T0348597

ORGANIC CHEMICAL
TOXICOLOGY OF FISHES

This is Volume 33 in the

FISH PHYSIOLOGY series
Edited by Anthony P. Farrell and Colin J. Brauner
Honorary Editors: William S. Hoar and David J. Randall

A complete list of books in this series appears at the end of the volume

ORGANIC CHEMICAL TOXICOLOGY OF FISHES

Edited by

KEITH B. TIERNEY

Department of Biological Sciences
University of Alberta
Edmonton, Alberta
Canada

ANTHONY P. FARRELL

Department of Zoology, and
Faculty of Land and Food Systems
The University of British Columbia
Vancouver, British Columbia
Canada

COLIN J. BRAUNER

Department of Zoology
The University of British Columbia
Vancouver, British Columbia
Canada

AMSTERDAM • BOSTON • HEIDELBERG • LONDON
NEW YORK • OXFORD • PARIS • SAN DIEGO
SAN FRANCISCO • SINGAPORE • SYDNEY • TOKYO
Academic Press is an imprint of Elsevier

Academic Press is an imprint of Elsevier
32 Jamestown Road, London NW1 7BY, UK
225 Wyman Street, Waltham, MA 02451, USA
525 B Street, Suite 1800, San Diego, CA 92101-4495, USA

The cover illustrates the diversity of effects an example synthetic organic water pollutant can have on fish. The chemical shown is 2,4-D, an herbicide that can be found in streams near urbanization and agriculture. The fish shown is one that can live in such streams: rainbow trout (*Oncorhynchus mykiss*). The effect shown on the left is the ability of 2,4-D (yellow line) to stimulate olfactory sensory neurons vs. control (black line) (measured as an electro-olfactogram; EOG). The effect shown on the right is the ability of 2,4-D to induce the expression of an egg yolk precursor protein (vitellogenin) in male fish. Image credits: chemical, Wikipedia; fish, Matthew Gilbert; EOG, Brian Blunt; well plate, Daniel Schlenk.

Notices
No responsibility is assumed by the publisher for any injury and/or damage to persons or property as a matter of products liability, negligence or otherwise, or from any use or operation of any methods, products, instructions or ideas contained in the material herein

Because of rapid advances in the medical sciences, in particular, independent verification of diagnoses and drug dosages should be made

British Library Cataloguing-in-Publication Data
A catalogue record for this book is available from the British Library

Library of Congress Cataloging-in-Publication Data
A catalog record for this book is available from the Library of Congress

ISBN: 978-0-12-398254-4
ISSN: 1546-5098

For information on all Academic Press publications
visit our website at **store.elsevier.com**

Typeset by MPS Limited, Chennai, India
www.adi-mps.com

Printed and bound by CPI Group (UK) Ltd, Croydon, CR0 4YY

Transferred to digital print 2012

Working together
to grow libraries in
developing countries

www.elsevier.com • www.bookaid.org

For Jeannie

CONTENTS

CONTRIBUTORS

The numbers in parenthesis indicate the pages on which the authors contributions begin.

BERNADITA F. ANULACION (53, 195), *Environmental and Fisheries Science Division, Northwest Fisheries Science Center, National Marine Fisheries Service, National Oceanic and Atmospheric Administration, Seattle, WA, USA*

MARY R. ARKOOSH (53, 195), *Environmental and Fisheries Science Division, Northwest Fisheries Science Center, National Marine Fisheries Service, National Oceanic and Atmospheric Administration, Newport, OR, USA*

NILADRI BASU (141), *Department of Environmental Health Sciences, University of Michigan, Ann Arbor, MI, USA*

DOUGLAS G. BURROWS (53), *Environmental Conservation Division, Northwest Fisheries Science Center, National Marine Fisheries Service, National Oceanic and Atmospheric Administration, Seattle, WA, USA*

TRACY K. COLLIER (195), *Science Director, Puget Sound Partnership, Tacoma, WA, USA*

DENIS A.M. DA SILVA (53), *Environmental and Fisheries Science Division, Northwest Fisheries Science Center, National Marine Fisheries Service, National Oceanic and Atmospheric Administration, Seattle, WA, USA*

DICK DE ZWART (483), *RIVM, Centre for Sustainability, Environment and Health, The Netherlands*

MARIE E. DELORENZO (309), *Center for Coastal Environmental Health and Biomolecular Research, National Ocean Service, National Oceanic and Atmospheric Administration, Charleston, SC, USA*

KRISTOFFER DALHOFF (371), *Faculty of Science, University of Copenhagen, Frederiksberg, Denmark*

JOSEPH P. DIETRICH (53, 195), *Environmental and Fisheries Science Division, Northwest Fisheries Science Center, National Marine Fisheries Service, National Oceanic and Atmospheric Administration, Newport, OR, USA*

MICHAEL H. FULTON (309), *Center for Coastal Environmental Health and Biomolecular Research, National Ocean Service, National Oceanic and Atmospheric Administration, Charleston, SC, USA*

MELISSA GLEDHILL (1), *Environment Canada, Pacific and Yukon Water Quality Monitoring, Vancouver, BC, Canada*

FRANK GOBAS (1), *Faculty of Environment, School of Resource & Environmental Management, Simon Fraser University, Burnaby, BC, Canada*

GREG G. GOSS (441), *Department of Biological Sciences and School of Public Health, Office of Environmental Nanosafety, University of Alberta, BC, Canada*

HAMID R. HABIBI (413), *Department of Biology, University of Calgary, Calgary, AB, Canada*

RICHARD D. HANDY (441), *Ecotoxicology Research and Innovation Centre, School of Biomedical and Biological Sciences, University of Plymouth, Plymouth, UK*

ALICE HONTELA (413), *Department of Biological Sciences, Alberta Water and Environmental Science Bldg., University of Lethbridge, Lethbridge, AB, Canada*

JOHN P. INCARDONA (195), *Environmental and Fisheries Science Division, Northwest Fisheries Science Center, National Marine Fisheries Service, National Oceanic and Atmospheric Administration, Seattle, WA, USA*

DAVID M. JANZ (141), *Department of Veterinary Biomedical Sciences, University of Saskatchewan, Saskatoon, SK, Canada*

LYNDAL L. JOHNSON (53, 195), *Environmental and Fisheries Science Division, Northwest Fisheries Science Center, National Marine Fisheries Service, National Oceanic and Atmospheric Administration, Seattle, WA, USA*

CHRISTOPHER J. KENNEDY (1, 257), *Department of Biological Sciences, Simon Fraser University, Burnaby, BC, Canada*

PETER B. KEY (309), *Center for Coastal Environmental Health and Biomolecular Research, National Ocean Service, National Oceanic and Atmospheric Administration, Charleston, SC, USA*

Tyson J. MacCormack (441), *Department of Chemistry and Biochemistry, Mount Allison University, Sackville, NB, Canada*

Mark S. Myers (53, 195), *Environmental and Fisheries Science Division, Northwest Fisheries Science Center, National Marine Fisheries Service, National Oceanic and Atmospheric Administration, Seattle, WA, USA*

Heather L. Osachoff (257), *Department of Biological Sciences, Simon Fraser University, Burnaby, BC, Canada*

Leo Posthuma (483), *RIVM, Centre for Sustainability, Environment and Health, The Netherlands*

Mark Sekela (1), *Environment Canada, Pacific and Yukon Water Quality Monitoring, Vancouver, BC, Canada*

Lesley K. Shelley (257), *Department of Biological Sciences, Simon Fraser University, Burnaby, BC, Canada*

Keith R. Solomon (371), *Centre for Toxicology, School of Environmental Sciences, University of Guelph, Guelph, ON, Canada*

Julann Spromberg (53), *Environmental and Fisheries Science Division, Northwest Fisheries Science Center, National Marine Fisheries Service, National Oceanic and Atmospheric Administration, Seattle, WA, USA*

Keith B. Tierney (1), *University of Alberta, Department of Biological Sciences, Edmonton, AB, Canada*

Glen Van Der Kraak (371), *Department of Integrative Biology, University of Guelph, Guelph, ON Canada*

David Volz (371), *Department of Environmental Health Sciences, Arnold School of Public Health, University of South Carolina, Columbia, SC, USA*

Gina M. Ylitalo (53, 195), *Environmental and Fisheries Science Division, Northwest Fisheries Science Center, National Marine Fisheries Service, National Oceanic and Atmospheric Administration, Seattle, WA, USA*

ABBREVIATIONS AND ACRONYMS

Abbreviations used throughout this volume are listed below. Standard abbreviations are not defined in the text. Nonstandard abbreviations are defined in each chapter and listed below for reference purposes.

11-KT	11-ketotestosterone
2,3,7,8 TCDD	2,3,7,8-tetrachlorodibenzo-p-dioxin
2,4,5-T	2,4,5-trichlorophenoxyacetic acid
2,4-D	2,4-dichlorophenoxyacetic acid
ABC	ATP binding cassette
AChE	acetylcholinesterase
ACTH	adrenocorticotropic hormone
ADME	absorption, distribution, metabolism (biotransformation), and excretion
Ag	silver
Ag-ENM-C	citrate-capped silver engineered nanomaterials
Ag-ENM-PVP	polyvinylpyrrilodone-capped silver engineered nanomaterials
AhR	aryl hydrocarbon receptor
AhRE	aryl hydrocarbon responsive element (interchangeable with xenobiotic responsive element; XRE)
AHTN	6-acetyl-1,1,2,4,4,7 hexamethyltetraline
ANSCO	Alaska north slope crude oil
AOP	adverse outcome pathway
AP-1	activator protein-1
APEs	alkylphenol ethoxylates
AR	androgen receptor
ARE	antioxidant response element (interchangeable with electrophile response element, EpRE)

ARNT	aryl hydrocarbon receptor nuclear translocator
BAF	bioaccumulation factor
BaP	benzo[a]pyrene
BCF	bioconcentration factor
BCRP	breast cancer resistance protein
BMF	biomagnification factor
BNF	β-naphthoflavone
BPA	bisphenol A
BSA	bovine serum albumin
BSAF	biota to sediment accumulation factor
C60	buckyball, carbon fullerene
CA	concentration addition
cAMP	cyclic adenosine monophosphate
CAR	constitutive androstane receptor
CBG	corticosteroid-binding globulin
CCAAT	cytidine-cytidine-adenosine-adenosine-thymidine
CEBPs	CCAAT/enhancer-binding proteins
CH	chlorinated hydrocarbon
CNT	carbon nanotube
CRH	corticotropin releasing hormone
Cu	copper
CYP	cytochrome P-450
CYP19A1	an isoform of aromatase, predominantly in the ovary
CYP19A2	an isoform of aromatase, predominantly in the brain
CYP1A	cytochrome P4501A
DCDD	dichlorodibenzo-p-dioxin
DDD	dichlorodiphenyldichloroethane
DDE	1,1-dichloro-2,2-bis(p-dichlorodiphenyl)ethylene
DDT	1,1,1-trichloro-2,2-di(4-chlorophenyl)ethane, or dichloro-diphenyltrichloroethane
DEHP	di(2-ethylhexyl) phthalate
DES	diethylstilbesterol
DHAA	dehydroabietic acid
DMBA	7,12-dimethylbenz[a]anthracene
DNA	deoxyribonucleic acid
Dpf	days post-fertilization
DWH	deepwater horizon
E1	estrone
E2	17β-estradiol
E3	estriol
EAC	endocrine active chemical

EC_{50}	effective concentration that causes a 50% change in a response
EDC	endocrine disrupting compound
EE2	17α-ethinylestradiol
EER	expected ecological risk (calculated with the ETX 2.0-software)
ENMs	engineered nanomaterials
EPA	USA Environmental Protection Agency
EPC	effect-and-probable cause (pie diagrams)
EpRE	electrophile response element (interchangeable with anti-oxidant response element, ARE)
ER	endoplasmic reticulum or estrogen receptor
EREs	estrogen-responsive elements
EROD	ethoxyresorufin-O-deethylase
ERR	estrogen-related receptor
ERRE	ERR-response element
EU	European Union
F1	the first filial generation comprised of offspring(s) resulting from a cross between strains of distinct genotypes
FA	fatty acid
FACs	fluorescent aromatic compounds
FCA	foci of cellular alteration
FCA/AHF	foci of cellular alteration or altered hepatocellular foci
FSH	follicle-stimulating hormone
FXR	farnesoid X receptor
GC/MS	gas chromatography/mass spectrometry
GFR	glomerular filtration rate
GH	growth hormone
GHR	growth hormone receptor
GnRH	gonadotropin-releasing hormone
GPER	G-protein linked estrogen receptor
GSH	glutathione (tripeptide; gamma-glutamyl-cysteinyl-glycine)
GSI	gonadosomatic index or gonadal somatic index ([gonad mass/body mass] × 100%)
GST	glutathione-S-transferase
GTH	gonadotropic hormone
HAHs	higher molecular weight polycyclic aromatic hydrocarbons
HC	hydrocarbon

HC5	hazardous concentration for 5% of the test species
HC5-NOEC	an HC5 exactly defined by using no observed effect concentrations in the SSD model used to derive it
HCB	hexachlorobenzene
HHCB	1,3,4,6,7,8-hexahydro-4,6,6,7,8,8-hexamethylcyclopenta (g)-2-benzopyran (aka galaxolide)
HI	hazard index (interchangeable with RCR or HQ)
HNF4A	hepatocyte nuclear factor 4 alpha
HP/MH	hepatomegaly/megalocytic hepatosis
HPA	axis hypothalamic-pituitary-adrenal axis
Hpf	hours post-fertilization
HPG	hypothalamus-pituitary-gonad
HPLC	high-performance liquid chromatography
HPT	hypothalamic-pituitary-thyroid
HQ	hazard quotient (interchangeable with RCR or HI)
HIS	hepatosomatic index ([liver mass/body mass] $\times$ 100%)
HydVac	hydropic vacuolation of biliary epithelial cells and hepatocytes
IC_{50}	inhibitory concentration that inhibits 50% of the response
ICP-OES	inductively coupled plasma emission spectrophotometry
IGF	insulin-like growth factor
kDa	kilodaltons
K_{OA}	octanol-air partition coefficient
K_{OC}	octanol-carbon partition coefficient
K_{OW}	octanol-water partitioning coefficient
LAHs	lower molecular weight polycyclic aromatic hydrocarbons
LC_{50}	lethal concentration that causes death of 50% of the population
LD_{50}	lethal dose that causes death of 50% of the population
LH	luteinizing hormone
LXR	liver X receptor
MDR	multidrug resistance
mER	membrane estrogen receptor
MFO	mixed function oxidase
mg l^{-1}	milligrams per liter
MoA	mode of action
mRNA	messenger ribonucleic acid
MRP	multidrug resistance protein
MRP1	multidrug resistance-associated protein
msPAF	multisubstance potentially affected fraction
MTCS	methyltriclosan
MWCNT	multi-walled carbon nanotubes

MXR	multixenobiotic resistance
NA	naphthenic acids
NCC	nanocrystalline cellulose
NCOA3	nuclear receptor coactivator 3
NFC	nanofibrillar cellulose
NF-κB	nuclear factor kappa B
ng l^{-1}	nanograms per liter
NOAA	national oceanic and atmospheric administration
NOAEL	no observable adverse effect level
NOEC	no observed effect concentration
NP	nonylphenol
NP/MH	nuclear pleomorphism/hepatocellular megalocytosis
NPE	nonylphenol ethoxylate
NSAID	non-steroidal anti-inflammatory drug
OC	organochlorine
OECD	organization for economic cooperation and development
OH-PBDEs	hydroxylated PBDEs
OP	organophosphate
PAA	polyacrylic acid
PAF	potentially affected fraction of species
PAH	polycyclic aromatic hydrocarbon
PAMP	pathogen-associated molecular patterns
PAPS	3'-phosphoadenosine-5'-phosphosulfate
PBDEs	polybrominated diphenyl ethers
PCB	polychlorinated biphenyl
PCDDs	polychlorinated dibenzodioxins
PCDFs	pentachlorodibenzo-p-furans
PEC	predicted environmental concentration
PeCDF	2,3,4,7,8-pentachlorodibenzo-p-furan
PERA	probabilistic ecological risk assessment
PFC	plaque-forming cell
P-gp	permeability glycoprotein
Phase I	biotransformation reactions—functionalization reactions, e.g. oxidation and reduction reactions
Phase II	biotransformation reactions—conjugation (synthetic) reactions
Phase III	membrane efflux transporters
PHH	polyhalogenated hydrocarbon
PME	pulp mill effluent
PNEC	predicted no effect concentration
POP	persistent organic pollutant
PPAR	peroxisome proliferator activated receptors

PPCP	pharmaceutical and personal care product
PRP	pattern recognition proteins
PRR	pattern recognition receptors
PSAMP	Puget Sound assessment and monitoring program
PVP	polyvinylpyrrolidone
PXR	pregnane X receptor
Pzp	pregnancy zone protein
RA	retinoic acid
RAR	retinoic acid receptor
RCR	risk characterization ratio or risk quotient (interchangeable with HI and HQ)
REACH	registration, evaluation, authorization, and restriction of chemical substances
ROS	reactive oxygen species
RXR	9-cis retinoic acid receptor
RyR	ryanodine receptor
SAM	S-adenosylmethione
SLC	solute carrier proteins
SLC15	solute organic oligopeptide transporter
SLC22	solute organic anion/cation/zwitterion transporter
SLC47	solute organic cation transporter
SLCO	solute organic anion transporter
SLP	solute transport proteins
SPEAR	species at risk approach
SP-ICP-MS	single-particle inductively coupled mass spectroscopy
SSBG	sex steroid-binding globulin
SSD	species sensitivity distribution
SSD-EC$_{50}$	an SSD model derived from EC$_{50}$ data
SSD-NOEC	an SSD model derived from no observed effect concentration data
ssPAF	single-substance potentially affected fraction
StAR	steroidogenic acute regulatory
STP	sewage treatment plant
SULT	sulfotransferases
SWCNT	single-walled carbon nanotubes
T3	triiodothyronine, a thyroid hormone
T4	thyroxine, a thyroid hormone
TAG	triacylglycerol
TAG:ST	triacylglycerol to structural lipids ratio
TBT	tributyl tin
TCDD	2,3,7,8-tetrachlorodibenzo-p-dioxin
TCS	triclosan

TEF	toxic equivalency factor
TEQ	toxic equivalent
TH	thyroid hormone
TiO_2	titanium dioxide
TMDL	total maximum daily loadings
TOC	total organic carbon
TPT	triphenyl tin
TRA	tissue residue approach
TRH	thyrotropin-releasing hormone
TSH	thyroid-stimulating hormone
UDP	uridine diphosphate
UDPGA	uridine diphosphate glucuronic acid
UGT	UDP-glucuronosyltransferase
UV	ultraviolet radiation
VEP	vitelline envelope protein
VTG	vitellogenin
WQC	water quality criteria
Ww	wet weight
WWTP	wastewater treatment plant
XRE	xenobiotic responsive element (interchangeable with aromatic hydrocarbon responsive element; AhRE)
ZnO	zinc oxide

PREFACE

To survive today's changing world, fish must cope with exposures to numerous chemical compounds, some synthetic and some natural, but often in complex mixtures. The fact that a compound may no longer be manufactured or legal is no safeguard for fish. Organic compounds can persist for several decades, be transported across and immune to international borders, and invariably end up in aquatic ecosystems where they exert extremely toxic effects many years after their release. Methods and technologies have been developed to better detect even the minutest quantities of dissolved organic compounds, and these developments have coincided with new methods to better assess the toxic impact of these chemicals on fishes. No longer is lethality testing the standard. Genetic, physiological and behavioural tests now examine sublethal toxicity, which gives chemical toxicity a mechanistic currency as well as a biological relevance. This volume draws on these new advances and summarizes and integrates the toxicity of organic chemicals to fishes.

By grouping chemicals according to their structural and functional characteristics, toxicity can be better integrated with the all-important fate and behavior of organic compounds within the environment, which determines bioavailability, and within the organism, which relates to toxicity. Indeed, organic compounds vary as widely in their behavior in the environment as they do in the organism. Of course, fishes like other biota are mostly water surrounded by cellular barriers of lipids—the cell membrane itself or an epithelial tissue with tight intercellular junctions. Thus, synthetic lipophilic compounds, for example, may have very low water solubility and will move into fish with ease, where they can "hijack" existing cellular machinery to alter physiology, and require complex biotransformation to increase their water solubility to facilitate excretion into the surrounding water. It is lipid solubility and the associated difficulty of excretion that make certain organic chemicals particularly prone to persist in fish and bioaccumulate to levels that exert toxicity.

There is public concern in both developed and developing nations about the unintentional exposure of fish to toxic organic compounds and the potential fish, human and ecosystem health impacts that this may have. People do not wish to consume contaminated fish nor do they want to destroy fish populations. Yet the regulations that govern environmental water quality vary considerably worldwide, and compliance is never complete. We hope that the information synthesized herein helps increase awareness and regulation of particularly toxic organic chemicals. With this in mind, the identification of emerging problems offers pathways to move forward in protecting aquatic systems and fish.

In sum, this book updates and collates the current literature on how the diverse classes of organic chemicals affect fish. It integrates diverse fields, from environmental sensing to toxicology, and acts as a guide for future research and the development of water quality guidelines.

A book of this magnitude and complexity requires the talents and dedication of many authors, who painstakingly worked with difficult deadlines and paid great attention to detail. For this we are very grateful. We thank the many reviewers for their valuable suggestions and the production staff at Elsevier, especially Pat Gonzalez and Kristi Gomez.

Keith B. Tierney
Anthony P. Farrell
Colin J. Brauner

1

ORGANIC CONTAMINANTS AND FISH

KEITH B. TIERNEY
CHRISTOPHER J. KENNEDY
FRANK GOBAS
MELISSA GLEDHILL
MARK SEKELA

1. CONTAMINANTS: A SHORT HISTORICAL PREAMBLE

The chemical nature of the environment can be modified by the very life contained within it, and other life must cope and adapt to these changes. Early in evolution, the changes that life brought to the environment were slow and progressive and could be accommodated through adaptation and natural selection by other organisms. For example, an increase in oxygen from primary productivity enabled the evolution of secondary consumers, and periodic oscillations in oxygen tension have seen the rise and fall of

1

Organic Chemical Toxicology of Fishes: Volume 33
FISH PHYSIOLOGY

large insect and air-breather aquatic animals. More central to this chapter is the chemical warfare between producer and consumer organisms that has lasted billions of years. For example, as plants evolved defense mechanisms, including toxic chemicals that served to limit their consumption, consumers developed detoxification mechanisms to counteract these toxins. Recently, this war took a different turn when a large group of consumers decided to go to war with plants, animals, and the environment in which they live. In a blink of evolutionary time, thousands of years for some and barely 50 years for most, humans have generated an arsenal of chemicals used to control their biotic and abiotic environments. This turned out to be problematic: As with any war, there is always collateral damage, some predictable, some not. In the context of this volume, the blink has been too short for most organisms to adapt to the chemical barrage.

The wave of chemicals introduced by humans commenced with simple elements and plant derivatives a few thousand years ago. The Chinese used sulfur as an antibacterial and fungicide at least 3000 years ago.[1] In the Middle East, arsenic was used for a variety of pest control purposes over 2000 years ago (Bentley and Chasteen, 2002). At some point in antiquity, humans discovered the power of plant-based agents for pest control. Perhaps the first plants used for their toxic properties were those of the strychnine family, specifically the seeds of fruit from the tree (*Strychnos nux-vomica*). The seeds contain an alkaloid known in India and China to be deadly to animal pests and were used as control agents. Other plants were recognized and harnessed for their pesticidal ability, including nightshade (nicotine), chrysanthemum (pyrethrin), and the root of derris (rotenone). Today, humans have very effectively hijacked, modified, and synthesized the chemical agents that plants evolved to cope with pests for their own purposes (see Chapter 6).

Ancient pest control agents, while still widely used today, represent only a small part of a chemical arsenal that now includes an array of synthetic compounds. In the mid-twentieth century, chemicals were created to kill broadleaf plants (weeds; see Chapter 7). The first synthetic weed killer, which was developed in the 1940s, is the infamous Agent Orange. This product was a blend of 2,4-dichlorophenoxyacetic acid (2,4-D) and 2,4,5-trichlorophenoxyacetic acid (2,4,5-T), two compounds that mimic plant growth hormones (auxins). Regrettably, an impurity of Agent Orange (2,3,7,8-tetrachlorodibenzo-*p*-dioxin; 2,3,7,8-TCDD) is a highly toxic, endocrine disrupting compound (EDC). In the next decade, another weed-killing compound was released: atrazine. This chemical blocks

[1] http://toxipedia.org

photosynthesis and has attracted considerable attention owing to its feminizing effects of wildlife (Hayes et al., 2006). In 1974, a compound was synthesized that blocked synthesis of specific amino acids in plants, glyphosate. While glyphosate was celebrated as a compound with very low nontarget species toxicity, the so-called inert ingredients of its formulations have generated serious concern (Tierney et al., 2007). Remarkably, all of these first-generation synthetic herbicides remain in use today, with the exception of 2,4,5-T, despite obvious unexpected environmental concerns.

At the same time that synthetic weed killers were first created, the first synthetic insecticides were created (1940s). First to emerge were the organophosphates (OPs), which were developed in Germany from the nerve gas used in World War I to kill troops in trench warfare. Following World War II, Americans began using parathion, the first of the OP revolutionary insecticides. The OP compounds represented the first synthetic, multi-taxa threat because of their toxic mechanism of action—an inhibition of the breakdown of the neuromuscular transmitter acetylcholine—a mechanism broadly conserved across the kingdom Animalia. The other class of synthetic insecticides to emerge in the 1940s was the organochlorines (OCs). The first OC was DDT—a chemical created much earlier (1874) but only later recognized as an insecticide. The mechanism of action of DDT was similar to that of parathion in that it was a neurotoxin (see Chapter 2). However, DDT possessed other, more insidious effects for other species in part because of its environmental persistence. DDT accumulated through the food chain, disrupted endocrine systems, and thinned bird eggshells to such a degree that embryos died and bird populations plummeted. The devastation of DDT was so total that it arguably spurred an entire discipline in the unwanted effects of synthetic agents on life, ecotoxicology (Carson, 1962). The nature of DDT helped form a new way to classify chemicals, as persistent organic pollutants (POPs). Other notable synthetic OC introductions include pentachlorophenol (1930s), lindane (1942), chlordane (1948), endosulfan (1954), and mirex (1955). Ironically, some introductions of new insecticides were primarily a result of natural selection of individual insects that possessed genes that conferred insecticide resistance and had flourished when other, more sensitive individuals had died during pesticide applications.

While herbicides and insecticides represent an explicit threat to life, they are by no means our most dangerous synthetic chemicals. The pesticide revolution was all about managing biology, but we also invented compounds for the purpose of managing heat. These synthetic compounds that are recalcitrant to heat have proven to be some the most resistant to degradation, and the most harming to life. These compounds are the polychlorinated biphenyls (PCBs), created in the late nineteenth century to serve as coolants for electrical transformers (see Chapter 2). PCBs represent a potent mix of persistence and toxicity: They are long lived and cause toxic effects at low concentrations, and so like DDT, they are considered POPs. While PCBs are no longer produced, their legacy remains strong. New hot spots (areas of high PCB concentration) continue to be found (EPA, 2010). Indeed, PCBs epitomize the naiveté of our "disposable culture." Discarded PCBs partitioned into life, bioaccumulated up the food chain, and caused health problems in mammals. While we no longer use PCBs, we do use structurally similar compounds (polybrominated diphenyl ethers; PBDEs) and have a new class of flame retardants (e.g., triphenyl phosphates).

The explosion of chemicals to control pests was paralleled by a revolution in chemicals to medicate or otherwise deal with human health issues. Included in these pharmaceuticals and personal care products (PPCPs) are various classes of painkillers, including opioids such as morphine (1827), nonsteroidal anti-inflammatory drugs (NSAIDs) such as aspirin (1899), and analgesics such as acetaminophen (1953); drugs to control human "bacterial pests," including antibiotics such as sulfonamides (1932) and penicillin (1945); mood-altering psychoactive drugs such as fluoxetine (i.e., Prozac; 1987); and ingredients in our hygiene-related products, which may include dyes, surfactants, soaps, and other antibacterial agents, such as triclosan (1972; see Chapter 8). Of the PPCPs, one that has received considerable attention owing to its potential impact on aquatic life is synthetic estrogen from birth control pills (ethinyl estradiol, EE2). It is released to the environment in municipal graywater/blackwater (Kvanli et al., 2008) and is toxic at very low (i.e., ng/L) concentrations (Palace et al., 2006). A chemical that may be grouped in PPCPs as it is used to "self-medicate" is caffeine. Although not strictly a PPCP, since it is a natural neurotoxin/neurostimulant intended for insects, it has been used since 3000 BC in China and is released into the environment in large quantities and can be found in fish (Wang and Gardinali, 2012). Together, highly complex mixtures of PPCPs can be found in the environment (Table 1.1; Figure 1.1).

Synthetic organic compounds will likely be used regardless of concerns over any collateral effects because of their overall value. There are some exceptions, however. Global regulations were enacted to ban PCB production in 2001, as were some OCs, for example, endrin and toxaphene (Stockholm Convention on Persistent Organic Pollutants).[2] Likewise, the European Union banned atrazine use in 2004 (Sass and Colangelo, 2006), owing to its groundwater presence and concern over its ability to harm

[2]www.pops.int

Table 1.1
Concentrations* of pharmaceuticals in surfacewater samples collected between June 2007 and February 2012 in the Okanagan River Basin, British Columbia, Canada

			Okanagan River upstream of Penticton WWTP	Okanagan River downstream of Penticton WWTP			Okanagan River downstream of Oliver, BC			Osoyoos Lake at the Canada/US Border		
Site			Coordinates 49.4846°N, 119.6048°W	Coordinates 49.4790°N, 119.5977°W			Coordinates 49.1145°N, 119.5661°W			Coordinates 49.0000°N, 119.4450°W		
			Sampling Period February 2012	January 2008-February 2012			June 2007-February 2012			June 2007-October 2010		
Drug Type	Chemical	Method Detection Limit	Concentration	Detection Frequency	Concentration Range	Median Detected Value	Detection Frequency	Concentration Range	Median Detected Value	Detection Frequency	Concentration Range	Median Detected Value
Acid inhibitor	Cimetidine	0.67	1.01	2/8	<0.639-59.7	30.6	1/10	<0.338-0.702	0.702	0/4	<0.663	—
Angiotensin receptor blocker	Valsartan	9.66	<4.24	3/3	15.9-60.1	26.1	0/3	<4.38	—	0/2	<4.48	—
Antibiotic	Azithromycin	3.23	<1.59	1/8	<1.58-2.86	2.86	1/10	<1.81-55.4	55.4	0/4	<1.68	—
Antibiotic	Cefotaxime	14.65	—	0/5	<13.3	—	1/7	<13.2-41.9	41.9	0/3	<12.6	—
Antibiotic	Ciprofloxacin	6.55	<8.57	1/8	<8.11-16.1	16.1	0/10	<11	—	0/4	<11	—
Antibiotic	Clarithromycin	1.41	<1.59	7/8	3.13-66.5	20.3	0/10	<1.64	—	0/4	<1.68	—
Antibiotic	Clinafloxacin	4.47	<35.5	0/8	<15.2	—	1/10	<12.6-40.8	40.8	1/4	<13.0-65.3	65.3
Antibiotic	Oxolinic acid	1.28	<1.35	0/8	<0.623	—	1/10	<0.749-2.56	2.56	1/4	<1.10-2.34	2.34
Antibiotic	Sulfamethoxazole	0.55	6.95	8/8	6.43-34.6	18.6	9/10	<2.34-7.82	5.56	4/4	1.81-3.98	2.25
Antibiotic	Trimethoprim	2.32	<1.85	5/8	<1.56-17.5	14.2	0/10	<1.81	—	0/4	<1.68	—
Antibiotic metabolite	Erythromycin-H2O	0.37	<2.44	6/7	1.56-12.6	5.6	6/9	0.67-4.4	0.755	3/4	<0.667-0.95	0.558
Anticholinergic	Benztropine	0.35	0.041	1/3	0.131-<0.626	0.131	1/3	0.229-<0.648	0.229	0/2	<0.336	—
Anticonvulsant	Carbamazepine	1.45	4.91	8/8	15.7-55.6	24.4	10/10	6.13-16.8	9.4	4/4	6.31-9.64	7.06
Antidepressant	Amitriptyline	0.71	<0.578	3/3	1.33-4.44	3.15	1/3	<0.353-1.1	1.1	1/2	<0.336-2.36	2.36
Amitriptyline metabolite	10-hydroxy-amitriptyline	0.15	<0.216	3/3	0.276-0.798	0.361	0/3	<0.181	—	0/2	<0.168	—
Antidepressant	Sertraline	0.30	<0.424	1/3	<0.45-0.599	0.599	0/3	<0.438	—	0/2	<0.448	—

(Continued)

Table 1.1 (continued)

Drug Type	Chemical	Method Detection Limit	Okanagan River upstream of Penticton WWTP 49.4846°N, 119.6048°W February 2012 Concentration	Okanagan River downstream of Penticton WWTP 49.4790°N, 119.5977°W January 2008–February 2012			Okanagan River downstream of Oliver, BC 49.1145°N, 119.5661°W June 2007–February 2012			Osoyoos Lake at the Canada/US Border 49.0000°N, 119.4450°W June 2007–October 2010		
				Detection Frequency	Concentration Range	Median Detected Value	Detection Frequency	Concentration Range	Median Detected Value	Detection Frequency	Concentration Range	Median Detected Value
Antihistamine	Ranitidine	0.99	<0.614	1/8	<0.639–17.9	17.9	0/10	<0.633	—	0/4	<0.663	—
Antihistamine, sedative	Diphenhydramine	1.08	<0.637	6/8	1.79–14.7	5.51	1/10	<0.725–1.69	1.69	0/4	<0.671	—
Anti-inflammatory	Naproxen	10.00	<3.18	2/8	<3.12–13.8	11.3	0/10	<3.29	—	0/4	<3.36	—
Bronchodilator	Albuterol	0.72	<0.307	3/8	<0.339–0.766	0.402	0/10	<0.316	—	0/4	<0.332	—
Ca channel blocker	Diltiazem	0.98	<0.318	5/7	<0.312–5.94	3.08	0/8	<0.329	—	0/3	<0.336	—
Diltiazem metabolite	Desmethyldiltiazem	0.60	<0.159	3/3	0.467–0.998	0.843	0/3	<0.164	—	0/2	<0.168	—
Ca channel metabolite	Norverapamil	0.10	<0.159	1/3	<0.169–0.202	0.202	0/3	<0.17	—	0/2	<0.168	—
Ca channel blocker	Verapamil	0.15	<0.159	1/3	<0.169–0.62	0.62	0/3	<0.164	—	0/2	<0.168	—
Ca channel blocker metabolite	Dehydronifedipine	1.27	<1.01	2/8	<0.676–1.61	1.3	1/10	<0.725–0.803	0.803	0/4	<0.671	—
Cholesterol drug	Gemfibrozil	1.56	<1.59	2/8	<1.56–11.9	7.76	0/10	<1.64	—	0/4	<1.68	—
Diabetes drug	Metformin	46.78	133	6/8	<128–469	272	6/10	69.3–263	138	2/4	81–<191	85.9
Diuretic	Furosemide	26.91	<43.1	1/4	<44.5–50.3	50.3	0/4	<48.3	—	0/2	<44.8	—
Diuretic	Triamterene	0.33	<0.307	4/4	1.55–4.7	3.92	1/4	<0.316–0.379	0.379	0/2	<0.332	—
Fungicide	Thiabendazole	0.19	<1.59	2/8	<1.58–3.58	2.66	0/10	<1.64	—	0/4	<1.68	—

Category	Compound											
Insecticide	DEET	0.20	3.32	3/3	3.82–6.1	4.51	3/3	4.80–6.50	5.25	2/2	9.82–11.7	10.8
Nicotine metabolite	Cotinine	1.45	4.16	4/8	<3.21–5.69	4.95	4/9	<3.45–5.26	4.4	2/4	<3.34–5.87	5.22
Painkiller (Opiate)	Codeine	3.85	<3.07	6/8	5.57–120	38.4	0/10	<3.16	–	0/4	<3.32	–
Painkiller (Opioid)	Hydrocodone	1.40	<1.53	1/4	<1.7–3.56	3.56	0/4	<1.58	–	0/2	<1.66	–
Painkiller (Opioid)	Oxycodone	0.63	<0.614	3/4	1.16–2.88	2.49	0/4	<0.633	–	0/2	<0.663	–
Psychostimulant	Amphetamine	2.14	<1.53	1/4	<1.58–1.93	1.93	0/4	<1.58	–	1/2	<3.85–5.7	5.7
Stimulant	Caffeine	28.47	<15.9	1/7	<15.8–32.4	32.4	2/9	<16.4–56	45.8	0/4	<16.8	–
Stimulant	Cocaine	0.17	<0.224	1/3	<0.169–0.81	0.81	1/3	<0.181–0.203	0.203	0/2	<0.168	–
Cocaine metabolite, analgesic	Benzoylecgonine	0.33	0.432	1/3	<0.338–3.48	3.48	0/3	<0.329	–	0/2	<0.336	–
Tranquilizer	Meprobamate	7.88	<5.16	1/3	<4.5–6.27	6.27	0/3	<4.4	–	0/2	<4.48	–
β-blocker	Atenolol	1.03	0.718	4/4	12.1–44	33.2	2/4	<0.732–0.946	0.925	0/2	<0.663	–
β-blocker	Metoprolol	2.83	<10.6	3/3	5.81–25.7	18.5	1/3	1.81–4.41	4.41	0/2	<1.68	–
β-blocker	Propranolol	1.04	<2.12	1/3	<2.25–4.57	4.57	0/3	<2.19	–	0/2	<2.24	–

*All concentrations and median values presented are in ng/L.

Case Study
Pharmaceuticals Up and Downstream of a Treatment Facility

Between June 2007 and February 2012, Environment Canada conducted a survey of pharmaceuticals in surface water collected from four locations in the Okanagan River Basin, British Columbia, Canada. A total of 23 samples were analyzed for trace levels of 72–118 individual compounds (Table 1.1). All analyses were performed by AXYS Analytical, Ltd. in Sidney, BC, and were based on the U.S. EPA method 1694, Pharmaceuticals and Personal Care Products in Water, Soil, Sediment, and Biosolids by HPLC/MS/MS.

All of the samples had measurable levels of pharmaceuticals and/or their breakdown products, and a total of 43 different compounds were detected between 2007 and 2012. Two compounds (carbamazepine and DEET) were detected in every sample for which they were analyzed. The site located downstream of the city of Penticton's wastewater treatment facility had the greatest number of compounds detected over the course of the study (40) compared to 20 different compounds detected in the Okanagan River at Oliver and only 10 detected in Osoyoos Lake. Only nine different compounds were detected in the sample collected upstream of the Penticton wastewater facility, compared to 33 compounds measured in the sample collected downstream of the facility on the same day.

The hypoglycemic drug metformin and the analgesic codeine were measured in the highest concentrations, at 469 and 120 ng/L, respectively. The anticonvulsant carbamazepine was detected in all 23 samples collected, and the antibiotic sulfamethoxazole was detected 22 times. These data represent the most current information on the presence and concentrations of pharmaceuticals in the ambient aquatic receiving environment in the Okanagan River Basin.

---

vertebrates. However, some compounds known to be overtly toxic to global ecology remain in use in specific regions. For example, some countries see the benefits of DDT in preventing the spreading of malaria as outweighing the negative environmental impact.

For reasons such as increased regulatory scrutiny and optics (i.e., public concern over chemical safety), there has been a change in the types of new chemicals developed. A trend of next generation "soft" pesticides (i.e., those that readily degrade in the environment) is that they are often tailored to resemble natural biological analogues, which brings the industry full circle

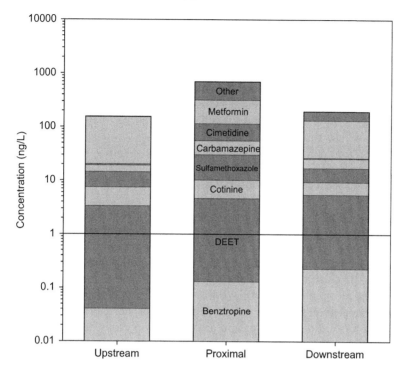

Figure 1.1. The relative concentrations of pharmaceuticals measured on February 9, 2012 upstream and downstream of the Penticton wastewater treatment outfall on the Okanagan River (site location details given in Table 1.1).

and back to its origins. There are relatively new insecticides based on the chrysanthemum (pyrethroids) and the nightshade (neonicotinoids), and new means of hijacking the physiology of pests: for example, compounds that mimic insect growth regulators and so affect the ability of insects to transition from juvenile to adult. These new control agents typically have relatively short half-lives (they readily hydrolyze or photolyze).

As the world has moved toward synthesizing less toxic control agents, an entirely new class of compounds has been produced, one for which toxicity remains almost completely unknown: nanomaterials (Slinn, 2006) (see Chapter 9). These new compounds already have application in industry, cosmetics, and drug and pesticide delivery. A concern with nanomaterials is that they are defined chiefly by the size of the molecule in one dimension (one-tenth of a micrometer—the size of DNA), not by any mechanism of action (MacCormack and Goss, 2008). However, it is debatable whether a class of compounds can be considered a threat before we actually know much about them, and for which environmental monitoring techniques are absent.

A major class of organic contaminants that we have not yet mentioned is one whose synthesis we are not typically responsible for, though we bear responsibility for their release: hydrocarbons (HCs), including polycyclic aromatic hydrocarbons (PAHs; see Chapter 4). Through our dependence on fossil-fuel-based energy, we have facilitated the release of HCs. Their release occurs not just through their burning, but also through their accidental dumping. Accidental oil releases can deposit vast amounts of HCs, and these releases continue to occur. The *Exxon Valdez* accident in 1989, which caused a marine release of 0.75 million barrels of crude oil, was surpassed in 2010 by the *Deepwater Horizon* uncapped wellhead release, which spewed perhaps 4.9 million barrels of crude oil into the Gulf of Mexico,[3] creating a 10,000-sq-km[4] surface spill and underwater plumes stretching 35 km[5] that blanketed the seafloor.[6] The impact of such spills remains unknown.

In the context of this book, the problem with the above chemicals is that they have found their way into the world's freshwater and saltwater (the hydrosphere). At the core of this problem are several associations that result in the tenet that all toxicants end up in the hydrosphere: Pesticides are often delivered in solutions of water and are sprayed on crops next to waterways or water bodies; our garbage laden with industrial compounds has been disposed of in or close to water; and our drugs and PPCPs go down the drain, mostly to treatment facilities where they may or may not be broken down before reaching the hydrosphere. Thus, fish receive exposures to a diverse array of synthetic chemicals, for which they likely will not have evolved defense systems (with the possible exception of HCs), sometimes all at once as complex mixtures and maybe continuously. Within mixtures there are the possibilities of additive or synergistic effects (see Chapter 10). With such a diversity of chemicals, predicting the outcome of mixture exposures is a daunting task, although a step toward doing so is provided at the end of this volume. Moving into the future, we sense the major concern that a larger number of people will be using a diminishing amount of the world's aquatic resources. Add to this concern an increasing array of synthetic chemicals entering water, and there is cause to worry about the future of aquatic, water-breathing vertebrates.

---

[3]www.pbs.org/newshour/rundown/2010/08/new-estimate-puts-oil-leak-at-49-million-barrels.html

[4]www.cbsnews.com/stories/2010/04/30/national/main6447428.shtml

[5]www.huffingtonpost.com/2010/05/27/gulf-oil-spill-new-plumes_n_591994.html

[6]www.npr.org/templates/story/story.php?storyId=129782098&ft=1&f=1007

1.1. Aquatic Contaminants: where They Come from

Civilizations continue to use water bodies as disposal sites, with the rationale being that the amount of water is sufficient to dilute or eliminate toxicity. We can detect water-soluble compounds in pristine environments hundreds of kilometers removed from human activity (Harris et al., 2008), and fat-soluble compounds in animals that have never been directly exposed to any contaminants (Ross, 2006). Overall, in the present world, the question is not whether an animal has been exposed, but rather how much an animal has been exposed to. The main driver in the broad distribution of organic contaminants is chemical persistence, which may have environmental half-lives in the decades.

Chemicals reach aquatic ecosystems from either point or nonpoint sources, with landscape features and meteorological conditions modifying their entry paths. Point sources indicate contaminant entry locations that may have been knowingly constructed and therefore are definable in time and space. They include end-of-pipe effluent discharges, such as treated municipal water, graywater (water used by humans but excluding waste), blackwater (with waste), or industrial discharges (e.g., pulp mill effluent [PME]). A common discharge is treated municipal blackwater that has undergone a multistep treatment process before being released. It may still contain a plethora of human-use chemicals that are uneconomical to remove completely, resulting in measurable concentrations of compounds such as carbamazepine (antiseizure drug) and ibuprofen (an NSAID) (Table 1.1). An important historic example of point-source pollution that was a major source of endocrine-disrupting chlorinated compounds was PME (Munkittrick et al., 1992).

Nonpoint contaminant sources are more difficult to define in time and space. These sources can be characterized by the processes by which contaminants are moved. Indeed, atmospheric transport may move compounds vast distances, such as the entire Pacific Ocean in the case of POPs, discovered in previously considered pristine ecosystems in British Columbia (Harris et al., 2008). Deposition can be divided into wet or dry routes. In dry deposition, compounds will settle out of the atmosphere by the action of gravity, or otherwise be deposited through forces such as turbulence or Brownian motion. In wet deposition, compounds leave the atmosphere to reach the hydrosphere via rain.

Rain events can lead to edge-of-field runoff of pesticides into nearby fish-bearing streams, even when the fields have riparian buffer strips. For example, rainfall of just 1 cm/d increased azinphos-methyl and chlorpyrifos concentrations in the Lourens River, South Africa, to detectable concentrations downstream of a 400 ha fruit orchard (Schulz et al., 2001). The more severe the rainfall, the more likely contaminants are to wash off and, as predicted, heavy rain increased azinphos-methyl and endosulfan concentrations in the Lourens River, South Africa, by as much as 40-fold (Schulz, 2001).

Rain events also transport sediments, which are sinks for lipophilic organic compounds (Bohm and During, 2010; Wu et al., 2004). For example, in Askim, Norway, surface water runoff from a rain event had fine particles bound with the fungicide propiconazole (Wu et al., 2004). While sediment can also be viewed as a transport vehicle for lipophilic contaminants (i.e., chaperone for contaminants) (Wu et al., 2004), once sediments settle, they may also represent a stable contaminant sink, one that can reveal contaminant signatures over time using sediment cores (Breedveld et al., 2010; Choi et al., 2011).

Contaminants reaching groundwater must pass through soil or rock. Factors such as vegetation (e.g., presence of large roots) and soil type (permeability) will affect how contaminants move into groundwater. Contaminated groundwater is a potential problem for drinking water and for fishes if it also feeds open water.[7] A commonly reported groundwater contaminant is atrazine, which has been found in groundwater of Canada (Masse et al., 1998), France (Garmouma et al., 1997), Germany (Dörfler et al., 1997), the United States (Blanchard and Donald, 1997), and Yugoslavia (Pucarević et al., 2002).

Spray drift is another nonpoint source of pesticides applied to crops using aerial spray. The amount of drift will depend on meteorological conditions (e.g., wind), the height of the spray nozzle, as well as characteristics of the spray aerosols (the bigger the droplet, the less likely to drift) (Gil and Sinfort, 2005). Chlorpyrifos and metalaxyl aerial applications in a vineyard in Italy were found to drift up to 24 m (Vischetti et al., 2008). Buffer strips, ranging from simple vegetative barriers to deep riparian areas with trees, shrubs, and grasses, are used to limit runoff and spray drift (Reichenberger et al., 2007). A 30 m buffer is expected to remove $\geq 85\%$ of contaminants (Zhang et al., 2010). Trees and shrubbery are important for preventing spray drift. A study on vineyards in the Palatinate region of Germany found that shrubs of height >1.5 m were capable of reducing drift by 72% (Ohliger and Schulz, 2010). Target-sensing sprayer technology, which includes "smart" sprayers that detect when they are near a tree, is also being used to reduce drift (Giles et al., 2011). Yet "hydrologic bypasses" such as rills (eroded channels) and ephemeral ditches can almost eliminate any buffering action (Reichenberger et al., 2007).

---

[7]water.usgs.gov/ogw/pubs/WRI004008/groundwater.htm

1.2. Aquatic Contaminants: Their Persistence

One of the principal ways to define contaminants is by their persistence. The POP grouping does just this. As mentioned previously, we are working to move away from synthesizing POPs. Even so, POPs are still found in fish, and their persistence translates to global distribution (Weber and Goerke, 2003). Because POPs are lipophilic/hydrophobic, they can be found in organisms with higher fat content (older organisms, higher on the food chain). In an Arctic study, PCB and mirex concentrations (1–7 ng/g and 0.4–2 ng/g lipid, respectively) were higher in blackfin icefish, which is piscivorous (fish eating), and lowest in mackerel icefish, which feed on krill (Weber and Goerke, 2003). Persistence must be viewed as being directly related to toxicity and thus as something to be avoided.

Almost 0.5 billion kg of insecticides were produced in 2007,[8] mostly for agricultural purposes. Many of these insecticides are next-generation (postchlorinated hydrocarbon) insecticides, and so they tend to be less persistent than their predecessors, but they still accumulate in the aquatic environment and its fishes (Chapter 6). For example, heavy use of OPs in fish-bearing waters that pass through agricultural lands resulted in concentrations reaching 1 µg/L such as the Nicomekl River, near Vancouver, BC (Tierney et al., 2008), the Wilmot River, Prince Edward Island (Purcell and Giberson, 2007), and two tributaries of the San Joaquin River in California's Central Valley (Ensminger et al., 2011). Although many types of pyrethroids are less persistent than OPs, they may still be found in the hydrosphere. For example, bifenthrin, esfenvalerate, lambda-cyhalothrin, and permethrin were all detected in 75% of sediment samples from streams and creeks in California's Central Valley (Weston et al., 2004). Clearly, current-use agricultural insecticides remain a concern for fish-bearing waters.

Over 1 billion kg of herbicides were produced in 2007,[9] twice that of insecticides, and again mostly for agricultural purposes and with some being persistent. Wide use of compounds such as 2,4-D and other phenoxy-family chemicals, atrazine and Roundup®, has resulted in concentrations in fish-bearing waterways (Gilliom et al., 2006; Harris et al., 2008; Tierney et al., 2011; Verrin et al., 2004) (see Chapter 7), as well as their "inert" formulants (Tierney et al., 2007). Unlike insecticides, urban use of herbicides for front lawns and golf courses has been identified as a potential concern for fish-bearing streams (Tierney et al., 2011). Thus, concern exists over the longer-term toxicity of many herbicides to fishes.

[8]http://www.epa.gov/pesticides/pestsales/07pestsales/usage2007.htm
[9]http://www.epa.gov/pesticides/pestsales/07pestsales/usage2007.htm

In discussions of chemical persistence, metal-containing compounds must not be omitted (metal speciation may change and toxicity with it, but metals cannot be broken down). Organic forms of metals (organometallics), that is, those bound to carbon such as organotins and organomercuries, may be highly persistent (see Chapter 3, as well as volumes 31A and 31B of this series for ionic forms of metals). Marine concentrations of tributyl tin (TBT) and triphenyl tin (TPT), used as antifouling coatings on (marine) boat hulls and fishing nets, extend to deep-sea Mediterranean fishes and may reach concentrations of 175–1700 ng/g wet wt (Borghi and Porte, 2002). A major concern is the accelerated release of Hg via mining activities, as Hg can be transported atmospherically great distances and converted by aquatic microorganisms into highly toxic organic forms (e.g., methyl and dimethyl-mercury) (Hintelmann, 2010), even in alpine lakes (Phillips et al., 2011).

2. CONTAMINANT TOXICOKINETICS IN FISHES

Chemical concentration at a target receptor site is the final determinant of the magnitude of a toxic response and is dependent on the chemical's kinetics (movement). Studies on chemical kinetics were historically performed with pharmaceuticals and were termed *pharmacokinetics*. However, with the evolution of toxicological sciences toward the inclusion of environmental contaminants and an increased concern for chemical effects on wildlife, *toxicokinetics* is now used as the appropriate term for the study of the kinetics of any foreign substance in a living organism. Toxicokinetics encompasses four time-dependent processes: chemical absorption, distribution, biotransformation (metabolism), and excretion (collectively abbreviated as ADME). These processes dictate the circulating and cellular levels of both endogenous and exogenous compounds and their physiological and toxicological activities. Toxicokinetic studies are concerned with the change in concentration of the chemical or its metabolites with time in circulatory fluids or other tissues; therefore, toxicokinetic data can be used to predict toxicity and to extrapolate data obtained in experimental animals to other species through mathematical models. Toxicokinetic models (both classical and physiologically based) have been well developed in fish, as our knowledge regarding toxic chemical ADME in this area has advanced rapidly (Stadnicka et al., 2012).

2.1. Uptake

The potential for xenobiotics to be taken up by fish exists in all epithelial cells that are in contact with the water and food, but it is highest in respiratory and gastrointestinal epithelia, which are designed for rapid and efficient uptake. Several epithelial tissue characteristics enhance absorption rates: a large surface area, short diffusion distance, and high blood perfusion. These characteristics vary greatly among fishes.

Xenobiotics cross into or through epithelia using several mechanisms that include simple passive diffusion, facilitated diffusion, filtration through membrane channels, active transport, and endocytosis (Figure 1.2). PCBs, dioxins, and legacy pesticides such as mirex and DDT are hydrophobic and lipophilic and have high log octanol water partition coefficient (log K_{OW}) values. As a result, they partition into membranes by simple passive diffusion, which is a first-order process dependent on pressure gradients. Cell membranes act as a barrier to most ions and nonionized hydrophilic molecules larger than 100 daltons (Schanker, 1961), but these may enter organisms along paracellular aqueous channels adapted for small polar solutes and ions (Diamond and Wright, 1969; Saarikoski et al., 1986). Xenobiotics may also commandeer epithelial carrier proteins involved in active transport and facilitated diffusion, which follow zero-order kinetics and are saturable, in contrast to diffusion. Endocytotic processes include pinocytosis and phagocytosis. The overall importance of these transport mechanisms for synthetic chemicals is likely limited, but can be a major route for some toxins of natural origin.

The overall importance of the dermal route in xenobiotic uptake in adult fish is probably limited, due to skin thickness, low surface area, and limited blood perfusion. However, an example of where skin uptake is significant is the approximately 20% uptake of a waterborne dose of sodium lauryl sulfate in goldfish (Tovell et al., 1975). Changes in skin thickness and scale presence will alter dermal absorption rates. Moreover, juvenile fish have a much larger skin surface area:gill surface area ratio, and the skin may be the principal route for oxygen uptake in larvae until the gills fully develop (Rombough, 2002). Thus, skin uptake of xenobiotics is a concern in small fish (Lien and McKim, 1993).

The relative importance to overall body burden of toxicants in fish of branchial uptake (from water) and gastrointestinal uptake (from food and water) are known (Clark et al., 1990; Erickson et al., 2006a, 2006b). Uptake by both routes can be viewed as transfer through a series of aqueous and lipid compartments (Hansch, 1969; Opperhuizen, 1986): First, the chemical must be delivered (in water or digestate) to the epithelium, and then it must move through several diffusion barriers, including the cell membrane, into the blood. The rate-limiting steps of uptake vary with log K_{OW}, creating a sigmoidal relationship between gill uptake and log K_{OW} (Erickson and McKim, 1990; McKim et al., 1985; Spacie and Hamelink, 1982; Yalkowski and Morozowich, 1980). Water-soluble chemicals with log $K_{OW} < 1$ are very

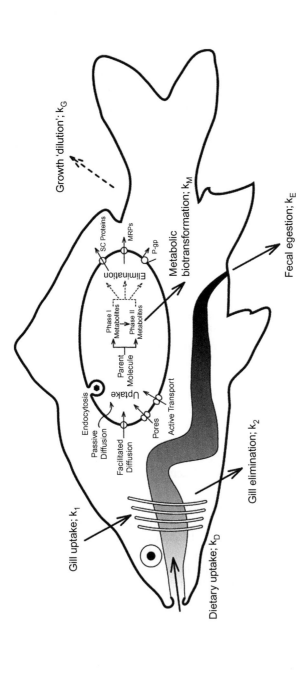

Figure 1.2. Chemical uptake, biotransformation, and elimination in fishes. Rate constants (k) of these major processes, as well as a model cell, are shown. Large arrows indicate organismal routes; smaller arrows indicate cellular routes; dashed arrows represent routes of potential excretion or "pseudo-excretion" (k_G). Shading corresponds to an increase in the polar nature of the gut and so an increase in the tendency of hydrophobic compounds to leave. P-gp: permeability glycoprotein; MRP: multidrug resistance protein; SC protein: solute carrier protein.

available to the gill epithelium since they dissolve readily in water. Their uptake, however, is very low (McKim et al., 1985) because they are not easily transferred from water into cells due to a very low lipid solubility (Hansch and Clayton, 1973). For chemicals with log K_{OW} values of 1 to 3, uptake increases with increasing log K_{OW} as they have sufficient water solubility to be delivered to the gill epithelium and sufficient lipid solubility to diffuse through cell membranes. Gill uptake efficiencies remain high and relatively constant for chemicals with log K_{OW} values between 4 and 6 (Qiao and Farrell, 1996). For hydrophobic chemicals (log $K_{OW} > 6$), delivery to the gill epithelium is limited by their very low water solubility. These very hydrophobic compounds can also self-associate into micelles that will reduce uptake rates through an apparent increase in their molecular weight and volume (McKim et al., 1985).

The gastrointestinal epithelium is an important uptake route for xenobiotics associated with food or in drinking water, and especially for very hydrophobic compounds dissolved in dietary lipids. These are taken up by passive diffusion, and the factors that are important in absorption in the gastrointestinal tract are similar to those that act at the gills. Hydrophobic chemicals, which are taken up by passive diffusion and governed by the same factors that act at the gills, can also bypass aqueous diffusional barriers in the gut by associating with fat droplets or fatty acid micelles (Ekelund, 1989) during pinocytosis (Addison, 1982; Bruggeman, 1982; Opperhuizen, 1986). Modification of gut contents during digestion may also alter the kinetics and processes involved in the transfer of chemicals from the lumen of the gut into the organism (described in *biomagnification*).

2.2. Distribution

Most xenobiotic translocation or distribution occurs through the circulatory systems of fish following absorption. Fish have two circulatory systems (primary; high blood pressure, high volume; secondary; low blood pressure, high volume (Farrell, 1993)); however, the contributions of each to xenobiotic distribution are unknown. The secondary circulation is specialized in fish scales, but its involvement in toxicant uptake is unknown.

The aqueous component of blood is the plasma, and many hydrophilic xenobiotics have sufficient water solubility to be dissolved and transported in this matrix. Hydrophobic compounds do not dissolve readily in plasma and associate with other constituents of fish blood (e.g., cholesterol, triglycerides), including macrocellular constituents (e.g., cell membranes) but most often plasma proteins. PCBs (McKim and Heath, 1983) and PAHs (Kennedy et al., 1991) bind more to plasma proteins than red blood cells of trout and toadfish.

Plasma proteins include albumins, globulins, and other immune system components, fibrinogen, regulatory proteins, clotting factors, transport proteins, and even unusual proteins such as ice-structuring proteins. Physicochemical characteristics of xenobiotics determine binding to individual proteins, but most typically through hydrophobic interactions. Albumin, the dominant plasma protein in fish (Perrier et al., 1977), contains a domain that attracts lipophilic substances (Goodman, 1958), and therefore xenobiotics will likely be associated with this or similar proteins. Even the sex steroid binding proteins (SSBPs) have been found to bind hydrophobic xenobiotics. For example, the xenobiotics nonylphenol and other alkyl phenols, bisphenol A, phthalates, and endosulfan bind to and are transported by SSBPs (Gale et al., 2004; Tollefsen, 2002).

The high molecular weight of protein-bound xenobiotics precludes easy exit from capillaries, making this fraction of xenobiotics not immediately available for distribution into the extravascular space and tissues. However, hydrophobic interactions are rapidly reversible, and the release of the protein-bound chemical occurs when the affinity of another biomolecule or tissue component is greater than that of the plasma protein. Once unbound, the chemical can easily diffuse between or through capillary endothelial cells until a steady state is reached between tissues and plasma proteins. Overall, tissue distribution depends mainly on the physicochemical properties of the chemical, blood flow delivery to a particular tissue, and the affinity of the chemical for tissue constituents. Regardless, they become distributed widely throughout the body, mainly according to the lipid content of a particular tissue (Clark et al., 1990), because of a high affinity for lipids as well as certain protein constituents. Redistribution among tissues occurs to some degree because hydrophobic/lipophilic chemicals readily diffuse through cell membranes, and so initial distribution patterns may change with time.

Water-soluble/charged chemicals do not readily pass through cell membranes and have a restricted distribution unless they are taken up by cell membrane transport proteins. Several toxicokinetic models have been developed for fish that adequately describe the movement of environmental chemicals between various tissue compartments using physiological parameters (e.g., such as tissue-specific blood flows) obtained from the literature, coupled with empirically derived tissue-specific partition coefficients (Law et al., 1991).

2.3. Biotransformation

Biotransformation is defined as the enzymatic conversion of a chemical from its parent form to a metabolite, an action that is distinct from physically and chemically mediated processes such as photolysis.

Biotransformation *ability* refers to combined transformations and reflects the fact that chemicals can undergo multiple transformations at varying rates, and that chemicals may be modified at a number of different molecular positions, resulting in a variety of metabolites.

Biotransformation ability determines the ultimate accumulation and excretion rates of chemicals, which affect the biological half-lives, persistence, and body burden, which in turn set the severity and duration of a toxic response. This is especially true for hydrophobic chemicals. For example, the recalcitrance of DDT to be biotransformed in mosquitofish (*Gambusia affinis*) results in a four order-of-magnitude difference in its bioaccumulation compared to that for the structurally related compound 2-bis(p-methylthio-phenyl)-1,1,1-trichloroethane which has similar lipophilic characteristics (Kapoor et al., 1973).

Fish are entirely reliant on the ability of a chemical to dissolve in water to be eliminated for effective excretion by the gills, gastrointestinal tract, and kidneys. Chemicals that have high log K_{OW} values either will not partition to the aqueous environment (e.g., water surrounding gills) or will be reabsorbed by the cells by back diffusion from excretory media (e.g., urine, bile, feces). Thus, another general purpose of biotransformation is to synthesize more polar hydrophilic metabolites to better utilize water-based excretory mechanisms. Such biotransformations introduce a hydrophilic moiety onto the molecule through functionalization (asynthetic, Phase I) or conjugation (synthetic, Phase II) reactions. Specifically, Phase I reactions introduce or expose a polar reactive functional group (e.g., OH, SH, NH2, COOH) in the parent molecule. Phase II reactions conjugate an endogenous molecule such as glucuronic acid, glutathione, sulfate, and taurine to the parent compound or its Phase I metabolite.

In the 1960s, pharmaceutical sciences promulgated the idea that fish did not oxidize or conjugate foreign chemicals (Lech and Bend, 1980). This idea persisted despite existing evidence for mixed-function oxidase-like activity in fish livers (Potter and O'Brien, 1964). Today, this falsehood has been exposed by the many reviews of Phase I and Phase II biotransformation among animal taxa. Animal groups are ranked in terms of biotransformation ability as mammals/birds > fish/reptiles > invertebrates. The major sites (high biotransformation enzyme concentrations) of biotransformation in fish are the liver (primary site), gastrointestinal (GI) tract, and gill. Thus, all sites of xenobiotic absorption are guarded, although these enzymes are found in virtually all fish tissues.

2.3.1. PHASE I REACTIONS

Phase I reactions are the predominant biotransformation pathway for most xenobiotics in fish. They provide polar functional groups to increase the water solubility of xenobiotics in preparation for excretion. These reactions involve microsomal mono-oxygenations, co-oxidations in the prostaglandin synthetase reaction, cytosolic and mitochondrial oxidations, and reduction reactions.

2.3.1.1. Oxidations. Oxidation reactions are the predominant metabolic pathway for many xenobiotics as well as endogenous compounds (e.g., steroids, vitamins, and fatty acids) in fish. Among the Phase I enzymes, cytochrome P450 ranks first in terms of catalytic versatility and the sheer number of xenobiotics it detoxifies or activates to reactive intermediates (Guengerich, 1987; Waterman and Johnson, 1991). The cytochrome P-450 (CYP)-dependent mixed-function oxidase (MFO) system consists of phospholipids (localized mainly to the endoplasmic reticulum (ER) and mitochondrial inner membranes), the enzyme cytochrome P450 (P450s), an electron transfer chain (NADPH-cytochrome P450 reductase in the ER, and redoxin reductase/redoxin in mitochondria), and cytochrome b5. In addition to xenobiotic degradation, P450 enzymes aid in the biosynthesis of critical signaling molecules used for the control of development and homeostasis. Monooxygenation reactions allow organisms to metabolize unknown compounds, in several reaction categories: hydroxylation of aliphatic or aromatic carbon compounds, the epoxidation of double bonds, heteroatom (*S-*, *N-*, and *I-*) oxygenation and *N*-hydroxylation, heteroatom (*O-*, *S-*, *N-*, and *Si-*) dealkylations, oxidative group transfers, cleavage of esters, and dehydrogenation reactions.

The first fish CYP to be cloned and sequenced was a CYP1A from trout and designated CYP1A1 (Heilmann et al., 1988). Due to its high diversity, a systematic classification of individual CYP forms has allocated P450 genes as isoforms, subfamilies, and families using amino acid sequence homology (Nelson, 1999). All CYPs have a similar structural fold with a highly conserved core, despite their occasionally minimal sequence similarity. The protein sequences within a given gene family are at least 40% identical (e.g., CYP1A1 and CYP1B1), and the sequences within a given subfamily are >55% identical (e.g., CYP2K7 and CYP2K16). Individual members of a family or subfamily are labeled again by Arabic numerals (e.g., CYP1A1, CYP1A2). In fish, 8 families (CYP1, CYP2, CYP3, CYP4, CYP11, CYP17, CYP19, and CYP26), 19 subfamilies, and at least 34 isoforms of P450 have been described (Arellano-Aguilar et al., 2009). However, the great increase in new CYP sequences has prompted the suggestion to move to a functional classification scheme (Kelly et al., 2006).

CYPs have a variety of endogenous functions in fish, and several families can even be detected during embryogenesis and organogenesis, particularly during neuronal and reproductive system development (Wang et al., 2006).

Xenobiotic substrates include most of the therapeutic drugs and environ-mental contaminants, antioxidants, dyes, and plant products such as flavorants and odorants. The enzymes in families 1–3 are mostly active in the biotransformation of xenobiotics, whereas the other families have important endogenous functions. CYP1A has been detected in most species examined, and the 1A1 isoform has a high affinity for many environmental contaminants (Livingstone et al., 2000) and is responsible for the oxidation of hundreds of different substrates.

P450 genes are regulated by both exogenous chemicals and endogenous factors involving as many as 70 nuclear receptors (Honkakoski and Negishi, 2000) that include the AhR (aryl hydrocarbon receptor), ARNT (aryl hydrocarbon receptor nuclear translocator), CAR (constitutive androstane receptor), and RAR (retinoic acid receptor) (Lewis, 2001). Induction of CYP P450 genes in fish differs from that for mammals, and substrate-specific induction is not necessarily equivalent. For example, the traditional P450 inducer phenobarbital in mammals does not induce similar isozymes in fish (Stuchal et al., 2006).

Biotransformation by cytochrome P450 does not always result in the detoxification of a chemical. For example, arene oxides that are highly reactive to cellular macromolecules such as DNA and protein can be formed during the cytochrome P450-dependent oxidation of aromatic hydrocarbons such as benzo[a]pyrene (Kennedy et al., 1991). Epoxide hydrolases can then generate vicinal diols with a *trans*-configuration (i.e., *trans*-1,2-dihydro-diols), which are less reactive and easier to excrete. Fish, like mammals, can produce the reactive metabolite benzo[a]pyrene 7,8-dihydrodiol-9,10-oxide (known as a bay-region diol epoxide) which is recognized as a tumorigenic metabolite (Varanasi et al., 1981).

2.3.1.2. Reduction Reactions. Certain metals (e.g., pentavalent arsenic) and xenobiotics containing an aldehyde, ketone, disulfide, sulfoxide, quinone, *N*-oxide, alkene, azo- or nitro-group can be reduced through reduction reactions catalyzed by P450, owing to the transfer of reducing equivalents of P450 reactions to the xenobiotic substrate instead of to molecular oxygen and usually under anaerobic or low-oxygen tension conditions. However, it is often difficult to ascertain whether the reaction has proceeded enzymatically or nonenzymatically by interaction with reducing agents (e.g., reduced glutathione, FAD, FMN or NAD[P]).

2.3.2. PHASE II REACTIONS

Phase II biotransformation reactions are biosynthetic and mediated by enzymes (e.g., transferases) that covalently bind endogenous cofactors to polar functional groups present on either parent compound or Phase I metabolite. These conjugation reactions have dual roles: generating intermediary endogenous metabolites and biotransformation of xenobiotics. The most common reactions include glucuronidation, sulfation, acetylation, methylation reactions, and conjugation with glutathione (mercapturic acid synthesis) or amino acids (e.g., glycine, taurine) (Paulson et al., 1986). In fish, conjugation with glycosides, sulfonate, and glutathione are the most common Phase II reactions. Hydrolysis reactions are also included here, although typically they are categorized as a Phase I reaction (Williams, 1959).

In these reactions, ATP is used to put either the substrate or cofactor into a "high-energy" state. In some cases, a xenobiotic or its metabolite already possesses the intrinsic reactivity to enzymatically react with a conjugating agent without the preformation of a high-energy intermediate (e.g., formation of glutathione conjugates). Glucuronidation, sulfonation, acetylation, and methylation reactions (i.e., glycoside, sulfonate, methyl-, and acetyl-group conjugates) involve high-energy cofactors; conjugation with amino acids involves "activated" xenobiotics. Phase II reactions result in a large increase in chemical hydrophilicity (exceptions are methylation and acetylation products) since the conjugated cofactors are usually ionized at physiological pH. Such reactions are usually considered as detoxication reactions. In mammals, however, bioactivation by Phase II reactions has been shown to occur. In addition, the conjugating moieties are readily recognized by efflux transport proteins in the secretory epithelial tissues, which aids in rapid excretion of the metabolite.

2.3.2.1. Hydrolysis Reactions. Broadly speaking, Phase II hydrolysis reactions conjugate the abundant cellular nucleophile (water) with a xenobiotic electrophile (Williams, 1959). Hydrolysis reactions are important for the disposition of PAH dihydrodiol epoxides, alkylating agents, and O-acetoxy esters of aromatic amines. The most significant hydrolyzing enzymes in fish and mammals are carboxylesterases, arylesterases, cholinesterases, serine endopeptidases, and epoxide hydrolases. The hydrolysis of xenobiotic esters and amides generates carboxylic acids, alcohols, and amines that are all susceptible to further metabolism by Phase II conjugation reactions. Epoxides are organic three-membered oxygen compounds that arise from the P450-mediated oxidative metabolism of PAH, for example, and are typically chemically reactive. Epoxide hydrolases catalyze the hydration of chemically reactive epoxides to their corresponding dihydrodiol products, which are less reactive.

2.3.2.2. Glucuronidation. Glucuronidation is a major pathway of xenobiotic transformation and of carbohydrate conjugation in fish.

Uridine diphosphate (UDP)-glucuronosyltransferases (UGTs), a family of endoplasmic reticulum-bound enzymes, accepts a very broad range of substrates, including a wide variety of industrial chemicals, agrochemicals, pesticides, and pharmaceuticals, such as potentially carcinogenic planar and bulky phenols, aromatic amines, and endogenous compounds (e.g., retinol, bilirubin, steroid hormones, thyroid hormones). The site of glucuronidation is usually an electron-rich nucleophilic heteroatom forming *O*-, *N*-, *S*-, or *C*-glucuronides of xenobiotics. In glucuronidation, the cofactor is the high-energy nucleotide uridine diphosphate (UDP)-glucuronic acid (UDPGA).

The UGT1 and UGT2 subfamilies in mammals belong to a superfamily of glycosyltransferases (Mackenzie et al., 2003). The UGT1 family in rat, for example, contains at least seven enzymes and belongs to the same subfamily designated UGT1A (UGT1A1, 1A2, 1A3, 1A5, 1A6, 1A7, and 1A8). UGT2 is subdivided into UGT2A (UGT2A1) and UGT2B (UGT2B1, 2B2, 2B3, 2B6, 2B8, and 2B12). Molecular evidence suggests at least 14 UGT genes in fish that may be closely related to the mammalian UGT1 and UGT2 families (George and Taylor, 2002). Amino acid sequences of UGT genes from both plaice (*Pleuronectes platessa*) and flounder (*Platichthys flesus*) indicate that these genes share greater similarity with mammalian UGT1 family genes (than UGT2) and have been designated UGT1B. UGT1B genes are also present in pufferfish (*Tetraodon nigroviridis*) and zebrafish (*Danio rerio*) (Leaver et al., 2007). There is some evidence of UGT-like genes identified in the zebrafish and pufferfish that have an unknown function and appear to be unique to fish (George and Taylor, 2002).

UGT activity is concentrated in the liver and intestines, though it is found in the ER of all tissues (Clarke et al., 1991). Under certain conditions, UDPGT activity is comparable in trout and rat liver (Lindström-Seppä et al., 1981). The *C*-terminus of all UDP-glucuronosyltransferases contains a membrane-spanning domain that anchors the enzyme in the endoplasmic reticulum. The enzyme faces the lumen of the endoplasmic reticulum, where it is ideally placed for direct access to products generated by cytochrome P450 and other microsomal Phase I enzymes. In *in vitro* enzyme assays, UGT activity in liver microsomes can be stimulated by detergents (Andersson et al., 1985), which disrupt the lipid bilayer of the endoplasmic reticulum and allow UDP-glucuronosyltransferases free access to UDP-glucuronic acid. Conversely, if concentrations are too high, they can inhibit UGTs by disrupting their interaction with phospholipids, which are important for catalytic activity.

The carboxylic acid moiety of glucuronic acid, which is ionized at physiologic pH, promotes excretion by increasing the aqueous solubility of the xenobiotic and by being recognized by the organic anion transport proteins, which secrete these metabolites into urine and bile. It is now generally recognized that glucuronides and other conjugates cannot leave cells without organic anion transporters (members of the ATP-binding cassette [ABC] superfamily) such as MRP2 (ABCC2) (König et al., 1999; Trauner and Boyer, 2003).

2.3.2.3. Sulfonation. Sulfotransferases (SULT) catalyze sulfate conjugation, or sulfonation reactions, whereby the sulfonate, not sulfate (i.e., SO_3^{2-} not SO_4^{2-}), group of 3'-phosphoadenosine-5'-phosphosulfate (PAPS) is readily transferred in a reaction involving nucleophilic attack of a phenolic oxygen or amine nitrogen on the sulfur atom, with the subsequent displacement of adensine-3',5'-diphosphate. SULT enzymes perform this transfer to hydroxyl (phenols, 1°, 2°, and 3° alcohols) and amine groups, and to a minor extent, thiols (Coughtrie, 2002). Similar to glucuronidation, sulfate conjugation reactions are an important biotransformation pathway in fish for endogenous (e.g., hormones, neurotransmitters, peptides and proteins, and bile acids) and xenobiotic compounds (e.g., pentochlorophenol, ethinylestradiol, naphthol) and their metabolites (e.g., phenolic metabolites of PAH and PCBs, ring hydroxylated metabolites of N-acetylaminofluorene). The major classes of SULTs that metabolize xenobiotics are: aryl sulfotransferases, alcohol sulfotransferases, tyrosine-ester sulfotransferases, and amine N-sulfotransferases. These reactions generally produce metabolites containing a highly water-soluble sulfuric acid ester (sulfate or sulfamate conjugates with pK_a values between 2 and 4 that exist as anions at physiological pH) that are generally less toxicologically active; however, sulfonation can bioactivate some chemicals (e.g., sulfonation of benzylic hydroxyl groups, formed from the hydroxylation of the methyl group in methylated PAHs, can alkylate DNA) (Watabe, 1983).

The sulfonate donor PAPS is synthesized in all cells but at low levels in most fish tissues and at relatively high activity in liver and intestine (James et al., 1997; Tong and James, 2000). Interestingly, sulfonation is a high-affinity, low-capacity conjugation reaction, which is opposite to glucuronidation. The rate of PAPS formation is believed to be relatively slow, with low basal levels. Therefore, when exposure to substrates is high, the importance of sulfonation to overall biotransformation may be low. Thus, at low phenol concentrations, sulfonation predominates, while glucuronidation predominates at high concentrations (Kobayashi et al., 1976, 1984). However, even with reports of low SULT activity in some fish (Parker et al., 1981), sulfonate conjugates can form the predominant class of xenobiotic metabolites compared to glucuronide conjugates in both fresh and marine species (Layiwola and Linnecar, 1981; Pritchard and Bend, 1984).

Multiple SULT enzymes have been identified in mammals, and the nomenclature system is similar to the nomenclature system developed for cytochrome P450 (Nagata and Yamazoe, 2000). The five known gene families (SULT1–SULT5) have subfamilies that are 40 to 65% identical. In rats, the individual enzymes are SULT1A1, 1B1, 1C1, 1C6, 1C7, 1D2, 1E2, 1E6, 2A1, 2A2, and 2A5 (Nagata and Yamazoe, 2000). Seven different SULTs have been cloned and sequenced from the zebrafish genome (Ohkimoto et al., 2003; Sugahara et al., 2003a, 2003b).

2.3.2.4. Methylation. Methylation is a common but minor xenobiotic biotransformation pathway that has received little attention in fish. It differs from other conjugation reactions in that it masks functional groups that may be otherwise used in other conjugation reactions, and it reduces water solubility and excretion potential. Methylation involves the transfer of methyl groups from one of two methyl donor substrates, S-adenosylmethione (SAM) or N5-methyltetrahydrofolic acid. The methyl group bound to the sulfonium ion in SAM has the characteristics of a carbonium ion and is transferred to xenobiotics and endogenous substrates by nucleophilic attack from an electron-rich heteroatom (*O, N,* or *S*). Consequently, the functional groups involved in methylation reactions are phenols, catechols, aliphatic and aromatic amines, *N*-heterocyclics, and sulfhydryl-containing compounds. Methylation is a common biochemical reaction of endogenous compounds (e.g., proteins, nucleic acids, phospholipids), but is not usually of quantitative significance for xenobiotics.

2.3.2.5. Acetylation. The primary site of acetylation is the liver. For xenobiotics containing an aromatic amine (R–NH2) group (e.g., 2-aminofluorene, aniline, quinolone), acetylation to aromatic amides (R–NH–COCH3) is known to be a major pathway in fishes (Kleinow et al., 1992). Xenobiotics containing primary aliphatic amines are rarely substrates for *N*-acetylation, a notable exception being cysteine conjugates, which are formed from glutathione conjugates and converted to mercapturic acids. Like methylation, *N*-acetylation masks an amine with a nonionizable group, so that many *N*-acetylated metabolites are less water soluble than the parent compound, although this is not always the case and many metabolites are more water soluble than the original chemical.

The *N*-acetylation of xenobiotics is catalyzed by acetyl CoA: amine *N*-acetyl-transferases and involves the activated conjugating intermediate acetyl CoA. The specific cofactor for the reaction, acetyl-CoA, is obtained mainly from the glycolysis pathway, via direct interaction of acetate and coenzyme A, or through the catabolism of fatty acids or amino acids.

2.3.2.6. Amino Acid Conjugation. Fish can conjugate amino acids with xenobiotics (e.g., 2,4-dichlorophenoxyacetic acid, 2,4,5-trichlorophenoxyacetic acid, the herbicide triclopyr, p-amino benzoic acid) and endogenous compounds (e.g., bile acids, phenylacetic acid) to enhance excretion. Water solubility increases because the conjugate is an anion at physiological pH and is recognized by organic anion transporters (Guarino et al., 1977; James and Pritchard, 1987). The amino group of amino acids are conjugated to carboxylic acid groups on xenobiotics following activation of the chemical to an acyl CoA intermediate (thioether) catalyzed by the enzyme acid:CoA ligase. The acyl CoA thioether subsequently reacts with the amino acid taurine in fish (James, 1976) to form an acylated amino acid conjugate, catalyzed by acyl-CoA-taurine aminoacyltransferase, which has its highest levels in the mitochondria of fish kidney and liver tissues (James, 1976). The reaction is categorized as a high-affinity, low-medium capacity system; however, it is not considered as effective as the main conjugation reactions since it has a relatively small substrate base.

2.3.2.7. Glutathione Conjugation and Mercapturic Acid Formation. Fish can also conjugate xenobiotics with the tripeptide glutathione (GSH; gamma-glutamyl-cysteinyl-glycine), although the process is fundamentally different from their conjugation with other amino acids and dipeptides. GSH in its reduced form also acts as an antioxidant due to its nucleophilic nature at the sulfur group and is essential in protection from reactive oxygen species. GSH reductase is constitutively active, and therefore GSH is found almost exclusively in its reduced form in the cell.

The conjugation of xenobiotics (e.g., PAH, PCBs, pesticides [alachlor, atrazine, DDT, lindane, methylparathion]), carcinogens and their metabolites (aflatoxin B_1, benzo[a]pyrene, 7,12-dimethylbenzanthracene), and endogenous molecules (prostaglandins, leukotrienes, and products of oxidation with lipids, nucleic acids, and proteins) with glutathione is catalyzed by a supergene family of mainly cytosolic enzymes (microsomal and mitochondrial forms also exist) termed glutathione S-transferases (GSTs). They are present in most tissues, with the highest concentrations found in the liver, intestine, and kidney, and can constitute 2–4% of total cytosolic protein in liver. These enzymes can also bind, store, and/or transport compounds that are not substrates for glutathione conjugation.

Substrates for GSTs include an enormous array of xenobiotics or their metabolites that share three common features: they are hydrophobic, they contain an electrophilic atom, and they react nonenzymatically with glutathione to some degree. In contrast to the amides formed by conjugation of xenobiotics to other amino acids, glutathione conjugates are thioethers, which form by nucleophilic attack of glutathione thiolate anion with an electrophilic carbon atom in the xenobiotic. Glutathione can also conjugate xenobiotics containing electrophilic heteroatoms (O, N, and S).

Arguably the most important function is to catalyze the initial reaction in the formation of N-acetylcysteine (mercapturic acid) derivatives through several steps, including the formation of a glutathione conjugate; GSH conjugation involves the formation of a thioether link between the GSH and electrophilic compound. The reaction can be considered a detoxification reaction by preventing the electrophilic compound from reacting with nucleophilic centers in other biomolecules. Following conjugation, the conversion of glutathione conjugates to mercapturic acids involves the sequential cleavage of glutamic acid (by the enzyme γ-glutamyl transpeptidase) and glycine (by the enzyme cysteinyl glycine dipeptidase or aminopeptidase M) from the glutathione moiety, followed by acetylation (by N-acetyltransferases) of the resulting cysteine conjugate, forming the N-acetyl derivative, mercapturic acid.

Many glutathione S-transferases have been cloned and sequenced, which has resulted in a mammalian nomenclature system (Mannervik et al., 1992) arranged into seven soluble GST classes designated alpha, mu, omega, pi, sigma, theta, and zeta, and one microsomal/mitochondrial kappa class. Although few GSTs have been fully characterized in fish, nucleotide and protein sequence data indicate that all classes identified in mammals have also been identified in fish (Schlenk et al., 2008). In 2005, a new class of piscine GSTs was identified as "rho-class" (Konishi et al., 2005). The highest levels in liver and intestine are the pi-class GSTs which have high activity toward B[a]P epoxides and diol epoxides which are important in detoxifying potentially carcinogenic PAH metabolites.

2.3.3. Regulation of Biotransformation

Evidence exists that all phases of xenobiotic metabolism in fish and mammals may be coordinately regulated by specific nuclear receptors such as PXR and CAR, and an ever-expanding repertoire of other orphan nuclear receptors (Maglich et al., 2002; Rae et al., 2001; Xie and Evans, 2001; Xie et al., 2003). For example, AhR-type induction of CYPs (Dixon et al., 2002; Leaver et al., 1993) and UGTs (Clarke et al., 1992; George and Taylor, 2002) by xenobiotic chemicals such as planar PAH and HAH (e.g., chlorinated biphenyls, dibenzo-p-dioxins, and dibenzofurans) is well known in fish. In this process, the aryl hydrocarbon receptor nuclear translocator (ARNT) and the AhR together form a transcription factor complex that binds specific genomic recognition sites (xenobiotic response elements, XRE) on DNA, altering the transcription of target genes that include both Phase I and Phase II biotransforming enzymes. Although reasonably well studied in mammals, how other orphan nuclear receptors in fish (e.g., constitutive androstane receptor [CAR], pregnane X receptor [PXR], retinoid X receptor [RXR]) sense endogenous (e.g., small lipophilic hormones), and exogenous (e.g. drugs, xenobiotics) compounds, and transfer this information into cellular processes by regulating the expression of their target genes, is still a relatively unexplored area. The regulation and induction of Phase II enzymes is less well understood, although several cis-acting regulatory elements exist, including the antioxidant response element (ARE/electrophile response element [EpRE]), xenobiotic responsive element (XRE/aromatic hydrocarbon response element [AhRE]), activator protein-1 (AP-1), and nuclear factor-kappa B (NF-kB) binding sites. However, despite their coregulation by the above nuclear receptors, CYPs, UGTs, GSTs, and conjugate transporters are also controlled individually in a complex species- and tissue-specific manner (Gonzalez, 1988; Mackenzie et al., 2003; Nebert and Gonzalez, 1987; Trauner and Boyer, 2003).

2.4. Excretion

The major sites of elimination in fish are the hepatobiliary system (fecal elimination), kidneys (renal), and gills (branchial) (Figure 1.2). Other minor excretory routes include milt (e.g., halogenated hydrocarbons), eggs (e.g., selenomethionine), alimentary elimination (via secretions such as mucous [e.g., DDT]), and scales (e.g., tetracyclines). Thus, uptake sites serve as excretion sites because of the bidirectionality for chemical flux, provided activity (concentration or fugacity) gradients favor outward movement. Membrane efflux transporters, filtration and secretion (Phase III mechanisms) of water-soluble compounds, and modification of chemicals by biotransformation reactions (Phase I and II mechanisms) are critical in chemical elimination and in setting body concentrations and burdens.

2.4.1. BRANCHIAL ELIMINATION

The hydrophobicity of a chemical governs elimination mainly through the gills. For example, the rate of excretion of individual hydrocarbons by the gills is inversely proportional to the size of the molecule and the log K_{OW} values (Thomas and Rice, 1981). As explained above for uptake, water-soluble, low log K_{OW} compounds dissolve in water but do not move easily through the cell membranes of the gill epithelial cells. Moderately hydrophobic chemicals can diffuse to the cell surface and into water, but are limited by water solubility. Yet if gill ventilation volume is very large and activity gradients are high, a significant amount of chemical could be eliminated via the gills. Low water-soluble, high log K_{OW} compounds are not eliminated by the gills to any significant degree.

Water quality also plays a role in gill elimination. For ionizable organic chemicals, the pH of the water at the gill boundary layer affects diffusional gradients. The efflux of weakly acidic chlorinated phenolic compounds was increased at water pH values > pKa values of each chemical. When phenols diffused through the membrane and reached the surface, they were ionized and moved easily into boundary water layer, thus maintaining outward gradients of the neutral unionized forms (Erickson et al., 2006b). This is often called ion trapping and is a mode of excretion of the broad-spectrum biocide pentachlorophenol at the surface of teleost gills (Kennedy and Law, 1986). Compounds such as pentachlorophenol (McKim et al., 1986), naphthalene and toluene (Thomas and Rice, 1981), and ethyl-m-amino-benzoate (MS222) (Hunn and Allen, 1974) undergo appreciable excretion through the gills.

Since rapid gill elimination of very hydrophobic chemicals is limited under most conditions, fish must rely on prior biotransformation to more water-soluble metabolites, which are usually eliminated by the liver and kidney. Increasing water solubility also mitigates the dilemma of reabsorption after renal/hepatobiliary excretion.

2.4.2. RENAL EXCRETION

Filtration and active tubular secretion are critical processes for renal excretion. Glomerular filtration is the major route to eliminate small xenobiotics dissolved in plasma, which pass ultra-filtrate in the kidney tubule lumen through glomerular pores. Weak acids may diffuse through tubular cells and be excreted via ion trapping in the urine. Numerous xenobiotics and their metabolites have been quantified in fish urine, including PAHs, PCBs, pesticides, and pharmaceuticals. Most fish species have kidneys that are glomerular and can utilize filtration for the excretion of xenobiotics with the appropriate characteristics. In some species such as the Antarctic notothenioids, evolution of aglomerularism is a well-known phenomenon suggested to represent an adaptation to reduce water efflux and prevent the loss of low-molecular-weight compounds such as antifreezes. Christiansen et al. (1996) showed that the elimination of the small polysaccharide laminaran in the glomerular Atlantic cod species *Gadus morhua* is solely via the urine, whereas this xenobiotic is eliminated only through the bile in the aglomerular polar cod *Boreogadus saida* (Christiansen et al., 1996). The mechanism by which this is achieved without filtration is unknown.

Renal active tubular secretion is via ATP-binding cassette (ABC) transporters and solute transport proteins (SLPs) that mediate the ATP-dependent transport of conjugates of lipophilic compounds (Phase II products). These are similar in kidney and liver/biliary systems in vertebrates and will therefore be discussed below.

2.4.3. HEPATOBILIARY EXCRETION

Hepatobiliary excretion is a major route of xenobiotic excretion in fish. The liver is in an advantageous position by receiving all of the blood flow from the stomach, intestines, and pyloric cecae via the hepatic portal vein before it can enter the general systemic circulation. The liver's efficient biotransformation and elimination of xenobiotics rely on high titers of biotransformation enzymes with high affinity for hydrophobic chemicals. The well-defined lobular structure of the mammalian liver is not a characteristic of fish (Gingerich, 1982). Instead, hepatocytes are arranged as tubules of cells, where a tubule is surrounded by five to seven hepatocytes, with their apices directed toward the central bile canaliculus or preductule. The tubules are surrounded by sinusoids that replace capillaries, which have a very porous endothelial lining. The cells produce bile, which is secreted into the central bile canliculus that flows into the gallbladder.

Although anatomically different, the process of elimination of xenobiotics or metabolites from hepatocytes to the bile is similar to that in mammals, but without the profound metabolic zonation of specialized areas of biotransforming enzyme typical of mammals (Schär et al., 1985). Enzymes tend to be more uniformly distributed in fish liver (Gingerich, 1982). Blood delivery to fish livers is similar to that of mammals, at least in terms of total cardiac output as in mammals (Farrell et al., 2001). Thus, any differences in clearance are more likely a function of metabolic rate and hence temperature/endothermy.

Liver metabolites (and the parent compound) can enter either the blood for excretion at the kidney or the bile for excretion in the feces. The transport of these compounds out of the hepatocytes and into the bile canaliculi is mediated mainly by ATPase efflux transport proteins. The term *Phase III* was coined to denote proteins involved in the elimination of xenobiotics (Ishikawa, 1992) due to the potential coregulation and cooperation with biotransformation. Those xenobiotics released to the bile are held in the gallbladder as a reservoir until release when the organism eats a meal and they exit the organism in fecal material. Biliary excretion is arguably the most important route of xenobiotic excretion in fish; there are hundreds of examples of environmental chemicals eliminated by this route.

2.4.4. TRANSPORTERS

Over the last 20 years, many membrane transport proteins have emerged as prime determinants of absorption, distribution, and elimination of xenobiotics in fish. Protein-mediated xenobiotic excretion was appreciated well before the cloning of transporters, and it has become increasingly recognized that transporters are also critical for xenobiotic absorption and distribution. The interplay between transporters, and the asymmetrical membrane distribution on both apical and basolateral membranes in epithelial cells, is central in determining the extent and direction of xenobiotic movement and ultimately contributes to the toxicokinetic profile of xenobiotics within an organism. Transporters that mediate vectorial transfer of substrates into the systemic circulation are termed *absorptive transporters*, regardless of whether they are influx or efflux transporters. In contrast, transporters that mediate vectorial transport of substrates from the systemic circulation into bile, urine, or gut lumen are termed *secretory transporters*. Xenobiotic transport proteins of this nature are currently grouped into two major classes: (1) ATP-binding cassette (ABC) transporters and solute carrier proteins (SLCs).

2.4.4.1. ABC Proteins.   ABC transporters contain ATP-binding domains that possess ATPase activity for translocating substrates (xenobiotics or their metabolites, endogenous molecules) across cell or organelle membranes. The three ABC subfamilies of (multidrug) transporters identified in fish (Lončar et al., 2010) are involved in the efflux of xenobiotics and endogenous molecules and include members of the P-glycoprotein (P-gp [ABCB]) family (Figure 1.2), the multidrug resistance protein (MRP [ABCC]) family (Figure 1.2), and the breast cancer resistance protein (BCRP [ABCG]) family. These transporters play a role in the epithelia of excretory organs, provide important components of permeability barriers (e.g., blood–brain barrier), and prevent the initial entry of environmental contaminants at absorption sites (e.g., gastrointestinal tract). Extremely low expression of all analyzed ABC genes were found in the gills of rainbow trout (Lončar et al., 2010). The ABC *tap* was also found in zebrafish liver following methyl mercury (MeHg) exposure (Gonzalez et al., 2005).

2.4.4.2. P-glycoprotein (ABCB Family).   The multidrug resistance (MDR) pump P-gp was the first transporter described (Juliano and Ling, 1976) and is to date the most widely studied efflux transporter in fish. This transporter has been recognized as the molecular basis for the resistance of tumors against a broad spectrum of unrelated cytostatic drugs in mammals. Kurelec described the ecotoxicological significance of xenobiotic transport, termed *multixenobiotic resistance* (MXR) (Kurelec, 1992; Kurelec et al., 1992). For fish living in polluted environments, the increased expression of MXR or other MDR-like transport mechanisms is of great protective value (e.g.,

P-gp and multidrug resistance proteins [MRPs] that restrict xenobiotic uptake) (Guarino et al., 1974).

The P-gp ABCB1 transports a broad variety of structurally unrelated compounds that are moderately hydrophobic and small in size, and contain positively charged domains; this includes endogenous molecules (e.g., progesterone, deoxycorticosterone), numerous drugs (e.g., doxorubicin, vinca alkaloids), and environmental chemicals (e.g., pesticides, PAH, nonylphenol ethoxylates) (Sturm and Segner, 2005). Functional efflux assays of P-gp activity exist for liver (Albertus and Laine, 2001), kidney (Van Aubel et al., 2000), intestine (Doi et al., 2001), the operculum (Karnaky et al., 1993), and brain capillaries (Miller et al., 2002). High-resolution structures of ABC proteins and the structure–function relationships of mammalian P-gp have been characterized in recent years, but its role in normal physiology or its mechanism of action is not known, and even less is known in fish. Their interaction with biotransformation systems, or other transporters (e.g., multidrug resistance protein [MRP]), is unclear. Although induction of CYP enzymes and P-gp is not coregulated, they may interact through CYP-mediated production of P-gp substrates or CYP-mediated or glutathione (GSH) depletion-mediated increases in intracellular reactive oxygen species.

The ABC superfamily is divided into seven subfamilies, ABCA to ABCG, and the individual transporters are identified by the name of the subfamily followed by a number. In winter flounder, *Pleuronectes americanus* (Chan et al., 1992) and other fish species, one gene with sequence homologies to mammalian ABCB genes was termed flounder genes fP-gp B, which show homology to both ABCB1 and ABCB2 genes but cannot be attributed to either class.

2.4.4.3. Multidrug Resistance-Associated Protein (ABCC Family). Water-soluble conjugated metabolites require export across cell membranes. This action is mainly accomplished by the MRP ABCC family in addition to affecting the transport of some chemicals in their parent form; as well, they prevent cellular uptake, as does P-gp. Their main role is in conjugate transport via recognition of the added endogenous moiety of the conjugated product. MRPs are located in both the basolateral and apical membranes (both directed toward the lumen of the tubule) of kidney tubule cells. Active transport of these conjugates and other like molecules is a significant route of excretion, but the process is saturable. Two tubular secretory processes exist, one for organic anions (acids) and the other for organic cations (bases). Genes and/or proteins of the MRP family are found in dogfish shark (*Squalus acanthias*) (Miller et al., 1998) and small skate (*Raja erinacea*) (Cai et al., 2003), and a partial cDNA sequence in red mullet represents the first

MRP-related protein sequence reported for a teleost fish species (*Mullus barbatus*) (Sauerborn et al., 2004).

2.4.4.4. Multixenobiotic Resistance Protein (ABCG Family). Breast cancer resistance protein (BCRP) or ABCG2 is a member of the ABCG family of transporters and is the only member of this subfamily involved in drug transport. Its substrate specificity is less broad than for ABCB and ABCC proteins in mammals (Litman et al., 2000; van Herwaarden et al., 2006). ABCG2 is known as a half transporter, which is different from the ABCB and ABCC transporter types, which are full transporters (the gene product constitutes one structural unit). ABCG2 consists of two protein molecules assembled to form an active homodimer.

2.4.4.5. Solute Carrier Proteins (SLC Family). Another family of export pumps transport of substrates from the basolateral extracellular fluid into the basolateral sides of cells (Figure 1.2). These proteins are important in renal and hepatobiliary elimination of xenobiotics or their metabolites. The SLC family is part of the major transport superfamily of proteins, typically considered to be uptake transporters, although there are examples of bidirectional transport. SLC transporters typically use secondary and tertiary active transport to move chemicals across biological membranes. In terms of toxicological relevance, these transporters are involved in the hepatic uptake of organic anions (e.g., conjugated metabolites), organic cations, organic solutes, and bile salts. SLCs include 43 families and 298 transporter genes in humans (Fredriksson et al., 2008). The transport activities for xenobiotics for approximately 19 of these gene families are known. They include the organic anion transporting polypeptide (SLCO), the oligopeptide transporter (SLC15), the organic anion/cation/zwitterion transporter (SLC22), and the organic cation transporter (SLC47) families. In fish, members of the SLC superfamily of proteins that have been identified at the basolateral membrane of renal tubules include the organic anion transporters (SLC21/SLCO) and organic cation/anion/zwitterion transporters (SLC22). SLCO exhibits a broad substrate specificity and is responsible for the cellular uptake of most amphiphilic organic anions. High levels of expression of SLC22 are seen in barrier and excretory tissues as well (Miller, 2002).

2.4.4.6. Regulation of Efflux Transporters. The regulation of efflux transporters is unclear in mammals and less clear in fish. Some evidence suggests that PXR, CAR, peroxisome proliferators activated receptors (PPAR), liver X receptor (LXR), and farnesoid X receptor (FXR) are all nuclear receptors that bind to specific DNA response elements, which produce effects on Phase III efflux transporters (Xu et al., 2005). PXR appears to regulate several efflux transporters, including P-gp, MRPs, and SLCOs. Common regulatory mechanisms may exist for xenobiotic metabolizing enzymes through pathways such as AhR/ARNT, although there is little, and even contradictory, evidence for coregulated pathways (Bard et al., 2002). However, since Phase II metabolites are transported out of cells by P-gp, MRPs, and SLCOs, Phase III transport processes may be coordinately regulated.

3. PREDICTING CONTAMINANT MOVEMENT IN THE ENVIRONMENT AND FISH

Predictive chemical fate models are useful when problems arise concerning chemical contamination of fish and other wildlife species. One problem occurs when assessing the impacts of chemical contamination resulting from a spill, point-source emission, or historic sediment contamination on fish or wildlife species. In such cases, chemical concentrations or quantities at the source (e.g., spill, effluent, water, sediments) may be known, but the resulting concentrations and associated risks in fish and other affected wildlife species will not be. A second problem emerges when the chemical concentration in fish or wildlife species is known through sampling programs, but it is unknown how the organisms acquired the chemical concentration and what actions need to be taken to eliminate concern. Both problems are examples of a missing cause–effect relationship between the chemical concentrations in environmental media such as water, air, sediments, and soil and those in the wildlife species of interest. This relationship can be complex. For example, a predatory fish absorbs contaminants from the water via its gills and/or ingest contaminates from ingested sediments and through predation on a variety of prey items, which acquire their contaminant burden via similar mechanisms as the predator. A better understanding of these difficulties comes from exploring the fundamental principles that control ADME of contaminants in single organisms and in food webs, which are recognized and formalized in certain bioaccumulation models for single organisms and food webs.

Current predictive models aim to describe some of the complexity and have been increasingly used to better characterize and understand the relationship between contaminant concentrations in abiotic environmental media and those in wildlife and humans. These models can be used for risk assessments, the derivation of total maximum daily loadings (TMDL) for impacted water bodies, the derivation of criteria for water and sediment quality criteria, as well as wildlife criteria, the development of target levels for water and sediment clean-up levels, and the screening of large numbers of chemicals for substances with high bioaccumulation potential. The following section discusses the main principles that can be invoked to develop an understanding of the relationship between chemical concentrations in organisms and hence help develop better strategies for environmental management with regard to chemical pollution.

3.1. Chemical Equilibrium

One of the most important chemical principles that can be used to better understand the uptake, bioaccumulation, and food-web transfer of contaminants in food webs is the natural tendency of chemicals to participate in a net transport between the organism and environment media that produces a chemical equilibrium. Equilibrium is thermodynamically defined as a situation in which the chemical potential μ or the fugacity f (Pa) or the activity (unitless) of the chemical in the organisms and environmental media (e.g., water) is the same (Mackay and Arnot, 2011), that is.

$$\mu_B = \mu_W \text{ or } f_B = f_W \text{ or } a_B = a_W \tag{1}$$

where μ_B and μ_W, f_B, and f_W and a_B and a_W are, respectively, the chemical potentials (μ), fugacities (f), and activities (a) in the organism or biota (B) and water (W). The fugacity of a chemical f (in units of Pascal) in a given phase is related to the molar concentration C (in mol m^{-3}) by the fugacity capacity Z (in mol m^{-3} Pa^{-1}) of the phase in which the chemical is solubilized (Mackay, 2001; Mackay and Arnot, 2011):

$$f = C/Z \tag{2}$$

The fugacity capacity Z is compound and phase specific and represents the capacity of that phase to sorb and retain a given chemical within its matrix.

The activity is an alternative way to express equilibrium (Mackay and Arnot, 2011) and is more suitable for chemicals of low volatility but relatively high water solubility and often appeals to aquatic toxicologists who deal with contaminants in water. The activity of a chemical a (unitless) in a given phase is related to the molar concentration C (in mol m^{-3}) through the chemical's solubility (S) in the medium in which the contaminant resides:

$$a = C/S \tag{3}$$

Net passive chemical transport occurs from the medium of high fugacity or activity (e.g., in the water) to the medium with the lower fugacity or activity (e.g., an organism) until the fugacities or activities of the contaminant in the media are equal and the contaminant is at equilibrium. At equilibrium, the chemical concentrations in both media (e.g., in the organism C_B and the water C_W [mol m^{-3}]) are related by the chemical's partition coefficient K_{BW} (unitless), that is

$$K_{BW} = C_B/C_W = Z_B/Z_W = S_B/S_W \qquad (4)$$

Application of the equilibrium principle allows a first-order assessment of the relationship of chemical contaminants between organisms and their environment. The equilibrium principle also reveals one of the main driving forces that controls the chemical distribution between environmental media and biota. In particular, it describes the contribution of the chemistry to the larger issue of chemical uptake and elimination. Chemical partitioning drives the chemical to achieve concentrations in the environment and the organisms that correspond to equal fugacities or equal activities or concentrations between the organism and the environmental media represented by the relevant partition coefficient(s). A chemical equilibrium is not always achieved as environmental and biological processes can interfere with the chemical's natural tendency to reach equilibrium, where fugacities or activities (but not concentrations) in the organism and its environment are equal. Chemical equilibrium models that have been widely used to describe the distribution of chemicals in evaluative environments are the level I fugacity models developed by Mackay and co-investigators (Mackay, 2001). These models are freely accessible (Chemistry, 2012) and relatively easy to use.

3.2. Mass Balance

The second principle that can be used to understand and describe the relationship between contaminant concentrations in organisms and those in environmental media is that of the chemical mass balance equation where the net flux of chemical N_B (in units of moles day^{-1}) into an organism is the sum of the chemical fluxes into and out of the organism (Arnot and Gobas, 2004; Gobas and Arnot, 2010). The net fluxes reflect the several chemical uptake and elimination processes acting together, as detailed above. For example, in fish, the most important uptake processes are dietary and respiratory via the gills and body surface area, whereas elimination is primarily via the respiratory surface, kidneys, and feces, and often involves metabolic transformation, as detailed earlier (Figure 1.2). Somatic growth is often referred to as growth dilution, although no chemical is actually excreted or transformed. An increase in body mass has a diluting effect on the chemical mass in the organisms, provided growth exceeds further uptake. A mass balance model for bioaccumulation in fish can therefore be formulated as (Arnot and Gobas, 2004):

$$N_B = V_B \cdot dC_B/dt = k_1 \cdot V_B \cdot C_W + k_{D,i} \cdot V_B \cdot \Sigma(P_i \cdot C_{D,i})$$
$$- (k_2 + k_E + k_M + k_G) \cdot V_B \cdot C_B \tag{5}$$

where k_1, k_D, k_2, k_E, and k_G are the rate constant (in units of d^{-1} if concentrations are in $mol.m^{-3}$) for chemical uptake via the respiratory area (k_1), uptake via food ingestion (k_D) and elimination via the respiratory area (k_2), excretion into egested feces (k_E), metabolic transformation (k_M), and growth dilution (k_G); P_i is the fraction of the diet consisting of prey item i, $C_{D,i}$ is the concentration of chemical (g/kg) in prey item i, k_2 is the rate constant (d^{-1}) for chemical elimination via the respiratory area (i.e., gills and skin), k_E is the rate constant (d^{-1}) for chemical elimination via excretion into egested feces, and k_M is the rate constant (d^{-1}) for metabolic transformation of the chemical. A similar approach has been followed to develop mass balance models for other species in both aquatic and terrestrial food webs (Alava et al., 2012; Armitage and Gobas, 2007; Gobas and Arnot, 2010; Kelly and Gobas, 2003).

Practical application of Eq. 5 to environmental pollution problems is often limited by access to time-dependent model input parameter values. Hence, the model is simplified by applying a steady-state assumption ($N_B = 0$), which turns Eq. 5 into a true mass balance:

$$C_B = (k_1.C_W + k_{D,i}.\Sigma(P_i.C_{D,i}))/(k_2 + k_E + k_M + k_G) \tag{6}$$

The steady-state assumption is reasonable for applications to field situations where organisms have been exposed to the chemical over a long period of time, often throughout their entire life since conception. It applies best to chemicals that are subject to relatively fast exchange kinetics as steady state is achieved rapidly or in situations where contamination levels have existed for long periods of time. It should be used with caution in situations where the exchange kinetics are very slow or subject to temporal changes in concentrations. One option to include the age of the animal is to apply the steady-state model to each age class independently.

3.3. Biomagnification

One of the key principles that controls contaminant distribution in food webs is the biomagnification effect. Biomagnification is the process resulting from dietary consumption that causes the fugacity and activity of the chemical in a predator to exceed that in its prey. This process can lead to food-web magnification of the chemical when this process occurs at each or several predator–prey interactions in a food web. Biomagnification in the food web can produce a 10,000- to 100,000-fold increase in fugacity, activity, or lipid normalized concentration of a contaminant (Kelly et al., 2007). Biomagnification is different from bioconcentration in that it produces an increase in thermodynamic potential, while bioconcentration is a chemical partitioning process that, if achieved (e.g., if there is no significant biotransformation, fecal excretion, or growth dilution), establishes a chemical equilibrium between the organism and water where the chemical potential in the organism is equal to that in the environment. Bioaccumulation refers to the combined process of bioconcentration and biomagnification.

Biomagnification is of ecotoxicological significance because it can cause organisms at higher trophic levels to be exposed to high concentrations, which can produce toxicological effects or high-risk levels. Biomagnification occurs as a result of food absorption and digestion in the gastrointestinal tract, which modifies the composition of the dietary matrix and reduces the volume of the dietary matrix in the intestinal tract. This gastrointestinal magnification process causes the ingested chemical to "concentrate" in the gastrointestinal tract (Gobas et al., 1998) (Figure 1.2). This concentration effect varies widely among different animals due to differences in diet composition and dietary assimilation efficiency. For example, in fish the gastrointestinal magnification factor is approximately 8, based on a 50% food-absorption efficiency and a fourfold drop in the diet's absorptive capacity for the chemical upon digestion (Clark et al., 1988). In birds, this factor is higher, for example, 50 in herring gulls (Clark et al., 1988), due to a higher assimilation efficiency of the diet. The gastrointestinal magnification process can cause chemicals to biomagnify in food webs if the rates of chemical elimination and metabolic transformation are low. The degree of biomagnification therefore depends on the extent of gastrointestinal magnification and the rate of chemical elimination and biotransformation. Equilibrium partitioning (which is the mechanism of bioconcentration), and gastrointestinal magnification (which causes biomagnification), are the two main forces that contribute to the process of contaminant bioaccumulation in organisms. Biotransformation, fecal egestion, and somatic growth (which involve processes such as egg and sperm production in fish, or lactation and parturition in mammals) act to reduce fugacities, activities, and concentrations in organisms. The net difference in contaminant fluxes associated with these processes produces the contaminant concentrations.

3.4. Energy Balance

While the mass balance principle ensures the conservation of mass in the model, it does not necessarily ensure that an energy balance is maintained. It is therefore important to ensure that feeding rate, fecal egestion rate, and animal growth rate are consistent (deBruyn and Gobas, 2006). This can be done by applying a general energy balance, that is,

$$IL = R + P \tag{7}$$

where I is energy ingestion, L is the sum of fecal and urinary losses, R is respiration, and P is production, all expressed in units of energy flux $(kJ\, d^{-1})$. These can be converted to mass fluxes $(g\, d^{-1})$ by energy–biomass interconversion ratios. This provides the opportunity to apply the model to a large variety of species for which bioenergetic efficiencies are known (deBruyn and Gobas, 2006) and assess which species are most likely to receive the highest body burdens of particular contaminants.

3.5. Bioaccumulation Models

Several food-web models exist. The most often used models for organic contaminants are those developed by Thomann and coworkers (Thomann, 1989), Gobas and coworkers (Alava et al., 2012; Armitage and Gobas, 2007; Arnot and Gobas, 2003; Gobas, 1993; Gobas and Arnot, 2010; Kelly and Gobas, 2003; Kelly et al., 2007) and Mackay and coworkers (Campfens and Mackay, 1997). The models are used for a range of purposes. For example, the Gobas model (Gobas, 1993), which was originally published for application to the Lake Ontario ecosystem, has been used for the development of water quality criteria and waste load allocations in the United States under the Great Lakes Water Quality Initiative. This model, updated in 2004, is now referred to as AQUAWEB v1.1 (Arnot and Gobas, 2004). It has been used for (i) the derivation of water quality criteria (Arnot and Gobas, 2004), (ii) sediment quality criteria and open-sea disposal scenarios (Alava et al., 2012), (iii) pesticide registration in the form of the Environmental Protection Agency's KABAM (EPA, 2009) and AGRO (Chemistry, 2012) models, (iv) ecosystem-wide exposure and risk assessments to support the derivation of total maximum daily loads (TMDLs) in the United States (Gobas and Arnot, 2010), and (v) chemical screening for bioaccumulative substances in Canada (Arnot and Gobas, 2003). The models developed can be freely downloaded from Chemistry (2012) and Group (2012).

4. QUANTIFYING CONTAMINANTS AND EXPOSURES

Quantifying contaminant concentration is essential to determining exposures and assessing biological effects, but precise quantification of an organism's exposure may be incredibly difficult. The challenges arise from the organism's behavior, the heterogeneity of the chemical in the environment (Figure 1.3), and the expense of chemical analyses, which are necessarily targeted to specific chemicals. Environmental contaminant survey studies often endeavor to measure contaminants within surface waters, which may capture water-soluble compounds, as well as sediments or biota, which may capture hydrophobic, lipophilic compounds. Groundwater can also be sampled, and different depths may be used to reflect a contaminant residency time (the deeper a contaminant's presence, theoretically the longer it has been around and will remain). The expense and challenge of measuring contaminant concentrations may mean that assessment is based on a single grab sample (Tierney et al., 2008, 2011). More comprehensive studies use repeated sampling at monitoring stations (e.g., Table 1.1).

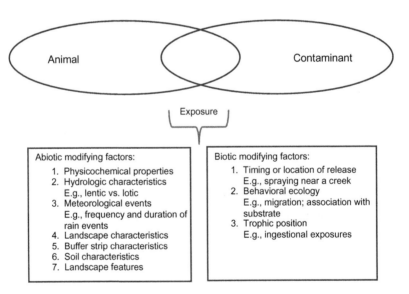

Figure 1.3. The extent to which organic chemicals overlap with life is determined by many abiotic and biotic factors. Some of these factors deal with human activities and how we shape the land, some deal with weather, and others deal with organismal ecology.

An indirect method to test for contaminant exposures is by evaluating the diversity of invertebrate communities. In streams impacted by pesticides, especially neurotoxic insecticides that are efficacious in arthropods, characteristic changes in species diversity and abundance may be evident. For example, ephemeroptera (mayfly) abundance was correlated with OP contamination in Western Cape River, South Africa (Bollmohr and Schulz, 2009), and a rapid, 10-fold decrease in abundance of aquatic insects was associated with a pulse of azinphos-methyl following a heavy rainfall in the Wilmot River, PEI, Canada (Gilliom et al., 2006). This approach is at the core of the SPEAR (SPEcies At Risk), which links environmental stressors with community composition (Rasmussen et al., 2012). Species traits of toxicological sensitivity, generation time, migratory ability, and emergence time are all factored into a predictive model (Liess and Ohe, 2005; Liess et al., 2008).

Land-use practice may be one of the best predictors of regional contaminant profile for (and thus exposure to) agricultural pesticides. Gilliom et al. (2006) proposed grouping agriculture by crops, such as corn and soybeans, wheat and alfalfa, and rice. These groupings had different geographic distributions and were associated with the use of different pesticides. The rice grouping was associated with use of molinate (a carbamate insecticide), while the corn and soybean grouping was associated with several pesticides, including metolachlor and atrazine (herbicides), and the wheat and alfalfa group was low in all pesticides.

Overall, determining some contaminant exposures is a challenge but not an impossible one. An important and topical example is HC mixtures, consisting of polycyclic aromatic PAHs and naphthenic acids (NAs), released by oil sands and their mining. Natural oil sands interface with river water, exposing oil sands and causing a local background level of HCs. HCs are also released by oil sands mining (through storage of HC-contaminated water on site, spills and aerial HC release from upgrading and fractionation of oil). Through sampling surface deposits on the snow, researchers were able to show that aerial deposition was increasing environmental HC contamination above background levels (Kelly et al., 2009). In this example and others, multiple environmental contaminant signals may be integrated to characterize exposures.

5. CONCLUSION

Humankind commenced the foray into organic compounds by harnessing compounds that already existed. As the ability to synthesize our own compounds developed, short-term thinking allowed the production of compounds resistant to degradation. We are now aware of the cost of our short-term thinking, but the legacy of our chemical mistakes continues. Compounds we created decades ago continue to harm life across the planet, especially aquatic life such as fishes in which many of these compounds end up. Although the persistence and toxicity of released compounds are decreasing, the number of compounds is increasing. The toxicities of these compounds remain largely unknown. We may ask, can we keep pace with change?

Although the understanding of the kinetics of xenobiotics in fishes lags far behind the knowledge base that exists for mammalian species, it is clear that many of the principles and mechanisms that govern chemical toxicokinetics are the same. However, evolution has provided some very different outcomes in fish species from different evolutionary lines and lifestyles. This information on biological fate, along with information on environmental fate, has afforded us the ability to place knowledge in a framework whereby we can now reliably model the movement and accumulation of chemicals into fish species. Such modeling has the potential to reduce the heavy costs that may be associated with contaminant monitoring. Future work is focused on model refinement, in the pursuit of linking toxicokinetic and other models to toxicodynamics, and predicting toxicity outcomes. There is much to do regarding the use of such models in a regulatory framework and in risk assessment, but the trend is moving in that direction.

ACKNOWLEDGMENT

The authors are grateful to Tony Farrell for his very thorough review of this chapter.

REFERENCES

Addison, R. F. (1982). Organochlorine compounds and marine lipids. *Progress in Lipid Research* **21**, 47–71.

Alava, J. J., Ross, P. S., Lachmuth, C., Ford, J. K. B., Hickie, B. E. and Gobas, F. A. P. C. (2012). Habitat-based PCB environmental quality criteria for the protection of endangered killer whales (*Orcinus orca*). *Environmental Science and Technology* **46**, 12655–12663.

Albertus, J. A. and Laine, R. O. (2001). Enhanced xenobiotic transporter expression in normal teleost hepatocytes: response to environmental and chemotherapeutic toxins. *Journal of Experimental Biology* **204**, 217–227.

Andersson, T., Koivusaari, U. and Förlin, L. (1985). Xenobiotic biotransformation in the rainbow trout liver and kidney during starvation. *Comparative Biochemistry and Physiology Part C: Comparative Pharmacology* **82**, 221–225.

Arellano-Aguilar, O., Montoya, R. M. and Garcia, C. M. (2009). Endogenous functions and expression of cytochrome P450 enzymes in teleost fish: A review. *Reviews in Fisheries Science* **17**, 541–556.

Armitage, J. M. and Gobas, F. A. P. C. (2007). A terrestrial food-chain bioaccumulation model for POPs. *Environmental Science and Technology* **41**, 4019–4025.

Arnot, J. A. and Gobas, F. A. P. C. (2003). A generic QSAR for assessing the bioaccumulation potential of organic chemicals in aquatic food webs. *QSAR and Combinatorial Science* **22**, 337–345.

Arnot, J. A. and Gobas, F. A. P. C. (2004). A food web bioaccumulation model for organic chemicals in aquatic ecosystems. *Environmental Toxicology and Chemistry* **23**, 2343–2355.

Bard, S. M., Woodin, B. R. and Stegeman, J. J. (2002). Expression of P-glycoprotein and cytochrome P450 1A in intertidal fish (*Anoplarchus purpurescens*) exposed to environmental contaminants. *Aquatic Toxicology* **60**, 17–32.

Bentley, R. and Chasteen, T. G. (2002). Arsenic curiosa and humanity. *Chemical Educator* **7**, 51–60.

Blanchard, P. E. and Donald, W. W. (1997). Herbicide contamination of groundwater beneath claypan soils in north-central Missouri. *Journal of Environmental Quality* **26**, 1612–1621.

Bohm, L. and During, R. A. (2010). Partitioning of polycyclic musk compounds in soil and aquatic environment-experimental determination of K-DOC. *Journal of Soils and Sediments* **10**, 708–713.

Bollmohr, S. and Schulz, R. (2009). Seasonal changes of macroinvertebrate communities in a Western Cape river, South Africa, receiving nonpoint-source insecticide pollution. *Environmental Toxicology and Chemistry* **28**, 809–817.

Borghi, V. and Porte, C. (2002). Organotin pollution in deep-sea fish from the northwestern Mediterranean. *Environmental Science and Technology* **36**, 4224–4228.

Breedveld, G., Skei, J. and Hauge, A. (2010). Contaminants in Norwegian fjord sediments: industrial history or future source? *Journal of Soils and Sediments* **10**, 151–154.

Bruggeman, W. A. (1982). Hydrophobic interactions in the aquatic environment. In *Handbook of Environmental Chemistry*, vol. 2b (ed. O. Hutzinger), pp. 29–49. Berlin Heidelberg: Springer.

Cai, S.-Y., Soroka, C. J., Ballatori, N. and Boyer, J. L. (2003). Molecular characterization of a multidrug resistance-associated protein, Mrp2, from the little skate. *American Journal of Physiology—Regulatory, Integrative and Comparative Physiology* **284**, R125–R130.

Campfens, J. and Mackay, D. (1997). Fugacity-based model of PCB bioaccumulation in complex aquatic food webs. *Environmental Science and Technology* **31**, 577–583.

Carson, R. (1962). *Silent Spring*. 1st Pub. Houghton Mifflin, Mariner Books.

Chan, K. M., Davies, P. L., Childs, S., Veinot, L. and Ling, V. (1992). P-glycoprotein genes in the winter flounder, *Pleuronectes americanus*: Isolation of two types of genomic clones carrying 3′ terminal exons. *Biochimica et Biophysica Acta (BBA)—Gene Structure and Expression* **1171**, 65–72.

Chemistry, C.C.f.E.M.a. (2012). http://www.trentu.ca/academic/aminss/envmodel/: Trent University, Ontario, Canada.

Choi, H. G., Moon, H. B., Choi, M. and Yu, J. (2011). Monitoring of organic contaminants in sediments from the Korean coast: Spatial distribution and temporal trends (2001–2007). *Marine Pollution Bulletin* **62**, 1352–1361.

Christiansen, J. S., Dalmo, R. A. and Ingebrigtsen, K. (1996). Xenobiotic excretion in fish with aglomerular kidneys. *Marine Ecology Progress Series* **136**, 303–304.

Clark, K. E., Gobas, F. and Mackay, D. (1990). Model of organic chemical uptake and clearance by fish from food and water. *Environmental Science and Technology* **24**, 1203–1213.

Clark, T., Clark, K., Paterson, S., Mackay, D. and Norstrom, R. J. (1988). Wildlife monitoring, modeling, and fugacity. Indicators of chemical contamination. *Environmental Science and Technology* **22**, 120–127.

Clarke, D. J., Burchell, B. and George, S. G. (1992). Differential expression and induction of UDP-glucuronosyltransferase isoforms in hepatic and extrahepatic tissues of a fish, Pleuronectes platessa: Immunochemical and functional characterization. *Toxicology and Applied Pharmacology* **115**, 130–136.

Clarke, D. J., George, S. G. and Burchell, B. (1991). Glucuronidation in fish. *Aquatic Toxicology* **20**, 35–56.

Coughtrie, M. W. H. (2002). Sulfation through the looking glass: recent advances in sulfotransferase research for the curious. *Pharmacogenomics Journal* **2**, 297–308.

deBruyn, A. M. H. and Gobas, F. A. P. C. (2006). A bioenergetic biomagnification model for the animal kingdom. *Environmental Science and Technology* **40**, 1581–1587.

Diamond, J. M. and Wright, E. M. (1969). Biological membranes: the physical basis of ION and nonelectrolyte selectivity. *Annual Review of Physiology* **31**, 581–646.

Dixon, T. J., Taggart, J. B. and George, S. G. (2002). Application of real time PCR determination to assess interanimal variabilities in CYP1A induction in the European flounder (*Platichthys flesus*). *Marine Environmental Research* **54**, 267–270.

Doi, A. M., Holmes, E. and Kleinow, K. M. (2001). P-glycoprotein in the catfish intestine: inducibility by xenobiotics and functional properties. *Aquatic Toxicology* **55**, 157–170.

Dörfler, U., Feicht, E. A. and Scheunert, I. (1997). S-triazine residues in groundwater. *Chemosphere* **35**, 99–106.

Ekelund, R. (1989). Bioaccumulation and biomagnification of hydrophobic persistent compounds as exemplified by hexachlorobenze. In *Chemicals in the Aquatic Environment. Advanced Hazard Assessment* (ed. L. Landner), pp. 128–149. New York: Springer-Verlag.

Ensminger, M., Bergin, R., Spurlock, F. and Goh, K. (2011). Pesticide concentrations in water and sediment and associated invertebrate toxicity in Del Puerto and Orestimba Creeks, California, 2007–2008. *Environmental Monitoring and Assessment* **175**, 573–587.

Erickson, R. J. and McKim, J. M. (1990). A simple flow-limited model for exchange of organic chemicals at fish gills. *Environmental Toxicology and Chemistry* **9**, 159–165.

Erickson, R. J., McKim, J. M., Lien, G. J., Hoffman, A. D. and Batterman, S. L. (2006a). Uptake and elimination of ionizable organic chemicals at fish gills: I. Model formulation, parameterization, and behavior. *Environmental Toxicology and Chemistry* **25**, 1512–1521.

Erickson, R. J., McKim, J. M., Lien, G. J., Hoffman, A. D. and Batterman, S. L. (2006b). Uptake and elimination of ionizable organic chemicals at fish gills: II. Observed and predicted effects of pH, alkalinity, and chemical properties. *Environmental Toxicology and Chemistry* **25**, 1522–1532.

Farrell, A. P. (1993). Cardiovascular system. In *The Physiology of Fishes* (ed. D. H. Evans), pp. 219–250. Boca Raton, FL: CRC Press.

Farrell, A. P., Thorarensen, H., Axelsson, M., Crocker, C. E., Gamperl, A. K. and Cech, J. J., Jr (2001). Gut blood flow in fish during exercise and severe hypercapnia. *Comparative Biochemistry and Physiology Part A: Molecular and Integrative Physiology* **128**, 549–561.

Fredriksson, R., Nordström, K. J. V., Stephansson, O., Hägglund, M. G. A. and Schiöth, H. B. (2008). The solute carrier (SLC) complement of the human genome: phylogenetic classification reveals four major families. *FEBS Letters* **582**, 3811–3816.

Gale, W. L., Patiño, R. and Maule, A. G. (2004). Interaction of xenobiotics with estrogen receptors α and β and a putative plasma sex hormone-binding globulin from channel catfish (*Ictalurus punctatus*). *General and Comparative Endocrinology* **136**, 338–345.

Garmouma, M., Blanchard, M., Chesterikoff, A., Ansart, P. and Chevreuil, M. (1997). Seasonal transport of herbicides (triazines and phenylureas) in a small stream draining an agricultural basin: Mélarchez (France). *Water Research* **31**, 1489–1503.

George, S. G. and Taylor, B. (2002). Molecular evidence for multiple UDP-glucuronosyl-transferase gene familes in fish. *Marine Environmental Research* **54**, 253–257.

Gil, Y. and Sinfort, C. (2005). Emission of pesticides to the air during sprayer application: a bibliographic review. *Atmospheric Environment* **39**, 5183–5193.

Giles, D. K., Klassen, P., Niederholzer, F. J. A. and Downey, D. (2011). "Smart" sprayer technology provides environmental and economic benefits in California orchards. *California Agriculture* **65**, 85–89.

Gilliom, R. J., Barbash, J. E., Crawford, C. G., Hamilton, P. A., Martin, J. D., Nakagaki, N., Nowell, L. H., Scott, J. C., Stackelberg, P. E., Thelin, G. P. et al. (2006). The quality of our nation's waters, pesticides in the nation's streams and ground water, 1992–2001, vol. 1291, p. 184: USGS.

Gingerich, W. H. (1982). Hepatic toxicology of fishes. In *Aquatic Toxicology* (ed. L. J. Weber), pp. 55–105. New York: Raven Press.

Gobas, F. A. P. C. (1993). A model for predicting the bioaccumulation of hydrophobic organic chemicals in aquatic food-webs: Application to Lake Ontario. *Ecological Modelling* **69**, 1–17.

Gobas, F. A. P. C. and Arnot, J. A. (2010). Food web bioaccumulation model for polychlorinated biphenyls in San Francisco Bay, California, USA. *Environmental Toxicology and Chemistry* **29**, 1385–1395.

Gobas, F. A. P. C., Wilcockson, J. B., Russell, R. W. and Haffner, G. D. (1998). Mechanism of biomagnification in fish under laboratory and field conditions. *Environmental Science and Technology* **33**, 133–141.

Gobas, F. A. P. C., Zhang, X. and Wells, R. (1993). Gastrointestinal magnification: The mechanism of biomagnification and food chain accumulation of organic chemicals. *Environmental Science and Technology* **27**, 2855–2863.

Gonzalez, F. J. (1988). The molecular biology of cytochrome P450s. *Pharmacological Reviews* **40**, 243–288.

Gonzalez, P., Dominique, Y., Massabuau, J. C., Boudou, A. and Bourdineaud, J. P. (2005). Comparative effects of dietary methylmercury on gene expression in liver, skeletal muscle, and brain of the zebrafish (*Danio rerio*). *Environmental Science and Technology* **39**, 3972–3980.

Goodman, D. S. (1958). The interaction of human serum albumin with long-chain fatty acid anions. *Journal of the American Chemical Society* **80**, 3892–3898.

Group, E. T. R. (2012). http://www.rem.sfu.ca/toxicology/index.htm School of Resource and Environmental Management, Simon Fraser University, Burnaby, BC, Canada.

Guarino, A. M., James, M. O. and Bend, J. R. (1977). Fate and distribution of the herbicides 2,4-dichlor-phenoxyacetic acid (2,4-D) and 2,4,5-trichlorophenoxyacetic acid (2,4,5-T) in the dogfish shark. *Xenobiotica* **7**, 623–631.

Guarino, A. M., Pritchard, J. B., Anderson, J. B. and Rall, D. P. (1974). Tissue distribution of [14C]DDT in the lobster after administration via intravascular or oral routes or after exposure from ambient sea water. *Toxicology and Applied Pharmacology* **29**, 277–288.

Guengerich, F. P. (1987). *Mammalian Cytochrome P450*. Boca Raton, FL: CRC Press.

Hansch, C. (1969). Quantitative approach to biochemical structure-activity relationships. *Accounts of Chemical Research* **2**, 232–239.

Hansch, C. and Clayton, J. M. (1973). Lipophilic character and biological activity of drugs II: The parabolic case. *Journal of Pharmaceutical Sciences* **62**, 1–21.

Harris, K. A., Dangerfield, N., Woudneh, M., Brown, T. G., Verrin, S. and Ross, P. S. (2008). Partitioning of current-use and legacy pesticides in salmon habitat in British Columbia, Canada. *Environmental Toxicology and Chemistry* **27**, 2253–2262.

Hayes, T. B., Case, P., Chui, S., Chung, D., Haefele, C., Haston, K., Lee, M., Pheng Mai, V., Marjuoa, Y., Parker, J., et al. (2006). Pesticide mixtures, endocrine disruption, and amphibian declines: are we underestimating the impact? *Environmental Health Perspectives* **114**, 70.

Heilmann, L. J., Sheen, Y. Y., Bigelow, S. W. and Nebert, D. W. (1988). Trout P4501A1: DNA and deduced protein sequence, expression in liver and evolutionary significance. *DNA* **7**, 379–387.

Hintelmann, H. (2010). Organomercurials. Their formation and pathways in the environment. In *Organometallics in Environment and Toxicology*, vol. 7 (eds. A. Sigel, H. Sigel and R. K. O. Sigel), pp. 365–401. Cambridge: Royal Soc Chemistry.

Honkakoski, P. and Negishi, M. (2000). Regulation of cytochrome P450 (CYP) genes by nuclear receptors. *Biochemical Journal* **347**, 321–337.

Hunn, J. B. and Allen, J. L. (1974). Movement of drugs across the gills of fishes. *Annual Review of Pharmacology* **14**, 47–54.

Ishikawa, T. (1992). The ATP-dependent glutathione S-conjugate export pump. *Trends in Biochemical Sciences* **17**, 463–468.

James, M. O. (1976). Taurine conjugation of carboxylic acids in some marine species. In *Conjugation Reactions in Drug Biotransformations* (ed. A. Aitio), pp. 121–129. Amsterdam: Elsevier.

James, M. O., Altman, A. H., Morris, K., Kleinow, K. M. and Tong, Z. (1997). Dietary modulation of phase 1 and phase 2 activities with benzo(a)pyrene and related compounds in the intestine but not the liver of the channel catfish, *Ictalurus punctatus*. *Drug Metabolism and Disposition* **25**, 346–354.

James, M. O. and Pritchard, J. B. (1987). *In vivo* and *in vitro* renal metabolism and excretion of benzoic acid by a marine teleost, the southern flounder. *Drug Metabolism and Disposition* **15**, 665–670.

Juliano, R. L. and Ling, V. (1976). A surface glycoprotein modulating drug permeability in Chinese hamster ovary cell mutants. *Biochimica et Biophysica Acta (BBA)—Biomembranes* **455**, 152–162.

Kapoor, I. P., Metcalf, R. L., Hirwe, A. S., Coats, J. R. and Khalsa, M. S. (1973). Structure activity correlations of biodegradability of DDT analogs. *Journal of Agricultural and Food Chemistry* **21**, 310–315.

Karnaky, K. J., Sluggs, K., French, S. and Willingham, S. C. (1993). Evidence for the multidrug transporter, P-glycoprotein, in the killifish, *Fundulus heteroclitus*. *Bulletin of the Mount Desert Island Biological Laboratory* **31**, 61–62.

Kelly, B. C. and Gobas, F. A. P. C. (2003). An arctic terrestrial food-chain bioaccumulation model for persistent organic pollutants. *Environmental Science and Technology* **37**, 2966–2974.

Kelly, B. C., Ikonomou, M. G., Blair, J. D., Morin, A. E. and Gobas, F. A. P. C. (2007). Food web-specific biomagnification of persistent organic pollutants. *Science* **317**, 236–239.

Kelly, E. N., Short, J. W., Schindler, D. W., Hodson, P. V., Ma, M., Kwan, A. K. and Fortin, B. L. (2009). Oil sands development contributes polycyclic aromatic compounds to the Athabasca River and its tributaries. *Proceedings of the National Academy of Sciences* **106**, 22346–22351.

Kelly, S. L., Lamb, D. C. and Kelly, D. E. (2006). Cytochrome P450 biodiversity and biotechnology. *Biochemical Society Transactions* **34**, 1159–1160.

Kennedy, C. J., Gill, K. A. and Walsh, P. J. (1991). *In-vitro* metabolism of benzo(a)pyrene in the blood of the gulf toadfish, *Opsanus beta*. *Marine Environmental Research* **31**, 37–53.

Kennedy, C. J. and Law, F. C. P. (1986). Toxicokinetics of chlorinated phenols in rainbow trout following different routes of chemical administration. *Canadian Technical Report of Fisheries and Aquatic Sciences* 124–125.

Kleinow, K. M., Beilfuss, W. L., Jarboe, H. H., Droy, B. F. and Lech, J. J. (1992). Pharmacokinetics, bioavailability, distribution, and metabolism of sulfadimethoxine in the rainbow trout (*Oncorhynchus mykiss*). *Canadian Journal of Fisheries and Aquatic Sciences* **49**, 1070–1077.

Kobayashi, K., Kimura, S. and Akitake, H. (1976). Studies on the metabolism of chlorophenols in fish. VII. Sulfate conjugation of phenol and PCP by fish livers. *Bulletin of the Japanese Society for the Science of Fish* **42**, 171–176.

Kobayashi, K., Kimura, S. and Akitake, H. (1984). Sulfate conjugation of various phenols by liver-soluble fraction of goldfish. *Bulletin of the Japanese Society for the Science of Fish* **50**, 833–837.

König, J., Nies, A. T., Cui, Y., Leier, I. and Keppler, D. (1999). Conjugate export pumps of the multidrug resistance protein (MRP) family: Localization, substrate specificity, and MRP2-mediated drug resistance. *Biochimica et Biophysica Acta (BBA)—Biomembranes* **1461**, 377–394.

Konishi, T., Kato, K., Araki, T., Shiraki, K., Takagi, M. and Tamaru, Y. (2005). A new class of glutathione S-transferase from the hepatopancreas of the red sea bream *Pagrus major*. *Biochemical Journal* **388**, 299–307.

Kurelec, B. (1992). The multixenobiotic resistance mechanism in aquatic organisms. *Critical Reviews in Toxicology* **22**, 23–43.

Kurelec, B., Krča, S., Pivčevic, B., Ugarković, D., Bachmann, M., Imsiecke, G. and Müller, W. E. G. (1992). Expression of P-glycoprotein gene in marine sponges. Identification and characterization of the 125 kDa drug-binding glycoprotein. *Carcinogenesis* **13**, 69–76.

Kvanli, D. M., Marisetty, S., Anderson, T. A., Jackson, W. A. and Morse, A. N. (2008). Monitoring estrogen compounds in wastewater recycling systems. *Water, Air, and Soil Pollution* **188**, 31–40.

Law, F. C. P., Abedini, S. and Kennedy, C. J. (1991). A biologically based toxicokinetic model for pyrene in rainbow trout. *Toxicology and Applied Pharmacology* **110**, 390–402.

Layiwola, P. J. and Linnecar, D. F. C. (1981). The biotransformation of [14C]phenol in some freshwater fish. *Xenobiotica* **11**, 167–171.

Leaver, M. J., Pirrit, L. and George, S. G. (1993). Cytochrome P450 1A1 cDNA from plaice (*Pleuronectes platessa*) and induction of P450 1A1 mRNA in various tissues by 3-methylcholanthrene and isosafrole. *Molecular Marine Biology and Biotechnology* **2**, 338–345.

Leaver, M. J., Wright, J., Hodgson, P., Boukouvala, E. and George, S. G. (2007). Piscine UDP-glucuronosyltransferase 1B. *Aquatic Toxicology* **84**, 356–365.

Lech, J. J. and Bend, J. R. (1980). Relationship between biotransformation and the toxicity and fate of xenobiotic chemicals in fish. *Environmental Health Perspectives* **34**, 115–131.

Lewis, V. (2001). *Guide to Cytochromes P450 Structure and Function*. London: Taylor and Francis.

Lien, G. J. and McKim, J. M. (1993). Predicting branchial and cutaneous uptake of 2,2′,5,5′-tetrachlorobiphenyl in fathead minnows (*Pimephales promelas*) and Japanese medaka (*Oryzias latipes*): Rate limiting factors. *Aquatic Toxicology* **27**, 15–31.

Liess, M. and Ohe, P. C. V. D. (2005). Analyzing effects of pesticides on invertebrate communities in streams. *Environmental Toxicology and Chemistry* **24**, 954–965.

Liess, M., Schäfer, R. B. and Schriever, C. A. (2008). The footprint of pesticide stress in communities—Species traits reveal community effects of toxicants. *Science of the Total Environment* **406**, 484–490.

Lindström-Seppä, P., Koivusaari, U. and Hänninen, O. (1981). Extrahepatic xenobiotic metabolism in North-European freshwater fish. *Comparative Biochemistry and Physiology Part C: Comparative Pharmacology* **69**, 259–263.

Litman, T., Brangi, M., Hudson, E., Fetsch, P., Abati, A., Ross, D. D., Miyake, K., Resau, J. H. and Bates, S. E. (2000). The multidrug-resistant phenotype associated with overexpression of the new ABC half-transporter, MXR (ABCG2). *Journal of Cell Science* **113**, 2011–2021.

Livingstone, D. R., Chipman, J. K., Lowe, D. M., Minier, C., Mitchelmore, C. L., Moore, M. N., Peters, L. D. and Pipe, R. K. (2000). Development of biomarkers to detect the effects of organic pollution on aquatic invertebrates: recent molecular, genotoxic, cellular and immunological studies on the common mussel (*Mytilus edulis* L.) and other mytilids. *International Journal of Environmental Pollution* **13**, 56–91

Lončar, J., Popović, M., Zaja, R. and Smital, T. (2010). Gene expression analysis of the ABC efflux transporters in rainbow trout (*Oncorhynchus mykiss*). *Comparative Biochemistry and Physiology Part C: Toxicology and Pharmacology* **151**, 209–215.

MacCormack, T. J. and Goss, G. G. (2008). Identifying and predicting biological risks associated with manufactured nanoparticles in aquatic ecosystems. *Journal of Industrial Ecology* **12**, 286–296.

Mackay, D. (2001). *Multimedia Environmental Models: The Fugacity Approach*. Boca Raton, FL: Lewis Publishers.

Mackay, D. and Arnot, J. A. (2011). The application of fugacity and activity to simulating the environmental fate of organic contaminants. *Journal of Chemical and Engineering Data* **56**, 1348–1355.

Mackenzie, P. I., Gregory, P. A., Gardner-Stephen, D. A., Lewinsky, R. H., Jorgensen, B. R., Nishiyama, T., Xie, W. and Radominska-Pandya, A. (2003). Regulation of UDP Glucuronosyltransferase Genes. *Current Drug Metabolism* **4**, 249–257.

Maglich, J. M., Stoltz, C. M., Goodwin, B., Hawkins-Brown, D., Moore, J. T. and Kliewer, S. A. (2002). Nuclear pregnane X receptor and constitutive androstane receptor regulate overlapping but distinct sets of genes involved in xenobiotic detoxification. *Molecular Pharmacology* **62**, 638–646.

Mannervik, B., Awasthi, Y. C., Board, P. G., Hayes, J. D., Di Ilio, C., Ketterer, B., Listowsky, I., Morgenstern, R., Muramatsu, M., Pearson, W. R., et al. (1992). Nomenclature for human glutathione transferases. *Biochemical Journal* **282** (Pt 1), 305–306.

Masse, L., Patni, N. K., Jui, P. Y. and Clegg, B. S. (1998). Groundwater quality under conventional and no tillage: II. atrazine, deethylatrazine, and metolachlor. *Journal of Environmental Quality* **27**, 877–883.

McKim, J., Schmieder, P. and Veith, G. (1985). Absorption dynamics of organic chemical transport across trout gills as related to octanol-water partition coefficient. *Toxicology and Applied Pharmacology* **77**, 1–10.

McKim, J. M. and Heath, E. M. (1983). Dose determinations for waterborne 2,5,2′,5′-[14C] tetrachlorobiphenyl and related pharmacokinetics in two species of trout (*Salmo gairdneri* and *Salvelinus fontinalis*): A mass-balance approach. *Toxicology and Applied Pharmacology* **68**, 177–187.

McKim, J. M., Schmieder, P. K. and Erickson, R. J. (1986). Toxicokinetic modeling of [14C]pentachlorophenol in the rainbow trout (*Salmo gairdneri*). *Aquatic Toxicology* **9**, 59–80.

Miller, D. S. (2002). Xenobiotic export pumps, endothelin signaling, and tubular nephrotoxicants—a case of molecular hijacking. *Journal of Biochemical and Molecular Toxicology* **16**, 121–127.

Miller, D. S., Graeff, C., Droulle, L., Fricker, S. and Fricker, G. (2002). Xenobiotic efflux pumps in isolated fish brain capillaries. *American Journal of Physiology—Regulatory, Integrative and Comparative Physiology* **282**, R191–R198.

Miller, D. S., Masereeuw, R., Henson, J. and Karnaky, K. J. (1998). Excretory transport of xenobiotics by dogfish shark rectal gland tubules. *American Journal of Physiology—Regulatory, Integrative and Comparative Physiology* **275**, R697–R705.

Munkittrick, K. R., McMaster, M. E., Portt, C. B., Kraak, G. J. V. D., Smith, I. R. and Dixon, D. G. (1992). Changes in maturity, plasma sex steroid levels, hepatic mixed-function oxygenase activity, and the presence of external lesions in Lake Whitefish (*Coregonus clupeaformis*) exposed to bleached kraft mill effluent. *Canadian Journal of Fisheries and Aquatic Sciences* **49**, 1560–1569.

Nagata, K. and Yamazoe, Y. (2000). Pharmacogenetics of sulfotransferase. *Annual Review of Pharmacology and Toxicology* **40**, 159–176.

Nebert, D. W. and Gonzalez, F. J. (1987). P450 genes: Structure, evolution, and regulation. *Annual Reviews of Biochemistry* **56**, 945–993.

Nelson, D. R. (1999). A second CYP26 P450 in humans and zebrafish: CYP26B1. *Archives of Biochemistry and Biophysics* **371**, 345–347.

Ohkimoto, K., Sugahara, T., Sakakibara, Y., Suiko, M., Liu, M.-J., Carter, G. and Liu, M.-C. (2003). Sulfonation of environmental estrogens by zebrafish cytosolic sulfotransferases. *Biochemical and Biophysical Research Communications* **309**, 7–11.

Ohliger, R. and Schulz, R. (2010). Water body and riparian buffer strip characteristics in a vineyard area to support aquatic pesticide exposure assessment. *Science of the Total Environment* **408**, 5405–5413.

Opperhuizen, A. (1986). *Bioconcentration of hydrophobic chemicals in fishAquatic Toxicology and Environmental Fate*, vol. 9. Baltimore, MD: American Society for Testing and Materials, (pp. 304–315).

Palace, V. P., Wautier, K. G., Evans, R. E., Blanchfield, P. J., Mills, K. H., Chalanchuk, S. M., Godard, D., McMaster, M. E., Tetreault, G. R., Peters, L. E., et al. (2006). Biochemical and histopathological effects in pearl dace (*Margariscus margarita*) chronically exposed to a synthetic estrogen in a whole lake experiment. *Environmental Toxicology and Chemistry* **25**, 1114–1125.

Parker, R. S., Morrissey, M. T., Moldeus, P. and Selivonchick, D. P. (1981). The use of isolated hepatocytes from rainbow trout (*Salmo gairdneri*) in the metabolism of acetaminophen. *Comparative Biochemistry and Physiology Part B: Comparative Biochemistry* **70**, 631–633.

Paulson, G. D., Caldwell, J., Hutson, D. H. and Menn, J. J. (1986). *Xenobiotic Conjugation Chemistry*. Washington, DC: American Chemical Society.

Perrier, H., Perrier, C., Peres, G. and Gras, J. (1977). The perchlorosoluble proteins of the serum of the rainbow trout (*Salmo gairdnerii* Richardson): Albumin like and hemoglobin binding fraction. *Comparative Biochemistry and Physiology Part B: Comparative Biochemistry* **57**, 325–327.

Phillips, V. J. A., St. Louis, V. L., Cooke, C. A., Vinebrooke, R. D. and Hobbs, W. O. (2011). Increased mercury loadings to western Canadian alpine lakes over the past 150 years. *Environmental Science and Technology* **45**, 2042–2047.

Potter, J. L. and O'Brien, R. D. (1964). Parathion activation by livers of aquatic and terrestrial vertebrates. *Science* **144**, 55–57.

Pritchard, J. B. and Bend, J. R. (1984). Mechanisms controlling the renal excretion of xenobiotics in fish: effects of chemical structure. *Drug Metabolism Reviews* **15**, 655–671.

Pucarević, M., Šovljanski, R., Lazić, S. and Marjanović, N. (2002). Atrazine in groundwater of Vojvodina Province. *Water Research* **36**, 5120–5126.

Purcell, L. A. and Giberson, D. J. (2007). Effects of an azinphos-methyl runoff event on macroinvertebrates in the Wilmot River, Prince Edward Island, Canada. *The Canadian Entomologist* **139**, 523–533.

Qiao, P. and Farrell, A. P. (1996). Uptake of hydrophobic xenobiotics by fish in water laden with sediments from the fraser river. *Environmental Toxicology and Chemistry* **15**, 1555–1563.

Rae, J. M., Johnson, M. D., Lippman, M. E. and Flockhart, D. A. (2001). Rifampin is a selective, pleiotropic inducer of drug metabolism genes in human hepatocytes: studies with cDNA and oligonucleotide expression arrays. *Journal of Pharmacology and Experimental Therapeutics* **299**, 849–857.

Rasmussen, J. J., Wiberg-Larsen, P., Baattrup-Pedersen, A., Friberg, N. and Kronvang, B. (2012). Stream habitat structure influences macroinvertebrate response to pesticides. *Environmental Pollution* **164**, 142–149.

Reichenberger, S., Bach, M., Skitschak, A. and Frede, H.-G. (2007). Mitigation strategies to reduce pesticide inputs into ground- and surface water and their effectiveness; A review. *Science of the Total Environment* **384**, 1–35.

Rombough, P. (2002). Gills are needed for ionoregulation before they are needed for O_2 uptake in developing zebrafish, *Danio rerio*. *Journal of Experimental Biology* **205**, 1787–1794.

Ross, P. S. (2006). Fireproof killer whales (*Orcinus orca*): Flame-retardant chemicals and the conservation imperative in the charismatic icon of British Columbia, Canada. *Canadian Journal of Fisheries and Aquatic Sciences* **63**, 224–234.

Saarikoski, J., Lindström, R., Tyynelä, M. and Viluksela, M. (1986). Factors affecting the absorption of phenolics and carboxylic acids in the guppy (*Poecilia reticulata*). *Ecotoxicology and Environmental Safety* **11**, 158–173.

Sass, J. B. and Colangelo, A. (2006). European Union bans atrazine, while the United States negotiates continued use. *International Journal of Occupational and Environmental Health* **12**, 260–267.

Sauerborn, R., Polancec, D. S., Zaja, R. and Smital, T. (2004). Identification of the multidrug resistance-associated protein (mrp) related gene in red mullet (*Mullus barbatus*). *Marine Environmental Research* **58**, 199–204.

Schanker, L. S. (1961). Mechanisms of drug absorption and distribution. *Annual Review of Pharmacology* **1**, 29–45.

Schär, M., Maly, I. P. and Sasse, D. (1985). Histochemical studies on metabolic zonation of the liver in the trout (*Salmo gairdneri*). *Histochemistry* **83**, 147–151.

Schlenk, D., Celander, M., Gallaher, E., George, S., James, M., Kullman, S., vanden Hurk, P. and Willett, K. (2008). Biotransformation in fishes. In *The Toxicology of Fishes* (eds. R. Di Giulio and D. Hinton), Boca Raton, FL: CRC Press.

Schulz, R. (2001). Comparison of spray drift- and runoff-related input of azinphos-methyl and endosulfan from fruit orchards into the Lourens River, South Africa. *Chemosphere* **45**, 543–551.

Schulz, R., Peall, S. K. C., Dabrowski, J. M. and Reinecke, A. J. (2001). Current-use insecticides, phosphates and suspended solids in the Lourens River, Western Cape, during the first rainfall event of the wet season. *Water SA* **27**, 65–70.

Slinn, J. (2006). Shaping the industrial century. The remarkable story of the evolution of the modern chemical and pharmaceutical industries. *Business History* **48**, 604–605.

Spacie, A. and Hamelink, J. L. (1982). Alternative models for describing the bioconcentration of organics in fish. *Environmental Toxicology and Chemistry* **1**, 309–320.

Stadnicka, J., Schirmer, K. and Ashauer, R. (2012). Predicting concentrations of organic chemicals in fish by using toxicokinetic models. *Environmental Science and Technology* **46**, 3273–3280.

Stuchal, L. D., Kleinow, K. M., Stegeman, J. J. and James, M. O. (2006). Demethylation of the pesticide methoxychlor in liver and intestine from untreated, methoxychlor-treated, and 3-methylcolanthrene-treated channel catfish (Ictalurus punctatus): Evidence for roles of CYP1 and CYP3 family isozymes. *Drug Metabolism and Disposition* **34**, 932–938.

Sturm, A. and Segner, H. (2005). P-glycoproteins and xenobiotic efflux transport in fish. In *Biochemistry and Molecular Biology of Fishes*, vol. 6 (eds. T. P. Mommsen and T. W. Moon), pp. 495–533. Amsterdam: New York: Elsevier Science B.V.

Sugahara, T., Liu, C.-C., Carter, G., Govind Pai, T. and Liu, M.-C. (2003a). cDNA cloning, expression, and functional characterization of a zebrafish SULT1 cytosolic sulfotransferase. *Archives of Biochemistry and Biophysics* **414**, 67–73.

Sugahara, T., Liu, C.-C., Pai, T. G., Collodi, P., Suiko, M., Sakakibara, Y., Nishiyama, K. and Liu, M.-C. (2003b). Sulfation of hydroxychlorobiphenyls. *European Journal of Biochemistry* **270**, 2404–2411.

Thomann, R. V. (1989). Bioaccumulation model of organic chemical distribution in aquatic food chains. *Environmental Science and Technology* **23**, 699–707.

Thomas, R. E. and Rice, S. D. (1981). Excretion of aromatic hydrocarbons and their metabolites by freshwater and saltwater *Dolly Varden* char. In *Biological Monitoring of Marine Pollutants* (eds. F. J. Vernberg, F. P. Thurberg, A. Calabrese and W. B. Vernberg), pp. 425–448. New York: Academic Press.

Tierney, K. B., Sampson, J. L., Ross, P. S., Sekela, M. A. and Kennedy, C. J. (2008). Salmon olfaction is impaired by an environmentally realistic pesticide mixture. *Environmental Science and Technology* **42**, 4996–5001.

Tierney, K. B., Sekela, M. A., Cobbler, C. E., Xhabija, B., Gledhill, M., Ananvoranich, S. and Zielinski, B. S. (2011). Evidence for behavioral preference towards environmental concentrations of urban-use herbicides in a model adult fish. *Environmental Toxicology and Chemistry* **30**, 2046–2054.

Tierney, K. B., Singh, C. R., Ross, P. S. and Kennedy, C. J. (2007). Relating olfactory neurotoxicity to altered olfactory-mediated behaviors in rainbow trout exposed to three currently-used pesticides. *Aquatic Toxicology* **81**, 55–64.

Tollefsen, K. E. (2002). Interaction of estrogen mimics, singly and in combination, with plasma sex steroid-binding proteins in rainbow trout (*Oncorhynchus mykiss*). *Aquatic Toxicology* **56**, 215–225.

Tong, Z. and James, M. O. (2000). Purification and characterization of hepatic and intestinal phenol sulfotransferase with high affinity for benzo[a]pyrene phenols from channel catfish, *Ictalurus punctatus*. *Archives of Biochemistry and Biophysics* **376**, 409–419.

Tovell, P. W. A., Howes, D. and Newsome, C. S. (1975). Absorption, metabolism and excretion by goldfish of the anionic detergent sodium lauryl sulphate. *Toxicology* **4**, 17–29.

Trauner, M. and Boyer, J. L. (2003). Bile salt transporters: molecular characterization, function, and regulation. *Physiological Reviews* **83**, 633–671.

U.S. EPA (2010). Hudson river PCBs site EPA phase 1 evaluation report. In *1 Appendices* (ed. L. B. Group), p. 272. Morristown, NJ: Prepared for U.S. EPA, Region 2 and U.S. Army Corps of Engineers.

U.S. EPA (2009). User's Guide and Technical Documentation of KABAM Version 1.0 (Kow (based) Aquatic BioAccumulation Model) http://www.epa.gov/oppefed1/models/water/kabam/kabam_user_guide.html: US Environmental Protection Agency

Van Aubel, R. A. M. H., Masereeuw, R. and Russel, F. G. M. (2000). Molecular pharmacology of renal organic anion transporters. *American Journal of Physiology—Renal Physiology* **279**, F216–F232.

van Herwaarden, A. E., Wagenaar, E., Karnekamp, B., Merino, G., Jonker, J. W. and Schinkel, A. H. (2006). Breast cancer resistance protein (Bcrp1/Abcg2) reduces systemic exposure of the dietary carcinogens aflatoxin B1, IQ and Trp-P-1 but also mediates their secretion into breast milk. *Carcinogenesis* **27**, 123–130.

Varanasi, U., Stein, J. E. and Hom, T. (1981). Covalent binding of benzo[a]pyrene to DNA in fish liver. *Biochemical and Biophysical Research Communications* **103**, 780–787.

Verrin, S. M., Begg, S. J. and Ross, P. S. (2004). Pesticide use in British Columbia and the Yukon: An assessment of types, applications, and risks to aquatic biota. In *Canadian Technical Report of Fisheries and Aquatic Sciences*, vol. 2517, p. 209: Fisheries and Oceans Canada.

Vischetti, C., Cardinali, A., Monaci, E., Nicelli, M., Ferrari, F., Trevisan, M. and Capri, E. (2008). Measures to reduce pesticide spray drift in a small aquatic ecosystem in vineyard estate. *Science of the Total Environment* **389**, 497–502.

Wang, J. and Gardinali, P. R. (2012). Analysis of selected pharmaceuticals in fish and the fresh water bodies directly affected by reclaimed water using liquid chromatography-tandem mass spectrometry. *Analytical and Bioanalytical Chemistry* **404**, 2711–2720.

Wang, L., Scheffler, B. E. and Willett, K. L. (2006). CYP1C1 messenger RNA expression is inducible by benzo[a]pyrene in *fundulus heteroclitus* embryos and adults. *Toxicological Sciences* **93**, 331–340.

Watabe, T. (1983). Metabolic activation of 7,12-dimethylbenzanthracene (DMBA) and 7-methylbenz[alpha]anthracene (7-MBA) by rat liver P-450 and sulfotransferase. *Journal of Toxicological Sciences* **8**, 119–131.

Waterman, M. R. and Johnson, E. F. (1991). *Cytochrome P450. Methods in Enzymology*. New York: Academic Press.

Weber, K. and Goerke, H. (2003). Persistent organic pollutants (POPs) in antarctic fish: levels, patterns, changes. *Chemosphere* **53**, 667–678.

Weston, D. P., You, J. and Lydy, M. J. (2004). Distribution and toxicity of sediment-associated pesticides in agriculture-dominated water bodies of California's central valley. *Environmental Science and Technology* **38**, 2752–2759.

Williams, R. (1959). *Detoxication Mechanisms: The Metabolism and Detoxication of Drugs, Toxic Substances, and Other Organic Compounds*. New York: John Wiley & Sons.

Wu, Q., Riise, G., Lundekvam, H., Mulder, J. and Haugen, L. E. (2004). Influences of suspended particles on the runoff of pesticides from an agricultural field at Askim, SE-Norway. *Environmental Geochemistry and Health* **26**, 295–302.

Xie, W. and Evans, R. M. (2001). Orphan nuclear receptors: The exotics of xenobiotics. *Journal of Biological Chemistry* **276**, 37739–37742.

Xie, W., Yeuh, M.-F., Radominska-Pandya, A., Saini, S. P. S., Negishi, Y., Bottroff, B. S., Cabrera, G. Y., Tukey, R. H. and Evans, R. M. (2003). Control of steroid, heme, and carcinogen metabolism by nuclear pregnane X receptor and constitutive androstane receptor. *Proceedings of the National Academy of Sciences* **100**, 4150–4155.

Xu, C., Li, C.-T. and Kong, A.-N. (2005). Induction of phase I, II and III drug metabolism/transport by xenobiotics. *Archives of Pharmacal Research* **28**, 249–268.

Yalkowski, S. H. and Morozowich, W. (1980). A physical chemical basis for the design of orally active prodrugs. In *Drug Design*, vol. 9 (ed. E. J. Ariens), pp. 122–185. New York: Academic Press.

Zhang, X., Liu, X., Zhang, M., Dahlgren, R. A. and Eitzel, M. (2010). A review of vegetated buffers and a meta-analysis of their mitigation efficacy in reducing nonpoint source pollution. *Journal of Environmental Quality* **39**, 76–84.

2

EFFECTS OF LEGACY PERSISTENT ORGANIC POLLUTANTS (POPS) IN FISH—CURRENT AND FUTURE CHALLENGES

LYNDAL L. JOHNSON

BERNADITA F. ANULACION

MARY R. ARKOOSH

DOUGLAS G. BURROWS

DENIS A. M. DA SILVA

JOSEPH P. DIETRICH

MARK S. MYERS

JULANN SPROMBERG

GINA M. YLITALO

1. Introduction
2. Transformations of POPs in the Aquatic Environment
3. Endocrine Disruption
   3.1. Disruption of Reproductive Function
   3.2. Disruption of Thyroid Function
   3.3. Disruption of the Hypothalamus-Adrenal-Pituitary Axis
   3.4. Disruption of Other Metabolic Processes
4. Effects on Early Development
   4.1. Maternal Transfer of POPs
   4.2. Direct Toxic Effects
   4.3. Impaired Maternal Transfer of Nutrients or Hormones
5. Transgenerational Effects
6. Effects on Growth, Condition, and Energy Reserves
7. Effects on Reproduction
   7.1. Onset of Puberty
   7.2. Fecundity in Females
   7.3. Fertility of Males
   7.4. Reproductive Behavior
8. Immunotoxicity
9. Neoplasia and Related Pathological Conditions
   9.1. Liver Glycogen Depletion and Hepatocyte Alterations
   9.2. Neoplasia and Other Toxicopathic Lesions

Organic Chemical Toxicology of Fishes: Volume 33
FISH PHYSIOLOGY

1. INTRODUCTION

Persistent organic pollutants (POPs) are ubiquitous environmental contaminants that are not readily degraded in the environment and can biomagnify in aquatic and marine food webs (Jones and de Voogt, 1999; Mackay and Fraser, 2000). POPs include a wide range of halogenated legacy contaminants [e.g., polychlorinated biphenyls (PCBs), DDTs, chlordanes, hexachlorobenzene (HCB)] and chemicals of emerging concern [e.g., polybrominated diphenyl ethers (PBDEs), perfluorinated compounds] that are lipophilic or proteophilic and can be found in relatively high concentrations in tissues of aquatic organisms, including fish and marine mammals (Schmidt, 1999; AMAP, 2004). Many of these chemicals were used as industrial compounds (e.g., PCBs, PBDEs) or pesticides (e.g., dieldrin, DDT). Other classes of POPs [e.g., polychlorinated dibenzofurans (PCDFs) and polychlorinated dibenzodioxins (PCDDs)] can be produced during natural disasters (e.g., forest fires, volcanic eruptions) or as by-products during wood pulp processing, as the synthesis of chlorinated chemicals, or as a result of incineration of chlorine-containing compounds (Safe, 1990; Fiedler, 1996; Srogi, 2008).

POPs enter aquatic environments directly (e.g., transformer spills, agricultural runoff) or indirectly via atmospheric transport and ocean currents (Breivik et al., 2004; Li and Macdonald, 2005). For example, high PCB levels have been measured in environmental samples (including fish tissues) collected downstream from electric capacitor manufacturing plants along the Hudson River as a result of PCB discharges into the river and nearby soil for more than 30 years (HRTC, 2002; Baldigo et al., 2006). DDTs can be transported from areas where they are still used to control disease vectors (e.g., malaria-carrying mosquitoes) to more remote regions of the world (e.g., the Arctic) via atmospheric transport or ocean currents (Hageman et al., 2006; Van den Berg, 2009; Letcher et al., 2010). Recently, studies have shown that POPs in seawater can be adsorbed on plastics (e.g., microplastic particles, plastic resin) and can be transported via ocean currents to various regions around the world (Rios et al., 2007; Teuten et al.,

2009). Marine organisms can then ingest POPs via these plastic vectors. In addition, migrating biota (e.g., fish, birds) have been shown to be sources of POPs to Arctic or subarctic aquatic ecosystems (Ewald et al., 1998; Evenset et al., 2007).

Many POPs have been banned for production and use in a number of countries around the world due to their toxic effects on wildlife and humans. For example, PCBs have not been produced in North America since 1977 (HealthCanada, 2001), whereas DDTs have been completely banned for use and production in North America since 2000 (CEC, 2003). On a more global scale, the Stockholm Convention on Persistent Organic Pollutants, an international environmental treaty, required that the member nations eliminate or restrict the production and use, or reduce unintentional use, of the following 12 POPs: aldrin, chlordane, dieldrin, endrin, heptachlor, HCB, mirex, toxaphene, PCBs, DDT, PCDDs, and PCDFs (UNEP, 2001). The Convention adopted amendments to the original document in 2009 and 2011, which included the addition of the following 10 POPs: alpha-hexachloro-cyclohexane, beta-hexachlorocyclohexane, chlordecone, hexabromobiphe-nyl, hexa- and hepta-bromodiphenyl ethers, lindane, pentachlorobenzene, tetra- and penta-bromodiphenyl ethers, perfluorooctane sulfonic acid, its salts and perfluorooctane sulfonyl fluoride, and technical endosulfan and its isomers (UNEP, 2009, 2011). The chemical structures of some of these compounds are shown in Figure 2.1. Although POP levels rapidly decreased in environmental samples in the years immediately following these bans, they continue to be detected in environmental samples around the world due to their persistence, limited metabolism, and capacity for long-range transport.

Since Jensen (1966) first reported measuring POPs in tissues of fish from Sweden, a large number of studies have been published in the peer-reviewed literature that report environmentally relevant concentrations of POPs determined in freshwater and marine species of fish, as well as their potential biological effects. These research efforts included determining concentra-tions in various tissues of fish, identifying the biological factors that influence POP concentrations, examining the fate and transport of these compounds in different ecosystems, or showing the effects of sublethal and chronic exposure of POPs to fish, particularly PCBs, including the toxic dioxin-like congeners (Safe, 1990). More recent studies have examined the effects of exposure to the PBDE-flame retardants (reviewed in Shaw and Kannan, 2009), and have determined temporal trends of various POPs in fish (Rigét et al., 2010). In this chapter we report on the toxic effects of a select group of POPs—PCBs, DDTs, chlordanes, HCHs, dieldrin, endo-sulfan, endrin, toxaphene, PCDDs, PCDFs, PBDEs—on fish, with a focus on studies conducted since 2000.

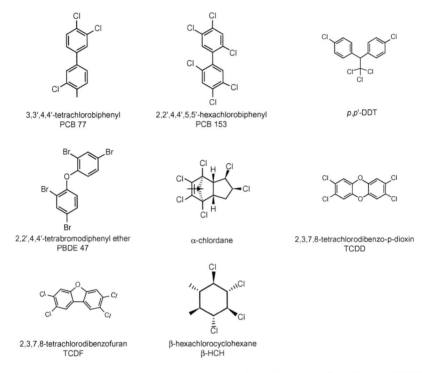

3,3',4,4'-tetrachlorobiphenyl
PCB 77

2,2',4,4',5,5'-hexachlorobiphenyl
PCB 153

p,p'-DDT

2,2',4,4'-tetrabromodiphenyl ether
PBDE 47

α-chlordane

2,3,7,8-tetrachlorodibenzo-p-dioxin
TCDD

2,3,7,8-tetrachlorodibenzofuran
TCDF

β-hexachlorocyclohexane
β-HCH

Figure 2.1. Chemical structure of major classes of persistent organic pollutants (POPs) discussed in this chapter.

2. TRANSFORMATIONS OF POPS IN THE AQUATIC ENVIRONMENT

Although many POPs found in environmental media (e.g., water, soil, sediment) are resistant to physical and chemical degradation, some of these compounds (e.g., lower chlorinated PCBs) have been shown to degrade, with photolysis and biodegradation being the primary processes (Sinkkonen and Paasivirta, 2000; Borja et al., 2005). Photolysis of POPs has been shown to occur (Fang et al., 2008), with hydroxy radical addition being the primary photolytic mechanism (Sinkkonen and Paasivirta, 2000). POPs in air and water samples are more likely to be degraded via photolysis compared to contaminated soils and sediments due to limited ultraviolet light penetration in these solid compartments (Hooper et al., 1990; Sinkkonen and Paasivirta, 2000). Because the majority of POPs are contained in soils and sediment in aquatic ecosystems, the primary degradation process of POPs is biodegradation (Jones and de Voogt, 1999). Both aerobic and anaerobic microorganisms biodegrade POPs, with a number of environmental factors (e.g., degree of chlorination, water solubility, temperature) affecting the accessibility and degradation rates (Abramowicz, 1990; Borja et al., 2005). For example, aerobic microorganisms degrade lightly chlorinated PCBs (e.g., containing four or fewer chlorine atoms) to their elements (e.g., carbon, hydrogen) via mineralization, but this process does not occur with the more highly chlorinated congeners (Borja et al., 2005). Anaerobic microorganisms dehalogenate POPs but cannot mineralize these compounds to their basic elements (Borja et al., 2005; He et al., 2006). Hydrolysis (chemical degradation) of POPs, on the other hand, is quite slow and is thus considered a negligible contributor to degradation of these compounds in the aquatic environment (Sinkkonen and Paasivirta, 2000).

   In addition to physical and chemical degradation, POPs can be biotransformed in the aquatic food web. Metabolites of POPs include hydroxylated as well as methylsulfone POPs (Klasson-Wehler et al., 1998; Letcher et al., 1998; Herman et al., 2001). POP metabolites have been measured in various tissues (e.g., blood, blubber, fat, bird eggs) of marine organisms (including fish), but, in many cases, it is not known if these compounds are formed as a result of *in vivo* metabolism or are obtained from the environment (Valters et al., 2005; Verreault et al., 2005; Buckman et al., 2006). Buckman et al. (2006) exposed juvenile rainbow trout (*Oncorhynchus mykiss*) to dietary PCBs for 30 days and reported measurable concentrations of hydroxylated PCBs in plasma, whereas these compounds were below the limit of quantitation in the food and in plasma of control fish, indicating metabolism. In another laboratory exposure study, metabolites of PBDE 209, were also measured in blood, kidney, and liver of juvenile rainbow trout exposed to various levels of this flame retardant (Feng et al., 2012). Certain POPs can biomagnify in aquatic organisms, with tissues of higher trophic level biota having higher concentrations of POPs compared to those determined in lower trophic level organisms in the same food web (Jones and de Voogt, 1999; Hoekstra et al., 2003; Kelly et al., 2007). In addition, the patterns of individual POPs can be altered through food webs. Muir et al. (1988) examined the proportions of PCB homologue groups measured in Arctic cod muscle, ringed seal blubber, and polar bear fat. The patterns of PCBs changed in each trophic level, with hexa- and heptachlorinated PCBs making up the bulk of the ΣPCBs in polar bear (ringed seal predator) fat, whereas penta- and hexachlorinated PCBs were the more abundant congeners found in the seal (cod predator) and Arctic cod tissues. These data indicate that polar bears more readily metabolized lower chlorinated PCB homologues compared to lower trophic level Arctic biota.

   Several studies have examined toxicokinetic processes in fish (see Kleinow et al., 2008). POPs are lipophilic ("fat-loving") chemicals that can be absorbed by fish across the gut, gills, and skin. Studies have shown that these compounds are then transported to tissues that contain high concentrations of lipids, particularly neutral lipids (e.g., triglycerides). Because POPs are not readily metabolized in fish, these lipophilic chemicals are then stored in the lipid-rich tissues (e.g., visceral fat, eggs). During food-limiting periods, the stored lipids are mobilized and used as an energy source. As a result, the POPs associated with the lipids can remain and concentrate in the tissue, or they can be transported to target sites where toxic effects can occur. A review on interactions between lipids and lipophilic contaminants in fish can be found in Elskus et al. (2005). A number of toxicokinetic models have been developed to estimate the bioaccumulation of organic contaminants in fish and aquatic food webs (Arnot and Gobas, 2003, 2004; Kleinow et al., 2008). These include both compartmental and physiologically based modeling approaches (Kleinow et al., 2008). For additional details of these modeling approaches and parameters, see Kleinow et al. (2008).

3. ENDOCRINE DISRUPTION

The endocrine system contains tissues that synthesize, store, and release their secretory products directly into the bloodstream, including signaling molecules synthesized by specific hormone-producing cells. These hormones regulate metabolism, growth and development, reproduction, and other physiological processes in organisms, so a well-functioning endocrine system is essential for health and survival. Some of the major endocrine glands and tissues in fish (Figure 2.2) include the pituitary and hypothalamus, thyroid follicles, pancreas, interrenal tissue, and gonads (testes and ovaries) (Bern, 1967; Pait and Nelson, 2002; Thomas, 2008). Several environmental contaminants, including POPs, are known to disrupt the endocrine system and affect the reproduction, development, and other hormonal functions of fish and wildlife. Although POPs have relatively low hormonal activity compared to chemicals specifically produced to mimic hormones (e.g., ethynylestradiol in birth control pills), they are more persistent in fish tissues and the aquatic environment and, thus, also pose a health risk.

There are a wide range of molecular interactions between POPs and the endocrine system of fish. The effects of these interactions can vary, depending on the chemical, fish species, sex, reproductive stage of the fish, exposure conditions, and other factors. Due to the complexity of the endocrine system, organisms exposed to POPs may suffer endocrine disruption owing to the effects of POPs on hormone synthesis, transport

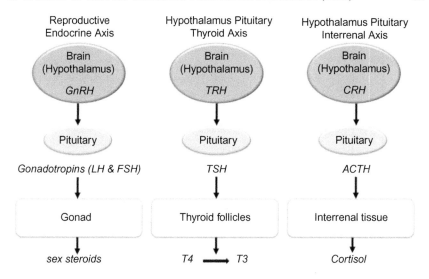

Figure 2.2. Simplified diagrams of reproductive, thyroid, and interrenal endocrine axes in teleost fish. Contaminants may disrupt these axes at the level of the brain by affecting releasing hormones (i.e., gonadotropin-releasing hormone, GnRH; thyrotropin-releasing hormone (TRH), and corticotropin-releasing hormone (CRH)); at the pituitary by interfering with the synthesis and release of gonadotropins, thyroid-stimulating hormone (TSH), or adrenocorticotropic hormone (ACTH); by direct toxic effects to gonads, thyroid follicles, or interrenal tissue; or by affecting the synthesis, release, metabolism, and clearance of the hormones that they produce.

and secretion, transformation, excretion or clearance, receptor recognition/ binding, and postreceptor response (U.S. EPA, 1997; Kendall et al., 2001; Thomas, 2008). Thus, the mechanisms of POP toxicity to the endocrine system can be very complex, with limited information on the details of the molecular interaction between these toxicants and the hormonal system, although this is an active area of research.

Many studies have examined the effects of POPs on reproductive hormones, as well as other components of the endocrine system (e.g., thyroid and interrenal hormones). The following reviews current studies demonstrating endocrine disruption in response to POPs (see Tables 2.1–2.4). More details on the effects of POPs on fish development, growth, reproduction, and survival will be given in subsequent sections of this chapter.

3.1. Disruption of Reproductive Function

Since the 1970s, various classes of POPs have been found to disrupt the reproductive endocrine system of female fish, resulting in low hatching

Table 2.1

Recent laboratory studies on the toxic effects of polychlorinated dibenzodioxins (PCDDs) and polychlorinated dibenzofurans (PCDFs) on fish. Age class of fish includes both sexes unless otherwise noted.

Effect Category	Compound(s)	Genus Species/ Common Name	Age/Sex	Effect	Concentration Range	Reference
Endocrine system	2,3,7,8-TCDD	*Oncorhynchus mykiss*/Rainbow trout and *Danio rerio*/Zebrafish	Adult females; ovaries with previtellogenic and vitellogenic oocytes	Inhibition of estrogen synthesis	0.08 to 0.8 ng TCDD/female/day in diet	Hutz et al. (2006)
Endocrine system/ Reproductive system	2,3,7,8-TCDD	*Danio rerio*/Zebrafish	Adult females, ovaries with previtellogenic and vitellogenic oocytes	Decline in ovosomatic index; ovarian necrosis; decreased serum estradiol and vitellogenin concentrations, inhibition of the transition from pre-vitellogenic stage follicles to vitellogenic stage follicles; induction of follicular atresia; reduced egg production and spawning success	0.08 to 0.8 ng TCDD/female/day in diet	King Heiden et al., 2005, 2006, 2008, 2012
Endocrine system/ Reproductive system	PCBs 118, 126; 2,3,7,8-tetrachlorodibenzo-dioxin; 2,3,4,7,8-pentachlorodibenzofuran	*Abramis brama*/ Seabream and *Cyprinus carpio*/ Carp	Cultured hepatocytes from adult male bream and carp	Inhibition of E2-induced vitellogenesis in hepatocytes	IC50 in bream: TCDD (0.02–0.09 nM); PCB126 (0.35–0.1 nM); PCDF (2.0–0.1) in carp: TCDD (0.01 nM); PCB126 (0.4 nM); no effect of PCB 118	Rankouhi et al. (2004)
Endocrine system/ Reproductive system	2,3,7,8-TCDD	*Gobiocypris rarus*/ Goby	Males exposed from embryo to adult	Feminization and overdevelopment of gonadal connective tissue	2–30 pg/L	Wu et al. (2001)
Growth/Survival	2,3,7,8-TCDD	*Oncorhynchus mykiss*/Rainbow trout	Adult females	Reduced survival of fry; no growth effects for adults at high dose	1.8 ng TCDD/kg ww fed for up to 300 days	Giesy et al. (2002)
Growth/Survival	PCDDs; PCDFs	*Sparus aurata*/ Gilthead seabream	Juveniles	No effect on fish growth	23 ng WHO/TEQ/kg fed for 300 days 5.5 pg WHO-TEQ/g ww in flesh	Ábalos et al. (2008)
Growth	PCB 126 and TCDD	*Fundulus heteroclitus*/ Mummichog	Larvae	EROD induction and reduced growth in larvae	80–1250 pg/g ww	Rigaud et al (2013)

Table 2.2

Recent laboratory studies on the toxic effects of polychlorinated biphenyls (PCBs) on fish. Age class of fish includes both sexes unless otherwise noted.

Effect Category	Compound(s)	Genus Species/ Common Name	Age/Sex	Effect	Concentration Range	Reference
Endocrine system	Aroclor 1242, 1254, 1260; PCBs 77, 87, 99, 101, 126, 153, 180, 183, 194, 202, 209	*Oncorhynchus mykiss*/Rainbow trout	Juveniles	Changes in some measures of thyroid status (e.g., liver deiodinase activity); no changes in circulating hormone levels	6000 ng/g lipid in diet; 28.5–30 µg/g lipid in carcass	Buckman et al. (2007)
Endocrine system	Aroclor 1254	*Salmo salar*/Atlantic salmon	Juveniles	Decreased production of plasma thyroid hormone triiodothyronine and cortisol in exposed smolts	≥ 1 ug/L Aroclor 1254	Lerner et al. (2007)
Endocrine system	Aroclor 1254	*Salvelinus alpinus*/ Arctic charr	Adults	Impaired production and regulation of cortisol by interrenal tissue	≥ 1 mg/kg body mass in diet Aroclor 1254	Aluru et al. (2004); Vijayan et al. (2006)
Endocrine system	Aroclor 1254; PCBs 47, 77, 153	*Micropogonias undulates*/Atlantic croaker	Juvenile to adult males	Reduced circulating levels of triiodothyronine and thyroxine (PCB 153)	Aroclor 1254: 0.2 and 1.0 mg/kg body wt/day; PCB 153: 0.1 and 1.0 mg/kg body wt/day; PCB 77: 0.1 mg/kg body wt/day; PCB 17: no effect at doses up to 1.0 mg/kg body wt/day	LeRoy et al. (2006)
Endocrine system	PCBs 47, 77, 153	*Micropogonias undulates*/Atlantic croaker	Adult males	Inhibition of hypothalamic tryptophan hydroxylase activity and gonadal growth by PCB77 in males; no effects observed for PCB 47 and PCB 153	PCBs 47 and 153: no effect at 1 mg/kg body wt for 30 days; PCB 77: effects at 0.01 mg/kg body wt for 15 days	Khan and Thomas (2006)
Endocrine system/ Reproductive system	PCBs 118, 126; 2,3,7,8-tetrachlorodibenzodioxin; 2,3,4,7,8-pentachlorodibenzofuran	*Abramis brama*/ Seabream and *Cyprinus carpio*/ Carp	Cultured hepatocytes from adult male bream and carp	Inhibition of E2-induced vitellogenesis in hepatocytes	IC50 in bream: TCDD (0.02–0.09 nM); PCB126 (0.35–0.1 nM); PCDF (2.0–0.1) in carp: TCDD (0.01 nM); PCB126 (0.4 nM).: no effect of PCB 118	Rankouhi et al. (2004)
Growth and development	Aroclor 1254	*Oncorhynchus mykiss*/Rainbow trout	Juveniles	Increased liver glycogen levels and hexokinase activity; decreased lactate dehydrogenase activity	10 mg/kg body weight	Wiseman and Vijayan (2011)

(Continued)

Table 2.2 (continued)

Effect Category	Compound(s)	Genus Species/Common Name	Age/Sex	Effect	Concentration Range	Reference
Growth and development	Aroclor 1254	*Micropogonias undulates*/Atlantic croaker	Fertilized eggs and larvae	Reduced growth rate and behavioral responses in larvae	0.66 µg Aroclor 1254 g−1 egg,	McCarthy et al. (2003)
Growth and development	Kanechlor-400, PCB 126	*Oryzias latipes*/Medaka	Adults	Changes in genes related to glycolysis and gluconeogenesis	0.01 ug PCB126/g b.w. daily	Nakayama et al. (2011)
Growth and development	PCBs 77, 126, 153	*Anguilla anguilla*/European eel	Adult females	Weight loss, lower metabolic rate and oxygen consumption	7 ug PCB126/kg body wt; 5 mg PCB153/kg body wt; 50 ug PCB77/kg body wt via I.P. injection	Van Ginneken et al. (2009)
Growth and development	PCBs 77, 126	*Oncorhynchus mykiss*/Rainbow trout	Adult females	Increased plasma glucose concentrations (PCB 126); changes in mitochondrial respiration in hepatocytes (PCB 77)	50 ug PCB 126/kg body wt; 1 nM PCB 77 and 126 for *in vitro* hepatocyte exposure	Nault et al. (2012)
Growth and development	PCB 126 and TCDD	*Fundulus heteroclitus*/Mummichog	Larvae	EROD induction and reduced growth in larvae	80–1250 pg/g wet wt	Rigaud et al. (2013)
Reproductive system	Aroclor 1254; PCBs 47, 153, 77	*Micropogonias undulates*/Atlantic croaker	Juvenile to adult males	Decline in serotonin production, inhibition of luteinizing hormone (LH) secretion reduction in gonadal growth; inhibition of hypothalamic TPH activity and gonadal growth	Aroclor 1254: 1 mg/kg body wt for 30 days in diet; PCBs 47 and 153: no effect at 1 mg/kg body wt for 30 days; PCB 77: effects at 0.01 mg/kg body wt for 15 days	Khan and Thomas (2000); Khan et al. (2001); Khan and Thomas (2006)
Reproductive system	Aroclor 1254	*Orechromis niloticus*/Nile tilapia	Larval exposure; effects in adults	Testicular and ovarian alterations and reduced plasma levels of thyroid hormones in early life stages, reproductive inhibition in adults	0.05 ug/g in diet	Coimbra and Reis-Henriques (2007)
Reproductive system	PCB mixtures	*Danio rerio*/Zebrafish	Juvenile to adult exposure; effects in adults	Increased atresia, reduced production of fertilized eggs per spawn and	500–2300 ng/g sum PCBs in diet	Daouk et al. (2011)

Immune system	Aroclor 1242, 1254, 1260 in mixture	*Limanda limanda* L./Dab	Adult females	increase production of poorly fertilized spawns	21–94 ug PCB/kg dry wt in sediment	Hutchinson et al. (2003)
Immune system	Aroclor 1248	*Ameiurus nebulosus*/Brown bullhead	Adults	No observed effect on lysozyme or ROS activity. Decrease in bactericidal activity and antibody production to pathogenic bacterium *Edwardsiella ictaluri*. Lower survival to *E. ictaluri*. Elevated proliferation to T-cell mitogens. Decreased levels of cortisol and triiodothyronine in plasma	ip injection of 50 ug–5 mg/kg body wt of Aroclor 1248	Iwanowicz et al. (2009)
Immune system	Aroclor 1254	*Salvelinus alpinus*/Arctic charr	Juveniles	Lower survival to pathogenic bacterium *Aeromonas salmonicida*	1 to 100 mg/kg body wt in diet; 130 to 8400 ng/g ww in fish muscle	Maule et al. (2005)
Immune system	Aroclor 1254	*Salmo salar*/Atlantic salmon	Juveniles	Elevated proliferation to T-cell mitogen	1 or 10 ug/L Aroclor 1254	Iwanowicz et al. (2005)
Immune system	Aroclor 1254	*Oncorhynchus tshawytscha*/Chinook salmon	Juveniles	Lower survival to pathogen pathogenic bacterium *Listonella anguillarum*	94 ng/g in sediment extract injected ip	Arkoosh et al. (2001)
Immune system	Aroclor 1254	*Oncorhynchus tshawytscha*/Chinook salmon	Juveniles	No effect on *Listonella anguillarum* disease susceptibility and antibody production	430–17,000 ug Aroclor 1254/kg food (wet wt); 50–980 ug/kg Aroclor 1254 in tissue (wet wt)	Powell et al. (2003)
Immune system	Aroclor 1254	*Oncorhynchus tshawytscha*/Chinook salmon	Juveniles	Reduced number of primary and secondary antibody forming cells	54 mg Aroclor 1254/kg wet weight via ip injection	Jacobson et al. (2003)
Immune system	Clophen A50	*Oncorhynchus mykiss*/Rainbow trout	Egg exposure; effects in juveniles	Lower survival to pathogenic bacterium *Flavobacterium psychrophilum* at the lower dose	0.4 or 2 ug Chophen A50/egg via microinjection	Ekman et al. (2004)
Immune system	PCB 126, PCB 153	*Lepomis macrochirus*/Bluegill sunfish	Adults	B-lymphocyte production altered; reduction in phagocytic respiratory burst and B- and T-lymphocyte proliferation	PCB 126: 0.01 and 1.0 ug/g BW); PCB 153: noncopl5.5.0 and 50.0 ug/g BW); both via ip injection	Duffy and Zelikoff (2006)
Immune system	PBC 126	*Oncorhynchus mykiss*/Rainbow trout	Head kidney cell cultures	Increased expression of IL-1β gene	0.001 nM to 1 uM; effects at concentrations $\geq$ 0.1 nM	Quabius et al. (2005)

(Continued)

Table 2.2 (continued)

Effect Category	Compound(s)	Genus Species/ Common Name	Age/Sex	Effect	Concentration Range	Reference
Immune system	PCB 126	*Ictalurus punctatus*/ Channel catfish	Juveniles	Low levels suppressed circulating plasma lysozyme, high levels enhanced the response	1 or 100 ppb in feed; effects at high dose	Burton et al. (2002)
Immune system	PCB 126	*Oryzias latipes*/ Japanese medaka	Juveniles and adults	Reduction in antibody forming cells	0.01 or 1.0 mg/g body weight via ip injection	Duffy et al. (2002)
Immune system	PCB 126	*Oryzias latipes*/ Japanese medaka	Juveniles and adults	Increase or reduction of ROS production depending upon the time examined post exposure	0.01 or 1.0 ug PCB 126/g body wt via ip injection; effects only at 1 ug/g dose?	Duffy et al. (2003)
Immune system	PCB 126	*Ictalurus punctatus*/ Channel catfish	Juveniles	Antibody response to *Listonella anguillarum* reduced; suppressed phagocyte oxidative burst activity	0.01 or 1.0 ug PCB 126/g BW via ip injection	Regala et al. (2001)
Immune system	PCB 126	*Oryzias latipes*/ Japanese medaka	Juveniles and adults	Reduced number of antibody forming cells	0.01 or 1.0 ug PCB 126/g BW via ip injection; effects only at 1 ug/g dose?	Duffy et al. (2003)

Table 2.3

Recent laboratory studies on the toxic effects of organochlorine (OC) pesticides on fish. Age class of fish includes both sexes unless otherwise noted.

Effect Category	Compound(s)	Genus Species/Common Name	Sex/Age	Effect	Concentration Range	Reference
Endocrine system	*o,p'*-DDT	*Oncorhynchus mykiss*/Rainbow trout	Pituitary cell culture	Stimulation of growth hormone and prolactin mRNA synthesis in pituitary gland culture	500 and 1000 nM	Elango et al. (2006)
Endocrine system	*o,p'*-DDT	*Cyprinus carpio*/Carp	Testicular microsomes	Enhanced ovarian synthesis of maturation-inducing hormones and altered hormone metabolism and clearance	100 μM–1 mM	Thibaut and Porte (2004)
Endocrine system/Reproductive system	*o,p'*-DDT and *p,p'*-DDE	*Paralichthys dentatus*/Summer flounder	Juvenile males	Reduced gonadosomatic indices and decreased plasma testosterone production, inhibition of spermatogenesis, altered gonadal development; no observed effects with *p,p'*-DDE	30, 60, and 120 mg; kg body wt via ip injection; 188–521 mg/g ww *o,p'*-DDT in liver and 200–372 mg/kg ww *p,p'*-DDE in liver	Mills et al. (2001) and Zaroogian et al. (2001)
Endocrine system	*p,p'*-DDE	*Salmo salar*/Atlantic salmon	Juveniles	Altered thyroid-stimulating hormone and thyroid receptor in salmon parr	10 ug/L in water	Mortensen and Arukwe (2006)
Endocrine system	*p,p'*-DDE	*Poecilia reticulata*/Guppy	Exposed birth to adulthood; effects in adults	Inhibition of gonopodium development, reduced sperm count, suppressed courtship behavior, delayed sexual maturation	0.01 and 0.1 ug/mg feed	Bayley et al. (2002)
Endocrine system	Endosulfan; *o,p'*-DDD	*Oncorhynchus mykiss*/Rainbow trout	Head kidney cell culture	Disruption of ACTH stimulated cortisol secretion in trout cells	19–366 uM endosulfan; 200 uM *o,p'*-DDD	Dorval et al. (2003) and Lacroix and Hontela (2003)
Endocrine system	Endosulfan	*Salmo salar*/Atlantic salmon	Hepatocyte cell culture	Induced vitellogenin production in hepatocyte cultures	0.001–100 uM; effects at 10–100 uM	Krovel al. (2010)

(Continued)

Table 2.3 (continued)

Effect Category	Compound(s)	Genus Species/Common Name	Sex/Age	Effect	Concentration Range	Reference
Endocrine system	Endosulfan	*Cichlasoma dimerus*	Adults	Gonadal impairment and alteration of pituitary hormone levels	0.1, 0.3 and 1 μg/L	Da Cuña et al. (2013)
Growth and development	*o,p′*-DDD	*Oncorhynchus mykiss*/Rainbow trout	Juveniles	Decreased liver glycogen reserves and plasma cortisol levels	5, 20, or 50 mg/kg *o,p′*-DDD via ip injection; 2.25 mg/kg in liver at highest dose	Benguira et al. (2002)
Growth and development	*p,p′*-DDE	*Parophrys vetulus*/English sole	Juvenile males and females	Reduced growth rate of juveniles	215–219 ug/g dry wt in feed	Rice et al. (2000)
Reproductive system	*o,p′*-DDE and *o,p′*-DDT	*Oncorhynchus mykiss*/Rainbow trout	Juveniles	Induction of plasma vitellogenin in sexually immature fish	45 mg/kg *o,p′*-DDT and 90 mg/kg *o,p′*-DDE by injection	Donohoe and Curtis (1996)
Reproductive system	*o,p′*-DDE	*Oryzias latipes*/Japanese medaka	Embryonic exposure; effects in adults	Reduced vitellogenic oocyte production, increased atresia, reduced gonadosomatic indices	0.5, 0.05, 0.005 ng/embryo via microinjection	Papoulias et al. (2003)
Reproductive system	*p,p′*-DDE	*Poecilia reticulata*/Guppy	Adult males	Reduced number of spermatogenetic cysts and an increased number of sperm bundles	0.1, 1, and 10 ug/mg food; effects at 10 ug/mg dose	Kinnberg and Toft (2003)
Reproductive system	*p,p′*-DDE	*Oryzias latipes*/Japanese medaka	Juvenile exposure; effects in adults	Decreased testis growth, induction of vitellogenin and choriogenin production in males, intersex	1, 5, 20, and 100 microg/L *p,p′*-DDE for two months	Zhang and Hu (2008)
Reproductive system	*p,p′*-DDE	*Poecilia reticulata*/Guppy	Adult males	No harmful effect of male fish siring young	0.01–1 mg *p,p′*-DDE/mg food	Kristensen et al. (2006)
Immune system	*o,p′*-DDE	*Oncorhynchus tshawytscha*/Chinook salmon	Egg stage exposure; effects in juveniles	Reduction in mitogenic response to LPS	Egg exposure 10 and 100 ppm in water; 0.53 ug/g lipid in fish at 10 ppm exposure	Milston et al. (2003)

System	Contaminant	Species	Life stage/Cells	Effect	Dose/Concentration	Reference
Immune system	p,p'-DDE	Oncorhynchus tshawytscha/Chinook salmon	Juveniles	Reduced cell viability and proliferation and increased apoptosis in splenic and head kidney leukocytes. Biphasic mitogenic response	In vitro exposure: 5-15 mg/L. In vivo exposure: 23-105 mg/kg ww fish; effects at doses ≥59 mg/kg ww	Misumi et al. (2005)
Immune system	Lindane and p,p'-DDE	Sparus aurata L./Gilthead seabream	Juvenile males	Upregulation of immune-related genes: Il-1β, TNFα, MHC11α, Mx, TLR9, IgML and TCRα	5 ng to 50 ug/ml	Cuesta et al. (2008)
Immune system	Lindane	Oncorhynchus mykiss/Rainbow trout	Blood leucocytes	Increases macrophage-activating factor in PBL	2.5–100 uM	Duchiron et al. (2002)
Immune system	Lindane	Cyprinus carpio/Carp	Adults	Suppression of immunoglobulin and antibody secreting cells	10 mg/kg body wt via ip injection	Studnicka et al. (2000)
Immune system	Lindane	Oncorhynchus mykiss/Rainbow trout	Rainbow trout phagocytes	Increased ROS production and intracellular calcium in head kidney phagocytes	2.5–200 uM	Betoulle et al. (2000)
Immune system	Endosulfan	Maccullochella peelii/Crimson-spotted rainbowfish, Bidyanus bidyanus/Silver perch, Macquaria ambigua/Golden perch, Maccullochella peelii/Murry cod	Head kidney cell cultures	Modulation of phagocytic response	0.1 to 10 mg/L; effects at 10 mg/L	Harford et al. (2005)
Immune system	Endosulfan	Oreochromis niloticus/Nile tilapia	Juvenile males	Nonspecific activation of macrophages followed by an increased synthesis of IL-2L factor	7 ug/L	Tellez-Banuelos et al. (2010)
Immune system	Toxaphene	Salvelinus alpinus/Arctic charr	Adults	Did not increase disease susceptibility to a cestode	10 mg/g ww in corn oil	Blanar et al. (2005)

Table 2.4

Recent laboratory studies on the toxic effects of polybrominated diphenyl ethers (PBDEs) on fish. Age class of fish includes both sexes unless otherwise noted.

Effect Category	Compound(s)	Genus Species/ Common Name	Age/Sex	Effect	Concentration Range	Reference
Endocrine system/Growth and development	PBDE 47	*Pimephales promelas*/ Fathead minnows	Adults	Decreased plasma thyroxine production; altered thyroid hormone at various levels of hypothalamic-pituitary-thyroid axis	2.4–12.3 μg PBDE-47/pair/day; 11. μg PBDE-47/g carcass in males and, 20. μg PBDE 47/g carcass in females from the low dose 64. μg/g carcass in males and 108 μg/g carcass in females for high dose treatment	Lema et al. (2008)
Endocrine system	PBDEs 47 and 209	*Danio rerio*/ Zebrafish	Larvae and juveniles	Altered triiodothyronine and thyroxine levels; upregulation of genes encoding corticotrophin-releasing hormone and thyroid-stimulating hormone	74.9 ng PBDE47/g wet wt in diet; 68.8 ng/PBDE 47 g ww in fish; 0.08–1.92 mg PBDE 209/L in water; 2400–39,000 ng PBDE 209/g ww in fish	Chen et al (2010) and Chen et al (2012a)
Endocrine system	PBDE 209	*Oncorhynchus mykiss*/Rainbow trout	Juvenile male and female	Altered thyroid hormones in plasma	50 to 1000 ng/g body wt via ip injection; 39–80 ng/g in trout liver	Feng et al. (2012)
Endocrine system	DE-71	*Danio rerio*/ Zebrafish	Embryo to adult males and females	Altered thyroxine levels; upregulation of transcription of corticotrophin-releasing hormone and thyroid-stimulating hormone genes	1–10 ug/L; effects at 3 and 10 ug/L, 7706–55,000 ng/g ww in adult females; 11,816–118,419 ng/g ww in adult males; 1689–13,701 ng/g wet weight in eggs	Yu et al. (2010) and Yu et al. (2011)
Endocrine system	3-OH-PBDE 47, 5-OH-PBDE 47, and 6-OH-PBDE 47	*Danio rerio*/ Zebrafish	Embryos	Upregulation of genes associated with thyroid hormone regulation and stress response	0.625 ppm	Usenko et al. (2012)
Endocrine system	PBDEs 47, 99 and 205: PBB-153 and technical Firemaster BP-6	*Oncorhynchus mykiss*/Rainbow trout	Hepatocyte culture	Reduced vitellogenin production in hepatocyte cultures	1–500 ug/L	Nakari and Pessala (2005)

Endpoint	Compound	Species	Life stage	Effect	Dose/Concentration	Reference
Endocrine system	PBDEs 47, 153 154	*Salmo salar* L./ Atlantic salmon	Hepatocyte cultures	Altered glucose homeostasis; induction of estrogen-responsive genes	0.01–100 uM	Softeland et al. (2011)
Endocrine system/Growth	PBDE 47	*Danio rerio*/ Zebrafish	Juveniles (prepubertal)	No effects on thyroid condition but some changes in body morphometry and mass	0.7–18 ug/g wet wt in diet; 6–130 ug/g lipid in fish	Torres et al. (2013)
Growth and development	PBDE 49	*Danio rerio*/ Zebrafish	Embryos and larvae	Mortality at 6 days post fertilization; severe dorsal curvature and cardiac toxicity	4–32 uM	McClain et al. (2012)
Growth and condition	PBDE 209	*Coregonus clupeaformis*/ Lake whitefish	Juvenile males and females	Altered growth rate	0.1–2 ug/g in food; growth effects at 2 ug/g	Kuo et al. (2010)
Growth and condition	PBDEs 28, 47, 100, 99, 154, 153, 183 and 209 as mixture	*Trematomus bernacchii*/ Emerald rock cod	Juveniles	Reduced total liver lipid, increased liver lipid peroxide and protein carbonyl concentrations	3.2 and 32 ng administered orally in capsules	Ghosh et al. (2013)
Reproductive system	PBDE 47	*Pimephales promelas*/ Fathead minnows	Adults	Reduced condition factor and decreased production of mature sperm	28.7 ug/pair/day; ~15 ug/g wet and 50 ug/g wet wt in carcasses of males and females	Muirhead et al. (2006)
Reproductive system/ Growth	PBDE 47	*Danio rerio*/ Zebrafish	Larval to adults	No significant variation in survival, body size and gonad histology; altered locomotion behavior	12–643 ng/g ww in food; 2.7–410 ng/g ww in fish tissue	Chou et al. (2010)
Reproductive system	PBDE 209	*Danio rerio*/ Zebrafish	Embryos to adults	Reduced ovarian growth; reduced testis development, and reduced male gamete quantity and quality	0.001 to 1 uM; embryos exposed in water with DMSO: ~1 ug/g; 32 ug/g ww and 4.7 ug/g ww in bodies of males and females in 1 uM treatment	He et al. (2011)

(Continued)

Table 2.4 (continued)

Effect Category	Compound(s)	Genus Species/ Common Name	Age/Sex	Effect	Concentration Range	Reference
Reproductive system	DE-71	*Platichthys flesus*/ European flounder and *Danio rerio*/ Zebrafish	Adult zebrafish and juvenile flounder	No histopathological changes found in reproductive organs; decreased egg production and reduced larval survival in zebrafish	Zebrafish: waterborne exposure at 5, 16, 50, 160, and 500 ug/L; 8.8–460 ug/g ww in tissue. Flounder: 0.007–700 ug/g TOC in sediment and 0.014 to 14,000 ug/g lipid in food; of 0.13 to 71 ug/g wet wt in tissue	Kuiper et al. (2008)
Reproductive system	PBDE 47	*Danio rerio*/ Zebrafish	Prepubertal males and females	No effects on gonadal development	0.7–18 ug/g wet wt in diet; 6–130 ug/g lipid in fish	Torres et al. (2013)
Immune system	Mixture (PBDE-47, 99, 100, 153, 154)	*Oncorhynchus tshawytscha*/ Chinook salmon	Juveniles	Biphasic response in disease susceptibility to *Listonella anguillarum*	270 and 1400 ng/g dry wt BDE sum in diet; 1600 ng/g lipid and 14,000 ng/g lipid in fish	Arkoosh et al. (2010)
Immune system	PBDE 47	*Oryzias melastigma*/ Medaka	Adults	Alteration of complement gene expression	290 and 580 ng/day in diet	Ye et al. (2012)
Immune system	PBDEs 47 and 99	*Salvelinus namaycush*/Lake trout	Thymocyte culture	Reduced viability, apoptosis, and necrosis of thymocytes	10 ug/L to 100 mg/L; effects at concentration > 1 mg/L	Birchmeier et al. (2005)

success, posthatch mortality, and reproductive disruption (reviewed in Kime, 1995; Monosson, 2000; Arukwe, 2001; Thomas, 2008). PCB exposure has frequently been associated with depressed sex steroid levels and associated reproductive dysfunction. Some PCB congeners, particularly the dioxin-like PCBs, have anti-estrogenic activity in fish. Rankouhi et al. (2004) found that in bream (*Abramis brama*) and carp (*Cyprinus carpio*) hepatocytes co-exposed to E2 and PCB 126, PCB 126 showed a concentration-dependent inhibition of E2-induced vitellogenesis.

Fish exposed to organochlorine (OC) pesticides also show reproductive disruption, including induction of plasma vitellogenin (VTG) levels in juvenile rainbow trout exposed to *o,p'*-DDT, *o,p'*-DDE, and *p,p'*-DDE (Donohoe and Curtis, 1996), and stimulation of growth hormone and prolactin mRNA synthesis in a rainbow trout pituitary gland culture, in a manner similar to that of estrogens (Elango et al., 2006). In another study, Thibaut and Porte (2004) observed enhanced ovarian synthesis of maturation-inducing hormones and altered hormone metabolism and clearance in *p,p'*-DDE-exposed carp. Endosulfan exposure was found to cause gonadal impairment and alteration of pituitary hormone levels in the cichlid, *Cichlasoma dimerus* (Da Cuña et al., 2013), as well as inducing VTG production in Atlantic salmon (*Salmo salar*) hepatocyte cultures (Krøvel et al., 2010). Some PBDEs (e.g., PBDEs 47 and 154) show weak estrogenic activity, based on increased VTG production or VTG gene induction in cell cultures of Atlantic salmon and rainbow trout hepatocytes (Nakari and Pessala, 2005; Søfteland et al., 2011).

The anti-estrogenic activity of PCDDs and PCDFs has been documented in mammals (Zacharewski and Safe, 1998), and there is evidence of similar effects in fish (King Heiden et al., 2012). The PCDD, 2,3,7,8-tetrachloro-dibenzo-*p*-dioxin (TCDD), the most potent aryl hydrocarbon receptor (AhR) agonist among the chlorinated aromatic hydrocarbons, has been used to represent this class of compounds in most studies on effect; PCDFs are thought to have a common mechanism of toxicity in fish (Wisk and Cooper, 1990) but have been less frequently studied. *In vitro* studies have described the anti-estrogenic potency of TCDD and 2,3,4,7,8-pentachlorodibenzo-*p*-furan (PeCDF), based on their ability to inhibit estrogen-induced vitellogenesis in fish hepatocytes (Smeets et al., 1999; Rankouhi et al., 2004). Similarly, rainbow trout (*Oncorhynchus mykiss*) and zebrafish (*Danio rerio*) exposed to TCDD *in vivo* showed inhibition of estrogen synthesis, potentially causing fertility defects (Hutz et al., 2006). A series of studies using zebrafish as a model species investigated TCDD effects on ovarian development (King Heiden et al., 2005, 2006, 2008, 2009). The studies showed that chronic dietary exposure led to decreased serum estradiol with associated inhibition of ovarian development.

POPs have also been associated with disrupted reproductive endocrine function in male fish (e.g., Kime, 1995, 1998; Arukwe, 2001; Thomas, 2008), and relatively recent studies have elucidated the mechanisms of action of some POPs, such as PCBs. For example, in male Atlantic croaker (*Micropogonias undulates*), exposure to Aroclor 1254 at 1 mg/kg body wt/day for 30 days during gonadal recrudescence caused a significant decline in serotonin, leading to an inhibition of luteinizing hormone secretion, and a consequent reduction in gonadal growth (Khan and Thomas, 2000, 2001; Khan et al., 2001). A subsequent study (Khan and Thomas, 2006) showed that the dioxin-like PCB 77 was more effective in inhibiting serotonin synthesis than noncoplanar congeners, PCB 47 and PCB 153, suggesting that dioxin-like PCBs may play an important role in disruption of the male reproductive endocrine system in fish. OC pesticides have also been found to impair the endocrine system of fish. Mills et al. (2001) and Zaroogian et al. (2001) reported depression of plasma testosterone in juvenile male summer flounder (*Paralichthys dentatus*) exposed to *o,p'*-DDT at concentrations resulting in liver concentrations of 188–521 µg/g ww, as well as induction of vitellogenin in the males. No such effects were elicited by comparable exposure to *p,p'*-DDE. A more limited number of studies have examined endocrine disruption in male fish exposed to PCDDs and PCDFs. Wu et al. (2001) observed that long-term TCDD exposure caused feminization and overdevelopment of gonadal connective tissue of the goby, *Gobiocypris rarus*. These studies illustrate the diversity of impacts that various POP classes can have on the reproductive endocrine system of fish across taxa.

Similar to field efforts conducted prior to 2000 (Kime, 1998; Monosson, 2000), recent field studies have reported reproductive endocrine alterations in fish exposed to mixtures of POPs. Wild white sturgeon (*Acipenser transmontanus*) from the Columbia River were examined for signs of reproductive endocrine disruption (Feist et al., 2005). Both *p,p'*-DDE and *p,p'*-DDD were consistently found at high levels, with mean liver and gonad DDT concentrations of 20.6 ug/g lipid and 12.5 ug/g lipid, respectively; concentrations of PCBs were lower, with mean levels of 1.9 ug/g lipid in liver and 1.6 ug/g lipid in gonad. In males, plasma androgens were negatively correlated with total DDT, total pesticides, and PCBs. Sol et al. (2008) compared reproductive parameters in male English sole from the Hylebos Waterway in Puget Sound—a Superfund site contaminated with a variety of chemicals, including PCBs, DDTs, polycyclic aromatic hydrocarbons (PAHs), and metals—with sole from an uncontaminated reference area and found significant negative correlations between body concentrations of PCBs, DDTs, and HCB and plasma 11-ketotestosterone levels. Gilroy et al. (2012) observed reduced 17β-estradiol (E2) levels and *in vitro* E2 production in wild female brown bullhead (*Ameiurus nebulosus*) and goldfish (*Carassius auratus*), from PCB-contaminated Wheatley Harbour in Lake Erie, Ontario, Canada, as well as reduced levels of testosterone and 11-keto-testosterone in males of both species.

In field and laboratory studies, endocrine disruption has been documented in fish exposed to bleached kraft mill effluent (reviewed in McMaster et al., 2006). In earlier studies, PCDDs and PCDFs in chlorinated effluent likely contributed to these abnormalities, but subsequent research has shown that process improvements that greatly decreased or eliminated PCDD releases from such plants have not eliminated the toxic effect on fish reproduction; the reproductive toxicity of pulp mill effluents is attributable, in part, to bioactive substances associated with wood-derived lignin and terpenoids (Hewitt et al., 2006).

3.2. Disruption of Thyroid Function

POPs have been shown to disrupt thyroid function in fish, with associated impacts on early development, metabolism, and growth (see Section 6 for details). Thyroxine (T4) and triiodothyronine (T3) are the principal thyroid hormones (THs) in fish. PCBs and especially their hydroxylated metabolites, as well as PBDEs, are structurally similar to T4. Their effects on thyroid hormone levels and functions are well documented in mammals (Boas et al., 2006), and some similar impacts, reviewed below, are also reported in fish.

The effects of PCBs on thyroid function in fish were first noted in the 1970s in coho and Chinook salmon from the Great Lakes; subsequently, a number of studies have confirmed that PCB mixtures and some individual PCB congeners can alter indices of thyroid status in fish (reviewed in Brown et al., 2004). More recently, Coimbra and Reis-Henriques (2007) found that tilapia (*Orechromis niloticus*) larvae exposed to the commercial PCB mixture, Aroclor 1254, had reduced plasma levels of THs in early life stages. Brown bullhead exposed to Aroclor 1248 had lower levels of T3 (Iwanowicz et al., 2009) compared to control fish. Schnitzler et al. (2011) demonstrated that environmentally relevant levels of PCBs affected the thyroid physiology of sea bass (*Dicentrarchus labrax*) by impairing TH synthesis and secretion.

Similar to PCBs, multiple laboratory studies have shown that exposure to PBDEs can alter thyroid function in fish species, including zebrafish, stickleback, and European flounder (Holm et al., 1993; Lema et al., 2007; Kuiper et al., 2008; Chou et al., 2010; Feng et al., 2012). While many studies document the effects of PBDEs on thyroid function at doses much higher than those typical of environmental samples, some effects have been documented at environmentally relevant concentrations. For example, Feng et al. (2012) observed alterations in the thyroid hormones of juvenile rainbow trout exposed to PBDE 209 through intraperitoneal injection at concentrations as low as 50 ng/g. Kuiper et al., (2008) documented a 10% decline in T4 levels in flounder with PBDE 47 concentrations of 51 ng/g muscle. It should be noted that there are species differences in thyroidal responses to PBDE congeners. For example, studies show that BDE-47 inhibits thyroid function in fathead minnow (Lema et al., 2008) but has little effect on thyroid condition in zebrafish (Chen et al., 2010; Torres et al., 2013).

Exposure to PBDEs appears to affect the hypothalamus-pituitary-thyroid axis at several levels, including synthesis of thyroid-stimulating hormone, conversion of T4 to T3, and synthesis of proteins regulating hormone supply to target tissues (Yu et al., 2010, 2011; Feng et al., 2012). In some cases such changes may affect not only exposed parents, but also their offspring. Yu et al. (2010, 2011) found that exposure to the PBDE mixture DE-71 not only caused changes in thyroid hormones in exposed parents, but in addition their progeny, even if they had no further exposure to DE-71, had increased plasma T3 and T4 concentrations and altered gene expression in the hypothalamic-pituitary-thyroid axis (HPT axis). However, dioxin impurities have been detected in some samples of DE-71 (Hanari et al., 2006), so caution must be taken in interpreting the results of exposure to this PBDE mixture. Additional examples of transgenerational effects of POPs on fish can be found in a later subsection of this chapter.

Thyroid function in fish may also be affected by OC pesticide exposure (Brown et al., 2004). Exposure to high doses (>1 mg/L in water or 5 mg/g in food) of mirex, DDTs, endrin, endosulfan, and lindane has been shown to alter thyroid follicle morphology and T3 and T4 levels in many fish species, including tilapia, catfish, mullet, coho salmon, rainbow trout, goldfish, perch, and medaka. More recently, Mortensen and Arukwe (2006) found that exposure to p,p'-DDE altered thyroid-stimulating hormone and thyroid receptor in Atlantic salmon parr. However, the potential of OC pesticides to alter thyroid function at environmentally realistic concentrations is less certain.

PCDDs and PCDFs, like coplanar PCBs, have been associated with changes in thyroid function, which may affect growth and metabolism (Brown et al., 2004). Administration of TCDD and PCDFs to mammals typically decreases plasma T4 and sometimes plasma T3, but effects in fish appear to be less consistent (Brown et al., 2004). While in some studies, exposure to these compounds altered plasma T3 or T4 levels, the doses required were often quite high, and the effects sometimes transient and variable depending on the species involved and the route of exposure. Some of these studies further suggest that teleost fish may in some cases be able to compensate for effects that might otherwise depress plasma T4 or T3 levels through changes in conversion of T4 to T3 or rates of glucuronidation and clearance of these hormones. Thus, the thyroid of fish may be less sensitive to the disrupting effects of dioxin-like compounds relative to mammals and birds. The same may be true of other POPs that affect thyroid function in fish (Brown et al., 2004); additional evidence for the ability of fish to maintain thyroid function homeostasis in the face of the impacts of anthropogenic chemicals is reviewed in Peter and Peter (2011).

The impacts of toxicants on thyroid function may also be lessened by the large amount of T4 stored in the thyroid follicle cells of teleosts. This reserve may make it difficult for compounds that inhibit T4 synthesis to deplete its stores, increasing the exposure time necessary for thyroidal responses to occur that would have observable effects on organismal function (see Carr and Patino, 2011). This phenomenon was demonstrated in zebrafish by Mukhi and Patiño (2007), who used a long-term exposure to perchlorate to induce a thyroid-mediated inhibitory effect on reproduction, but it has not been specifically investigated with POPs.

Differences in the potencies of various POPs in altering thyroid function in fish may also contribute to variation in thyroid responses to POPs exposure. LeRoy et al. (2006) exposed Atlantic croaker to Aroclor 1254 or one of three individual congeners (planar PCB 77 or ortho-substituted PCB 47 and PCB 153) in the diet for 30 days to investigate the effects of different classes of PCBs on circulating concentrations of T4 and T3, and found that PCB 153 significantly lowered levels of both T3 and T4, while impacts of PCB 47 and PCB 77 were less consistent. These results suggest that the ortho-substituted PCBs may be greater contributors to the deleterious effects of Aroclor 1254 on thyroid status than planar PCBs. Similarly, PBDE congeners also differ in thyroid toxicity. Chen et al. (2010, 2012a) found altered thyroid hormone levels in juvenile zebrafish exposed to PBDE 209, but not PBDE 47. In contrast to these findings, rainbow trout held at 8, 12, or 16 degrees C and exposed to environmentally relevant concentrations of PCBs for 30 days, followed by a depuration phase, showed changes in some measures of thyroid status (e.g., liver deiodinase activity), but the responses did not differ among PCB mixtures containing different proportions of dioxin-like and non-dioxin-like PCBs, and circulating hormone concentrations did not change significantly (Buckman et al., 2007). Responses may also differ among life stages. For example, Lerner et al. (2007) observed little effect on thyroid hormone levels in juvenile Atlantic salmon exposed to PCB 1254 as yolk sac larvae, but juvenile fish exposed as smolts had significantly decreased production of plasma T3.

Recently, hydroxylated PBDEs (OH-PBDEs) have emerged as a topic of concern because of reports of their natural production in the environment as well as metabolites in fish (Feng et al., 2012). Usenko et al. (2012) studied the effects of three hydroxylated PBDEs (3-OH-PBDE 47, 5-OH-PBDE 47, and 6-OH-PBDE 47) on embryonic zebrafish and found that all three congeners upregulated genes associated with thyroid hormone regulation and stress response.

Although most studies on the effects of POPs on thyroid function in fish have been conducted in the laboratory, there are some cases in which thyroid hormone alterations have been observed in wild fish from sites contaminated with these compounds. Weis et al. (2003) found significantly elevated T4 levels and a trend of reduced levels of T3 in mummichog (*Fundulus heteroclitus*) from a PCB-contaminated site in New Jersey. This finding suggests that the pollutants may have affected conversion from T4 to T3. In a field study at PCB-contaminated Wheatley Harbour (Lake Erie, Ontario, Canada), perturbations in the thyroid status in brown bullhead and goldfish, including lower plasma concentrations of thyroid hormones and/or elevated liver deiodinase activity, were found (Gilroy et al., 2012). The mean ΣPCB concentration in affected brown bullhead was approximately 250 ng/g wet wt. Schnitzler et al. (2012) investigated thyroid function in wild sea bass collected near several estuaries in Europe and found that, particularly in fish exposed to high PCB concentrations, changes in thyroid hormone metabolism led to an increased conversion of T4 to T3 and reduced thyroid hormone excretion. Song et al. (2012) found elevated serum levels of thyroid-stimulating hormone and depressed levels of T4 in crucian carp (*Carassius auratus*) from a PBDE-contaminated site—a river close to an electronic waste site in China. The average muscle PBDE concentration in affected fish was 236 ng/g wet wt.

3.3. Disruption of the Hypothalamus-Adrenal-Pituitary Axis

The hypothalamus-adrenal-pituitary (HPA) axis in fish can also be affected by POPs, interfering with their ability to react normally to environmental stressors (Hontela, 1998). For example, brown bullhead exposed to Aroclor 1248 had lower levels of cortisol compared to control fish (Iwanowicz et al., 2009). Arctic charr (*Salvelinus alpinus*) exposed to environmentally relevant concentrations of PCBs found in the Arctic environment had impaired regulation of cortisol production by interrenal tissue, which could lead to a deficiency in adaptation to stress (Aluru et al., 2004; Vijayan et al., 2006). Stouthart et al. (1998) found that PCB 126 exposure during embryonic development at concentrations of 3.26 to 326 ng/L altered whole-body adrenocorticotropic hormone (ACTH) and cortisol levels in carp embryos, suggesting that PCB induced a stress response in these larvae. However, cortisol levels declined over time in exposed larvae to levels below those in controls. Juvenile fish are also sensitive to the effects of PCBs on the HPA axis. Lerner (2007) observed that Atlantic salmon exposed to Aroclor 1254 as smolts had significantly decreased plasma cortisol compared to smolts that had been exposed to PCBs as yolk-sac larvae. OC pesticides may also disrupt the HPA axis and interfere with cortisol secretion and normal stress response. Endosulfan and *o,p'*-DDT have both been shown to disrupt ACTH-stimulated cortisol secretion by rainbow trout cells *in vitro*, though at relatively high exposure concentrations (see Dorval et al., 2003; Lacroix and Hontela, 2003 and references therein). There is also some evidence for PBDE effects on the adrenal function, as Chen et al. (2012a) and Yu et al. (2010, 2011) report changes in expression of genes encoding corticotrophin-releasing hormones in zebrafish exposed to PBDE 209 or the PBDE mixture DE-71.

3.4. Disruption of Other Metabolic Processes

There is some evidence that endocrine-disrupting compounds, including some POPs, may interfere with mechanisms involved in weight control. These "obesogens" inappropriately regulate and promote lipid accumulation and adipogenesis, and may act through a variety of mechanisms (see review by Grun and Blumberg, 2009). Some of these compounds, such as tributyl tin (TBT), interact with the peroxisome proliferator-activated receptors (PPARα, δ and γ) and the 9-cis retinoic acid receptor (RXR), which play an important role in the regulation of lipid biosynthesis and storage. Other compounds may affect weight through disregulation of sex steroids or produce changes in the hypothalamic-pituitary-adrenal and hypothalamus-pituitary-thyroid axes. A number of studies suggest that tributyl tin may function as an obesogen in both mammals (Grun and Blumberg, 2009) and fish (Meador et al., 2011; Tingaud-Sequeira et al., 2011). However, while there is some evidence linking exposure to certain PCB congeners and organochlorine pesticides with obesity and adipose tissue accumulation in humans and rodents (Dirinck et al., 2011; Lee et al., 2012; Wahlang et al., 2013), such a relationship has not yet been confirmed in fish.

4. EFFECTS ON EARLY DEVELOPMENT

Fish are generally considered to be most sensitive to chemical contaminants during the early life stages. Exposure to POPs during early development in oviparous fish species can result in a broad spectrum of effects, such as reduced hatching success and embryo mortality, abnormal development and malformation of larvae, and limited survival (Westerlund et al., 2000; Foekema et al., 2012; Rigaud et al., 2013). Among the more significant aspects of developmental toxicity discussed here are the transfer of maternally derived POPs; the direct toxic effects of specific types of POPs; and finally, impaired maternal transfer of nutrients or hormones due to POPs exposure (see Tables 2.1–2.4).

4.1. Maternal Transfer of POPs

POPs measured in the embryonic and early developmental stages of fish are considered to be primarily of maternal origin (Elskus et al., 2005; Kleinow et al., 2008). When eggs are released into the water column at spawning, the chorion and vitelline envelope provide some protection for the embryo against exposure to waterborne contaminants (Finn, 2007). During reproduction, lipid mobilization occurs in female teleosts to support egg development, providing a route through which lipophilic POPs, accumulated in other organs high in lipids, such as liver, are transferred to the gonads and eventually to eggs. The lipid content, reproductive stage, and reproductive life history of the mother, as well as the overall spawning and reproductive strategy of the species, can influence the redistribution and maternal transfer of POPs. The transfer of maternally derived POPs, therefore, varies dramatically from species to species (Niimi, 1983; Elskus et al., 2005). Several researchers have observed that pre-spawning migration causes a magnification of POP concentrations in body tissues, female gonads, and roe of Pacific salmon (DeBruyn et al., 2004; DeBruyn and Gobas, 2006), with potential consequences for egg viability.

The selective transfer and retention of contaminants also occur and are regulated by physicochemical properties such as lipophilicity, water solubility, and chemical hydrophobicity (log Kow) of compounds (Nyholm et al., 2008; Arnoldsson et al., 2012; Daley et al., 2012; Foekema et al., 2012). For example, Serrano et al. (2008) found significant correlation between log Kow of OC pesticides and PCB transfer ratios from the liver to eggs in seabream (*Sparus aurata*). Higher log Kow compounds had lower transfer rates from liver to eggs. Recently, Daley et al. (2009, 2012) provided evidence for bioamplification of POPs during transfer from tissue of female fish to oocytes, eggs, and embryos when adult lipid stores are depleted, as well as in eggs and embryos themselves as yolk is depleted. In a larval fish bioaccumulation study, PCB accumulation peaked at the end of the yolk-sac stage, when lipid reserves are spent, with the highest levels for compounds with log Kow > 5.0 (Foekema et al., 2012).

4.2. Direct Toxic Effects

Studies characterizing the effects of POP exposure on early development have involved different oviparous fish species and exposure routes, including amended diets fed to adults and direct waterborne or injection delivery routes to embryos. Exposures with single compounds, technical mixtures, or combinations of different types of POPs have been used. A brief review of the lethal and sublethal effects of POPs (PCBs, PBDEs, PCDD/Fs, and OC pesticides) on early development is provided here.

A comprehensive discussion of the toxicity of PCDD/Fs in the early development of fish is found in Cook et al. (2003), Tillitt et al. (2008), and King-Heiden et al. (2012). Although the first two studies focus on lake trout from the North American Great Lakes, their findings are applicable to, and have been confirmed in other species globally, including marine fish (Yamauchi et al., 2006; King Heiden et al., 2012). In brief, exposure to dioxin-like POPs during early development induces a classic response, mediated through activation of AhR, known as the "blue sac disease." This syndrome includes edema, anemia, hemorrhage, ischemia, impaired heart and vasculature development, impaired bone formation, and jaw malformations, all of which are associated with arrested growth and development.

Recent studies with dioxin-like POPs have established more accurate threshold effect concentrations in both laboratory fish and naturally occurring species of concern. In a chronic exposure study, Geisy et al. (2002) reported a reduced survival of rainbow trout fry from females fed as little as 1.8 ng TCDD per kg feed. Another study using zebrafish as a model species (King Heiden et al., 2005) showed that chronic dietary exposure resulting in the accumulation of as little as 1.1 ng TCDD/g female impacted offspring health. Maternal transfer resulted in the accumulation of 0.094–1.2 ng/g TCDD in eggs and larvae, sufficient to induce larval TCDD toxicity. Similarly, Palstra et al. (2006) reported significant impacts on the development and survival of European eel (*Anguilla anguilla* L.) embryos whose parents had high concentrations of dioxin-like compounds in muscle and gonad tissues. The disrupting effects occurred at levels below 4 pg TEQ per g gonad, which is below the European Union eel consumption standard (Anonymous 2001). In a field study with walleye (*Sander vitreus*) in Lake Ontario, Canada, Johnston et al. (2005) found no effects on early life stage survival in the offspring of females whose ova TEQs ranged from 3.0 to 14.0 pg TCDD TEQ/g wet wt. However, they also noted that the toxic effects may have been offset by larger size and more favorable fatty acid (FA) composition of eggs of the largest females, which had the highest TCDD concentrations. Additionally, they observed that the toxic effects of these contaminants may not have had time to express themselves within the time frame of their study, which extended only through the embryonic and early post-hatch period. Yamauchi et al. (2006) characterized the early life stage toxicity of TCDD in red seabream (*Pagrus major*). In this species, the EC(egg)50s for yolk sac edema, underdeveloped fins, and spinal deformity were 0.170, 0.240, and 0.340 ng/g, respectively. The LC(egg)50 was 0.360 ng/g embryo, indicating that this species is one of the most sensitive fishes to TCDD toxicity. Based on these and other studies, the estimated no-observable-adverse-effect level (NOAEL) for these effects is 5 pg TCDD toxicity equivalence/g egg (Cook et al., 2003). Newer studies have also examined the toxicity of brominated dioxins, compounds that are commonly found in fish from the Baltic Sea (Norman Haldén et al., 2011). Preliminary findings suggest that they affect embryo viability in ways similar to TCDD.

PCBs have frequently been associated with endocrine disruption and reproductive toxicity in fish, with dioxin-like congeners eliciting effects similar to those associated with TCDD (e.g., Tillitt et al., 2008). For example, Grimes et al. (2008) used zebrafish to investigate the developmental toxicity of the model dioxin-like PCB, PCB 126, and observed cardiac and jaw abnormalities very similar to those documented in fish exposed to TCDD. The effects of dioxin-like PCBs on flatfish metamorphosis have also been examined. Foekema et al. (2008) conducted a prolonged early life stage test in which newly fertilized sole (*Solea solea*) eggs were placed in water containing varying concentrations of PCB 126 (0.1 to 1000 ng/L) until 4, 8, 10, or 15 days postfertilization (dpf). For embryos that survived to the free-feeding stage (12 dpf), the LC_{50} values were between 39 and 82 ng/L, depending on exposure duration, with increasing mortality rates a week beyond this stage. In addition, surviving fish developed pericardial and yolk-sac edema. When the tests were extended through the completion of metamorphosis, LC_{50} values were much lower, 1.7–3.7 ng/L. The internal dosages of these larvae revealed a LD_{50} of 1 ng TEQ/g lipid, which is within the same order of magnitude as TEQ levels found in fish from highly polluted areas. A similar exposure with larval summer flounder to PCB 126 reported median LD_{50}s for pre- and early metamorphic flounder ranging from 30 to 220 ng/g wet wt and a single sublethal dose of PCB 126 (15 ng/g) delayed metamorphic progress and caused abnormal gastric gland morphology (Soffientino et al., 2010).

Other investigations have studied the early developmental toxicity of dioxin-like PCBs as compared to TCDD. Rigaud et al. (2013) exposed mummichog embryos to various doses of PCB 126 and TCDD and found larval malformation and reduced growth at the highest doses (1280 pg/g ww for TCDD and >2500 for PCB 126). While PCB 126 was less potent in terms of causing mortality, malformations, and reductions in body size compared to TCDD, its relative potency in mummichog, based on induction of ethoxyresorufin-O-deethylase (EROD) activity, was unusually high, 0.71 compared to those (0.001 to 0.005 range) in other species. These findings highlight the variable sensitivities among fish species to these compounds.

Commercial PCB mixtures and non-dioxin-like PCBs have also been associated with developmental toxicity in fish, inducing reduced hatching of embryos and reduced or abnormal growth. Örn et al. (1998) exposed zebrafish to a mixture of 20 PCBs in three different dose levels (0.008, 0.08, and 0.4 ug of each congener per g of feed dry wt). A reproduction study performed with exposed females and unexposed males after 9 weeks revealed that median survival time for larvae was only 7.7 days in the high-dose group as compared with 14 days in controls. Atlantic croaker larvae obtained from females fed a diet amended with Aroclor 1254 were examined for growth and survival (McCarthy et al., 2003). Results showed developmental delays and reduced growth rates in these maternally dosed larvae at concentrations of 0.66 ng per egg. In fact, one study suggested that certain non-dioxin-like PCBs may have higher embryo toxicity than PCB 126. Westerlund et al. (2000) exposed zebrafish embryos to PCBs 60, 104, 112, 126, 143, 172, 184, and 190, and two hydroxylated PCBs (3'-OH-PCB 61 and 4'-OH-PCB 30). The two PCBs with estrogenic activity, PCB 104 and 4'-OH-PCB, caused the highest embryo mortalities, greater than the mortality associated with PCB 126.

Although not as extensively studied as PCDDs and PCBs, OC pesticides also appear to cause developmental toxicity in fish. For example, zebrafish embryos exposed directly to dissolved-phase endosulfan exhibited neurotoxic effects (Stanley et al., 2009). Fish exposed to certain OC pesticides have demonstrated effects on the development of the reproductive system and sex differentiation. Several studies examining the effects of early life exposure of fish to o,p'-DDT have shown this DDT isomer to be estrogenic (Edmunds et al., 2000; Metcalfe et al., 2000). Metcalfe et al. (2000) found that exposure of early life stages of Japanese medaka to o,p'-DDT at concentrations as low as 5 ug/L in water caused an intersex condition of the gonad (testis–ova) in male medaka. If female medaka were exposed to comparable concentrations of o,p'-DDT and the contaminant was transferred to embryos through maternal transfer, no testis–ova were observed in the offspring, although there was some delay in offspring time to hatch. When treated fish reached adulthood, the females showed more advanced development of oocytes and males showed greater induction of vitellogenin when treated with 17β-estradiol than control males. Edmunds et al. (2000) reported male-to-female sex reversal in the Japanese medaka embryos injected with o,p'-DDT at an embryonic dose of $\sim$230 ng/egg. Sex-reversed males (XY females) had male pigmentation accompanied by female secondary sexual characteristics. Interestingly, these fish had active ovaries and breeding success comparable to normal females. In another study, male juvenile Japanese medaka exposed to 1, 5, 20, and 100 ug/L p,p'-DDE for two months had increased hepatosomatic index (HSI) and decreased gonadosomatic index (GSI). Intersex was found in the 100 ug/L p,p'-DDE exposure group (Zhang and Hu, 2008). Exposure to p,p'-DDE also upregulated gene expression of VTG, choriogenins, and estrogen receptor alpha in fish liver. While these studies demonstrate the estrogenic activity and potential impact of DDTs on sex differentiation in fish, the doses were substantially higher than those typically found in environmental samples.

Mixtures of OC pesticides with other POPs have also been linked to developmental toxicity in fish. For example, striped bass embryos from the San Francisco Estuary, contaminated with PCBs, PBDEs, and legacy organochlorine and current-use pesticides, showed significantly slower growth when compared to controls, and their yolk sacs contained less yolk and were more rapidly absorbed (Ostrach et al., 2008). The majority of the exposed larvae exhibited yolk-sac edema, and brain and liver were also slower to develop. Lyche et al. (2013) exposed zebrafish embryos to a naturally occurring mixture of POPs. The mixture, obtained from liver extracts of burbot collected from two Norwegian lakes, contained environmentally relevant concentrations of PBDEs, PCBs, and DDTs. Changes in the transcription of genes regulating endocrine signaling, metabolism, reproduction, and immune function were observed in exposed embryos.

Several recent laboratory exposure studies have established the developmental toxicity of PBDEs, although at exposure concentrations substantially higher than those typically found in the environment. Zebrafish embryos exposed directly to PBDE 47 dissolved in water at doses ranging from 100 to 5000 ug/L showed a number of effects including delayed hatching, reduced growth posthatching, and morphological abnormalities in larvae (Lema et al., 2007). Similar effects were observed in zebrafish exposed to PBDE 49 (McClain et al., 2012); exposure at concentrations of 4–32 μM (1944–15,552 ug/L) caused dose-dependent mortality at 6 days postfertilization, as well as severe dorsal curvature and cardiac toxicity.

Developmental abnormalities in fish embryos and larvae due to parental exposure and maternal transfer of PBDEs may occur at much lower exposure concentrations. For example, Yu et al. (2011) exposed zebrafish embryos to the PBDE mixture DE-71 at concentrations of 1, 3, and 10 ug/L for 5 months until sexual maturation. Decreased hatching, altered thyroid hormone levels, and inhibition of growth in the offspring of exposed parents were observed; these conditions were exacerbated when F1 embryos and larvae continued to be treated with DE-71 at comparable concentrations. He et al. (2011) exposed zebrafish to deca-PBDE (PBDE 209), the only commercial PBDE mixture still allowed for use. They found that parental exposure to PBDE 209 at doses ranging from 0.001 to 1 μM (0.959 to 959 ug/L) led to delayed hatch and motor neuron development in offspring, as well as some behavioral abnormalities (see Section 7). Several others have reported alterations in motor neuron development and cardiac development in zebrafish exposed to PBDEs 47 and 49 (Chen et al., 2012b; McClain et al., 2012), as well as behavioral abnormalities discussed further in Section 8. As noted in Section 3, exposure of larvae and developing fish to high levels of OH-PBDEs has been shown to induce oxidative stress and disrupt the cholinergic system and thyroid function, leading to disrupted development (Feng et al., 2012; Song et al., 2012; Usenko et al., 2012). However, it is unknown if these same effects would occur at environmentally relevant levels.

Although most developmental toxicity studies of PBDEs have been conducted in zebrafish, some work has been done with other species. Mhadhbi et al. (2012) determined LC_{50} values for turbot embryos exposed to PBDEs 47 and 99. The values for embryos and larvae, respectively, were 27.35 and 14.13 μg L^{-1} for BDE-47 and 38.28 and 29.64 μg L^{-1} for PBDE 99, concentrations that were considerably higher than concentrations of dissolved PBDEs typically measured in aquatic systems. However, some developmental abnormalities have been reported as a result of early life stage exposure at lower concentrations (~ 1 ug/L). Timme-Laragy et al. (2006) exposed mummichog embryos to the PBDE mixture, DE-71, and observed delayed hatching and tail curvature malformation; the tail malformation possibly resulted from perturbation of the thyroid hormones. The larvae also displayed some behavioral abnormalities, such as lower activity and reduced startle response (see Section 7).

4.3. Impaired Maternal Transfer of Nutrients or Hormones

In oviparous fish, lipids for energy reserves as well as nutrients and hormones essential for development are transferred to developing embryos from the mother through deposition into the egg. Thiamine (vitamin B1), retinal (a form of vitamin A), and FAs, including eicosapentaenoic acid and arachidonic acid, essential for normal larval development, are among the nutrients transferred during vitellogenesis. Hormones including thyroid hormone and cortisol are also transferred, as well as antibodies and immune proteins. There is some evidence that exposure to POPs may interfere with the normal transfer of these important substances to developing embryos.

Retinal is the main retinoid stored in oviparous eggs of fish, amphibians, and reptiles, reaching the oocytes in association with VTGs, the yolk precursor proteins (Levi et al., 2012). During early development, it is metabolized to retinoic acid (RA), which regulates genes involved in cell proliferation, differentiation, and tissue function and is therefore essential for normal embryonic development. Several classes of POPs, including some OC pesticides, PCBs, and TCDD, can interfere with retinoid metabolism and signaling (Novák et al., 2008), and exposure to dioxin-like POPs may deplete retinoid stores in fish (reviewed in Rolland, 2000). Such effects could interfere with retinal transfer to oocytes and contribute to deformities in larvae.

Thiamine deficiency can also cause developmental defects in fish, such as the M74 syndrome, a constellation of developmental deformities observed in Baltic salmon and salmon from the Great Lakes. It has been hypothesized that maternal exposure to POPs could contribute to the development of this thiamine deficiency in eggs and embryos, but this has not yet been resolved (reviewed in Vuori and Nikinmaa, 2007; Tillitt et al., 2008).

As previously described, maternal lipid provisioning to embryos is critical for their development. FA concentrations and composition in female muscle are correlated with FA content in eggs, with egg FA concentrations decreasing as females lose condition (Garrido et al., 2007). Thus effects on female lipid content or metabolism caused by exposure to POPs could significantly impact egg quality, including the lipid reserves available to larvae and the FA composition required for normal growth.

During early development, innate immunity proteins and antibodies are also transferred from the maternal circulation to the oocytes. Maternally derived hormones are also transferred, including cortisol, which can have immunosuppressive effects (Li and Leatherland, 2012). Thus actions of POPs that increase stress in females and alter the cortisol content of eggs may affect the immune systems of early embryos. Moreover, females that are immunocompromised due to POP effects may be less able to contribute to the immunocompetence of embryos. Additionally, thyroid hormones, T4 and T3, transferred during vitellogenesis, are present in high quantities in fish eggs and play an important role in early development (Monteverdi and Di Giulio, 2000; Power et al., 2001). Thus the effects of POPs on thyroid function in adult females could interfere with the transfer of these vital hormones to eggs. Based on the study findings, there is increasing evidence that exposure to POPs impairs the maternal transfer of nutrients and hormones to early life stages of fish.

5. TRANSGENERATIONAL EFFECTS

In recent years a great deal of interest has been shown in the transgenerational effects of chemical contaminants that can be mediated through epigenetic mechanisms (Skinner et al., 2010, 2011). Epigenetics is the study of mitotically or meiotically heritable changes in gene function that occur without a change in the DNA sequence (Vandegehuchte and Janssen, 2011). The most well-studied epigenetic mechanisms for regulation of gene expression are DNA methylation and histone modification (Head et al., 2012). In mammals, DNA methylation regulates transcription of genes, alters chromosomal positioning, influences X-chromosome inactivation, controls imprinted genes, and represses parasitic (e.g., viral) DNAs (Skinner et al., 2010, 2011). Such changes occur in both somatic and germ cells; studies in mammals have revealed unique epigenetic profiles in sperm that may affect the male contribution to embryogenesis (Skinner et al., 2010, 2011). Environmental influences, including exposure to toxic chemicals, can alter DNA methylation and epigenetic programming in both germ and somatic cells (Skinner et al., 2010, 2011). When epigenetic mutations occur in somatic cells, they can produce disease in the exposed individual, but they will not be transmitted to the next generation. However, if the germ line is permanently modified, this will affect not only exposed individuals and their developing offspring, but also subsequent generations (Skinner et al., 2012).

In mammals, alterations in DNA methylation and changes in epigenetic programming are associated with multiple disease states, including cancers (Sadikovic et al., 2008; Ellis et al., 2009), male infertility (Anway et al., 2005; Anway and Skinner, 2008), prostate and kidney disease (Anway and Skinner, 2008), and several cognitive disorders including autism (Schanen, 2006; Gräff and Mansuy, 2009). However, not all of these diseases are transgenerational phenomena. Skinner et al. (2010) define transgenerational epigenetic effects as those in which direct exposure of the parent to the stressor results in an altered phenotype that is transmitted through multiple generations in the absence of direct exposure of those generations to the stressor. These are distinguished from multiple generation effects, such as those that occur when a gestating female is exposed to a chemical contaminant, thereby exposing the developing embryo (F1 generation) and its germ line, which will generate the F2 generation (Skinner et al., 2010). As defined by Skinner et al. (2010), true transgenerational events must extend to the F3 generation and beyond, demonstrating that the effects persist in the absence of any exposure to the stressor.

Research on humans and rodents shows that exposure to POPs can cause epigenetic modifications to somatic- and germ-cell lines, potentially leading to transgenerational effects. Several studies have linked exposure to PCBs, DDTs, or PBDEs to epigenetic changes such as hypomethylation or changes in histone modification of DNA in brain or liver of young rodents (Desaulniers et al., 2009; Shutoh et al., 2009; Casati et al., 2012; Woods et al., 2012). In some cases, such changes are linked to disease states; for example, Woods et al. (2012) found that perinatal PBDE exposure and associated epigenetic DNA changes in brain tissue of mice were linked to cognitive and behavioral impairments. Such associations have also been observed in epidemiological studies. Rusiecki et al. (2008) and Kim et al. (2010) found a significant and inverse relationship between plasma concentrations of DDTs, PCBs, chlordanes, mirex, and total POPs and blood DNA methylation in Greenland Inuit and in healthy Koreans. Mitchell et al. (2012) found a relationship between exposure to PCBs and PBDEs and DNA methylation patterns in individuals with autism. All of these studies provide evidence for the involvement of epigenetic modification in the etiology of POPs-related diseases.

Evidence for epigenetic transgenerational effects of POPs is provided by Manikkam et al. (2012), who found that when F0 female rats were exposed to TCDD during the period of embryonic gonadal sex determination, not only the F1 and F2 generations, but the subsequent F3 generation, which had no environmental exposure, exhibited early-onset female puberty, spermatogenic cell apoptosis, and reduction in the ovarian primordial follicle pool size. Differential DNA methylation of the F3 generation sperm promoter epigenome was also observed, supporting an epigenetic mechanism for these abnormalities. However, the dose of TCDD used was quite high, so it is uncertain how often this would occur in natural populations.

As yet, we have only limited information on contaminant-related epigenetic and transgenerational effects in fish. Recent studies indicate that DNA methylation and related epigenetic mechanisms are involved in developmental processes in fish as well as mammals, including epigenetic programming in the male germ line during sperm maturation (Carrell, 2011; Moran and Perez-Figueroa, 2011). There is also evidence of the transgenerational effects of other stressors in fish, such as hyperglycemia and exposure to anoxic conditions, that may be mediated through epigenetic mechanisms (Ho and Burggren, 2012; Olsen et al., 2012). Changes in DNA methylation in liver neoplasms have also been observed in fish, both in laboratory-reared zebrafish in which tumors were induced with a model carcinogen (Mirbahai et al., 2011a) and in wild dab that developed neoplasms as a result of environmental exposures (Mirbahai et al., 2011b). Several studies also report changes in DNA methylation in somatic or germ cells in fish exposed to heavy metals, tributyltin, and synthetic and natural estrogens (Contractor et al., 2004; Bagnyukova et al., 2007; Aniagu et al., 2008; Wang et al., 2009; Strömqvist et al., 2010). These studies indicate that epigenetic mechanisms occur in fish and may play a role in disease etiology, and that these mechanisms may be affected by chemical contaminants and other environmental factors.

Limited studies have reported multigenerational effects of POP exposure in fish. King-Heiden et al. (2009) used zebrafish to determine whether early life stage exposure to TCDD induces toxicity in adult zebrafish and their offspring. They found that offspring of parents exposed to TCDD during early life stages showed a 25% increase in mortality compared with the F1 of control fish and reduced egg production (30–50% of control). Similarly, Yu et al. (2011) characterized the disrupting effects of long-term exposure on the thyroid endocrine system in adult fish and their progeny following parental exposure to PBDEs. In both studies, adverse effects resulting from TCDD or PBDE exposure during early life stages for one generation of zebrafish were sufficient to cause adverse health and reproductive effects on a second generation of zebrafish. Persistence of such effects into the F3 generation would establish that these exposures led to a true transgenerational phenotype. However, this has not yet been established. In summary, mammalian studies as well as preliminary whole life-cycle exposure in fish suggest that some of the toxic effects of POPs in fish could be mediated through epigenetic mechanisms, including modifications to the germ line that could result in transgenerational effects. However, definitive evidence for these effects is still lacking.

6. EFFECTS ON GROWTH, CONDITION, AND ENERGY RESERVES

Normal growth and adequate energy reserves are critical to the survival of juvenile fish and have a strong influence on adult reproductive potential. Prior to 2000, studies showed that POP exposure can alter growth rates and condition in fish, particularly fish exposed to high levels (>1–2 ug/L) of PCBs, PCDDs, and OC pesticides (e.g., Mauck et al., 1978; Eisler, 1986a; Eisler and Belisle, 1996; Scott and Sloman, 2004). The effects of low environmentally relevant concentrations are less consistent, with some reports of low POP levels actually stimulating fish growth (e.g., Bengtsson, 1980). Exposure to low concentrations of POPs may also have neurological effects that impair foraging ability, leading to reduced growth (see Section 6 for details). Recently, growth studies have been expanded to include additional fish species (including wild fish from contaminated sites), as well as the effects of POPs on lipid content and energy metabolism. In this section we review recent studies on the effects of major classes of POPs on the growth and condition of fish (see Tables 2.1–2.4).

Exposure to TCDD and other dioxin-like compounds can reduce fish growth, particularly if exposures occur during early development (Giesy et al., 2002 and references therein). However, effects on growth and condition are found less frequently in adults and older juveniles, especially at environmentally realistic exposure concentrations. For example, no effects on growth or condition were found for adult rainbow trout exposed to TCDD concentrations as high as 90 ng TCDD/kg ww in food for up to 300 days (Giesy et al., 2002). Similarly, Ábalos et al. (2008) exposed gilthead seabream (*Sparus aurata*) to feed containing PCDDs and PCDFs at 23 ng TCDD-TEQ/kg for over 300 days, but found no effect on fish growth. In contrast, mixed effects on growth and condition have been reported in field studies. Hodson et al. (1992) observed no change in condition, but did detect an increase in carcass lipid content in white sucker exposed to PCDDs and PCDFs in bleached kraft mill effluent, while Adams et al. (1992) reported increased growth and lipid storage in redbreast sunfish (*Lepomis auritus*) exposed to dioxin-contaminated effluent.

Changes in energy metabolism in fish exposed to PCBs have been reported. For example, Nault et al. (2012) found increased plasma glucose concentrations in rainbow trout exposed to PCB 126 (50 ug per kg) through intraperitoneal injection, as well as some changes in mitochondrial respiration in trout hepatocytes exposed to PCB 77. Similarly, changes in growth and metabolic status have been reported in fish exposed to PCB technical mixtures (e.g., Aroclor 1254) and non-dioxin-like PCBs, particularly juvenile fish exposed to these compounds in food and water (Eisler, 1986b; Anderson et al., 2003; Vijayan et al., 2006). PCB exposure in Japanese medaka induced changes in genes related to glycolysis and gluconeogenesis, indicating that PCB exposure might enhance glucose production via gluconeogenesis, perhaps as a response to increased glucose consumption (Nakayama et al., 2011). Liver metabolic performance may also be altered in ways that may render fish less able to meet the enhanced energy demands of other environmental stressors (Wiseman and Vijayan, 2011). Several studies also document reduced growth of fish larvae exposed to PCBs during early development at more environmentally realistic concentrations; these studies are reviewed in Section 4.

Additional studies have documented negative correlations between PCB exposure and growth and body condition in wild fish populations. Khan (2011) found that marine sculpin, *Myoxocephalus scorpius*, exposed to PCBs near a shipping terminal in Newfoundland, Canada, had decreased body condition in comparison to sculpin from a reference site upcurrent from the contaminated terminal. Tillitt et al. (2003) report reduced weight gain and poor survival of offspring of largemouth bass from the PCB-contaminated Housatonic River, in Massachusetts in the United States, while Anderson et al. (2003) saw glycogen depletion in smallmouth bass from PCB-contaminated Kalamazoo River in Michigan. In contrast, Rypel and Bayne (2010) found that the lipid-corrected PCB concentrations correlated positively with the growth rate of four of six species of fish residing in the contaminated Logan Martin Reservoir system in Alabama.

The potential of DDTs and other OC pesticides to reduce fish growth was established in the late 1960s and 1970s (e.g., Macek, 1968; Mayer et al., 1975), though typically at concentrations higher than those currently measured in most environmental samples. More recent studies have documented the effects of DDTs and other OCs on the growth of flatfish, white sturgeon, guppies, and perch (Rice et al., 2000; Bayley et al., 2002; Feist et al., 2005; Linderoth et al., 2006). For example, Rice et al. (2000) observed reduced growth rates in juvenile English sole fed a diet of polychaete worms reared on sediment contaminated with 4.2–4.8 ug/g p,p'-DDE dry wt, with p,p'-DDE concentrations of 215–219 ug/g dry wt in tissue. Additional studies suggest the effects of DDTs, especially the estrogenic DDT isomer, o,p'-DDT, on growth and energy storage (Benton et al., 1994; Benguira et al., 2002).

Characterizing the effects of PBDEs on fish growth and condition has become a recent area of interest. Similar to fish exposed to dioxin-like compounds, reduced growth has also been documented in larval and juvenile fish exposed to various PBDE congeners in the diet (Chen et al., 2010; Kuo et al., 2010). There is also some evidence that PBDEs affect metabolic function (Søfteland et al., 2011). Ghosh et al. (2013) studied the impact of PBDEs on the Antarctic fish emerald rock cod (*Trematomus bernacchii*) and found that PBDE-exposed fish had lower total liver lipid, increased liver lipid peroxide and protein carbonyl concentrations, as well as changes in activities of several antioxidant enzymes. These findings suggested possible oxidative stress and disruption of the low basal metabolic rates and basal oxygen-holding capacity of this fish species that might lead to decreases in growth performance.

Although most studies focus on the potential of POPs to reduce growth and condition, a few reports highlight the potential for POPs to stimulate growth and weight gain under certain conditions. Berg et al. (2011) examined the effects of dietary exposure to environmentally realistic mixtures of POPs extracted from Norwegian burbot (*Lota lota*) on the growth of zebrafish. While these mixtures contained PCBs, DDTs, and PBDEs, PBDEs dominated the mixture, as the ratios of concentrations of these compounds were 6:10:270 for PCBs, DDTs, and PBDEs, respectively. Surprisingly, exposure was found to have effects on body weight. In the first generation, body weight was significantly higher in exposed fish compared to controls, while in the next generation the same exposures were associated with a decrease in body weight. As part of the same study, Lyche et al. (2011) examined the effects of exposure to this mixture on genes associated with weight homeostasis. They found altered regulation of three groups of genes in the exposed fish: regulators of weight homeostasis (PPARs, glucocorticoids, CEBPs, estradiol), steroid hormone function (glucocorticoids, estradiol, NCOA3), and insulin signaling (HNF4A, CEBPs, PPARγ). Similarly, fasting European eel exposed to a PCB mixture over a 27-day exposure period lost less weight compared to the unexposed controls (van Ginneken et al., 2009). In addition, the standard metabolic rate and oxygen consumption during swimming were significantly lower in the PCB-exposed eels compared to the unexposed controls. These results suggest that exposure to POPs may affect the endocrine regulation of the metabolism in a way that promotes increased weight gain and obesity.

As discussed in Section 3, POP exposure has been associated with changes in thyroid function, which could affect growth and metabolism. However, some studies suggest that compensatory mechanisms that help to maintain thyroid system homeostasis may protect fish, and thus these metabolic changes may not always result in biologically meaningful effects on energy balance and growth (Brown et al., 2004; Peter and Peter, 2011). For example, Brown et al. (2004) cite several studies in which treatment with PCB 126 led to temporary fluctuations in T4 levels in fish, but the thyroid system was capable of compensation and no changes in growth or condition occurred.

7. EFFECTS ON REPRODUCTION

Some classes of POPs (e.g., DDTs, PCBs, PCDDs) have long been recognized as reproductive toxicants in fish and wildlife (Kime, 1995, 1998; van der Oost et al., 2003; White and Birnbaum, 2009; Letcher et al., 2010). More recent studies focus on the endocrine-disrupting effects of POPs, impacts of individual compounds, and the multigenerational and transgenerational effects of these chemicals. POPs may affect the reproductive cycle in a variety of ways, influencing gonadal development and the onset of puberty, female fecundity and male fertility, and reproductive behavior (see Tables 2.1–2.4).

7.1. Onset of Puberty

Exposure to POPs may influence the onset of puberty in laboratory mammals and humans (Rogan and Ragan, 2007; Roy et al., 2009; Bell et al., 2010; Schell and Gallo, 2010). In both laboratory rodents and in epidemiological studies of human populations, exposure to PCDDs, PCDFs, and dioxin-like PCBs generally delays the onset of puberty in both sexes, whereas exposure to certain POPs with potential estrogenic activity is associated with early puberty, particularly in females.

Surprisingly few studies have examined the effects of POPs on the age of sexual maturation in fish. Bayley et al. (2002) reported reduced body weight and delayed sexual maturation in male guppies exposed to p,p'-DDE in the diet from birth to adulthood. However, the concentrations (10 and 100 ug/g in food) used in this experiment were high compared to DDT levels typically measured in prey of fish in the natural environment. Early onset of puberty was observed in both male and female zebrafish exposed through the diet, from posthatching to adulthood, to a POP mixture containing PBDEs, PCBs, DDTs, and other OC pesticides at environmentally relevant concentrations (Lyche et al., 2011). How these compounds may have interacted to produce this effect is uncertain, but of the chemicals measured in the exposed fish, concentrations of PBDEs were particularly high, with a maximum concentration of 6500 ng/g lipid. In a field study, Johnson et al. (1998a) observed early sexual maturation in wild female English sole from the Hylebos Waterway, an urban and industrial site in Puget Sound, Washington. While a variety of pollutants were present at the site, this phenomenon was correlated most strongly with liver PCB concentrations.

7.2. Fecundity in Females

In contrast to effects on age at sexual maturation, the effects of POPs on ovarian development, egg quality, and fecundity in female fish have been studied extensively. Both field and laboratory investigations conducted prior to 2000 (Niimi, 1983; Walker and Peterson, 1994; Monosson, 2000) indicate that these chemicals can disrupt reproductive processes and reduce egg quantity and quality in female fish.

In fish, early life stage exposure to PCDDs leads to reduced survival and a wide range of developmental defects (reviewed in Tillitt et al., 2008; King Heiden et al., 2012) (see Section 4 of this chapter). Recently, research efforts have focused on the effects of TCDD on gonadal development and egg production, as well as the molecular mechanisms of toxicity. Possible transgenerational effects of dioxin-like POPs have also been a focus of recent work (see Section 5). For example, a series of studies using zebrafish as a model species investigated TCDD effects on ovarian development (King Heiden et al., 2005, 2006, 2008, 2009). The studies showed that chronic dietary exposure resulting in the accumulation of as little as 0.6 ng TCDD/g fish led to a decline in the ovosomatic index, and 10% of the females showed signs of ovarian necrosis following accumulation of approximately 3 ng/g TCDD. Subsequently, the effects of TCDD on ovarian follicle development and vitellogenesis were investigated. Chronic dietary exposure to as little as 0.08 ng TCDD per fish per day and with accumulations of 0.5–1 ng TCDD/g fish led to decreased serum estradiol and vitellogenin concentrations, inhibition of the transition from pre-vitellogenic stage follicles to vitellogenic stage follicles, and induction of follicular atresia. These changes in ovarian development were associated with decreased egg production and spawning success. Molecular studies suggested that TCDD could also alter ovarian function by disrupting additional signaling pathways such as glucose and lipid metabolism (King Heiden et al., 2008).

As discussed in Section 3, PCDDs and PCDFs in bleached kraft pulp mill effluent have been associated with reproductive dysfunction in fish (McMaster et al., 2006). Subsequent studies, however, have shown that other bioactive substances in pulp mill effluents may also contribute to these impairments (Hewitt et al., 2006).

Endocrine disruption and reproductive toxicity have frequently been associated with PCB exposure in fish, with dioxin-like congeners eliciting effects similar to those associated with TCDD (Monosson, 2000; King Heiden et al., 2012). For example, Orn et al. (1998) observed reduced body, liver, and ovary weights, as well as a reduced liver and ovary somatic index, among zebrafish exposed to a mixture of 20 PCBs in three different dose levels. Egg production was also reduced in all three exposure groups. Histologically, females in both the intermediate and high-dose groups contained a reduced number of mature oocytes.

Studies since 2000 have focused on the effects of non-dioxin-like PCB congeners, including those suspected of estrogenic activity. For example, Daouk et al. (2011) performed a life-cycle dietary exposure of zebrafish to PCB mixtures containing primarily non-dioxin-like congeners in proportions and concentrations similar to those typically found in environmental samples. Exposure to the PCB-contaminated diet led to increased atresia and a decline in the proportion of maturing follicles in the ovaries of females, as well as a reduction in the number of fertilized eggs per spawn and an increase in the number of poorly fertilized spawns. Whole-body tissue concentrations at which these effects were observed were as low as 200 ng/g dry wt in juvenile and postspawning females and 350 ng/g dry wt in gravid females. These results demonstrate that exposure to non-dioxin-like PCBs at environmentally realistic concentrations can cause significant reproductive toxicity in female fish. There is also some evidence that exposure to PCBs during larval development can lead to pathology in adults. Coimbra and Reis-Henriques (2007) exposed tilapia larvae to the PCB mixture Aroclor 1254 and found that this exposure resulted in ovarian alterations when these larvae reached adulthood. The same authors also found that exposure of adult tilapia to Aroclor 1254 inhibited reproduction.

Other studies have focused on POPs with suspected estrogenic activity, such as the *o,p'*-substituted DDTs. Papoulias et al. (2003) conducted an egg injection study with medaka to examine the effects of *o,p'*-DDE on sexual differentiation and development. Following exposure at 0.005 ng/embryo or 5 ng/g egg, a concentration within the range measured in fish eggs sampled in the 1990s (Miller, 1993; Munn and Gruber, 1997), females at maturation exhibited few vitellogenic oocytes, increased atresia, and lower gonadosomatic indices than unexposed controls.

Although information on the reproductive toxicity of PBDEs in female fish is limited, evidence shows that these POPs can disrupt the reproductive endocrine axis (see Section 3), and reproductive impairment has been reported at relatively high PBDE body concentrations. Muirhead et al. (2006) studied the effects of PBDE 47 on reproductive function in fathead minnows and observed a cessation of spawning at body concentrations of 50–60 ug/g wet wt. He et al. (2011) exposed zebrafish to waterborne PBDE 209 at concentrations ranging from 0.001 to 1 uM (0.959 to 959 ug/L) for five months posthatching, and observed reduced ovarian growth in females at body concentrations in the 2–30 ug/g wet wt range. However, when Kuiper et al. (2008) exposed European flounder and zebrafish to a range of concentrations of the commercial PBDE mixture, DE-71, with fish accumulation concentrations and PBDE congener patterns similar to those observed in wild fish, no histopathological changes were observed in reproductive organs. Interestingly, some data suggest that those PBDE congeners that affect the thyroid system are the most likely to affect the reproductive system (Torres et al., 2013). This report is consistent with the close relationship that has been demonstrated between thyroid endocrine status, gonadal sex differentiation, pubertal development, and reproduction in zebrafish and other fish species (Blanton and Specker, 2007; Mukhi and Patiño, 2007; Mukhi et al., 2007; Sharma et al., 2013).

A few recent studies have examined the reproductive effects of female fish exposed to mixtures of POPs. Kraugerud et al. (2012) examined the effects of a mixture of PCBs, PBDEs, and DDTs on gonadal morphology in female zebrafish. In female zebrafish exposed to the higher doses with body POPs concentrations of 6488 ng/g lipid PBDEs, 1313 ng/g lipid PCBs, 618 ng/g lipid DDTs, and 280 ng/g lipid HCBD (Lyche et al., 2011), there was a significant decrease in late vitellogenic follicle stages in ovaries. In addition, granulosa cells showed increased apoptosis and decreased proliferation, results somewhat similar to those of Daouk et al. (2011). There was also some indication of reduced vitellogenin synthesis in exposed females, based on vitellogenin immunostaining and hepatocyte proliferation in the liver.

Reproductive toxicity has been observed in female fish in natural populations as well. Abnormal ovarian development, atresia, reduced gonad size, reduced fecundity, and poor hatching success and fry survival have also been reported in wild fish populations from areas contaminated with POPs since the early 1980s in both marine and freshwater species (reviewed in Kime, 1995; Johnson et al., 1998a, 1998b; Monosson, 2000; Thomas, 2008). More recent studies confirm a relationship between environmental POP exposure and impaired reproduction in both marine and freshwater female fish. Bugel et al. (2010, 2011) examined the reproductive health of killifish from Newark Bay, New Jersey, an area contaminated with high concentrations of PCDDs, PCDFs, and PCBs as well as other chemical contaminants (e.g., PAHs). Fish from POP-contaminated sites exhibited a variety of reproductive abnormalities, including decreased gonad to body weight ratios, reduced mature ovarian follicles and VTG production, and inhibition of ovarian development. Those females that did mature produced fewer viable eggs. Similarly, Hinck et al. (2007) observed low GSI and gonadal abnormalities in common carp, black bass (*Micropterus* spp.), and channel catfish (*Ictalurus punctatus*) from urbanized and agricultural areas in the Colorado River Basin. Concentrations of DDTs (>0.5 ug/g ww), PCBs (>0.11 microg/g ww), and TCDD-EQs (>5 pg/g ww) were elevated in affected fish. Mayon et al. (2006) found impaired reproductive responses, including a higher proportion of atretic oocytes in females, in chub from sites in Belgium contaminated with PCBs and OC pesticides.

7.3. Fertility of Males

As with females, the reproductive toxicity of POPs to male fish has been demonstrated in a number of studies conducted prior to 2000 (e.g., Niimi, 1983; Walker and Peterson, 1994; Kime, 1995; 1998; Monosson, 2000; Pait and Nelson, 2002). More recent studies elucidate the effects of POPs with suspected estrogenic and anti-estrogenic activity, and provide more insight into the mechanisms through which these contaminants affect reproductive function.

Several studies have investigated the effects of different DDT isomers on reproductive function in male fish. Mills et al. (2001) and Zaroogian et al. (2001) examined the effects of o,p'-DDT and p,p'-DDE on gonadal development in laboratory-raised juvenile male summer flounder. In these studies, flounder were injected with the test chemicals at doses of 30 and 60 mg/kg, resulting in o,p'-DDT concentrations of 188–521 ug/g ww and p,p'-DDE concentrations of 200–373 ug/g ww in liver. Fish treated with o,p'-DDT showed lower GSI, inhibition of spermatogenesis, and regression of the testes. Additionally, these researchers observed induction of vitellogenin in the males, indicative of the estrogenic activity of this DDT isomer. In contrast, exposure to p,p'-DDE has no effect on the reproductive development of male flounder. Similarly, in an egg injection study with medaka, Papoulias et al. (2003) found that o,p'-DDE exposure led to reduced testis size at concentrations as low as 0.005 ng/embryo or 5 ng/g egg; however, no testicular pathology was observed. These studies clearly demonstrate the estrogenic effects of o,p'-DDTs, although there is some uncertainty about whether comparable effects would occur in natural fish populations exposed to DDTs in the environment.

In contrast to Mills et al. (2001) and Zaroogian et al. (2001), several authors report reproductive toxicity in male fish exposed to p,p'-DDE. Bayley et al. (2002) observed delayed puberty, inhibited gonopodium development, and reduced sperm counts in male guppies exposed from birth to adulthood at concentrations of 10 to 100 ug/g in food, while Kinnberg and Toft (2003) found that exposure of adult male guppies (*Poecilia reticulata*) to p,p'-DDE at concentrations of 100 to 10,000 ug/g food in the diet led to a reduced number of spermatogenetic cysts and an increased number of spermatozuegmata, or sperm bundles, in the ducts, effects indicative of blocked spermatogonial mitosis. In a waterborne medaka exposure study, decreased testis growth and vitellogenin and choriogenin induction were observed in males exposed to p,p'-DDE at concentrations >1 ug/L for two months, as well as intersex at a concentration of 100 ug/L (Zhang and Hu, 2008). However, these exposure concentrations are quite high in comparison to p,p'-DDE concentrations typically measured in fish prey or in the water column. While less information is available on other OC pesticides, induction of testis–ova and induction of vitellogenin in males were observed in medaka chronically exposed to waterborne lindane (B-hexachlorocyclohexane) (Wester and Canton, 1986).

PCBs in mixtures as well as specific congeners have been associated with reproductive impairment, including poor gonadal growth, in male fish, in part through their effects on the serotoninergic systems in fish brains that regulate reproductive hormones (Khan et al., 2001; Khan and Thomas, 2006; see Section 3). DDTs and PCBs may also affect sperm release in male fish. Njiwa et al. (2004) conducted a long-term toxicity study with zebrafish exposed to DDTs and Aroclor 1254 and a 1:1 mixture of these two in the water column. Sperm count and activity were reduced significantly at 5 and 50 ug/L after 1 month of exposure for PCBs or DDTs and after 2 weeks of exposure to the mixture. This study suggests that DDTs and PCBs could synergistically affect sperm quality and quantity. However, these exposure concentrations are quite high as compared to those typically found in the water column in the environment or permitted concentrations under current regulations. In the United States, for example, the ambient water quality criterion for DDTs for the protection of aquatic life is 0.001 ug/L (U.S. EPA, 1980).

Information on the effects of PBDEs on reproductive function in male fish is limited. There is some evidence that PBDEs can affect the male reproductive endocrine axis (see Section 3) and reduce fertility. In male fathead minnows given a daily dose of PBDE 47, resulting in body concentrations of $\sim$15 ug/g wet wt, male condition factor was significantly reduced, and the quantity of mature sperm was reduced by more than 50% compared to controls (Muirhead et al., 2006). He et al. (2011) found that 150-day aqueous exposure (from hatching to adulthood) to PBDE 209 in zebrafish at doses ranging from 0.959 to 959 ug/L led to reduced testis development, and reduced male gamete quantity and quality. Effects on sperm quality were observed in the lowest dose group, where body concentrations were $\sim$0.5 ug/g wet wt, while the other effects were observed at body concentrations in the 1–5 ug/g range. These findings suggest a potential for impacts on sperm quality in male fish at environmentally relevant exposures to PBDEs.

Older field studies conducted prior to 2000 have reported poor reproductive success in male fish exposed to POPs (Kime, 1998; Pait and Nelson, 2002). These findings have been confirmed in more recent work. In a monitoring study of the Elbe River in the Czech Republic, for example, Randak et al. (2009) observed reduced gonad size and histological abnormalities in testes of chub from sites with high levels of a variety of POPs, including PCBs, DDTs, HCH, and HCB. Both EROD induction and vitellogenin induction were observed in these fish, indicating exposure to chemicals interacting with the Ah receptor such as dioxin-like PCBs, as well as compounds with estrogenic activity. Average concentrations of PCBs in chub from the sampling sites were in the 2000–8000 ng/g lipid range, while DDT values were in the 1300–6500 ng/g lipid range. Feist et al. (2005) examined wild white sturgeon from the Columbia River (Oregon) for signs of reproductive endocrine disruption. Metabolites of p,p'-DDT (i.e., p,p'-DDE and p,p'-DDD) were consistently found at high levels in fish, with mean DDT concentrations of 20.6 ug/g lipid in liver and 12.5 ug/g lipid in gonad; concentrations of PCBs were somewhat lower, with mean levels of 1.9 ug/g lipid in liver and 1.6 ug/g lipid in gonad. In males, gonad size was negatively correlated with DDTs, OC pesticides, and PCBs. Fish with the highest contaminant levels had the highest prevalence of gonadal abnormalities and the lowest gonad size. In another study, on marine fish, nesting plainfin midshipman *Porichthys notatus* were collected from areas with low and high contamination on Vancouver Island, British Columbia. Contaminants in the area included PCBs and dioxins, as well as metals and PAHs. Males in the contaminated areas had more testicular asymmetry, sperm with shorter heads, and fewer live eggs in their nests (Sopinka et al., 2012). Sol et al. (2008) found significant negative correlations between body concentrations of several classes of POPs (including PCBs, DDTs, and HCB) and gonadosomatic index in male English sole from Puget Sound in the United States.

7.4. Reproductive Behavior

Several studies reviewed in Scott and Sloman (2004) suggest that exposure to POPs affects the courtship behavior of guppies and chichlids (Matthiessen and Logan, 1984; Schröder and Peters, 1988a, 1988b; Baatrup and Junge, 2001). In more recent studies, Bayley et al. (2002) exposed male guppies to *p,p'*-DDE from birth to adulthood at concentrations of 10 and 100 ug/g in food. At adulthood, they found changes in male sexual characteristics such as a reduction in display coloration, inhibited gonopodium development, and suppressed courtship behavior in exposed fish. In a subsequent study, however, when male guppies exposed to *p,p'*-DDE competed against unexposed males for the opportunity to fertilize females, *p,p'*-DDE exposure produced no significant harmful effect on success in siring young, although the highest sublethal dose tested was only a factor of 10 below the dose producing 100% mortality (Kristensen et al., 2006). The high doses of *p,p'*-DDE needed to induce effects, and the conflicting results of these studies, suggest limited anti-androgenic potency for *p,p'*-DDE.

8. IMMUNOTOXICITY

A properly functioning immune system is an important fitness trait that is critical for both individual survival and population productivity (Segner et al., 2012a). An immune system altered by contaminant exposure, either alone (Arkoosh et al., 2000, 2001, 2010), or in conjunction with other stressors (Jacobson et al., 2003), can result in an increase in disease susceptibility. This increase in disease susceptibility can lead to potential population-level effects (Arkoosh et al., 1998a; Loge et al., 2005; Spromberg and Meador, 2005).

The immune system discriminates self from nonself or altered self in order to protect the host from internal insults such as neoplasia and external insults such as environmental chemicals (e.g., POPs) and micro- and macroparasites. The process of immunity that occurs once a fish encounters an insult is complex. The immune system in fish (Figure 2.3) can be divided broadly into three categories: integumental barriers, innate immunity, and

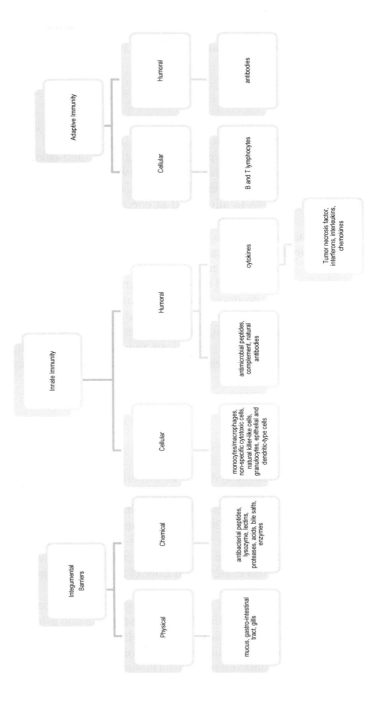

Figure 2.3. Simplified diagram of the teleost immune system. The immune system in teleosts can be divided broadly into three categories: integumental barriers, innate immunity, and adaptive immunity (Magnadottir 2006; Renshaw and Trede 2012; Segner et al. 2012b; Uribe et al. 2011; Whyte 2007). POPs have the potential to affect the function of these immune categories.

adaptive immunity (Magnadóttir, 2006; Whyte, 2007; Uribe et al., 2011; Renshaw and Trede, 2012; Segner et al., 2012b).

Both physical and chemical integumental barriers act as the first line of defense against the pathogen by keeping the pathogen from attaching, invading, or multiplying (Bayne and Gerwick, 2001; Ellis, 2001). However, if the pathogen does compromise the host tissue, various humoral and cellular components of the innate immune system interact to create an inflammatory response (Whyte, 2007). Innate immunity responds nonspecifically to an insult and occurs much faster than adaptive immunity but is of a limited duration. The innate immune system is activated by the interaction of pattern recognition receptors (PRR), with pathogen-associated molecular patterns (PAMP) that are present on most pathogens.

Adaptive immunity is required only if the host's innate defenses are overwhelmed or bypassed. Adaptive immunity is slower than innate immunity but is pathogen specific and longer lasting. Adaptive immunity results in a specific response from a limited repertoire of genes via gene rearrangement. Cells responsible for the specific nature of the adaptive immune response are B- and T-lymphocytes. B-lymphocytes can differentiate into antibody producing plasma cells or produce long-lasting memory cells (Magnadottir, 2010). The critical humoral component produced from the plasma cells is the antibody. The antibody binds to its target and allows for the initiation of effector functions such as phagocytosis and complement activation to help clear the host of the pathogen.

Immune factors have been detected early in ontogeny. Components of both innate and adaptive immunity have been detected prior to hatching. Transfer of immune factors such as lysozymes, lectin, cathelicidin, and complement occurs from mother to offspring (Swain and Nayak, 2009; Zhang et al., 2013). A physical factor protecting fish early in ontogeny is the chorion. The chorion is a membrane that surrounds the eggs and acts as a barrier against infection (Murray, 2012). Therefore, the vulnerable eggs are dependent on maternal transfer of immune-relevant factors and the chorion for protection against infectious diseases while their immune system is developing and maturing (Figure 2.4; Swain and Nayak, 2009; Zhang et al., 2013).

The various components of the fish immune system are known to be sensitive to contaminants. For example, the mononuclear phagocytic system is an important component of the innate immune system in fish (Bols et al., 2001). This system houses phagocytes and endothelial cells and acts as a filtering system for the blood and as a barrier to the outside of the host. As a result, the mononuclear phagocytic system may be exposed directly to contaminants. In addition, circulating immune cells in the peripheral blood are susceptible to contaminants in both their active and metabolized form

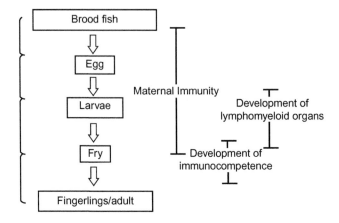

Figure 2.4. Ontogeny of immunity relative to the development of immunocompetence at different stages of growth. Figure reprinted from Swain and Nayak (2009).

(Rice, 2001). Once exposed, the immune process can be affected by acting on a number of these potential immune targets (Segner et al., 2012b), resulting in either suppression or hypersensitivity of the response (Rice, 2001; Rice and Arkoosh, 2002).

Understanding of immune processes in teleost fish has progressed rapidly due to interest in fish such as salmonids, carp, and catfish used in aquaculture settings and development of vaccines and immunostimulants (Carlson and Zelikoff, 2008). Although the exact mechanisms by which contaminants modulate the immune system are unclear, immune cells and tissues may be directly affected as described earlier or indirectly affected due to their interactions with the endocrine system (Rice and Arkoosh, 2002; Milston et al., 2003; Casanova-Nakayama et al., 2011).

The immunotoxicity caused by POPs has been assessed by examining the functionality of various components of the immune system or by examining the susceptibility of the host to a pathogen. Immunotoxicological studies have examined macrophage function (engulfment and production of reactive oxygen species), expression of innate humoral factors such as cytokines (tumor necrosis factor, interferons, interleukins, chemokines), mitogenic response of lymphocytes with B and T cell mitogens, activity of antibodies to a specific antigen, and the number of antibody forming cells to an antigen (Bols et al., 2001; Segner et al., 2012a). While important in examining the immunotoxicity of POPs, these assays do not directly indicate how toxicity may be expressed in the fish (Segner et al., 2012a). A more integrated and holistic approach to examining the immunotoxicity of a chemical is examining the host's susceptibility to a pathogen (Arkoosh et al., 2005). With the disease challenge assay, immunotoxicity is examined with an activated rather than a resting immune system (Köllner et al., 2002). Recently, molecular tools have been developed to help determine how contaminants may change immune-related gene expression during an active infection (Eder et al., 2008; Jin et al., 2010). Understanding how contaminants alter gene expression may ultimately lead to a more mechanistic interpretation of how they modulate the immune system. Extrapolating changes due to contaminants from these molecular and/or individual levels to alterations in the population or community is at the heart of ecotoxicology (Newman and Clements, 2008).

Recent studies (Tables 2.1–2.4) and reviews (Bols et al., 2001; Rice, 2001b; Rice and Arkoosh, 2002; Burnett, 2005; Rollins-Smith et al., 2007; Carlson and Zelikoff, 2008) have shown that various legacy POPs under controlled laboratory conditions can alter the immune system of fish as well as their susceptibility to disease. In general, studies examining the effects of environmental contaminants on the immune system in fish have examined insecticides, heavy metals, and surfactants. Fewer studies have examined the effects of POPs on the teleost immune system (Bols et al., 2001). There is little recent work on fish, with the most biologically potent halogenated aromatic hydrocarbon, TCDD, constituting an important research gap (Carlson and Zelikoff, 2008). Fish exposed to POPs under controlled conditions (see Tables 2.1–2.4) in the laboratory demonstrate altered immune functions, an increase in disease susceptibility, and altered immune-gene expression, depending on the POP.

A few studies have examined immunotoxicity in wild fish from contaminated bodies of water. These studies have focused primarily on fish from Puget Sound, the Great Lakes, and the Elizabeth River in Virginia. Fish exposed to these contaminated water bodies were found to have a suppressed secondary *in vitro* antibody response, increased disease suscept-ibility, reduced mitogenesis, and inhibition of macrophage function relative to fish from reference sites (Weeks and Warinner, 1986; Arkoosh et al., 1991, 1998b; Faisal et al., 1991).

The advantage of field exposure studies is that the fish are exposed to POPs in an environmentally relevant fashion. However, environmental variables such as other contaminants present in the water body (e.g., PAHs) and varying temperature, salinity, and diets may also contribute to or influence the response. A mesocosm approach in examining the immuno-toxicity of POPs would help to reduce the contribution of these other potentially interfering environmental variables and yet allow the fish to exist in a more natural setting and bridge both laboratory and field exposure studies (Grinwis et al., 2000).

9. NEOPLASIA AND RELATED PATHOLOGICAL CONDITIONS

This section focuses on important, widely recognized, and well-documented pathological effects of exposure to POPs, particularly effects in liver, including hepatic neoplasia. In adult fish, the liver is the organ commonly affected by exposure to POPs, although lesions in other organs can be caused by exposure to these compounds (reviewed in Hinton, 1993; Myers and Fournie, 2002). This is because, in teleosts, the liver is the primary organ of metabolism/detoxification where relatively high levels of lipophilic POPs can accumulate (Hinton, 1993). The pathological effects of POP exposure on reproductive and immune organs, as well as pathological conditions in embryos and larvae exposed to POPs, are presented in separate sections of this chapter.

9.1. Liver Glycogen Depletion and Hepatocyte Alterations

In juvenile and adult fish exposed to TCDD, hepatocellular glycogen depletion is a characteristic early response observed in several fish species, including lake trout, rainbow trout, yellow perch, carp, and zebrafish (Mac, 1986; Spitsbergen et al., 1988a, 1988b; Ferguson, 1989; van der Weiden et al., 1994; Zodrow et al., 2004). Hepatocytes of adult female rainbow trout exposed dietarily for 320 d to environmentally relevant doses of TCDD (Walter et al., 2000) displayed glycogen depletion, increased mitosis, nuclear and cellular pleomorphism, margination and clumping of chromatin, single-cell necrosis, and especially prominent nucleoli; these features were essentially absent in control fish. Hepatic glycogen depletion, as well as hepatocyte proliferation, have also been found in various species of fish exposed to PCB mixtures and dioxin-like PCBs (Grinwis et al., 2001; Hinton et al., 1978; Nestel and Budd, 1975; Rhodes et al., 1985).

Although investigations on the pathologic effects of PBDE exposure in fish are limited, a few examples of liver pathology associated with PBDE exposure are available. Pronounced hepatocellular lipid accumulation was observed in juvenile lake trout exposed dietarily for 56 d to a mixture of 13 PBDEs (Tomy et al., 2004). In contrast, chronic exposure of juvenile European flounder (dietary and sediment exposure, 101 d) and zebrafish (waterborne exposure, 30 d) to DE-71 at environmentally relevant doses did not result in any major histopathological changes (Kuiper et al., 2008). In a recent Norwegian field study, Berg et al. (2013) found that PBDE levels in burbot (*Lota lota*) from a polluted lake were 300 times higher than those from the reference lake, and the overall prevalence of histopathological lesions was significantly higher than in the reference lake fish.

However, these differences were not corroborated with statistical analyses linking these pathological conditions specifically to PBDE exposure and are likely the result of exposure to the ambient mixture of PBDEs, PCBs, and DDTs in the polluted lake.

9.2. Neoplasia and Other Toxicopathic Lesions

Many studies have reported neoplastic and nonneoplastic toxicopathic (having a likely etiology involving exposure to toxicants) liver lesions involved in the histogenesis of liver neoplasia in fish (Myers et al., 1987; Boorman et al., 1997; reviewed in Myers et al., 2003; Wolf and Wolfe, 2005), including lesions induced by experimental exposure to dioxin-like POPs, or statistically associated with environmental exposure to POPs in wild fish (reviewed in Myers and Fournie, 2002). However, most ecoepidemiological studies in wild fish have shown the strongest and most consistent associations with exposure to PAHs rather than POPs in species such as English sole from the Pacific Coast (Myers et al., 2003) and mummichog from the Elizabeth River and Chesapeake Bay (Vogelbein et al., 1990; Vogelbein and Unger, 2006).

While PAH exposure is most strongly and consistently associated with neoplastic and preneoplastic lesions in wild fish, the environments they inhabit are contaminated by a mixture of other chemicals, including POPs, which may contribute to the risk of neoplasia. In multiyear, multisite U.S. Pacific Coast studies on toxicopathic hepatic lesions in English sole (*Parophrys vetulus*), starry flounder (*Platichthys stellatus*), and white croaker (*Genyonemus lineatus*), relatively consistent, significant associations were found between PCBs in sediment, liver, and stomach contents and the prevalence of several categories of lesions, including liver neoplasms, preneoplastic foci of cellular alteration (FCA), hepatocellular nuclear pleomorphism/hepatocellular megalocytosis (NP/HM), and nonneoplastic proliferative lesions (Myers et al., 1998a, 1998b; Stehr et al., 1998) at contaminated sites. A similar study using logistic regression techniques in English sole strictly from Puget Sound sites in Washington State documented that PCB levels in liver were a significant risk factor explaining 8.5% of the variation in toxicopathic hepatic lesions, in comparison to 53.6% for high-molecular-weight PAHs in sediments (Myers et al., 2003). Other U.S. studies have also related PCB exposure to neoplasms and liver lesions involved in the neoplastic process in wild fish, including walleye (*Stizostedium vitreum vitreum*) from Lake Michigan (Barron et al., 2000), smallmouth bass from Kalamazoo River, Michigan (Anderson et al., 2003), and Atlantic tomcod from the Hudson Bay estuary in New York (Klauda et al., 1981). However, in similar studies with winter flounder on the U.S.

Atlantic Coast (Johnson et al., 1993; Chang et al., 1998), no toxicopathic liver lesions were statistically associated with PCB exposure. European studies on wild or mesocosm-exposed European flounder from the Dutch (Vethaak et al., 1996, 2009; Vethaak and Wester, 1996) and German (Koehler, 2004) coastal waters of the North Sea have also associated exposure to PCBs, including specific congeners such as PCB 126 and PCB 153, with the development of neoplasms and related lesions in fish. In European flounder studies, the current view is that exposure of wild fish to PCBs and other POPs is probably a more important risk factor in hepatocarcinogenesis than are genotoxic PAHs (Koehler, 2004; Vethaak et al., 2009). These associations between neoplasia-related lesions and exposure to POPs are consistent with extensive evidence from experimental studies with model fish species. Although these compounds do not bind readily to DNA and are not mutagenic (nongenotoxic), they are strongly mitogenic and cytotoxic. They therefore act mainly as strong epigenetic promoters of hepatocarcinogenesis (see references in Myers et al., 1987, 2003; Safe, 1994) by provoking cell proliferation and regenerative growth following hepatotoxicity, and stimulating and promoting the growth of previously initiated preneoplastic cells (Koehler, 2004).

The literature (especially recent studies) on the effects of experimental exposure to DDT and its metabolites in fish liver is not extensive. However, DDT is also considered a probable nongenotoxic promoter of hepatocarcinogenesis in fish (Halver, 1967; Metcalfe, 1989; Hawkins et al., 1995) and is considered hepatotoxic to fish (reviewed in Walsh and Ribelin, 1975). Moreover, DDT exposure in mice induced marked hepatocellular cytomegaly and karyomegaly (Frith and Ward, 1979), lesions that are morphologically similar to hepatic lesions (e.g., NP/MH) detected in several species of wild fish (Myers and Fournie, 2002). However, in a more recent study (Zaroogian et al., 2001) in which summer flounder were dorsal sinus-injected with either DDT or DDE, no histological effects were shown in either liver or kidney.

Field studies on toxicopathic liver lesions in Pacific Coast marine and estuarine fishes (Myers et al., 1998b; Stehr et al., 1998) have documented significant associations between the prevalence of neoplasms, FCA, HP/MH, HydVac, and nonneoplastic proliferative lesions and DDT concentrations in sediments, fish stomach contents, and liver tissue in English sole, starry flounder, and white croaker. In similar field studies on winter flounder from the U.S. East Coast (Johnson et al., 1993; Chang et al., 1998; Myers et al., 1998b), DDT exposure was associated with increased risk of HydVac and FCA. However, in all of these studies there were also correlations with PAHs, non-DDT pesticides, and PCBs, so the causal role of DDTs in lesion induction is uncertain. Another field study that attempted to relate histopathological lesions in tigerfish, *Hydrocynus vittatus*, from a highly contaminated area in the Phongola River in South Africa with exposure to this pesticide failed to correlate any lesions in the liver, kidney, or gill with DDT exposure (McHugh et al., 2011).

Other OC pesticides, including dieldrin, chlordane, and lindane together with its metabolites have also been linked to liver toxicity in fish in laboratory exposure studies, inducing lesions such as vacuolar degeneration, hypertrophy, hepatocellular pleomorphism, coagulative necrosis, hepatic lipid accumulation, and spongiotic edema (Mathur, 1962, 1976; Couch, 1975; Wester and Canton, 1986; Braunbeck et al., 1990). Dieldrin has a slight co-carcinogenic or promotional effect in rainbow trout hepatocarcinogenesis, although it is not a carcinogen when administered alone (Hendricks et al., 1979; Donohoe et al., 1998). While chlordane is generally considered a nongenotoxic, nonmutagenic, epigenetic promoter of hepatic carcinogenesis in mice (Williams and Weisburger, 1986; ATSDR, 1997), its hepatocarcinogenicity in fish has not been demonstrated.

Field studies in fish have also associated toxicopathic liver lesions involved in hepatocarcinogenesis with exposure to dieldrin and chlordane in English sole, starry flounder, and white croaker (Myers et al., 1994; Stehr et al., 1998) and with chlordane in winter flounder (Johnson et al., 1993), including an especially strong association between chlordane and HydVac in winter flounder. However, laboratory studies in which winter flounder were exposed to high levels of chlordane did not induce this unique degenerative lesion (Moore, 1991). We are not aware of any field studies that have specifically attributed or associated any pathological effects in fish to lindane exposure.

10. EFFECTS ON BEHAVIOR, INCLUDING FORAGING, AGGRESSION, AND PREDATOR AVOIDANCE

POPs can affect fish behavior, including their susceptibility to predation and foraging ability (reviewed in Kasumyan, 2001; Scott and Sloman, 2004; Amiard-Triquet, 2009; Weis and Candelmo, 2012; Tierney et al. 2010). Studies conducted prior to 2000 in fish species, including goldfish, rainbow trout, and medaka, showed that relatively short-term exposure (2–3 days) to waterborne chlordane, DDTs, endosulfan, and PCBs at concentrations in the 1–2 ug/L range disrupted predator avoidance behaviors, including schooling, and swimming, and decreased the likelihood of survival in encounters with predators (Scott and Sloman, 2004). In such studies, exposure to high POP concentrations typically leads to cessation of feeding and foraging behavior due to lack of appetite or energy to engage in behaviors related to prey search and eating; exposure to lower levels of POPs is associated with structural and functional changes in the sensory systems critical for prey identification and capture and predator avoidance (Kasumyan, 2001). Recently, the behavioral effects of sublethal, environmentally realistic POPs concentrations relevant to wild fish populations have been of greatest interest. Additional areas of recent concern are the biochemical bases for behavioral effects, and the behavioral effects of contaminants of emerging concern, such as PBDEs.

Two major mechanisms through which POP exposure appears to influence fish behavior are via changes in levels of brain neurotransmitters (e.g., serotonin) and through altered thyroid function. Both DDTs and PCBs alter brain serotonin, dopamine, and norepinephrine levels (Khan and Thomas, 2000, 2006; Khan et al., 2001), causing changes in spontaneous activity, schooling, and locomotor activity (reviewed in Weis et al., 2001). Altered thyroid function associated with exposure to PCBs and PBDEs has also been shown to influence spontaneous activity and feeding in fish. Hypothyroidism could lead to general sluggishness and reduced activity, while hyperthyroidism would tend to lead to hyperactivity, possibly enhancing vulnerability to predation by making an animal more conspicuous (Castonguay and Cyr, 1998; Weis et al., 2001).

Fish may be especially sensitive to the behavioral toxicity of POPs during early development. Carvalho et al. (2004), Couillard et al. (2011), and Rigaud et al. (2013) have studied the effects of exposure to dioxins and dioxin-like PCBs during early development on rainbow trout and mummichog. Carvalho et al. (2004) examined the behavior of larval rainbow trout hatched from TCDD-injected eggs. In these fish, biochemical and morphological changes in the brain and eyes were associated with impaired visual and motor function and reductions in prey capture ability. The investigators proposed visual impairment as a possible mechanism leading to reduced ability to capture prey in TCDD-exposed rainbow trout. Couillard et al. (2011) conducted a similar experiment with PCB126, exposing fertilized eggs of mummichog to this compound by topical application. They also found that prey capture ability was reduced in a dose-responsive fashion. Rigaud et al. (2013) compared the potency of PCB126 to TCDD in inducing behavioral toxicity in mummichog embryos and found that, in contrast to results with rainbow trout, PCB126 was nearly as potent as TCDD in inducing behavioral abnormalities. Different gene-expression patterns were also observed between PCB126 and TCDD exposed embryos, as PCB126 appeared to induce antioxidant responses while TCDD did not. In contrast to dioxinlike PCBs, the behavioral toxicity of non-dioxin-like PCBs has not been studied extensively in fish.

There is some evidence that exposure to these PCBs (e.g., PCBs 28, 52, 101, 138, 153, and 180) may increase anxious and aggressive behaviors in mice (Haave et al., 2011; Elnar et al., 2012), but it is unknown whether they have similar effects on fish.

Recent research efforts have focused on the behavioral effects of PBDEs, as they are known to affect brain development and behavior in mammals (Costa and Giordano, 2007; Herbstman et al., 2010) and appear to have similar effects in fish. A number of recent studies have demonstrated the behavioral and neurological effects of various PBDE congeners (e.g., BDE 47, 49, 209) and mixtures in the model laboratory fish, zebrafish (Chou et al., 2010; Chen et al., 2012b, 2012c, 2012d; McClain et al., 2012). All of these researchers reported behavioral deficits in exposed fish, such as reduced motor activity, slower swimming speed, and altered swimming behavior in response to light. Chen et al (2012b) also observed inhibited axonal growth of primary and secondary motor neurons in zebrafish with movement abnormalities, suggesting that the deficits were related to abnormal neurological development. McClain et al. (2012) found cardiac abnormalities, which might also contribute to reduced swimming speed and activity. Two researchers (He et al., 2011; Chen et al., 2012d) have examined the effects of parental exposure to PBDEs on behavior of the offspring. He et al. (2011) found that parents exposed to PBDE 209 at doses ranging from 0.001 to 1 μM (or ∼1 to 1000 ug/L) produced offspring with delayed hatch and slower motor neuron development, loose muscle fiber, slow locomotion behavior in normal conditions, and hyperactivity when subjected to light-dark photoperiod stimulation. Similarly, Chen et al. (2012d) exposed male and female adult zebrafish to the penta-BDE mixture DE-71 in water at concentrations of 0.16, 0.8, or 4.0 μg/L for 150 days and found slower swimming speed and altered responses to light and dark in the offspring. Exposed offspring also showed reduced expression of three genes critical for normal brain and neurological development (e.g., myelin basic protein, synapsin IIa, α1-tubulin), as well as reduced acetocholinesterase (AChE) gene expression and activity. Both He et al. (2011) and Chen et al. (2012d) also documented accumulation of PBDEs in the eggs and larvae of exposed parents. These two studies indicate that maternal exposure to and transfer of PBDEs to offspring can lead to developmental and behavioral abnormalities.

Much of the evidence for the behavioral effects of POPs comes from laboratory exposure studies, but effects on behavior have also been observed in field populations (Weis et al., 2001, 2003; Candelmo et al., 2010; Sopinka et al., 2010; Weis and Candelmo, 2012). Weis and colleagues have studied the behavior of mummichog and juvenile bluefish (*Pomatomus saltatrix*) from industrialized estuaries in New Jersey contaminated with PCBs and other chemicals (Weis et al., 2001, 2003; Candelmo et al., 2010; Weis and Candelmo, 2012). They found that mummichog from the contaminated site had a slower rate of spontaneous activity and a reduced rate of prey capture, and were more vulnerable to predation by blue crabs (*Callinectes sapidus*) than mummichog from reference areas (Weis et al., 2001). While mummichog were exposed to a variety of contaminants, PCBs were among the chemicals to show a significant negative association with prey capture ability (Weis et al., 2003). These behavioral impairments in contaminant-exposed mummichog appeared to be related to altered thyroid function. Young-of-the-year bluefish (*Pomatomus saltatrix*) from another industrialized site in the same area were significantly smaller than reference fish and had elevated levels of PCBs in their tissues. Also, a much lower percentage of fish had food in their stomachs than is typical of young bluefish, suggesting reduced feeding (Candelmo et al., 2010). Interestingly, the levels of PCBs and DDTs in prey fish found in bluefish stomachs were higher than levels in fish of the same species caught in trawls and seines. This finding suggested that the more contaminated fish were easier for predators to capture. In a subsequent laboratory study, the investigators fed young-of-the-year bluefish on diets of prey fish collected from the reference and contaminated sites. They found that bluefish fed prey from the contaminated site had elevated levels of PCBs and pesticides in tissues and displayed significantly reduced feeding, spontaneous activity, growth, and poorer schooling behavior compared to bluefish fed with uncontaminated prey. Bluefish fed contaminated prey also showed thyroid follicle abnormalities similar to those seen in earlier studies with mummichogs from contaminated New Jersey sites, as well as reduced levels of the metabolites of the neurotransmitter dopamine. In another study, Sopinka et al. (2010) compared aggression and dominance hierarchy formation in round gobies, *Neogobius melanostomus*, from polluted and cleaner sites in Lake Ontario and found that while stable hierarchies formed between pairs of fish from the cleaner sites, dominance was less obvious among fish from the more contaminated areas.

11. RISK ASSESSMENT CHALLENGES

The chemical characteristics inherent in defining a chemical as a POP are the same ones that pose specific difficulties when determining a chemical's risk to fish populations. High lipophilicity and long half-lives predict that these chemicals can settle in sediment and soils for long periods, and when they enter biota they will bioaccumulate and biomagnify up the food chain.

In an individual fish, POPs accumulate in lipid-rich tissues as lipid reserves, where they can be mobilized during times of stress (migration, low food supply) or maternally transferred. The chemical stability and biological persistence of these compounds leads to known exposures of individuals, but linking these interactions to population-level effects presents challenges related to the life history of the organism being exposed, the toxicity of individual compounds, and additional stressors on the population (Vasseur and Cossu-Leguille, 2006; Kramer et al., 2011).

11.1. Vulnerability of Different Life Stages, Populations, and Species

Each species evolves a life history strategy adapted to sustaining viable populations in their natural habitat. Resource availability and survival potential influence the time and energy invested in each life stage (i.e., for fish, embryo, larval, juvenile, and adult stages). It should be clarified here that the term *strategy* denotes "the way individuals allocate resources to reproduction under different circumstances" and "is a heritable trait, which is subject to natural selection in the same manner as any other characteristic determining the production of successful offspring" (Ware, 1984). The life history strategy develops as a set of trade-offs in the allocation of energy and resources between growth, current reproduction, future reproduction, and survival. Energy allocation trade-offs between growth and reproduction attempt to maximize both individual survival and the individual's reproductive contribution to the population (Roff, 1984).

Adaptations to local environmental factors, such as seasonal conditions and severity of stochastic impacts, occur over many generations, leading to the evolution of many variations in life history strategies. Each strategy provides for continuity of generations under normal environmental conditions; however, some strategies may confer greater or lesser susceptibility to anthropogenic perturbations that fall outside of the species' adaptive history. A species that reproduces multiple times each year may more easily compensate for a one-time impact than a species that reproduces annually. This illustrates how a population's response to an impact, relative to either a nonimpacted population or a similarly impacted population, can differ depending on the various life history and survivorship characteristics exhibited (Spromberg and Birge, 2005; Spromberg and Meador, 2005; Raimondo et al., 2006). Life stage, life span, and time to first reproduction all contribute to determining how a population will respond to a stressor (Caswell, 2001; Stark et al., 2004; Raimondo et al., 2006). Fish with a short time to reproduction may accumulate a lower contaminant concentration to pass on to eggs than those that take many years to reach reproductive maturity.

Another concern in predicting the risk of population-level impacts is the occurrence of differential sensitivity exhibited by different life stages to the same contaminant. For example, exposures of freshwater stages of Atlantic salmon to DDT resulted in over 90% mortality of underyearlings, but only 50–70% mortality of parr (Elson, 1967). Thus if laboratory toxicity tests used to determine risk were conducted on parr, the potential risks to field populations could be underestimated. Issues with differential sensitivity can also occur across species. Pesticide runoff events resulted in a shift in community structure when brook trout (*Salvelinus fontinalis*) populations declined relative to rainbow trout in a river on Prince Edward Island, Canada (Gormley et al., 2005). In this case, the young-of-the-year fish of both species also displayed greater sensitivity, resulting in a shift in age structure of the populations for several years (Gormley et al., 2005). As these examples illustrate, selection of appropriate toxicity data for estimating risk requires attention to how extrapolation across species and across ages is handled and how this affects the uncertainty of the risk assessment.

Sublethal effects that were mentioned in previous sections in this chapter demonstrate ways that various POPs can influence an individual's ability to grow, capture food, avoid predation, fight disease, or reproduce, which can alter the demographic rates of survival and reproduction and potentially influence population persistence and viability. These effects are often observed in a laboratory setting, but the prevalence in the field is uncertain. Hylland and others (2006) reviewed evidence collected in the North Sea and identified correlations between POPs and embryonic aberrations and disease prevalence. Direct impacts from these effects on the populations were unclear, but they encouraged a precautionary approach until further information is known (Hylland et al., 2006). Similarly, POPs have been characterized as endocrine disruptors, but directly connecting specific changes in vitellogenin production, intersex, or fertilization success to changes in a wild population's growth rate requires a level of data that is often lacking (Jobling and Tyler, 2003).

11.2. Adaptations

The persistence of POPs in the environment and biota creates a consistent, if heterogeneous, selection pressure on exposed populations. The selective pressure over generations allows for the possibility of genetic adaptation to occur. Populations inherently have distributions of individuals possessing a variety of alleles and mutations that reflect past selection pressures and are able to respond to current stressors. For example, New Bedford Harbor and other U.S. Atlantic urban coastal populations of killifish have adapted to high concentrations of PCDDs, other dioxin-like compounds, and PCBs by having reduced responsiveness in the aryl-hydrocarbon receptor (AHR) pathway (Nacci et al., 1999, 2010). This allows persistence of the populations in conditions that would be lethal to fish expressing typical levels of Ah-receptor pathway responsiveness (Nacci et al., 1999). Although each of these populations shows tolerance to dioxin-like compounds and passes this tolerance to the F1 generation, the extent of tolerance and the mechanism of adaptation appear to differ and have arisen separately in each population (Nacci et al., 2010). Some associated costs with these adaptations would suggest selective bottlenecks linked to low genetic diversity, but these tolerant killifish populations express high genetic diversity (McMillan et al., 2006). Other costs could include reduced resistance to other stressors such as disease or low oxygen levels (Meyer and Di Giulio, 2003).

Populations of Atlantic tomcod residing downstream of industrial sites highly contaminated with PCBs, PCDDs, and PCDFs have also developed resistance to these contaminants by undergoing rapid evolution (Wirgin et al., 2011). The form of resistance was a difference in the sequence of the AHR-2 allele that produced an AHR with a lower binding affinity to TCDD (Wirgin et al., 2011). The frequency of this allele in the resistant population approached 99%, while its frequency was less than 5% in other populations along the Atlantic coast of North America (Wirgin et al., 2011).

In light of the potential for rapid selection, determining the risk associated with POPs for a particular population must consider the timing of the exposure and the potential adaptation that could or may have already occurred. Populations with a high intrinsic maximum growth rate (rmax) and demographic and genotypic plasticity may be better able to respond to directional pressures posed by POPs exposures if adaptive traits can emerge and can allow sufficient population productivity to counter the stressor and minimize the extinction risk (Kinnison and Hairston, 2007).

11.3. Assessment Tools: Exposure Models, Adverse Outcome Pathways, Tissue Residue Approach

Although data are often lacking on how species of interest will respond to particular stressors, risk assessments must still be completed to inform regulatory and policy decisions (Sappington et al., 2011). In these situations, assessments often rely on models to link chemical exposure to risks to populations. An extensive review of models of exposure and effects is beyond the scope of this chapter, but a few areas will be discussed. Many methods have been employed to model the exposure of organisms to contaminants, including POPs. These have evolved from steady-state models of passive uptake from the water column based primarily on the chemical properties (Mackay, 1982) to more complex models of trophic uptake and bioaccumulation incorporating biological and ecological dynamics (Gobas, 1993; Morrison et al., 1997; Loizeau et al., 2001; Gobas et al., 2009). A review by Cowan-Ellsberry et al. (2009) encouraged the appropriate selection of exposure models to assist in interpreting monitoring data and combine fate, long-range transport, and bioaccumulation data and to foster analysis of time trend monitoring to assess the long-term risk of these compounds.

Once exposure has been established, several methodologies may be used to assess potential effects. Adverse outcome pathways (AOPs) link an initiating event at the molecular level across levels of biological organization to an adverse outcome on the individual fitness (Ankley et al., 2010). AOPs use known toxicity pathways and translate that chemical-specific data into information that can have demographic relevance for risk assessments that estimate population or ecosystem impacts (Kramer et al., 2011; Villeneuve and Garcia-Reyero, 2011). Once an AOP is established for a species and life stage, it could be applied to other chemicals that share that mechanism of action. For example, dioxin and dioxin-like compounds (e.g., coplanar PCBs) share the aryl hydrocarbon (Ah) receptor pathway and can use the same conceptual framework (Ankley et al., 2010). Each compound will express its own binding affinity and potency, but can trigger the same pathway. Similarly, endocrine-disrupting compounds that bind to the estrogen receptor can initiate known AOPs and predictable responses. The known AOP may not be the only mechanism of action for a particular compound; rather, it may be just the one that has been identified, and this could underestimate risk. This method was used to link PCB and PCDD exposures in Lake Ontario with early life stage mortality in lake trout and corresponding reductions in population abundance (Kramer et al., 2011).

The tissue residue approach (TRA) relates observed toxicological responses to the amount of chemical in an organism's body (Sappington et al., 2011). This approach focuses on the amount of chemical at the site of action in the fish's tissues to estimate risk rather than the amount in the surrounding water or sediment and incorporates all methods of uptake and is applicable to chemicals that bioaccumulate. POPs bioaccumulate and are highly lipophilic, so the portion of chemical stored in a fish's lipids will not be available to act on other biological molecules. This makes lipid stores protective until they are used, at which time any stored chemicals can mobilize to the site of action (Meador et al., 2002, 2006). For this reason, whole-body tissue residues need to be normalized to the amount of lipid in the organism at that life stage (Meador et al., 2002; Beckvar et al., 2005; Sappington et al., 2011). A common application of TRAs is to establish tissue residues below which fish are not expected to experience adverse effects, as has been done for PCBs and DDT (Meador et al., 2002; Beckvar et al., 2005). The biological effects threshold for juvenile salmonids exposed to PCBs was determined by calculating the 10th percentile whole-body residue of 15 laboratory studies examining the biological responses of salmonids to PCBs (Meador et al., 2002). Other TRAs utilize data on specific tissues to estimate effects (Sappington et al., 2011). The choice of methodology will depend on the data available, the chemical's mode and mechanism of action, and the species and life stage sensitivities (Meador, 2006; Sappington et al., 2011). This tool can be very powerful for estimating the risk of toxicological effects from exposures to bioaccumulating chemicals such as POPs (Sappington et al., 2011).

11.4. Complex Mixtures and Multiple Stressors

POPs are globally distributed and tend to concentrate in environmental compartments (i.e., attached to sediments or in the lipids of biota), creating complex mixtures. Many of these compounds share similar structures and toxicity pathways and are classified together, such as the DDTs and OC pesticides, or the PCBs and dioxin-like compounds. While these compounds may share a mode of action, their potency is often different. The calculation of toxic equivalency factors (TEFs) aids in expressing the toxicity of each compound relative to the typical or ideal compound. The position of the chlorine atoms on PCB congeners determines the binding affinity to the AHR. TEFs can be calculated relative to TCDD and then summed across all the congeners in a mixture to provide an assessment of toxicity (Meador et al., 2002). This method can produce a good estimate of risk due to the mixture for a particular response end-point, such as AHR activity, but omits any other modes of action that some compounds in the class may exhibit.

Other complications arising from complex mixtures of POPs are similar to those of any compound. The same biological target could be altered by different compounds in additive, antagonistic, synergistic, or potentiating ways, thus making it difficult to identify and establish causal links (Vasseur and Cossu-Leguille, 2006). Some POPs can induce P450 enzymes in the liver that metabolize endogenous hormones, while other POPs may simultaneously alter their production through a different pathway. The resulting levels of circulating hormone may be very different from what is expected due to the known toxicity pathways for those compounds acting alone.

Multiple stressors influence population and community dynamics, and interactions between stressors, such as harvesting, climate change, eutrophication, disease prevalence, and contaminants, can prove difficult to differentiate (Hylland et al., 2006; Vasseur and Cossu-Leguille, 2006).

Climate change poses particular concerns related to the transport of POPs and the changes in physiological responses of organisms experiencing toxicant stress with heat stress or other climatic stressors (Moe et al., 2013). Population dynamics may compensate for some stressors, depending on the timing and extent, but others, while small individually, may combine with other stressors to push a population into decline. Parsing out specific contributions will require attention to the community and ecosystem complexities present as well as the uncertainty surrounding the risk estimates (Barnthouse et al., 2007).

ACKNOWLEDGMENTS

We thank Drs. Nathanial Scholz and Walton Dickhoff of the Northwest Fisheries Science Center for providing helpful comments on this document. We greatly appreciate the excellent reviews provided by Dr. Larry Curtis, Professor, Department of Environmental and Molecular Toxicology, Oregon State University and Dr. Reynaldo Patiño, Research Professor, Department of Natural Resources Management, Texas Tech University.

REFERENCES

Ábalos, M., Abad, E., Estévez, A., Solé, M., Buet, A., Quirós, L., Piña, B. and Rivera, J. (2008). Effects on growth and biochemical responses in juvenile gilthead seabream (*Sparus aurata*) after long-term dietary exposure to low levels of dioxins. *Chemosphere* **73**, S303–S310.

Abramowicz, D. A. (1990). Aerobic and anaerobic biodegradation of PCBs: a review. *Biotechnology* **10**, 241–251.

Adams, S. M., Crumby, W. D., Greeley, M. S., Jr., Shugart, L. R. and Saylor, C. F. (1992). Responses of fish populations and communities to pulp mill effluents: A holistic assessment. *Ecotoxicology and Environmental Safety* **24**, 347–360.

Aluru, N., Jorgensen, E. H., Maule, A. G. and Vijayan, M. M. (2004). PCB disruption of the hypothalamus-pituitary-interrenal axis involves brain glucocorticoid receptor downregulation in anadromous Arctic charr. *American Journal of Physiology-Regulatory, Integrative and Comparative Physiology* **287**, R787–R793.

AMAP (2004). *AMAP Assessment 2002: Persistent Organic Pollutants in the Arctic*. Oslo, Norway: Arctic Monitoring and Assessment Program (AMAP), 310 p.

Amiard-Triquet, C. (2009). Behavioral disturbances: the missing link between sub-organismal and supra-organismal responses to stress? Prospects based on aquatic research. *Human and Ecological Risk Assessment: An International Journal* **15**, 87–110.

Anderson, M. J., Cacela, D., Beltman, D., Teh, S. J., Okihiro, M. S., Hinton, D. E., Denslow, N. and Zelikoff, J. T. (2003). Biochemical and toxicopathic biomarkers assessed in smallmouth bass recovered from a polychlorinated biphenyl-contaminated river. *Biomarkers* **8**, 371–393.

Aniagu, S. O., Williams, T. D., Allen, Y., Katsiadaki, I. and Chipman, J. K. (2008). Global genomic methylation levels in the liver and gonads of the three-spine stickleback (*Gasterosteus aculeatus*) after exposure to hexabromocyclododecane and 17-β oestradiol. *Environment International* **34**, 310–317.

Ankley, G. T., Bennett, R. S., Erickson, R. J., Hoff, D. J., Hornung, M. W., Johnson, R. D., Mount, D. R., Nichols, J. W., Russom, C. L., Schmieder, P. K., Serrrano, J. A., Tietge, J. E. and Villeneuve, D. L. (2010). Adverse outcome pathways: A conceptual framework to support ecotoxicology research and risk assessment. *Environmental Toxicology and Chemistry* **29**, 730–741.

Anonymous (2001). Council regulation (EC) No. 2375/2001, November 29, 2001.

Anway, M. D., Cupp, A. S., Uzumcu, M. and Skinner, M. K. (2005). Epigenetic transgenerational actions of endocrine disruptors and male fertility. *Science* **308**, 1466–1469.

Anway, M. D. and Skinner, M. K. (2008). Epigenetic programming of the germ line: effects of endocrine disruptors on the development of transgenerational disease. *Reproductive BioMedicine Online* **16**, 23–25.

Arkoosh, M., Casillas, E., Clemons, E., Huffman, P., Kagley, A., Collier, T. and Stein, J. (2000). Increased susceptibility of juvenile chinook salmon to infectious disease after exposure to chlorinated and aromatic compounds found in contaminated urban estuaries. *Marine Environmental Research* **50**, 470–471.

Arkoosh, M. R., Boylen, D., Dietrich, J., Anulacion, B. F., Ylitalo, G. M., Bravo, C. F., Johnson, L. L., Loge, F. J. and Collier, T. K. (2010). Disease susceptibility of salmon exposed to polybrominated diphenyl ethers (PBDEs). *Aquatic Toxicology* **98**, 51–59.

Arkoosh, M. R., Boylen, D., Stafford, C. L., Johnson, L. L. and Collier, T. K. (2005). Use of disease challenge assay to assess immunotoxicity of xenobiotics. In *Techniques in Aquatic Toxicology* (ed. G. K. Ostrander), pp. 19–38. New York: Taylor and Francis Group.

Arkoosh, M. R., Casillas, E., Clemons, E., Kagley, A. N., Olson, R., Reno, P. and Stein, J. E. (1998a). Effect of pollution on fish diseases: Potential impacts on salmonid populations. *Journal of Aquatic Animal Health* **10**, 182–190.

Arkoosh, M. R., Casillas, E., Clemons, E., McCain, B. and Varanasi, U. (1991). Suppression of immunological memory in juvenile chinook salmon (*Oncorhynchus tshawytscha*) from an urban estuary. *Fish and Shellfish Immunology* **1**, 261–277.

Arkoosh, M. R., Casillas, E., Huffman, P., Clemons, E., Evered, J., Stein, J. E. and Varanasi, U. (1998b). Increased susceptibility of juvenile chinook salmon from a contaminated estuary to vibrio anguillarum. *Transactions of the American Fisheries Society* **127**, 360–374.

Arkoosh, M. R., Clemons, E., Huffman, P., Kagley, A. N., Casillas, E., Adams, N., Sanborn, H. R., Collier, T. K. and Stein, J. E. (2001). Increased susceptibility of juvenile chinook salmon to vibriosis after exposure to chlorinated and aromatic compounds found in contaminated urban estuaries. *Journal of Aquatic Animal Health* **13**, 257–268.

Arnoldsson, K., Haldén, A. N., Norrgren, L. and Haglund, P. (2012). Retention and maternal transfer of environmentally relevant polybrominated dibenzo-p-dioxins and dibenzofurans, polychlorinated dibenzo-p-dioxins and dibenzofurans, and polychlorinated biphenyls in zebrafish (*Danio rerio*) after dietary exposure. *Environmental Toxicology and Chemistry* **31**, 804–812.

Arnot, J. A. and Gobas, F. A. P. C. (2003). A generic QSAR for assessing the bioaccumulation potential of organic chemicals in aquatic food webs. *QSAR and Combinatorial Science* **22**, 337–345.

Arnot, J. A. and Gobas, F. A. P. C. (2004). A food web bioaccumulation model for organic chemicals in aquatic ecosystems. *Environmental Toxicology and Chemistry* **23**, 2343–2355.

Arukwe, A. (2001). Cellular and molecular responses to endocrine-modulators and the impact on fish reproduction. *Marine Pollution Bulletin* **42**, 643–655.

ATSDR (1997). Chlordane. ATSDRs toxicological profiles on CD-ROM. Boca Raton, FL: Agency for Toxic Substances and Disease Registry (ATSDR).

Baatrup, E. and Junge, M. (2001). Antiandrogenic pesticides disrupt sexual characteristics in the adult male guppy (*Poecilia reticulata*). *Environmental Health Perspectives* **109**, 1063–1070.

Bagnyukova, T. V., Luzhna, L. I., Pogribny, I. P. and Lushchak, V. I. (2007). Oxidative stress and antioxidant defenses in goldfish liver in response to short-term exposure to arsenite. *Environmental and Molecular Mutagenesis* **48**, 658–665.

Baldigo, B. P., Sloan, R. J., Smith, S. B., Denslow, N. D., Blazer, V. S. and Gross, T. S. (2006). Polychlorinated biphenyls, mercury, and potential endocrine disruption in fish from the Hudson River, New York. *Aquatic Sciences* **68**, 206–228.

Barnthouse, L. W., Munns, W. R., Jr. and Sorensen, M. T. (2007). *Population-Level Ecological Risk Assessment*. Boca Raton, FL: CRC Press, 376 p.

Barron, M. G., Anderson, M. J., Cacela, D., Lipton, J., Teh, S. J., Hinton, D. E., Zelikoff, J. T., Dikkeboom, A. L., Tillitt, D. E., Holey, M. and Denslow, N. (2000). PCBs, liver lesions, and biomarker responses in adult Walleye (*Stizostedium vitreum vitreum*) collected from Green Bay, Wisconsin. *Journal of Great Lakes Research* **26**, 250–271.

Bayley, M., Junge, M. and Baatrup, E. (2002). Exposure of juvenile guppies to three antiandrogens causes demasculinization and a reduced sperm count in adult males. *Aquatic Toxicology* **56**, 227–239.

Bayne, C. J. and Gerwick, L. (2001). The acute phase response and innate immunity of fish. *Developmental and Comparative Immunology* **25**, 725–743.

Beckvar, N., Dillon, T. M. and Read, L. B. (2005). Approaches for linking whole-body fish tissue residues of mercury or DDT to biological effects thresholds. *Environmental Toxicology and Chemistry* **24**, 2094–2105.

Bell, D. R., Clode, S., Fan, M. Q., Fernandes, A., Foster, P. M. D., Jiang, T., Loizou, G., MacNicoll, A., Miller, B. G., Rose, M., Tran, L. and White, S. (2010). Interpretation of studies on the developmental reproductive toxicology of 2,3,7,8-tetrachlorodibenzo-p-dioxin in male offspring. *Food and Chemical Toxicology* **48**, 1439–1447.

Bengtsson, B.-E. (1980). Long-term effects of PCB (Clophen A50) on growth, reproduction and swimming performance in the minnow, Phoxinus phoxinus. *Water Research* **14**, 681–687.

Benguira, S., Leblond, V. S., Weber, J.-P. and Hontela, A. (2002). Loss of capacity to elevate plasma cortisol in rainbow trout (*Oncorhynchus mykiss*) treated with a single injection of o,p′-dichlorodiphenyldichloroethane. *Environmental Toxicology and Chemistry* **21**, 1753–1756.

Benton, M. J., Nimrod, A. C. and Benson, W. H. (1994). Evaluation of growth and energy storage as biological markers of DDT exposure in sailfin mollies. *Ecotoxicology and Environmental Safety* **29**, 1–12.

Berg, V., Lyche, J. L., Karlsson, C., Stavik, B., Nourizadeh-Lillabadi, R., Hårdnes, N., Skaare, J. U., Alestrøm, P., Lie, E. and Ropstad, E. (2011). Accumulation and effects of natural mixtures of persistent organic pollutants (POP) in zebrafish after two generations of exposure. *Journal of Toxicology and Environmental Health, Part A* **74**, 4074–4023.

Berg, V., Zerihun, M. A., Jørgensen, A., Lie, E., Dale, O. B., Skaare, J. U. and Lyche, J. L. (2013). High prevalence of infections and pathological changes in burbot (*Lota lota*) from a polluted lake (Lake Mjøsa, Norway). *Chemosphere* **90**, 1711–1718.

Bern, H. A. (1967). Hormones and endocrine glands of fishes: Studies of fish endocrinology reveal major physiologic and evolutionary problems. *Science* **158**, 455–462.

Betoulle, S., Duchiron, C. and Deschaux, P. (2000). Lindane increases in vitro respiratory burst activity and intracellular calcium levels in rainbow trout (*Oncorhynchus mykiss*) head kidney phagocytes. *Aquatic Toxicology* **48**, 211–221.

Birchmeier, K. L., Smith, K. A., Passino-Reader, D. R., Sweet, L. I., Chernyak, S. M., Adams, J. V. and Omann, G. M. (2005). Effects of selected polybrominated diphenyl ether flame retardants on lake trout (*Salvelinus namaycush*) thymocyte viability, apoptosis, and necrosis. *Environmental Toxicology and Chemistry* **24**, 1518–1522.

Blanar, C. A., Curtis, M. A. and Chan, H. M. (2005). Growth, nutritional composition, and hematology of Arctic charr (*Salvelinus Alpinus*) exposed to toxaphene and tapeworm

(*Diphyllobothrium dendriticum*) larvae. *Archives of Environmental Contamination and Toxicology* **48**, 397–404.

Blanton, M. L. and Specker, J. L. (2007). The hypothalamic-pituitary-thyroid (HPT) axis in fish and its role in fish development and reproduction. *Critical Reviews in Toxicology* **37**, 97–115.

Boas, M., Feldt-Rasmussen, U., Skakkebæk, N. E. and Main, K. M. (2006). Environmental chemicals and thyroid function. *European Journal of Endocrinology* **154**, 599–611.

Bols, N. C., Brubacher, J. L., Ganassin, R. C. and Lee, L. E. J. (2001). Ecotoxicology and innate immunity in fish. *Developmental and Comparative Immunology* **25**, 853–873.

Boorman, G. A., Botts, S., Bunton, T. E., Fournie, J. W., Harshbarger, J. C., Hawkins, W. E., Hinton, D. E., Jokinen, M. P., Okihiro, M. S. and Wolfe, M. J. (1997). Diagnostic criteria for degenerative, inflammatory, proliferative nonneoplastic and neoplastic liver lesions in Medaka (*Oryzias latipes*): Consensus of a National Toxicology Program Pathology Working Group. *Toxicologic Pathology* **25**, 202–210.

Borja, J., Taleon, D. M., Auresenia, J. and Gallardo, S. (2005). Polychlorinated biphenyls and their biodegradation. *Process Biochemistry* **40**, 1999–2013.

Braunbeck, T., Görge, G., Storch, V. and Nagel, R. (1990). Hepatic steatosis in zebra fish (*Brachydanio rerio*) induced by long-term exposure to γ-hexachlorocyclohexane. *Ecotoxicology and Environmental Safety* **19**, 355–374.

Breivik, K., Alcock, R., Li, Y.-F., Bailey, R. E., Fiedler, H. and Pacyna, J. M. (2004). Primary sources of selected POPs: Regional and global scale emission inventories. *Environmental Pollution* **128**, 3–16.

Brown, S. B., Adams, B. A., Cyr, D. G. and Eales, J. G. (2004). Contaminant effects on the teleost fish thyroid. *Environmental Toxicology and Chemistry* **23**, 1680–1701.

Buckman, A. H., Fisk, A. T., Parrott, J. L., Solomon, K. R. and Brown, S. B. (2007). PCBs can diminish the influence of temperature on thyroid indices in rainbow trout (*Oncorhynchus mykiss*). *Aquatic Toxicology* **84**, 366–378.

Buckman, A. H., Wong, C. S., Chow, E. A., Brown, S. B., Solomon, K. R. and Fisk, A. T. (2006). Biotransformation of polychlorinated biphenyls (PCBs) and bioformation of hydroxylated PCBs in fish. *Aquatic Toxicology* **78**, 176–185.

Bugel, S. M., White, L. A. and Cooper, K. R. (2010). Impaired reproductive health of killifish (*Fundulus heteroclitus*) inhabiting Newark Bay, NJ, a chronically contaminated estuary. *Aquatic Toxicology* **96**, 182–193.

Bugel, S. M., White, L. A. and Cooper, K. R. (2011). Decreased vitellogenin inducibility and 17β-estradiol levels correlated with reduced egg production in killifish (*Fundulus heteroclitus*) from Newark Bay, NJ. *Aquatic Toxicology* **105**, 1–12.

Burnett, K. G. (2005). Impact of environmental toxicants and natural variables on the immune system of fishes. In *Biochemistry and Molecular Biology of Fishes* (eds. T. P. Mommsen and T. W. Moon), pp. 231–253. St. Louis: Elsevier Science.

Burton, J. E., Dorociak, I. R., Schwedler, T. E. and Rice, C. D. (2002). Circulating lysozyme and hepatic CYP1A activities during a chronic dietary exposure to tributyltin (TBT) and 3,3′,4,4′,5-pentachlorbiphenyl (PCB-126) mixtures in channel catfish, *Ictalurus punctatus*. *Journal of Toxicology and Environmental Health, Part A* **65**, 589–602.

Candelmo, A., Deshpande, A., Dockum, B., Weis, P. and Weis, J. (2010). The effect of contaminated prey on feeding, activity, and growth of young-of-the-year bluefish, *Pomatomus saltatrix*, in the Laboratory. *Estuaries and Coasts* **33**, 1025–1038.

Carlson, E. and Zelikoff, J. T. (2008). The immune system of fish: A target organ of toxicity. In *The Toxicology of Fishes* (eds. R. T. DiGiulio and D. E. Hinton), pp. 489–529. Boca Raton, FL: CRC Press.

Carr, J. A. and Patiño, R. (2011). The hypothalamus-pituitary-thyroid axis in teleosts and amphibians: endocrine disruption and its consequences to natural populations. *General and Comparative Endocrinology* **170**, 299–312.

Carrell, D. T. (2011). Epigenetic marks in zebrafish sperm: Insights into chromatin compaction, maintenance of pluripotency, and the role of the paternal genome after fertilization. *Asian Journal of Andrology* **13**, 620–621.

Carvalho, P. S. M., Noltie, D. B. and Tillitt, D. E. (2004). Biochemical, histological and behavioural aspects of visual function during early development of rainbow trout. *Journal of Fish Biology* **64**, 833–850.

Casanova-Nakayama, A., Wenger, M., Burki, R., Eppler, E., Krasnov, A. and Segner, H. (2011). Endocrine disrupting compounds: Can they target the immune system of fish? *Marine Pollution Bulletin* **63**, 412–416.

Casati, L., Sendra, R., Colciago, A., Negri-Cesi, P., Berdasco, M., Esteller, M. and Celotti, F. (2012). Polychlorinated biphenyls affect histone modification pattern in early development of rats: A role for androgen receptor-dependent modulation? *Epigenomics* **4**, 101–112.

Castonguay, M. and Cyr, D. G. (1998). Effects on temperature on spontaneous and thyroxine-stimulated locomotor activity of Atlantic cod. *Journal of Fish Biology* **53**, 303–313.

Caswell, H. (2001). *Matrix population models: Construction, analysis and interpretation.* Sunderland, MA: Sinauer Associates, 722 p.

CEC (2003). *DDT no longer used in North America.* Montréal, Québec: Commission for Environmental Cooperation of North America.

Chang, S., Zdanowicz, V. S. and Murchelano, R. A. (1998). Associations between liver lesions in winter flounder (*Pleuronectes americanus*) and sediment chemical contaminants from northeast United States estuaries. *ICES Journal of Marine Science: Journal du Conseil* **55**, 954–969.

Chen, T.-H., Cheng, Y.-M., Cheng, J.-O., Chou, C.-T., Hsiao, Y.-C. and Ko, F.-C. (2010). Growth and transcriptional effect of dietary 2,2′,4,4′-tetrabromodiphenyl ether (PBDE-47) exposure in developing zebrafish (*Danio rerio*). *Ecotoxicology and Environmental Safety* **73**, 377–383.

Chen, Q., Yu, L., Yang, L. and Zhou, B. (2012a). Bioconcentration and metabolism of decabromodiphenyl ether (BDE-209) result in thyroid endocrine disruption in zebrafish larvae. *Aquatic Toxicology* **110–111**, 141–148.

Chen, X., Huang, C., Wang, X., Chen, J., Bai, C., Chen, Y., Chen, X., Dong, Q. and Yang, D. (2012b). BDE-47 disrupts axonal growth and motor behavior in developing zebrafish. *Aquatic Toxicology* **120–121**, 35–44.

Chen, L., Huang, C., Hu, C., Yu, K., Yang, L. and Zhou, B. (2012c). Acute exposure to DE-71: Effects on locomotor behavior and developmental neurotoxicity in zebrafish larvae. *Environmental Toxicology and Chemistry* **31**, 2338–2344.

Chen, L., Yu, K., Huang, C., Yu, L., Zhu, B., Lam, P. K. S., Lam, J. C. W. and Zhou, B. (2012d). Prenatal transfer of polybrominated diphenyl ethers (PBDEs) results in developmental neurotoxicity in zebrafish larvae. *Environmental Science and Technology* **46**, 9727–9734.

Chou, C.-T., Hsiao, Y.-C., Ko, F.-C., Cheng, J.-O., Cheng, Y.-M. and Chen, T.-H. (2010). Chronic exposure of 2,2′,4,4′-tetrabromodiphenyl ether (PBDE-47) alters locomotion behavior in juvenile zebrafish (*Danio rerio*). *Aquatic Toxicology* **98**, 388–395.

Coimbra, A. M. and Reis-Henriques, M. A. (2007). Tilapia larvae Aroclor 1254 exposure: Effects on gonads and circulating thyroid hormones during adulthood. *Bulletin of Environmental Contamination and Toxicology* **79**, 488–493.

Contractor, R. G., Foran, C. M., Li, S. and Willett, K. L. (2004). Evidence of gender- and yissue-dpecific promoter methylation and the potential for ethinyl estradiol-induced changes in Japanese Medaka (*Oryzias Latipes*) estrogen receptor and aromatase genes. *Journal of Toxicology and Environmental Health, Part A* **67**, 1–22.

Cook, P. M., Robbins, J. A., Endicott, D. D., Lodge, K. B., Guiney, P. D., Walker, M. K., Zabel, E. W. and Peterson, R. E. (2003). Effects of aryl hydrocarbon receptor-mediated early life stage toxicity on lake trout populations in Lake Ontario during the 20th century. *Environmental Science and Technology* **37**, 3864–3877.

Costa, L. G. and Giordano, G. (2007). Developmental neurotoxicity of polybrominated diphenyl ether (PBDE) flame retardants. *NeuroToxicology* **28**, 1047–1067.

Couch, J. A. (1975). Histopathologic effects of pesticides and related chemicals in the livers of fishes. In *The Pathology of Fishes* (eds. W. E. Ribelin and G. Migaki), pp. 559–584. Madison: University of Wisconsin Press.

Couillard, C. M., Légaré, B., Bernier, A. and Dionne, Z. (2011). Embryonic exposure to environmentally relevant concentrations of PCB126 affect prey capture ability of Fundulus heteroclitus larvae. *Marine Environmental Research* **71**, 257–265.

Cowan-Ellsberry, C. E., McLachlan, M. S., Arnot, J. A., MacLeod, M., McKone, T. E. and Wania, F. (2009). Modeling exposure to persistent chemicals in hazard and risk assessment. *Integrated Environmental Assessment and Management* **5**, 662 and 679.

Cuesta, A., Meseguer, J. and Ángeles Esteban, M. (2008). Effects of the organochlorines p,p′-DDE and lindane on gilthead seabream leucocyte immune parameters and gene expression. *Fish and Shellfish Immunology* **25**, 682 and 688.

Da Cuña, R. H., Pandolfi, M., Genovese, G., Piazza, Y., Ansaldo, M. and Lo Nostro, F. L. (2013). Endocrine disruptive potential of endosulfan on the reproductive axis of Cichlasoma dimerus (*Perciformes, Cichlidae*). *Aquatic Toxicology* **126**, 299–305.

Daley, J. M., Leadley, T. A. and Drouillard, K. G. (2009). Evidence for bioamplification of nine polychlorinated biphenyl (PCB) congeners in yellow perch (*Perca flavascens*) eggs during incubation. *Chemosphere* **75**, 1500–1505.

Daley, J. M., Leadley, T. A., Pitcher, T. E. and Drouillard, K. G. (2012). Bioamplification and the selective depletion of persistent organic pollutants in chinook salmon larvae. *Environmental Science and Technology* **46**, 2420–2426.

Daouk, T., Larcher, T., Roupsard, F., Lyphout, L., Rigaud, C., Ledevin, M., Loizeau, V. and Cousin, X. (2011). Long-term food-exposure of zebrafish to PCB mixtures mimicking some environmental situations induces ovary pathology and impairs reproduction ability. *Aquatic Toxicology* **105**, 270–278.

DeBruyn, A. M. H. and Gobas, F. A. P. C. (2006). A bioenergetic biomagnification model for the animal kingdom. *Environmental Science and Technology* **40** (1581–1587).

DeBruyn, A. M. H., Ikonomou, M. G. and Gobas, F. A. P. C. (2004). Magnification and toxicity of PCBs, PCDDs, and PCDFs in upriver-migrating Pacific salmon. *Environmental Scienceand Technology* **38** (6217–6224).

Desaulniers, D., Xiao, G.-h., Lian, H., Feng, Y.-L., Zhu, J., Nakai, J. and Bowers, W. J. (2009). Effects of mixtures of polychlorinated niphenyls, methylmercury, and organochlorine pesticides on hepatic DNA methylation in prepubertal female Sprague-Dawley rats. *International Journal of Toxicology* **28**, 294–307.

Dirinck, E., Jorens, P. G., Covaci, A., Geens, T., Roosens, L., Neels, H., Mertens, I. and Van Gaal, L. (2011). Obesity and persistent organic pollutants: possible obesogenic effect of organochlorine pesticides and polychlorinated biphenyls. *Obesity (Silver Spring)* **19**, 709–714.

Donohoe, R., Zhang, M., Siddens, Q., Carpenter, L. K., Hendricks, H. M., Curtis, J. D. and LR (1998). Modulation of 7,12-dimethylbenz[a]anthracene disposition and hepatocarcinogenesis by dieldrin and chlordecone in rainbow trout. *Journal of Toxicology and Environmental Health, Part A* **54**, 227–242.

Donohoe, R. M. and Curtis, L. R. (1996). Estrogenic activity of chlordecone, o,p'-DDT and o,p'-DDE in juvenile rainbow trout: induction of vitellogenesis and interaction with hepatic estrogen binding sites. *Aquatic Toxicology* **36**, 31–52.

Dorval, J., Leblond, V. S. and Hontela, A. (2003). Oxidative stress and loss of cortisol secretion in adrenocortical cells of rainbow trout (*Oncorhynchus mykiss*) exposed in vitro to endosulfan, an organochlorine pesticide. *Aquatic Toxicology* **63**, 229–241.

Duchiron, C., Betoulle, S., Reynaud, S. and Deschaux, P. (2002). Lindane increases macrophage-activating factor production and intracellular calcium in rainbow trout (*Oncorhynchus mykiss*) leukocytes. *Ecotoxicology and Environmental Safety* **53**, 388–396.

Duffy, J. E., Carlson, E., Li, Y., Prophete, C. and Zelikoff, J. T. (2002). Impact of polychlorinated biphenyls (PCBs) on the immune function of fish: age as a variable in determining adverse outcome. *Marine Environmental Research* **54**, 559–563.

Duffy, J. E., Carlson, E. A., Li, Y., Prophete, C. and Zelikoff, J. T. (2003). Age-related differences in the sensitivity of the fish immune response to a coplanar PCB. *Ecotoxicology* **12**, 251–259.

Duffy, J. E. and Zelikoff, J. T. (2006). The relationship between noncoplanar PCB-induced immunotoxicity and hepatic CYP1A induction in a fish model. *Journal of Immunotoxicology* **3**, 39–47.

Eder, K. J., Clifford, M. A., Hedrick, R. P., Köhler, H.-R. and Werner, I. (2008). Expression of immune-regulatory genes in juvenile chinook salmon following exposure to pesticides and infectious hematopoietic necrosis virus (IHNV). *Fish and Shellfish Immunology* **25**, 508–516.

Edmunds, J. S. G., McCarthy, R. A. and Ramsdell, J. S. (2000). Permanent and functional male-to-female sex reversal in d-rR strain medaka (*Oryzias latipes*) following egg microinjection of o,p'-DDT. *Environmental Health Perspectives* **108**, 219–224.

Eisler, R. (1986a). Polychlorinated biphenyls hazards to fish, wildlife and invertebrates: A synoptic review. Laurel, MD: U.S Fish and Wildlife Biological Report 85(1.7)—Contaminant Hazard Reviews Report No. 7.

Eisler, R. (1986b). Dioxin hazards to fish, wildlife and invertebrates: A synoptic review. Laurel, MD: U.S Fish and Wildlife Biological Report 85 (1.8)—Contaminant Hazard Reviews Report No. 8.

Eisler, R. and Belisle, A. (1996). Planar PCBs hazards to fish, wildlife and invertebrates: A synoptic review. Laurel, MD: U.S Fish and Wildlife Biological Report 31—Contaminant Hazard Reviews Report No. 31.

Ekman, E., Åkerman, G., Balk, L. and Norrgren, L. (2004). Impact of PCB on resistance to Flavobacterium psychrophilum after experimental infection of rainbow trout Oncorhynchus mykiss eggs by nanoinjection. *Diseases of Aquatic Organisms* **60**, 31–39.

Elango, A., Shepherd, B. and Chen, T. T. (2006). Effects of endocrine disrupters on the expression of growth hormone and prolactin mRNA in the rainbow trout pituitary. *General and Comparative Endocrinology* **145**, 116–127.

Ellis, A. E. (2001). Innate host defense mechanisms of fish against viruses and bacteria. *Developmental and Comparative Immunology* **25**, 827–839.

Ellis, L., Atadja, P. W. and Johnstone, R. W. (2009). Epigenetics in cancer: Targeting chromatin modifications. *Molecular Cancer Therapeutics* **8**, 1409–1420.

Elnar, A. A., Diesel, B., Desor, F., Feidt, C., Bouayed, J., Kiemer, A. K. and Soulimani, R. (2012). Neurodevelopmental and behavioral toxicity via lactational exposure to the sum of six indicator non-dioxin-like-polychlorinated biphenyls (Σ6 NDL-PCBs) in mice. *Toxicology* **299**, 44–54.

Elskus, A. A., Collier, T. K. and Monosson, E. (2005). Interactions between lipids and persistent organic pollutants in fish. In *Biochemistry and Molecular Biology of Fishes* (eds. T. P. Mommsen and T. W. Moon), pp. 119–152. St. Louis, MO: Elsevier Science.

Elson, P. F. (1967). Effects on wild young salmon of spraying DDT over New Brunswick Forests. *Journal of the Fisheries Research Board of Canada* 24, 731–767.

Evenset, A., Carroll, J., Christensen, G. N., Kallenborn, R., Gregor, D. and Gabrielsen, G. W. (2007). Seabird guano is an rfficient conveyer of persistent organic pollutants (POPs) to Arctic lake ecosystems. *Environmental Science and Technology* 41, 1173–1179.

Ewald, G., Larsson, P., Linge, H., Okla, L. and Szarzi, N. (1998). Biotransport of organic pollutants to an inland Alaska lake by migrating sockeye salmon (*Oncorhynchus nerka*). *Arctic* 51, 40–47.

Faisal, M., Marzouk, M. S. M., Smith, C. L. and Huggett, R. J. (1991). Mitogen induced proliferative responses of lymphocytes from spot (*Leiostomus Xanthurus*) exposed to polycyclic aromatic hydrocarbon contaminated environments. *Immunopharmacology and Immunotoxicology* 13, 311–327.

Fang, L., Huang, J., Yu, G. and Wang, L. (2008). Photochemical degradation of six polybrominated diphenyl ether congeners under ultraviolet irradiation in hexane. *Chemosphere* 71, 258–267.

Feist, G. W., Webb, M. A. H., Gundersen, D. T., Foster, E. P., Schreck, C. B., Maule, A. G. and Fitzpatrick, M. S. (2005). Evidence of detrimental effects of environmental contaminants on growth and reproductive physiology of white sturgeon in impounded areas of the Columbia River. *Environmental Health Perspectives* 113, 1675–1682.

Feng, C., Xu, Y., Zhao, G., Zha, J., Wu, F. and Wang, Z. (2012). Relationship between BDE 209 metabolites and thyroid hormone levels in rainbow trout (*Oncorhynchus mykiss*). *Aquatic Toxicology* 122–123, 28–35.

Ferguson, H. W. (1989). *Systemic Pathology of Fish.* Ames, IA: Iowa State University Press, 263 p.

Fiedler, H. (1996). Sources of PCDD/PCDF and impact on the environment. *Chemosphere* 32, 55–64.

Finn, R. N. (2007). The physiology and toxicology of salmonid eggs and larvae in relation to water quality criteria. *Aquatic Toxicology* 81, 337–354.

Foekema, E. M., Deerenberg, C. M. and Murk, A. J. (2008). Prolonged ELS test with the marine flatfish sole (*Solea solea*) shows delayed toxic effects of previous exposure to PCB 126. *Aquatic Toxicology* 90, 197–203.

Foekema, E. M., Fischer, A., Parron, M. L., Kwadijk, C., de Vries, P and Murk, A. J. (2012). Toxic concentrations in fish early life stages peak at a critical moment. *Environmental Toxicology and Chemistry* 31, 1381–1390.

Frith, C. and Ward, J. (1979). A morphologic classification of proliferative and neoplastic hepatic lesions in mice. *Journal of Environmental Pathology and Toxicology* 3, 329–351.

Garrido, S., Rosa, R., Ben-Hamadou, R., Cunha, M. E., Chícharo, M. A. and van der Lingen, C. D. (2007). Effect of maternal fat reserves on the fatty acid composition of sardine (*Sardina pilchardus*) oocytes. *Comparative Biochemistry and Physiology Part B: Biochemistry and Molecular Biology* 148, 398–409.

Ghosh, R., Lokman, P. M., Lamare, M. D., Metcalf, V. J., Burritt, D. J., Davison, W. and Hageman, K. J. (2013). Changes in physiological responses of an Antarctic fish, the emerald rock cod (*Trematomus bernacchii*), following exposure to polybrominated diphenyl ethers (PBDEs). *Aquatic Toxicology* 128–129, 91–100.

Giesy, J. P., Jones, P. D., Kannan, K., Newsted, J. L., Tillitt, D. E. and Williams, L. L. (2002). Effects of chronic dietary exposure to environmentally relevant concentrations to 2,3,7,8-tetrachlorodibenzo-p-dioxin on survival, growth, reproduction and biochemical responses of female rainbow trout (*Oncorhynchus mykiss*). *Aquatic Toxicology* 59, 35–53.

Gilroy, È. A. M., McMaster, M. E., Parrott, J. L., Hewitt, L. M., Park, B. J., Brown, S. B. and Sherry, J. P. (2012). Assessment of the health status of wild fish from the Wheatley Harbour area of Concern, Ontario, Canada. *Environmental Toxicology and Chemistry* **31**, 2798–2811.

Gobas, F. A. P. C. (1993). A model for predicting the bioaccumulation of hydrophobic organic chemicals in aquatic food-webs: application to Lake Ontario. *Ecological Modelling* **69**, 1–17.

Gobas, F. A. P. C., de Wolf, W., Burkhard, L. P., Verbruggen, E. and Plotzke, K. (2009). Revisiting bioaccumulation criteria for POPs and PBT assessments. *Integrated Environmental Assessment and Management* **5**, 624–637.

Gormley, K., Teather, K. and Guignion, D. (2005). Changes in salmonid communities associated with pesticide runoff events. *Ecotoxicology* **14**, 671–678.

Gräff, J. and Mansuy, I. M. (2009). Epigenetic dysregulation in cognitive disorders. *European Journal of Neuroscience* **30**, 1–8.

Grimes, A. C., Erwin, K. N., Stadt, H. A., Hunter, G. L., Gefroh, H. A., Tsai, H.-J. and Kirby, M. L. (2008). PCB126 exposure disrupts zebrafish ventricular and branchial but not early neural crest development. *Toxicological Sciences* **106**, 193–205.

Grinwis, G. C. M., van den Brandhof, E. J., Engelsma, M. Y., Kuiper, R. V., Vaal, M. A., Vethaak, A. D., Wester, P. W. and Vos, J. G. (2001). Toxicity of PCB-126 in European flounder (*Platichthys flesus*) with emphasis on histopathology and cytochrome P4501A induction in several organ systems. *Archives of Toxicology* **75**, 80–87.

Grinwis, G. C. M., Vethaak, A. D., Wester, P. W. and Vos, J. G. (2000). Toxicology of environmental chemicals in the flounder (*Platichthys flesus*) with emphasis on the immune system: field, semi-field (mesocosm) and laboratory studies. *Toxicology Letters* **112–113**, 289–301.

Grun, F. and Blumberg, B. (2009). Endocrine disrupters as obesogens. *Molecular and Cellular Endocrinology* **304**, 19–29.

Haave, M., Bernhard, A., Jellestad, F., Heegaard, E., Brattelid, T. and Lundebye, A.-K. (2011). Long-term effects of environmentally relevant doses of 2,2′,4,4′,5,5′ hexachlorobiphenyl (PCB153) on neurobehavioural development, health and spontaneous behaviour in maternally exposed mice. *Behavioral and Brain Functions* **7**, 3.

Hageman, K. J., Simonich, S. L., Campbell, D. H., Wilson, G. R. and Landers, D. H. (2006). Atmospheric deposition of current-use and historic-use pesticides in snow at national parks in the western United States. *Environmental Science and Technology* **40**, 3174–3180.

Halver, J. (1967). Crystalline aflatoxin and other vectors for trout hepatoma. In: Halver, J., Mitchell, I., editors. *Trout Hepatoma Research Conference Papers*, Research Rept. 70. Washington. pp. 78–102.

Hanari, N., Kannan, K., Miyake, Y., Okazawa, T., Kodavanti, P. R. S., Aldous, K. M. and Yamashita, N. (2006). Occurrence of polybrominated biphenyls, polybrominated dibenzo-p-dioxins, and polybrominated dibenzofurans as impurities in commercial polybrominated diphenyl ether mixtures. *Environmental Science and Technology* **40**, 4400–4405.

Harford, A. J., O'Halloran, K. and Wright, P. F. A. (2005). The effects of in vitro pesticide exposures on the phagocytic function of four native Australian freshwater fish. *Aquatic Toxicology* **75**, 330–342.

Hawkins, W. E., Walker, W. W. and Overstreet, R. M. (1995). Carcinogenicity tests using aquarium fish. *Toxicology Mechanisms and Methods* **5**, 225–263.

He, J., Robrock, K. R. and Alvarez-Cohen, L. (2006). Microbial reductive debromination of polybrominated diphenyl ethers (PBDEs). *Environmental Science and Technology* **40**, 4429–4434.

He, J., Yang, D., Wang, C., Liu, W., Liao, J., Xu, T., Bai, C., Chen, J., Lin, K., Huang, C. and Dong, Q. (2011). Chronic zebrafish low dose decabrominated diphenyl ether (BDE-209) exposure affected parental gonad development and locomotion in F1 offspring. *Ecotoxicology* **20**, 1813–1822.

Head, J. A., Dolinoy, D. C. and Basu, N. (2012). Epigenetics for ecotoxicologists. *Environmental Toxicology and Chemistry* **31**, 221–227.

HealthCanada (2001). PCBs. In: Canada, H., editor. Ottawa, ON: Health Canada-Management of Toxic Substances Division.

Hendricks, J. D., Putnam, T. and Sinnhuber, R. (1979). Effect of dietary dieldrin on aflatoxin-B1 carcinogenesis in rainbow trout (*Salmo gairdneri*). *Journal of Environmental Pathology and Toxicology* **2**, 719–728.

Herbstman, J. B., Sjodin, A., Kurzon, M., Lederman, S. A., Jones, R. S., Rauh, V., Needham, L. L., Tang, D., Niedzwiecki, M., Wang, R. Y. and Perera, F. (2010). Prenatal exposure to PBDEs and neurodevelopment. *Environmental Health Perspectives* **118**, 712–719.

Herman, D. P., Effler, J. I., Boyd, D. T. and Krahn, M. M. (2001). An efficient clean-up method for the GC–MS determination of methylsulfonyl-PCBs/DDEs extracted from various marine mammal tissues. *Marine Environmental Research* **52**, 127–150.

Hewitt, M. L., Parrott, J. L. and McMaster, M. E. (2006). A decade of research on the environmental impacts of pulp and paper mill effluents in Canada: Sources and characteristics of bioactive substances. *Journal of Toxicology and Environmental Health, Part B* **9**, 341–356.

Hinck, J. E., Blazer, V. S., Denslow, N. D., Echols, K. R., Gross, T. S., May, T. W., Anderson, P. J., Coyle, J. J. and Tillitt, D. E. (2007). Chemical contaminants, health indicators, and reproductive biomarker responses in fish from the Colorado River and its tributaries. *Science of the Environment* **378**, 376–402.

Hinton, D. E. (1993). Toxicologic histopatholoy of fishes: A systemic approach and overview. In *Pathobiology of Marine and Estuarine Organisms* (eds. J. A. Couch and J. W. Fournie), pp. 177–215. Boca Raton, FL: CRC Press.

Hinton, D. E., Klaunig, J. E. and Lipsky, M. M. (1978). PCB-induced alterations in telost liver: A model for environmental disease in fish. *Marine Fisheries Review* **40**, 47–50.

Ho, D. H. and Burggren, W. W. (2012). Parental hypoxic exposure confers offspring hypoxia resistance in zebrafish (*Danio rerio*). *The Journal of Experimental Biology* **215**, 4208–4216.

Hodson, P. V., Thivierge, D., Levesque, M.-C., McWhirter, M., Ralph, K., Gray, B., Whittle, D. M., Carey, J. H. and Van Der Kraak, G. (1992). Effects of bleached kraft mill effluent on fish in the St. Maurice River, Quebec. *Environmental Toxicology and Chemistry* **11**, 1635–1651.

Hoekstra, P. F., O'Hara, T. M., Fisk, A. T., Borgå, K., Solomon, K. R. and Muir, D. C. G. (2003). Trophic transfer of persistent organochlorine contaminants (OCs) within an Arctic marine food web from the southern Beaufort–Chukchi Seas. *Environmental Pollution* **124**, 509–522.

Holm, G., Norrgren, L., Andersson, T. and Thurén, A. (1993). Effects of exposure to food contaminated with PBDE, PCN or PCB on reproduction, liver morphology and cytochrome P450 activity in the three-spined stickleback, *Gasterosteus aculeatus*. *Aquatic Toxicology* **27**, 33–50.

Hontela, A. (1998). Interrenal dysfunction in fish from contaminated sites: In vivo and in vitro assessment. *Environmental Toxicology and Chemistry* **17**, 44–48.

Hooper, S. W., Pettigrew, C. A. and Sayler, G. S. (1990). Ecological fate, effects and prospects for the elimination of environmental polychlorinated biphenyls (PCBs). *Environmental Toxicology and Chemistry* **9**, 655–667.

HRTC (2002). *Hudson River Natural Resource Damage and Assessment Plan.* Albany, New York: Hudson River Trustee Council–New York State Department of Environmental Conservation, National Oceanic and Atmospheric Administration, U.S. Fish and Wildlife Service, 81 p.

Hutchinson, T. H., Field, M. D. R. and Manning, M. J. (2003). Evaluation of non-specific immune functions in dab, *Limanda limanda L.*, following short-term exposure to sediments contaminated with polyaromatic hydrocarbons and/or polychlorinated biphenyls. *Marine Environmental Research* **55**, 193–202.

Hutz, R. J., Carvan, M. J., III, Baldridge, M. G., Conley, L. K. and King Heiden, T. (2006). Environmental toxicants and effects on female reproductive function. *Trends in Reproductive Biology* **2**, 1–11.

Hylland, K., Beyer, J., Berntssen, M., Klungsøyr, J., Lang, T. and Balk, L. (2006). May organic pollutants affect fish populations in the North Sea? *Journal of Toxicology and Environmental Health, Part A* **69**, 125–138.

Iwanowicz, L. R., Blazer, V. S., McCormick, S. D., VanVeld, P. A. and Ottinger, C. A. (2009). Aroclor 1248 exposure leads to immunomodulation, decreased disease resistance and endocrine disruption in the brown bullhead, *Ameiurus nebulosus. Aquatic Toxicology* **93**, 70–82.

Iwanowicz, L. R., Lerner, D. T., Blazer, V. S. and McCormick, S. D. (2005). Aqueous exposure to Aroclor 1254 modulates the mitogenic response of Atlantic salmon anterior kidney T-cells: Indications of short- and long-term immunomodulation. *Aquatic Toxicology* **72**, 305–314.

Jacobson, K. C., Arkoosh, M. R., Kagley, A. N., Clemons, E. R., Collier, T. K. and Casillas, E. (2003). Cumulative effects of natural and anthropogenic stress on immune function and disease resistance in juvenile chinook salmon. *Journal of Aquatic Animal Health* **15**, 1–12.

Jensen, S. (1966). Reoprt of a new chemical hazard. *New Scientist* **15**, 612.

Jin, Y., Chen, R., Liu, W. and Fu, Z. (2010). Effect of endocrine disrupting chemicals on the transcription of genes related to the innate immune system in the early developmental stage of zebrafish (*Danio rerio*). *Fish and Shellfish Immunology* **28**, 854–861.

Jobling, S. and Tyler, C. R. (2003). Endocrine disruption in wild freshwater fish. *Pure and Applied Chemistry* **75**, 2219–2234.

Johnson, L., Sol, S., Ylitalo, G., Hom, T., French, B., Olson, O. P. and Collier, T. (1998a). Reproductive injury in English sole (*Pleuronectes vetulus*) from the Hylebos Waterway, Commencement Bay, Washington. *Journal of Aquatic Ecosystem Stress and Recovery* **6**, 289–310.

Johnson, L. L., Misitano, D., Sol, S. Y., Nelson, G. M., French, B., Ylitalo, G. M. and Hom, T. (1998b). Contaminant effects on ovarian development and spawning success in rock sole from Puget Sound, Washington. *Transactions of the American Fisheries Society* **127**, 375–392.

Johnson, L. L., Stehr, C. M., Olson, O. P., Myers, M. S., Pierce, S. M., Wigren, C. A., McCain, B. B. and Varanasi, U. (1993). Chemical contaminants and hepatic lesions in winter flounder (*Pleuronectes americanus*) from the northeast coast of the United States. *Environmental Science and Technology* **27**, 2759–2771.

Johnston, T. A., Miller, L. M., Whittle, D. M., Brown, S. B., Wiegand, M. D., Kapuscinski, A. R. and Leggetta, W. C. (2005). Effects of maternally transferred organochlorine contaminants on early life survival in a freshwater fish. *Environmental Toxicology and Chemistry* **24**, 2594–2602.

Jones, K. C. and de Voogt, P. (1999). Persistent organic pollutants (POPs): state of the science. *Environmental Pollution* **100**, 209–221.

Kasumyan, A. O. (2001). Effects of chemical pollutants on foraging behavior and sensitivity to fish food stimuli. *Journal of Ichthyology* **41**, 76–81.

Kelly, B. C., Ikonomou, M. G., Blair, J. D., Morin, A. E. and Gobas, F. A. (2007). Food web-specific biomagnification of persistent organic pollutants. *Science* **317**, 236–239.

Kendall, R. J., Anderson, T. A., Baker, R. J., Bens, C. M., Carr, J. A., Chiodo, L. A., Cobb, G. P., III, Dickerson, R. L., Dixon, K. R., Frame, L. T., Hooper, M. J., Martin, C. F., McMurry, S. T., Patino, R., Smith, E. E. and Theodorakis, C. W. (2001). Ecotoxicology. In *Casarett and Doull's Toxicology—The Basic Science of Poisons* (ed. C. D. Klaassen), 6th ed, pp. 1013–1045. McGraw-Hill.

Khan, I. A., Mathews, S., Okuzawa, K., Kagawa, H. and Thomas, P. (2001). Alterations in the GnRH–LH system in relation to gonadal stage and Aroclor 1254 exposure in Atlantic croaker. *Comparative Biochemistry and Physiology Part B: Biochemistry and Molecular Biology* **129**, 251–259.

Khan, I. A. and Thomas, P. (2000). Lead and Aroclor 1254 disrupt reproductive neuroendocrine function in Atlantic croaker. *Marine Environmental Research* **50**, 119–123.

Khan, I. A. and Thomas, P. (2001). Disruption of neuroendocrine control of luteinizing hormone secretion by Aroclor 1254 involves inhibition of hypothalamic tryptophan hydroxylase activity. *Biology of Reproduction* **64**, 955–964.

Khan, I. A. and Thomas, P. (2006). PCB congener-specific disruption of reproductive neuroendocrine function in Atlantic croaker. *Marine Environmental Research* **62** (Supplement 1), S25–S28.

Khan, R. A. (2011). Chronic exposure and decontamination of a marine sculpin (*Myoxocephalus scorpius*) to polychlorinated biphenyls using selected body indices, blood values, histopathology, and parasites as bioindicators. *Archives of Environmental Contamination and Toxicology* **60**, 479–485.

Kim, K. Y., Kim, D. S., Lee, S. K., Lee, I. K., Kang, J. H., Chang, Y. S., Jacobs, D. R., Steffes, M. and Lee, D. H. (2010). Association of low-dose exposure to persistent organic pollutants with global DNA hypomethylation in healthy Koreans. *Environmental Health Perspectives* **118**, 370–374.

Kime, D. E. (1995). The effects of pollution on reproduction in fish. *Reviews in Fish Biology and Fisheries* **5**, 52–95.

Kime, D. E. (1998). *Endocrine Disruption in Fish*. Boston, MA: Kluwer Academic Publishers, 396 p.

King Heiden, T., Carvan, M. J. and Hutz, R. J. (2006). Inhibition of follicular development, vitellogenesis, and serum 17β-estradiol concentrations in zebrafish following chronic, sublethal dietary exposure to 2,3,7,8-tetrachlorodibenzo-p-dioxin. *Toxicological Sciences* **90**, 490–499.

King Heiden, T., Hutz, R. J. and Carvan, M. J. (2005). Accumulation, tissue distribution, and maternal transfer of dietary 2,3,7,8,-tetrachlorodibenzo-p-dioxin: Impacts on reproductive success of zebrafish. *Toxicological Sciences* **87**, 497–507.

King Heiden, T. C., Mehta, V., Xiong, K. M., Lanham, K. A., Antkiewicz, D. S., Ganser, A., Heideman, W. and Peterson, R. E. (2012). Reproductive and developmental toxicity of dioxin in fish. *Molecular and Cellular Endocrinology* **354**, 121–138.

King Heiden, T. C., Spitsbergen, J., Heideman, W. and Peterson, R. E. (2009). Persistent adverse effects on health and reproduction caused by exposure of zebrafish to 2,3,7,8-tetrachlorodibenzo-p-dioxin during early development and gonad Differentiation. *Toxicological Sciences* **109**, 75–87.

King Heiden, T. C., Struble, C. A., Rise, M. L., Hessner, M. J., Hutz, R. J. and Carvan, M. J., III (2008). Molecular targets of 2,3,7,8-tetrachlorodibenzo-p-dioxin (TCDD) within the zebrafish ovary: Insights into TCDD-induced endocrine disruption and reproductive toxicity. *Reproductive Toxicology* **25**, 47–57.

Kinnberg, K. and Toft, G. (2003). Effects of estrogenic and antiandrogenic compounds on the testis structure of the adult guppy (*Poecilia reticulata*). *Ecotoxicology and Environmental Safety* **54**, 16–24.

Kinnison, M. T. and Hairston, N. G. (2007). Eco-evolutionary conservation biology: contemporary evolution and the dynamics of persistence. *Functional Ecology* **21**, 444–454.

Klasson-Wehler, E., Bergman, Å., Athanasiadou, M., Ludwig, J. P., Auman, H. J., Kannan, K., Berg, M. V. D., Murk, A. J., Feyk, L. A. and Giesy, J. P. (1998). Hydroxylated and methylsulfonyl polychlorinated biphenyl metabolites in albatrosses from Midway Atoll, North Pacific Ocean. *Environmental Toxicology and Chemistry* **17**, 1620–1625.

Klauda, R., Peck, T. and Rice, G. (1981). Accumulation of polychlorinated biphenyls in atlantic tomcod (*Microgadus tomcod*) collected from the Hudson River estuary, New York. *Bulletin of Environmental Contamination and Toxicology* **27**, 829–835.

Kleinow, K. M., Nichols, J. W., Hayton, W. L., McKim, J. M. and Barron, M. G. (2008). Toxicokinetics in fishes. In *The Toxicology of Fishes* (eds. R. T. DiGiulio and D. E. Hinton), pp. 55–152. Boca Raton, FL: CRC Press.

Koehler, A. (2004). The gender-specific risk to liver toxicity and cancer of flounder (*Platichthys flesus (L.)*) at the German Wadden Sea coast. *Aquatic Toxicology* **70**, 257–276.

Köllner, B., Wasserrab, B., Kotterba, G. and Fischer, U. (2002). Evaluation of immune functions of rainbow trout (*Oncorhynchus mykiss*)—how can environmental influences be detected? *Toxicology Letters* **131**, 83–95.

Kramer, V. J., Etterson, M. A., Hecker, M., Murphy, C. A., Roesijadi, G., Spade, D. J., Spromberg, J. A., Wang, M. and Ankley, G. T. (2011). Adverse outcome pathways and ecological risk assessment: Bridging to population-level effects. *Environmental Toxicology and Chemistry* **30**, 64–76.

Kraugerud, M., Doughty, R. W., Lyche, J. L., Berg, V., Tremoen, N. H., Alestrøm, P., Aleksandersen, M. and Ropstad, E. (2012). Natural mixtures of persistent organic pollutants (POPs) suppress ovarian follicle development, liver vitellogenin immunostaining and hepatocyte proliferation in female zebrafish (*Danio rerio*). *Aquatic Toxicology* **116–117**, 16–23.

Kristensen, T., Baatrup, E. and Bayley, M. (2006). p,p'-DDE fails to reduce the competitive reproductive fitness in Nigerian male guppies. *Ecotoxicology and Environmental Safety* **63**, 148–157.

Krøvel, A. V., Søfteland, L., Torstensen, B. E. and Olsvik, P. A. (2010). Endosulfan in vitro toxicity in Atlantic salmon hepatocytes obtained from fish fed either fish oil or vegetable oil. *Comparative Biochemistry and Physiology Part C: Toxicology and Pharmacology* **151**, 175–186.

Kuiper, R. V., Vethaak, A. D., Cantón, R. O. F., Anselmo, H., Dubbeldam, M., van den Brandhof, E.-J., Leonards, P. E. G., Wester, P. W. and Van den Berg, M. (2008). Toxicity of analytically cleaned pentabromodiphenylether after prolonged exposure in estuarine European flounder (*Platichthys flesus*), and partial life-cycle exposure in fresh water zebrafish (*Danio rerio*). *Chemosphere* **73**, 195–202.

Kuo, Y.-M., Sepúlveda, M., Hua, I., Ochoa-Acuña, H. and Sutton, T. (2010). Bioaccumulation and biomagnification of polybrominated diphenyl ethers in a food web of Lake Michigan. *Ecotoxicology* **19**, 623–634.

Lacroix, M. and Hontela, A. (2003). The organochlorine o,p'-DDD disrupts the adrenal steroidogenic signaling pathway in rainbow trout (*Oncorhynchus mykiss*). *Toxicology and Applied Pharmacology* **190**, 197–205.

Lee, D. H., Lind, L., Jacobs, D. R., Jr., Salihovic, S., van Bavel, B. and Lind, P. M. (2012). Associations of persistent organic pollutants with abdominal obesity in the elderly: The prospective investigation of the vasculature in Uppsala seniors (PIVUS) study. *Environment International* **40**, 170–178.

Lema, S. C., Dickey, J. T., Schultz, I. R. and Swanson, P. (2008). Dietary exposure to 2,2′,4,4′-tetrabromodiphenyl ether (PBDE-47) alters thyroid status and thyroid hormone–regulated gene transcription in the pituitary and brain. *Environmental Health Perspectives* **116**, 1694–1699.

Lema, S. C., Schultz, I. R., Scholz, N. L., Incardona, J. P. and Swanson, P. (2007). Neural defects and cardiac arrhythmia in fish larvae following embryonic exposure to 2,2′,4,4′-tetrabromodiphenyl ether (PBDE 47). *Aquatic Toxicology* **82**, 296–307.

Lerner, D. T., Björnsson, B. T. and McCormick, S. D. (2007). Effects of aqueous exposure to polychlorinated biphenyls (Aroclor 1254) on physiology and behavior of smolt development of Atlantic salmon. *Aquatic Toxicology* **81**, 329–336.

LeRoy, K. D., Thomas, P. and Khan, I. A. (2006). Thyroid hormone status of Atlantic croaker exposed to Aroclor 1254 and selected PCB congeners. *Comparative Biochemistry and Physiology Part C: Toxicology and Pharmacology* **144**, 263–271.

Letcher, R. J., Bustnes, J. O., Dietz, R., Jenssen, B. M., Jørgensen, E. H., Sonne, C., Verreault, J., Vijayan, M. M. and Gabrielsen, G. W. (2010). Exposure and effects assessment of persistent organohalogen contaminants in arctic wildlife and fish. *Science of the Total Environment* **408**, 2995–3043.

Letcher, R. J., Norstrom, R. J. and Muir, D. C. G. (1998). Biotransformation versus bioaccumulation: Sources of methyl sulfone PCB and 4,4′-DDE metabolites in the polar bear food chain. *Environmental Science and Technology* **32**, 1656–1661.

Levi, L., Ziv, T., Admon, A., Levavi-Sivan, B. and Lubzens, E. (2012). Insight into molecular pathways of retinal metabolism, associated with vitellogenesis in zebrafish. *American Journal of Physiology—Endocrinology and Metabolism* **302**, E626–E644.

Li, M. and Leatherland, J. F. (2012). The interaction between maternal stress and the ontogeny of the innate immune system during teleost embryogenesis: implications for aquaculture practice. *Journal of Fish Biology* **81**, 1793–1814.

Li, Y. F. and Macdonald, R. W. (2005). Sources and pathways of selected organochlorine pesticides to the Arctic and the effect of pathway divergence on HCH trends in biota: a review. *Science of the Total Environment* **342**, 87–106.

Linderoth, M., Hansson, T., Liewenborg, B., Sundberg, H., Noaksson, E., Hanson, M., Zebühr, Y. and Balk, L. (2006). Basic physiological biomarkers in adult female perch (Perca fluviatilis) in a chronically polluted gradient in the Stockholm recipient (Sweden). *Marine Pollution Bulletin* **53**, 437–450.

Loge, F. J., Arkoosh, M. R., Ginn, T. R., Johnson, L. L. and Collier, T. K. (2005). Impact of environmental stressors on the dynamics of disease transmission. *Environmental Science and Technology* **39**, 7329–7336.

Loizeau, V., Abarnou, A., Cugier, P., Jaouen-Madoulet, A., Le Guellec, A. M. and Menesguen, A. (2001). A model of PCB bioaccumulation in the sea bass food web from the Seine Estuary (Eastern English Channel). *Marine Pollution Bulletin* **43**, 242–255.

Lyche, J. L., Grześ, I. M., Karlsson, C., Nourizadeh-Lillabadi, R., Berg, V., Kristoffersen, A. B., Skåre, J. U., Alestrøm, P. and Ropstad, E. (2013). Parental exposure to natural mixtures of POPs reduced embryo production and altered gene transcription in zebrafish embryos. *Aquatic Toxicology* **126**, 424–434.

Lyche, J. L., Nourizadeh-Lillabadi, R., Karlsson, C., Stavik, B., Berg, V., Skåre, J. U., Alestrøm, P. and Ropstad, E. (2011). Natural mixtures of POPs affected body weight gain and induced transcription of genes involved in weight regulation and insulin signaling. *Aquatic Toxicology* **102**, 197–204.

Mac, M. (1986). *Mortality of lake trout swim-up fry from southwestern Lake Michigan: Documentation and hepatic structural analysis.* PhD Thesis *[PhD Thesis].* Laramie: University of Wyoming.

Macek, K. J. (1968). Growth and resistance to stress in brook trout fed sublethal levels of DDT. *Journal of the Fisheries Research Board of Canada* 25, 2443–2451.

Mackay, D. (1982). Correlation of bioconcentration factors. *Environmental Science and Technology* 16, 274–278.

Mackay, D. and Fraser, A. (2000). Bioaccumulation of persistent organic chemicals: mechanisms and models. *Environmental Pollution* 110, 375–391.

Magnadottir, B. (2010). Immunological control of fish diseases. *Marine Biotechnology* 12, 361–379.

Magnadóttir, B. (2006). Innate immunity of fish (overview). *Fish and Shellfish Immunology* 20, 137–151.

Manikkam, M., Guerrero-Bosagna, C., Tracey, R., Haque, M. M. and Skinner, M. K. (2012). Transgenerational actions of environmental compounds on reproductive Disease and identification of epigenetic biomarkers of ancestral exposures. *PLoS One* 7, e31901.

Mathur, D. S. (1962). Studies on the histopathological changes induced by DDT in the liver, kidney, and intestine of certain fishes. *Experientia* 18, 506–509.

Mathur, D. S. (1976). Histopathological changes in the liver of fishes resulting from exposure to dieldrin and lindane. In *Animal, Plant, and Microbial Toxins* (eds. A. Ohsaka, K. Hayashi, Y. Sawai, R. Murata, M. Funatsu, et al.), pp. 547–552. New York: Springer.

Matthiessen, P. and Logan, J. M. (1984). Low concentration effects of endosulfan insecticide on reproductive behaviour in the tropical cichlid fish Sarotherodon mossambicus. *Bulletin of Environmental Contamination and Toxicology* 33, 575–583.

Mauck, W. L., Mehrle, P. M. and Mayer, F. L. (1978). Effects of the polychlorinated biphenyl Aroclor® 1254 on growth, survival, and bone development in brook trout (*Salvelinus fontinalis*). *Journal of the Fisheries Research Board of Canada* 35, 1084–1088.

Maule, A. G., Jørgensen, E. H., Vijayan, M. M. and Killie, J.-E. A. (2005). Aroclor 1254 exposure reduces disease resistance and innate immune responses in fasted Arctic charr. *Environmental Toxicology and Chemistry* 24, 117–124.

Mayer, F. L., Mehrle, P. M. and Dwyer, W. P. (1975). Toxaphene effects on reproduction, growth, and mortality of brook trout. Ecological Research Series Report EPA 600/3-75-013. *Environmental Protection Agency* 42.

Mayon, N., Bertrand, A., Leroy, D., Malbrouck, C., Mandiki, S. N. M., Silvestre, F., Goffart, A., Thomé, J.-P. and Kestemont, P. (2006). Multiscale approach of fish responses to different types of environmental contaminations: A case study. *Science of the Total Environment* 367, 715–731.

McCarthy, I. D., Fuiman, L. A. and Alvarez, M. C. (2003). Aroclor 1254 affects growth and survival skills of Atlantic croaker *Micropogonias undulatus* larvae. *Marine Ecology Progress Series* 252, 295–301.

McClain, V., Stapleton, H. M., Tilton, F. and Gallagher, E. P. (2012). BDE 49 and developmental toxicity in zebrafish. *Comparative Biochemistry and Physiology Part C: Toxicology and Pharmacology* 155, 253–258.

McHugh, K. J., Smit, N. J., Van Vuren, J. H. J., Van Dyk, J. C., Bervoets, L., Covaci, A. and Wepener, V. (2011). A histology-based fish health assessment of the tigerfish. *Hydrocynus vittatus* from a DDT-affected area. *Physics and Chemistry of the Earth, Parts A/B/C* 36, 895–904.

McMaster, M. E., Evans, M. S., Alaee, M., Muir, D. C. G. and Hewitt, L. M. (2006). Northern rivers ecosystem initiative: Distribution and effects of contaminants. *Environmental Monitoring and Assessment* 113, 143–165.

McMillan, A., Bagley, M., Jackson, S. and Nacci, D. (2006). Genetic diversity and structure of an estuarine fish (*Fundulus heteroclitus*) indigenous to sites associated with a highly contaminated urban harbor. *Ecotoxicology* **15**, 539–548.

Meador, J. (2006). Rationale and procedures for using the tissue-residue approach for toxicity assessment and determination of tissue, water, and sediment quality guidelines for aquatic organisms. *Human and Ecological Risk Assessment: An International Journal* **12**, 1018–1073.

Meador, J. P., Collier, T. K. and Stein, J. E. (2002). Use of tissue and sediment-based threshold concentrations of polychlorinated biphenyls (PCBs) to protect juvenile salmonids listed under the U.S. Endangered Species Act. *Aquatic Conservation: Marine and Freshwater Ecosystems* **12**, 493–516.

Meador, J. P., Sommers, F. C., Cooper, K. A. and Yanagida, G. (2011). Tributyltin and the obesogen metabolic syndrome in a salmonid. *Environmental Research* **111**, 50–56.

Metcalfe, C. D. (1989). Tests for predicting carcinogenicity in fish. *CRC Reviews: Aquatic Sciences* **1**, 111–129.

Metcalfe, T. L., Metcalfe, C. D., Kiparissis, Y., Niimi, A. J., Foran, C. M. and Benson, W. H. (2000). Gonadal development and endocrine responses in Japanese medaka (*Oryzias latipes*) exposed to o,p′-DDT in water or through maternal transfer. *Environmental Toxicology and Chemistry* **19**, 1893–1900.

Meyer, J. N. and Di Giulio, R. T. (2003). Heritable adaptation and fitness costs in killifish (*Fundulus heteroclitus*) inhabiting a polluted estuary. *Ecological Applications* **13**, 490–503.

Mhadhbi, L., Fumega, J., Boumaiza, M. and Beiras, R. (2012). Acute toxicity of polybrominated diphenyl ethers (PBDEs) for turbot (*Psetta maxima*) early life stages (ELS). *Environmental Science and Pollution Research* **19**, 708–717.

Miller, M. A. (1993). Maternal transfer of organochlorine compounds in salmonines to their eggs. *Canadian Journal of Fisheries and Aquatic Sciences* **50**, 1405–1413.

Mills, L. J., Gutjahr-Gobell, R. E., Haebler, R. A., Borsay Horowitz, D. J., Jayaraman, S., Pruell, R. J., McKinney, R. A., Gardner, G. R. and Zaroogian, G. E. (2001). Effects of estrogenic (o,p′-DDT; octylphenol) and anti-androgenic (p,p′-DDE) chemicals on indicators of endocrine status in juvenile male summer flounder (*Paralichthys dentatus*). *Aquatic Toxicology* **52**, 157–176.

Milston, R. H., Fitzpatrick, M. S., Vella, A. T., Clements, S., Gundersen, D., Feist, G., Crippen, T. L., Leong, J. and Schreck, C. B. (2003). Short-term exposure of chinook salmon (*Oncoryhnchus tshawytscha*) to o,p′-DDE or DMSO during early life-history stages causes long-term humoral immunosuppression. *Environmental Health Perspectives* **111**, 1601–1607.

Mirbahai, L., Williams, T. D., Zhan, H., Gong, Z. and Chipman, J. K. (2011a). Comprehensive profiling of zebrafish hepatic proximal promoter CpG island methylation and its modification during chemical carcinogenesis. *BMC Genomics* **12**, 3.

Mirbahai, L., Yin, G., Bignell, J. P., Li, N., Williams, T. D. and Chipman, J. K. (2011b). DNA methylation in liver tumorigenesis in fish from the environment. *Epigenetics* **6**, 1319–1333.

Misumi, I., Vella, A. T., Leong, J.-A. C., Nakanishi, T. and Schreck, C. B. (2005). p,p′-DDE depresses the immune competence of chinook salmon (*Oncorhynchus tshawytscha*) leukocytes. *Fish and Shellfish Immunology* **19**, 97–114.

Mitchell, M. M., Woods, R., Chi, L.-H., Schmidt, R. J., Pessah, I. N., Kostyniak, P. J. and LaSalle, J. M. (2012). Levels of select PCB and PBDE congeners in human postmortem brain reveal possible environmental involvement in 15q11-q13 duplication autism spectrum disorder. *Environmental and Molecular Mutagenesis* **53**, 589–598.

Moe, S. J., De Schamphelaere, K., Clements, W. H., Sorensen, M. T., Van den Brink, P. J. and Liess, M. (2013). Combined and interactive effects of global climate change and toxicants on populations and communities. *Environmental Toxicology and Chemistry* **32**, 49–61.

Monosson, E. (2000). Reproductive and developmental effects of PCBs in fish: a synthesis of laboratory and field studies. *Reviews in Toxicology* **3**, 25–75.

Monteverdi, G. H. and Di Giulio, R. T. (2000). Vitellogenin association and oocytic accumulation of thyroxine and 3,5,3'-triiodothyronine in gravid Fundulus heteroclitus. *General and Comparative Endocrinology* **120**, 198–211.

Moore, M. (1991). *Vacuolation, proliferation, and neoplasia in the liver of Boston Harbor winter flounder (*Pseudopleuronectes americanus*). Technical Report 91–28 [PhD Thesis].* Woods Hole, MA: Woods Hole Oceanographic Institute, 268 p.

Moran, P. and Perez-Figueroa, A. (2011). Methylation changes associated with early maturation stages in the Atlantic salmon. *BMC Genetics* **12**, 86.

Morrison, H. A., Gobas, F. A. P. C., Lazar, R., Whittle, D. M. and Haffner, G. D. (1997). Development and verification of a benthic/pelagic food web bioaccumulation model for PCB congeners in western Lake Erie. *Environmental Science and Technology* **31**, 3267–3273.

Mortensen, A. S. and Arukwe, A. (2006). The persistent DDT metabolite, 1,1-dichloro-2,2-bis (p-chlorophenyl)ethylene, alters thyroid hormone-dependent genes, hepatic cytochrome P4503A, and pregnane × receptor gene expressions in atlantic salmon (*Salmo salar*) parr. *Environmental Toxicology and Chemistry* **25**, 1607–1615.

Muir, D. C. G., Norstrom, R. J. and Simon, M. (1988). Organochlorine contaminants in arctic marine food chains: accumulation of specific polychlorinated biphenyls and chlordane-related compounds. *Environmental Science and Technology* **22**, 1071–1079.

Muirhead, E. K., Skillman, A. D., Hook, S. E. and Schultz, I. R. (2006). Oral exposure of PBDE-47 in gish: Toxicokinetics and reproductive effects in Japanese medaka (*Oryzias latipes*) and Fathead Minnows (*Pimephales promelas*). *Environmental Science and Technology* **40**, 523–528.

Mukhi, S. and Patiño, R. (2007). Effects of prolonged exposure to perchlorate on thyroid and reproductive function in zebrafish. *Toxicological Sciences* **96**, 246–254.

Mukhi, S., Torres, L. and Patiño, R. (2007). Effects of larval–juvenile treatment with perchlorate and co-treatment with thyroxine on zebrafish sex ratios. *General and Comparative Endocrinology* **150**, 486–494.

Munn, M. D. and Gruber, S. J. (1997). The relationship between land use and organochlorine compounds in streambed sediment and fish in the Central Columbia Plateau, Washington and Idaho, USA. *Environmental Toxicology and Chemistry* **16**, 1877–1887.

Murray, D. S. (2012). The role of physical structure and micronutrient provisioning in determining egg quality and performance in fish. PhD thesis. http://theses.gla.ac.uk/3563/.

Myers, M., Stehr, C., Olson, O., Johnson, L., McCain, B., Chan, S.-L. and Varanasi, U. (1994). Relationships between toxicopathic hepatic lesions and exposure to chemical contaminants in English sole (*Pleuronectes vetulus*), starry flounder (*Platichthys stellatus*), and white croaker (*Genyonemus lineatus*) from selected marine sites on the Pacific Coast, USA. *Environmental Health Perspectives* **102**, 200–215.

Myers, M. S. and Fournie, J. W. (2002). Histopathological biomarkers as integrators of anthropogenic and environmental stressors. In *Biological Indicators of Aquatic Ecosystem Stress* (ed. S. M. Adams), pp. 221–287. Bethesda, MD: American Fisheries Society.

Myers, M. S., Johnson, L. L. and Collier, T. K. (2003). Establishing the causal relationship between polycyclic aromatic hydrocarbon (PAH) exposure and hepatic neoplasms and neoplasia-related liver lesions in English sole (*Pleuronectes vetulus*). *Human and Ecological Risk Assessment: An International Journal* **9**, 67–94.

Myers, M. S., Johnson, L. L., Hom, T., Collier, T. K., Stein, J. E. and Varanasi, U. (1998a). Toxicopathic hepatic lesions in subadult English sole (*Pleuronectes vetuls*) from Puget

Sound, Washington, DC: Relationships with other biomarkers of contaminant exposure. *Marine Environmental Research* **45**, 47–67.

Myers, M. S., Johnson, L. L., Olson, O. P., Stehr, C. M., Horness, B. H., Collier, T. K. and McCain, B. B. (1998b). Toxicopathic hepatic lesions as biomarkers of chemical contaminant exposure and effects in marine bottomfish species from the Northeast and Pacific Coasts, USA. *Marine Pollution Bulletin* **37**, 92–113.

Myers, M. S., Rhodes, L. D. and McCain, B. B. (1987). Pathologic anatomy and patterns of occurrence of hepatic neoplasms, putative preneoplastic lesions, and other idiopathic hepatic conditions in English sole (*Parophrys vetulus*) from Puget Sound, Washington. *Journal of the National Cancer Institute* **78**, 333–363.

Nacci, D., Champlin, D. and Jayaraman, S. (2010). Adaptation of the estuarine fish *Fundulus heteroclitus* (Atlantic killifish) to polychlorinated biphenyls (PCBs). *Estuaries and Coasts* **33**, 853–864.

Nacci, D., Coiro, L., Champlin, D., Jayaraman, S., McKinney, R., Gleason, T. R., Munns, W. R., Jr, Specker, J. L. and Cooper, K. R. (1999). Adaptations of wild populations of the estuarine fish Fundulus heteroclitus to persistent environmental contaminants. *Marine Biology* **134**, 9–17.

Nakari, T. and Pessala, P. (2005). In vitro estrogenicity of polybrominated flame retardants. *Aquatic Toxicology* **74**, 272–279.

Nakayama, K., Sei, N., Handoh, I. C., Shimasaki, Y., Honjo, T. and Oshima, Y. (2011). Effects of polychlorinated biphenyls on liver function and sexual characteristics in Japanese medaka (*Oryzias latipes*). *Marine Pollution Bulletin* **63**, 366–369.

Nault, R., Al-Hameedi, S. and Moon, T. W. (2012). Effects of polychlorinated biphenyls on whole animal energy mobilization and hepatic cellular respiration in rainbow trout, Oncorhynchus mykiss. *Chemosphere* **87**, 1057–1062.

Nestel, H. and Budd, J. (1975). Chronic oral exposure of rainbow trout (*Salmo gairdneri*) to a polychlorinated biphenyl (Aroclor 1254): pathological effects. *Canadian Journal of Comparative Medicine* **39**, 208–215.

Newman, M. C. and Clements, W. H. (2008). *Ecotoxicology: A Comprehensive Treatment*. Boca Raton, FL: CRC Press, 880 p.

Niimi, A. J. (1983). Biological and toxicologtcal effects of environmental contaminants in fish and their eggs. *Canadian Journal of Fisheries and Aquatic Sciences* **40**, 306–312.

Njiwa, J. R. K., Müller, P. and Klein, R. (2004). Binary mixture of DDT and Arochlor 1254: Effects on sperm release by *Danio rerio*. *Ecotoxicology and Environmental Safety* **58**, 211–219.

Norman Haldén, A., Arnoldsson, K., Haglund, P., Mattsson, A., Ullerås, E., Sturve, J. and Norrgren, L. (2011). Retention and maternal transfer of brominated dioxins in zebrafish (*Danio rerio*) and effects on reproduction, aryl hydrocarbon receptor-regulated genes, and ethoxyresorufin-O-deethylase (EROD) activity. *Aquatic Toxicology* **102**, 150–161.

Novák, J., Benišek, M. and Hilscherová, K. (2008). Disruption of retinoid transport, metabolism and signaling by environmental pollutants. *Environment International* **34**, 898–913.

Nyholm, J. R., Norman, A., Norrgren, L., Haglund, P. and Andersson, P. L. (2008). Maternal transfer of brominated flame retardants in zebrafish (*Danio rerio*). *Chemosphere* **73**, 203–208.

Olsen, A. S., Sarras, M. P., Leontovich, A. and Intine, R. V. (2012). Heritable transmission of diabetic metabolic memory in zebrafish correlates with DNA hypomethylation and aberrant gene expression. *Diabetes* **61**, 485–491.

Örn, S., Andersson, P. L., Förlin, L., Tysklind, M. and Norrgren, L. (1998). The impact on reproduction of an orally administered mixture of selected PCBs in zebrafish (*Danio rerio*). *Archives of Environmental Contamination and Toxicology* **35**, 52–57.

Ostrach, D. J., Low-Marchelli, J. M., Eder, K. J., Whiteman, S. J. and Zinkl, J. G. (2008). Maternal transfer of xenobiotics and effects on larval striped bass in the San Francisco Estuary. *Proceedings of the National Academy of Sciences* **105**, 19354–19359.

Pait, A. S. and Nelson, J. O. (2002). *Endocrine disruption in fish: An assessment of recent research and results. NOAA Tech. Memo. NOS NCCOS CCMA 149.* Silver Spring, MD: NOAA, NOS, Center for Coastal Monitoring and Assessment, 55 p.

Palstra, A. P., Ginneken, V. J. T., Murk, A. J. and Thillart, G. E. E. J. M. (2006). Are dioxin-like contaminants responsible for the eel (*Anguilla anguilla*) drama? *Naturwissenschaften* **93**, 145–148.

Papoulias, D. M., Villalobos, S. A., Meadows, J., Noltie, D. B., Giesy, J. P. and Tillitt, D. E. (2003). In ovo exposure to o,p'-DDE affects sexual development but not sexual differentiation in Japanese medaka (*Oryzias latipes*). *Environmental Health Perspectives* **111**, 29–32.

Peter, V. S. and Peter, M. C. S. (2011). The interruption of thyroid and interrenal and the inter-hormonal interference in fish: Does it promote physiologic adaptation or maladaptation? *General and Comparative Endocrinology* **174**, 249–258.

Powell, D. B., Palm, R. C., Skillman, A. and Godtfredsen, K. (2003). Immunocompetence of juvenile chinook salmon against Listonella anguillarum following dietary exposure to Aroclor® 1254. *Environmental Toxicology and Chemistry* **22**, 285–295.

Power, D. M., Llewellyn, L., Faustino, M., Nowell, M. A., Björnsson, B. T., Einarsdottir, I. E., Canario, A. V. M. and Sweeney, G. E. (2001). Thyroid hormones in growth and development of fish. *Comparative Biochemistry and Physiology Part C: Toxicology and Pharmacology* (130), 447–459.

Quabius, E. S., Krupp, G. and Secombes, C. J. (2005). Polychlorinated biphenyl 126 affects expression of genes involved in stress-immune interaction in primary cultures of rainbow trout anterior kidney cells. *Environmental Toxicology and Chemistry* **24**, 3053–3060.

Raimondo, S., McKenney, C. L. and Barron, M. G. (2006). Application of perturbation simulations in population risk assessment for different life history strategies and elasticity patterns. *Human and Ecological Risk Assessment: An International Journal* **12**, 983–999.

Randak, T., Zlabek, V., Pulkrabova, J., Kolarova, J., Kroupova, H., Siroka, Z., Velisek, J., Svobodova, Z. and Hajslova, J. (2009). Effects of pollution on chub in the River Elbe, Czech Republic. *Ecotoxicology and Environmental Safety* **72**, 737–746.

Rankouhi, T. R., Sanderson, J. T., van Holsteijn, I., van Leeuwen, C., Vethaak, A. D. and van den Berg, M. (2004). Effects of natural and synthetic estrogens and various environmental contaminants on vitellogenesis in fish primary hepatocytes: Comparison of bream (*Abramis brama*) and carp (*Cyprinus carpio*). *Toxicological Sciences* **81**, 90–102.

Regala, R. P., Rice, C. D., Schwedler, T. E. and Dorociak, I. R. (2001). The effects of tributyltin (TBT) and 3,3',4,4',5-pentachlorobiphenyl (PCB-126) mixtures on antibody responses and phagocyte oxidative burst activity in channel catfish, Ictalurus punctatus. *Archives of Environmental Contamination and Toxicology* **40**, 386–391.

Renshaw, S. A. and Trede, N. S. (2012). A model 450 million years in the making: Zebrafish and vertebrate immunity. *Disease Models and Mechanisms* **5**, 38–47.

Rhodes, L., Casillas, E., McKnight, B., Gronlund, W., Myers, M., Olson, O. P. and McCain, B. (1985). Interactive effects of cadmium, polychlorinated biphenyls, and fuel oil on experimentally exposed English sole (*Parophrys vetulus*). *Canadian Journal of Fisheries and Aquatic Sciences* **42**, 1870–1880.

Rice, C. A., Myers, M. S., Willis, M. L., French, B. L. and Casillas, E. (2000). From sediment bioassay to fish biomarker—connecting the dots using simple trophic relationships. *Marine Environmental Research* **50**, 527–533.

Rice, C. D. (2001). Fish immunotoxicology: Understanding the mechanisms of action. In *Target organ toxicity in marine and freshwater teleosts* (eds. D. Schlenk and W. H. Benson), pp. 96–138. London: Taylor and Francis Group.

Rice, C. D. and Arkoosh, M. R. (2002). Immunological indicators of environmental stress and disease susceptibility in fishes. In *Biological Indicators of Stress in Aquatic Ecosystem Stress* (ed. S. M. Adams), pp. 96–138. Bethesda, MD: American Fisheries Society Publication.

Rigaud, C., Couillard, C. M., Pellerin, J., Légaré, B., Gonzalez, P. and Hodson, P. V. (2013). Relative potency of PCB126 to TCDD for sublethal embryotoxicity in the mummichog (*Fundulus heteroclitus*). *Aquatic Toxicology* **128–129**, 203–214.

Rigét, F., Bignert, A., Braune, B., Stow, J. and Wilson, S. (2010). Temporal trends of legacy POPs in Arctic biota, an update. *Science of the Total Environment* **408**, 2874–2884.

Rios, L. M., Moore, C. and Jones, P. R. (2007). Persistent organic pollutants carried by synthetic polymers in the ocean environment. *Marine Pollution Bulletin* **54**, 1230–1237.

Roff, D. A. (1984). The evolution of life history parameters in teleosts. *Canadian Journal of Fisheries and Aquatic Sciences* **41**, 989–1000.

Rogan, W. J. and Ragan, N. B. (2007). Some evidence of effects of environmental chemicals on the endocrine system in children. *International Journal of Hygiene and Environmental Health* **210**, 659–667.

Rolland, R. (2000). A review of chemically-induced alterations in thyroid and vitamin A status from field studies of wildlife and fish. *Journal of Wildlife Diseases* **36**, 615–635.

Rollins-Smith, L. A., Rice, C. D. and Grasman, K. A. (2007). Amphibian, fish and bird immunotoxicology. In *Target Organ Toxicology Series: Immunotoxicology and Immuno-pharmacology* (eds. R. Luebke, R. House and I. Kimber), pp. 385–402. Boca Raton, FL: CRC Press.

Roy, J. R., Chakraborty, S. and Chakraborty, T. R. (2009). Estrogen-like endocrine disrupting chemicals affecting puberty in humans—a review. *Medical Science Monitor* **15**, RA137–145.

Rusiecki, J. A., Baccarelli, A., Bollati, V., Tarantini, L., Moore, L. E. and Bonefeld-Jorgensen, E. C. (2008). Global DNA hypomethylation is associated with high serum-persistent organic pollutants in Greenlandic Inuit. *Environmental Health Perspectives* **116**, 1547–1552.

Rypel, A. L. and Bayne, D. R. (2010). Do fish growth rates correlate with PCB body burdens? *Environmental Pollution* **158**, 2533–2536.

Sadikovic, B., Yoshimoto, M., Al-Romaih, K., Maire, G., Zielenska, M. and Squire, J. A. (2008). In vitro analysis of integrated global high-resolution DNA methylation profiling with genomic imbalance and gene expression in osteosarcoma. *PLoS One* **3**, e2834.

Safe, S. (1990). Polychlorinated biphenyls (PCBs), dibenzo-p-dioxins (PCDDs), dibenzofurans (PCDFs), and related compounds: Environmental and mechanistic considerations which support the development of Toxic Equivalency Factors (TEFs). *Critical Reviews in Toxicology* **21**, 51–88.

Safe, S. H. (1994). Polychlorinated biphenyls (PCBs): Environmental impact, biochemical and toxic responses, and implications for risk assessment. *Critical Reviews in Toxicology* **24**, 87–149.

Sappington, K. G., Bridges, T. S., Bradbury, S. P., Erickson, R. J., Hendriks, A. J., Lanno, R. P., Meador, J. P., Mount, D. R., Salazar, M. H. and Spry, D. J. (2011). Application of the tissue residue approach in ecological risk assessment. *Integrated Environmental Assessment and Management* **7**, 116–140.

Schanen, N. C. (2006). Epigenetics of autism spectrum disorders. *Human Molecular Genetics* **15**, R138–R150.

Schell, L. M. and Gallo, M. V. (2010). Relationships of putative endocrine disruptors to human sexual maturation and thyroid activity in youth. *Physiology and Behavior* **99**, 246–253.

Schmidt, C. (1999). Spheres of influence: No POPs. *Environmental Health Perspectives* **107**, A24–25.

Schnitzler, J. G., Celis, N., Klaren, P. H. M., Blust, R., Dirtu, A. C., Covaci, A. and Das, K. (2011). Thyroid dysfunction in sea bass (*Dicentrarchus labrax*): Underlying mechanisms and effects of polychlorinated biphenyls on thyroid hormone physiology and metabolism. *Aquatic Toxicology* **105**, 438–447.

Schnitzler, J. G., Klaren, P. H. M., Bouquegneau, J.-M. and Das, K. (2012). Environmental factors affecting thyroid function of wild sea bass (*Dicentrarchus labrax*) from European coasts. *Chemosphere* **87**, 1009–1017.

Schröder, J. and Peters, K. (1988a). Differential courtship activity of competing guppy males (*Poecilia reticulata peters; pisces: Poeciliidae*) as an indicator for low concentrations of aquatic pollutants. *Bulletin of Environmental Contamination and Toxicology* **40**, 396–404.

Schröder, J. and Peters, K. (1988b). Differential courtship activity and alterations of reproductive success of competing guppy males (*Poecilia reticulata Peters; Pisces: Poeciliidae*) as an indicator for low concentrations of aquatic pollutants. *Bulletin of Environmental Contamination and Toxicology* **41**, 385–390.

Scott, G. R. and Sloman, K. A. (2004). The effects of environmental pollutants on complex fish behaviour: integrating behavioural and physiological indicators of toxicity. *Aquatic Toxicology* **68**, 369–392.

Segner, H., Moller, A. M., Wenger, M. and Casanova-Nakayama, A. (2012a). Fish immunotoxicology: Research at the crossroads of immunology, ecology and toxicology. In *Interdisciplinary Studies on Environmental Chemistry—Environmental Pollution and Ecotoxicology* (eds. M. Kawaguchi, K. Misaki, H. Sato, T. Yokokawa, T. Itai, et al.), pp. 1–12. Tokyo: TERRAPUB.

Segner, H., Wenger, M., Möller, A., Köllner, B. and Casanova-Nakayama, A. (2012b). Immunotoxic effects of environmental toxicants in fish—how to assess them? *Environmental Science and Pollution Research* **19**, 2465–2476.

Serrano, R., Blanes, M. A. and López, F. J. (2008). Maternal transfer of organochlorine compounds to oocytes in wild and farmed gilthead sea bream (*Sparus aurata*). *Chemosphere* **70**, 561–566.

Sharma, P. and Patiño, R. (2013). Regulation of gonadal sex ratios and pubertal development by the thyroid endocrine system in zebrafish (*Danio rerio*). *General and Comparative Endocrinology* **184**, 111–119.

Shaw, S. D. and Kannan, K. (2009). Polybrominated diphenyl ethers in marine ecosystems of the American continents: Foresight from current knowledge. *Reviews on Environmental Health* 157.

Shutoh, Y., Takeda, M., Ohtsuka, R., Haishima, A., Yamaguchi, S., Fujie, H., Komatsu, Y., Maita, K. and Harada, T. (2009). Low dose effects of dichlorodiphenyltrichloroethane (DDT) on gene transcription and DNA methylation in the hypothalamus of young male rats: Implication of hormesis-like effects. *The Journal of Toxicological Sciences* **34**, 469–482.

Sinkkonen, S. and Paasivirta, J. (2000). Degradation half-life times of PCDDs, PCDFs and PCBs for environmental fate modeling. *Chemosphere* **40**, 943–949.

Skinner, M., Manikkam, M., Haque, M., Zhang, B. and Savenkova, M. (2012). Epigenetic transgenerational inheritance of somatic transcriptomes and epigenetic control regions. *Genome Biology* **13**, R91.

Skinner, M. K., Manikkam, M. and Guerrero-Bosagna, C. (2010). Epigenetic transgenerational actions of environmental factors in disease etiology. *Trends in Endocrinology and Metabolism: TEM* **21**, 214–222.

Skinner, M. K., Manikkam, M. and Guerrero-Bosagna, C. (2011). Epigenetic transgenerational actions of endocrine disruptors. *Reproductive Toxicology* **31**, 337–343.

Smeets, J. M. W., Rankouhi, T. R., Nichols, K. M., Komen, H., Kaminski, N. E., Giesy, J. P. and van den Berg, M. (1999). In vitro vitellogenin production by carp (*Cyprinus carpio*) hepatocytes as a screening method for determining (anti)estrogenic activity of xenobiotics. *Toxicology and Applied Pharmacology* **157**, 68–76.

Soffientino, B., Nacci, D. E. and Specker, J. L. (2010). Effects of the dioxin-like PCB 126 on larval summer flounder (*Paralichthys dentatus*). *Comparative Biochemistry and Physiology Part C: Toxicology and Pharmacology* **152**, 9–17.

Søfteland, L., Petersen, K., Stavrum, A.-K., Wu, T. and Olsvik, P. A. (2011). Hepatic in vitro toxicity assessment of PBDE congeners BDE47, BDE153 and BDE154 in Atlantic salmon (*Salmo salar L.*). *Aquatic Toxicology* **105**, 246–263.

Sol, S., Johnson, L., Boyd, D., Olson, O. P., Lomax, D. and Collier, T. (2008). Relationships between anthropogenic chemical contaminant exposure and associated changes in reproductive parameters in male English sole (*Parophrys vetulus*) collected from Hylebos Waterway, Puget Sound, Washington. *Archives of Environmental Contamination and Toxicology* **55**, 627–638.

Song, Y., Wu, N., Tao, H., Tan, Y., Gao, M., Han, J., Shen, H., Liu, K. and Lou, J. (2012). Thyroid endocrine dysregulation and erythrocyte DNA damage associated with PBDE exposure in juvenile crucian carp collected from an e-waste dismantling site in Zhejiang Province, China. *Environmental Toxicology and Chemistry* **31**, 2047–2051.

Sopinka, N., Marentette, J. and Balshine, S. (2010). Impact of contaminant exposure on resource contests in an invasive fish. *Behavioral Ecology and Sociobiology* **64**, 1947–1958.

Sopinka, N. M., Fitzpatrick, J. L., Taves, J. E., Ikonomou, M. G., Marsh-Rollo, S. E. and Balshine, S. (2012). Does proximity to aquatic pollution affect reproductive traits in a wild-caught intertidal fish? *Journal of Fish Biology* **80**, 2374–2383.

Spitsbergen, J. M., Kleeman, J. M. and Peterson, R. E. (1988a). 2,3,7,8-Tetrachlorodibenzo-p-dioxin toxicity in yellow perch (*Perca flavescens*). *Journal of Toxicology and Environmental Health* **23**, 359–383.

Spitsbergen, J. M., Kleeman, J. M. and Peterson, R. E. (1988b). Morphologic lesions and acute toxicity in rainbow trout (*Salmo Gairdneri*) treated with 2,3,7,8-tetrachlorodibenzo-p-dioxin. *Journal of Toxicology and Environmental Health* **23**, 333–358.

Spromberg, J. A. and Birge, W. J. (2005). Modeling the effects of chronic toxicity on fish populations: The influence of life-history strategies. *Environmental Toxicology and Chemistry* **24** (1), 532–1540.

Spromberg, J. A. and Meador, J. P. (2005). Relating results of chronic toxicity responses to population-level effects: Modeling effects on wild chinook salmon populations.. *Integrated Environmental Assessment and Management* **1**, 9–21.

Srogi, K. (2008). Levels and congener distributions of PCDDs, PCDFs and dioxin-like PCBs in environmental and human samples: A review. *Environmental Chemistry Letters* **6**, 1–28.

Stanley, K. A., Curtis, L. R., Massey Simonich, S. L. and Tanguay, R. L. (2009). Endosulfan I and endosulfan sulfate disrupts zebrafish embryonic development. *Aquatic Toxicology* **95**, 355–361.

Stark, J. D., Banks, J. E. and Vargas, R. (2004). How risky is risk assessment: The role that life history strategies play in susceptibility of species to stress. *Proceedings of the National Acadamy of Science U S A* **101**, 732–736.

Stehr, C. M., Johnson, L. L. and Myers, M. S. (1998). Hydropic vacuolation in the liver of three species of fish from the U.S. West Coast: Lesion description and risk assessment associated with contaminant exposure. *Diseases of Aquatic Organisms* **32**, 119–135.

Stouthart, X. J. H. X., Huijbregts, M. A. J., Balm, P. H. M., Lock, R. A. C. and Wendelaar Bonga, S. E. (1998). Endocrine stress response and abnormal development in carp (*Cyprinus carpio*) larvae after exposure of the embryos to PCB 126. *Fish Physiology and Biochemistry* **18**, 321–329.

Strömqvist, M., Tooke, N. and Brunström, B. (2010). DNA methylation levels in the 5′ flanking region of the vitellogenin I gene in liver and brain of adult zebrafish (*Danio rerio*)—Sex and tissue differences and effects of 17α-ethinylestradiol exposure. *Aquatic Toxicology* **98**, 275–281.

Studnicka, B. M., Siwicki, A. K., Morand, M., Rymuszka, A., Bownik, A. and Terech-Majewska, E. (2000). Modulation of nonspecific defence mechanisms and specific immune responses after suppression induced by xenobiotics. *Journal of Applied Ichthyology* **16**, 1–7.

Swain, P. and Nayak, S. K. (2009). Role of maternally derived immunity in fish. *Fish and Shellfish Immunology* **27**, 89–99.

Tellez-Bañuelos, M. C., Santerre, A., Casas-Solis, J. and Zaitseva, G. (2010). Endosulfan increases seric interleukin-2 like (IL-2L) factor and immunoglobulin M (IgM) of Nile tilapia (*Oreochromis niloticus*) challenged with Aeromona hydrophila. *Fish and Shellfish Immunology* **28**, 401–405.

Teuten, E. L., Saquing, J. M., Knappe, D. R. U., Barlaz, M. A., Jonsson, S., Björn, A., Rowland, S. J., Thompson, R. C., Galloway, T. S., Yamashita, R., Ochi, D., Watanuki, Y., Moore, C., Viet, P. H., Tana, T. S., Prudente, M., Boonyatumanond, R., Zakaria, M. P., Akkhavong, K., Ogata, Y., Hirai, H., Iwasa, S., Mizukawa, K., Hagino, Y., Imamura, A., Saha, M. and Takada, H. (2009). Transport and release of chemicals from plastics to the environment and to wildlife. *Philosophical Transactions of the Royal Society B: Biological Sciences* **364**, 2027–2045.

Thibaut, R. and Porte, C. (2004). Effects of endocrine disrupters on sex steroid synthesis and metabolism pathways in fish. *The Journal of Steroid Biochemistry and Molecular Biology* **92**, 485–494.

Thomas, P. (2008). The endocrine system. In *The Toxicology of Fishes* (eds. R. T. Di Giulio and D. E. Hinton), p. 1071. Boca Raton, FL: CRC Press.

Tierney, K. B., Baldwin, D. H., Hara, T. J., Ross, P. S., Scholz, N. L. and Kennedy, C. J. (2010). Olfactory toxicity in fishes. *Aquatic Toxicology* **96**, 2–26.

Tillitt, D., Papoulias, D. and Buckler, D. (2003). *Fish reproductive health assessment in PCB contaminated regions of the Housatonic River, Massachusetts, USA: Investigations of causal linkages between PCBs and fish health. Final Report of Phase II Studies, Interagency Agreement 1448-50181-99-H-008.* Columbia, MO: U.S. Geological Survey Columbia Environmental Research Center, 219 p.

Tillitt, D. E., Cook, P. M., Giesy, J. P., Heideman, W. and Peterson, R. E. (2008). Reproductive impairment of Great Lakes Lake Trout by dioxin-like chemicals. In *The Toxicology of Fishes* (eds. R. T. DiGiulio and D. E. Hinton), pp. 819–876. Boca Raton, FL: CRC Press.

Timme-Laragy, A. R., Levin, E. D. and Di Giulio, R. T. (2006). Developmental and behavioral effects of embryonic exposure to the polybrominated diphenylether mixture DE-71 in the killifish (*Fundulus heteroclitus*). *Chemosphere* **62**, 1097–1104.

Tingaud-Sequeira, A., Ouadah, N. and Babin, P. J. (2011). Zebrafish obesogenic test: A tool for screening molecules that target adiposity. *Journal of Lipid Research* **52**, 1765–1772.

Tomy, G. T., Palace, V. P., Halldorson, T., Braekevelt, E., Danell, R., Wautier, K., Evans, B., Brinkworth, L. and Fisk, A. T. (2004). Bioaccumulation, biotransformation, and biochemical effects of brominated diphenyl ethers in juvenile Lake Trout (*Salvelinus namaycush*). *Environmental Science and Technology* **38**, 1496–1504.

Torres, L., Orazio, C. E., Peterman, P. H. and Patiño, R. (2013). Effects of dietary exposure to brominated flame retardant BDE-47 on thyroid condition, gonadal development and growth of zebrafish. *Fish Physiology and Biochemistry*. http://dx.doi.org/10.1007/s10695-012-9768-0

UNEP (2001). Final Act of the Conference of Plenipotentiaries on the Stockholm Convention on Persistent Organic Pollutants. Geneva, Switzerland: United Nations Environment Program. 43 p.

UNEP (2009). Stockholm Convention on Persistent Organic Pollutants, Adoption of Amendments of Annexes A, B and C. Geneva, Switzerland: United Nations Environment Program. 63 p.

UNEP (2011). Stockholm Convention on Persistent Organic Pollutants, Adoption of an Amendment of Annex A. Geneva, Switzerland: United Nations Environment Program. 63 p.

Uribe, C., Folch, H., Enriquez, R. and Moran, G. (2011). Innate and adaptive immunity in teleost fish: A review. *Veterinarni Medicina* **56**, 486–503.

Usenko, C. Y., Hopkins, D. C., Trumble, S. J. and Bruce, E. D. (2012). Hydroxylated PBDEs induce developmental arrest in zebrafish. *Toxicology and Applied Pharmacology* **262**, 43–51.

U.S. EPA (1980). Ambient Water Quality Criteria for DDT. EPA Report #440/5-80-038. Washington, DC: U.S. Environmental Protection Agency, Office of Water Regulations and Standards Division.

U.S. EPA (1997). Special Report on Environmental Endocrine Disruption: an Effects Assessment and Analysis. US Environmental Protection Agency, Washington, DC: EPA Report no. EPA/630/R-96/012: 111 pp.

Valters, K., Li, H., Alaee, M., D'Sa, I., Marsh, G., Bergman, Å. and Letcher, R. J. (2005). Polybrominated diphenyl ethers and hydroxylated and methoxylated brominated and chlorinated analogues in the plasma of fish from the Detroit River. *Environmental Science and Technology* **39**, 5612–5619.

van den Berg, H. (2009). Global status of DDT and its alternatives for use in vector control to prevent disease. *Environmental Health Perspectives* **117**, 1656–1663.

van der Oost, R., Beyer, J. and Vermeulen, N. P. E. (2003). Fish bioaccumulation and biomarkers in environmental risk assessment: a review. *Environmental Toxicology and Pharmacology* **13** (57–149),

van der Weiden, M. E. J., Bleumink, R., Seinen, W. and van den Berg, M. (1994). Concurrence of P450 1A induction and toxic effects in the mirror carp (*Cyprinus carpio*), after administration of allow dose of 2,3,7,8-tetrachlorodibenzo-p-dioxin. *Aquatic Toxicology* **29**, 147–162.

van Ginneken, V., Palstra, A., Leonards, P., Nieveen, M., van den Berg, H., Flik, G., Spanings, T., Niemantsverdriet, P., van den Thillart, G. and Murk, A. (2009). PCBs and the energy cost of migration in the European eel (*Anguilla anguilla L.*). *Aquatic Toxicology* **92**, 213–220.

Vandegehuchte, M. and Janssen, C. (2011). Epigenetics and its implications for ecotoxicology. *Ecotoxicology* **20**, 607–624.

Vasseur, P. and Cossu-Leguille, C. (2006). Linking molecular interactions to consequent effects of persistent organic pollutants (POPs) upon populations. *Chemosphere* **62**, 1033–1042.

Verreault, J., Letcher, R. J., Muir, D. C. G., Chu, S., Gebbink, W. A. and Gabrielsen, G. W. (2005). New organochlorine contaminants and metabolites in plasma and eggs of glaucous gulls (*Larus hyperboreus*) from the Norwegian Arctic. *Environmental Toxicology and Chemistry* **24**, 2486–2499.

Vethaak, A., Jol, J., Meijboom, A., Eggens, M., Rheinallt, T., Wester, P., van de Zande, T., Bergman, A., Dankers, N., Ariese, F., Baan, R., Everts, J., Opperhuizen, A. and Marquenie, J. (1996). Skin and liver diseases induced in flounder (*Platichthys flesus*) after longterm exposure to contaminated sediments in large-scale mesocosms. *Environmental Health Perspectives* **104**, 1218–1229.

Vethaak, A. and Wester, P. (1996). Diseases of flounder *Platichthys flesus* in Dutch coastal and estuarine waters, with particular reference to environmental stress factors. II. Liver histopathology. *Diseases of Aquatic Organisms* **26**, 99–116.

Vethaak, A. D., Jol, J. G. and Pieters, J. P. F. (2009). Long-term trends in the prevalence of cancer and other major diseases among flatfish in the southeastern North Sea as indicators of changing ecosystem health. *Environmental Science and Technology* **43**, 2151–2158.

Vijayan, M. M., Aluru, N., Maule, A. G. and Jørgensen, E. H. (2006). Fasting augments PCB impact on liver metabolism in anadromous Arctic char. *Toxicological Sciences* **91**, 431–439.

Villeneuve, D. L. and Garcia-Reyero, N. (2011). Vision and strategy: Predictive ecotoxicology in the 21st century. *Environmental Toxicology and Chemistry* **30**, 1–8.

Vogelbein, W. K., Fournie, J. W., Van Veld, P. A. and Huggett, R. J. (1990). Hepatic neoplasms in the mummichog Fundulus heteroclitus from a creosote-contaminated Site. *Cancer Research* **50**, 5978–5986.

Vogelbein, W. K. and Unger, M. A. (2006). Liver carcinogenesis in a non-migratory fish:The association with polycyclic aromatic hydrocarbon exposure. *Bulletin of European Association of Fish Pathologists* **26**, 11–20.

Vuori, K. A. M. and Nikinmaa, M. (2007). M74 syndrome in Baltic salmon and the possible role of oxidative stresses in its development: Present knowledge and perspectives for future studies. *AMBIO: A Journal of the Human Environment* **36**, 168–172.

Wahlang, B., Falkner, K. C., Gregory, B., Ansert, D., Young, D., Conklin, D. J., Bhatnagar, A., McClain, C. J. and Cave, M. (2013). Polychlorinated biphenyl 153 is a diet-dependent obesogen that worsens nonalcoholic fatty liver disease in male C57BL6/J mice. *Journal of Nutritional Biochemistry* 22. DOI:10.1016/j.jnutbio.2013.01.009. Epub 2013 Apr.

Walker, M. K. and Peterson, R. E. (1994). Aquatic toxicity of dioxins and related chemicals. In *Dioxins and Health* (ed. A. Schecter). New York: Plenum Press.

Walsh, A. and Ribelin, W. (1975). The pathology of pesticide poisoning. In *Pathology of Fishes* (eds. W. Ribelin and G. Migaki), pp. 515–557. Madison: University of Wisconsin Press.

Walter, G. L., Jones, P. D. and Giesy, J. P. (2000). Pathologic alterations in adult rainbow trout, Oncorhynchus mykiss, exposed to dietary 2,3,7,8-tetrachlorodibenzo-p-dioxin. *Aquatic Toxicology* **50**, 287–299.

Wang, Y., Wang, C., Zhang, J., Chen, Y. and Zuo, Z. (2009). DNA hypomethylation induced by tributyltin, triphenyltin, and a mixture of these in Sebastiscus marmoratus liver. *Aquatic Toxicology* **95**, 93–98.

Ware, D. (1984). Fitness of different reproductive strategies in teleost fishes. In *Fish Reproduction* (eds. G. Potts and R. Wootton). London: Academic Press.

Weeks, B. A. and Warinner, J. E. (1986). Functional evaluation of macrophages in fish from a polluted estuary. *Veterinary Immunology and Immunopathology* **12**, 313–320.

Weis, J. S. and Candelmo, A. C. (2012). Pollutants and fish predator/prey behavior: A review of laboratory and field approaches. *Current Zoology* **58**, 9–20.

Weis, J. S., Samson, J., Zhou, T., Skurnick, J. and Weis, P. (2003). Evaluating prey capture by larval mummichogs (*Fundulus heteroclitus*) as a potential biomarker for contaminants. *Marine Environmental Research* **55**, 27–38.

Weis, J. S., Smith, G., Zhou, T., Santiago-Bass, C. and Weis, P. (2001). Effects of contaminants on behavior: Biochemical mechanisms and ecological consequences. *BioScience* **51**, 209–217.

Wester, P. W. and Canton, J. H. (1986). Histopathological study of Oryzias latipes (medaka) after long-term β-hexachlorocyclohexane exposure. *Aquatic Toxicology* **9**, 21–45.

Westerlund, L., Billsson, K., Andersson, P. L., Tysklind, M. and Olsson, P.-E. (2000). Early life-stage mortality in zebrafish (*Danio rerio*) following maternal exposure to polychlorinated biphenyls and estrogen. *Environmental Toxicology and Chemistry* **19**, 1582–1588.

White, S. S. and Birnbaum, L. S. (2009). An overview of the effects of dioxins and dioxin-like compounds on vertebrates, as documented in human and ecological epidemiology. *Journal of Environmental Science and Health, Part C* **27**, 197–211.

Whyte, S. K. (2007). The innate immune response of finfish—A review of current knowledge. *Fish and Shellfish Immunology* **23**, 1127–1151.

Williams, G. and Weisburger, J. (1986). Chemical carcinogens. In *Casarett and Doull's Toxicology: The Basic Science of Poisons* (eds. J. Doull, C. D. Klaassen and M. Amdur), 3rd ed. New York: Macmillan Publishing.

Wirgin, I., Roy, N. K., Loftus, M., Chambers, R. C., Franks, D. G. and Hahn, M. E. (2011). Mechanistic basis of resistance to PCBs in Atlantic tomcod from the Hudson River. *Science* **331**, 1322–1325.

Wiseman, S. and Vijayan, M. M. (2011). Aroclor 1254 disrupts liver glycogen metabolism and enhances acute stressor-mediated glycogenolysis in rainbow trout. *Comparative Biochemistry and Physiology Part C: Toxicology and Pharmacology* **154**, 254–260.

Wisk, J. D. and Cooper, K. R. (1990). Comparison of the toxicity of several polychlorinated dibenzo-p-dioxins and 2,3,7,8-tetrachlorodibenzofuran in embryos of the Japanese medaka (*Oryzias latipes*). *Chemosphere* **20**, 361–377.

Wolf, J. C. and Wolfe, M. J. (2005). A brief overview of nonneoplastic hepatic toxicity in fish. *Toxicologic Pathology* **33**, 75–85.

Woods, R., Vallero, R. O., Golub, M. S., Suarez, J. K., Ta, T. A., Yasui, D. H., Chi, L.-H., Kostyniak, P. J., Pessah, I. N., Berman, R. F. and LaSalle, J. M. (2012). Long-lived epigenetic interactions between perinatal PBDE exposure and Mecp2308 mutation. *Human Molecular Genetics* **21**, 2399–2411.

Wu, W. Z., Li, W., Xu, Y. and Wang, J. W. (2001). Long-term toxic impact of 2,3,7,8-tetrachlorodibenzo-p-dioxin on the reproduction, sexual differentiation, and development of different life stages of *Gobiocypris rarus and Daphnia magna*. *Ecotoxicology and Environmental Safety* **48**, 293–300.

Yamauchi, M., Kim, E.-Y., Iwata, H., Shima, Y. and Tanabe, S. (2006). Toxic effects of 2,3,7,8-tetrachlorodibenzo-p-dioxin (TCDD) in developing red seabream (*Pagrus major*) embryo: An association of morphological deformities with AHR1, AHR2 and CYP1A expressions. *Aquatic Toxicology* **80**, 166–179.

Ye, R., Lei, E. Y., Lam, M. W., Chan, A. Y., Bo, J., Merwe, J., Fong, A. C., Yang, M. S., Lee, J. S., Segner, H., Wong, C. C., Wu, R. S. and Au, D. T. (2012). Gender-specific modulation of immune system complement gene expression in marine medaka Oryzias melastigma following dietary exposure of BDE-47. *Environmental Science and Pollution Research* **19**, 2477–2487.

Yu, L., Deng, J., Shi, X., Liu, C., Yu, K. and Zhou, B. (2010). Exposure to DE-71 alters thyroid hormone levels and gene transcription in the hypothalamic–pituitary–thyroid axis of zebrafish larvae. *Aquatic Toxicology* **97**, 226–233.

Yu, L., Lam, J. C. W., Guo, Y., Wu, R. S. S., Lam, P. K. S. and Zhou, B. (2011). Parental transfer of polybrominated diphenyl ethers (PBDEs) and thyroid endocrine disruption in zebrafish. *Environmental Science and Technology* **45**, 10652–10659.

Zacharewski, T. and Safe, S. H. (1998). Antiestrogenic activity of TCDD and related compounds. In *Reproductive and Developmental Toxicology* (ed. K. S. Korach), p. 722. New York: Marcel Dekker.

Zaroogian, G., Gardner, G., Borsay Horowitz, D., Gutjahr-Gobell, R., Haebler, R. and Mills, L. (2001). Effect of 17β-estradiol, o,p'-DDT, octylphenol and p,p'-DDE on gonadal development and liver and kidney pathology in juvenile male summer flounder (*Paralichthys dentatus*). *Aquatic Toxicology* **54**, 101–112.

Zhang, Z. and Hu, J. (2008). Effects of p,p' -DDE exposure on gonadal development and gene expression in Japanese medaka (*Oryzias latipes*). *Journal of Environmental Sciences* **20**, 347–352.

Zhang, S. C., Wang, Z. P. and Wang, H. M. (2013). Maternal immunity in fish. *Developmental and Comparative Immunology* **39**, 72–78.

Zodrow, J. M., Stegeman, J. J. and Tanguay, R. L. (2004). Histological analysis of acute toxicity of 2,3,7,8-tetrachlorodibenzo-p-dioxin (TCDD) in zebrafish. *Aquatic Toxicology* **66**, 25–38.

3

ORGANOMETAL(LOID)S

NILADRI BASU

DAVID M. JANZ

1. Introduction
2. Organic Mercury
   2.1. The Chemicals
   2.2. Environmental Situations of Concern
   2.3. Toxicokinetics
   2.4. Potential for Bioconcentration or Biomagnification
   2.5. Mechanism(s) of Toxicity
   2.6. Organism Impacts
   2.7. Interactions with Other Toxic Agents
   2.8. Knowledge Gaps and Future Directions
3. Organoselenium
   3.1. The Chemicals
   3.2. Environmental Situations of Concern
   3.3. Toxicokinetics
   3.4. Potential for Bioconcentration or Biomagnification
   3.5. Mechanism(s) of Toxicity
   3.6. Organism Impacts
   3.7. Interactions with Other Toxic Agents
   3.8. Knowledge Gaps and Future Directions
4. Organoarsenicals
   4.1. The Chemicals
   4.2. Environmental Situations of Concern
   4.3. Toxicokinetics
   4.4. Potential for Bioconcentration or Biomagnification
   4.5. Mechanism(s) of Toxicity and Organism Impacts
   4.6. Knowledge Gaps and Future Directions
5. Organotin Compounds
   5.1. The Chemicals
   5.2. Environmental Situations of Concern
   5.3. Toxicokinetics
   5.4. Potential for Bioconcentration or Biomagnification
   5.5. Mechanism(s) of Toxicity and Organism Impacts
   5.6. Knowledge Gaps and Future Directions
6. Other Organometal(loid)s
   6.1. Organolead Compounds

Organic Chemical Toxicology of Fishes: Volume 33
FISH PHYSIOLOGY

1. INTRODUCTION

Organometals are compounds in which a metal(loid) is bonded with carbon (Dopp et al., 2004; Thayer, 2010). Owing to unique physical and chemical properties, certain organometal(loid)s have been favored in many industrial and chemical processes. In addition, natural biogeochemical processes may facilitate the conversion of inorganic elements into organometal(loid)s. Whether derived from sources that are anthropogenic, natural, or a combination of both, organometal(loid)s exist in a number of chemical forms, and they are widely dispersed in the environment.

The risks posed by organometal(loid)s are driven by a combination of their environmental levels, target organism bioavailability, and intrinsic hazard potential. Organometal(loid) risk is largely driven by anthropogenic activity, which may promote releases of organic metal(loid)s or their inorganic precursors into the environment. For example, alkylmercurials were historically used as fungicidal seed dressings for grain (e.g., wheat, barley, rye), and this use resulted in the intoxication and death of granivorous birds in Scandinavia and in humans that consumed the grains in Iraq (Grandjean et al., 2010). Another example is tributyltin, a proven endocrine disruptor that until recently was intentionally and widely used as an antifouling pesticide on ocean-faring ships (Gadaja and Janscó, 2010). In other cases, the discharge of inorganic metal(loid)s (which may subsequently be methylated in the environment), or their organic derivative, occur as by-products of anthropogenic activities. Concerning arsenic, a seminal paper by Challenger et al. (1933) demonstrated that fungi could methylate inorganic arsenic. For mercury (Hg), the elemental form that is released from coal-fired power plants or artisanal gold-mining operations may find its way into aquatic sediments where it becomes alkylated into methylmercury (MeHg) (Jensen and Jernelov, 1967).

A key feature in organometal(loid) toxicology is biological methylation or alkylation. This process refers to the transfer of carbon-containing groups to a metal(loid). The basic equation is:

$$M^{+n} + CH_3^- \rightarrow CH_3M^{+n-1}$$

where M refers to a metal(loid).

Biological methylation is usually enzyme mediated and facilitated by anaerobic prokaryotic microbes in anoxic zones of aquatic sediments, although other organisms, such as algae and fungi, have been shown to facilitate biomethylation (Bentley and Chasteen, 2002). The two major methyl donors are methylcobalamine (vitamin B12) derivatives and s-adenosylmethionine, but others have been found. Biomethylation has been shown for a number of metal(loid)s—As, Bi, Cd, Hg, Ge, Pb, Se, Sn, Te, Tl—though the underlying mechanisms are not necessarily established for all of them (Jenkins, 2010). Nonetheless, with advents in analytical chemistry and molecular biology, the ability to quantify and study organometal(loid)s is rapidly improving (Dopp et al., 2004). Since metal speciation is a primary determinant of bioavailability and toxicity, knowing the "total" amount of any metal(loid) provides only limited toxicological information. Thus, an understanding of the various chemical forms that comprise the total is critical in understanding risk.

The biomethylation of inorganic metal(loid)s can affect environmental fate and toxicity in a number of ways. In terms of physical and chemical properties, the resulting metal(loid)-carbon bond shows little polarity and thus may have low chemical reactivity (Thayer, 2010). The solubility of organometal(loid)s is also affected, with aqueous solubility generally decreasing with the increasing number (or size) of organic ligands. Mobility is enhanced for many elements, as organometal(loid)s are usually better absorbed by organisms and can undergo biomagnification through food webs. In terms of toxicodynamics, the cellular targets, uptake mechanisms, and health effects in fish following exposure to organometal(loid)s can vary widely depending on the chemical itself and other factors.

The purpose of this chapter is to review and synthesize consequences associated with exposure of fish to the most relevant organometal(loid)s. Although others have reviewed the biogeochemistry (Dopp et al., 2004; Thayer, 2010), a focused review concerning the potential physiological impacts on fish has not, to our knowledge, been published. This gap is surprising, considering the fact that biomethylation of inorganic metal (loids) often occurs in aquatic environments and that many organometal (loid)s or their inorganic precursors are directly discharged into waterways. Although several organometal(loid)s exist, here we focus mainly on MeHg and organoselenium, as well as organoarsenicals and organotin. These are perhaps the most relevant organometal(loid)s in terms of environmental concern and research findings. In this chapter we also make brief mention of other organometal(loid)s, including alkyl derivatives of lead, tellurium, antimony, and germanium. For the primary compounds, we detail their chemistry, sources and uses, regulatory status, and environmental situations of concern. Pertinent issues related to their toxicokinetics and toxicological consequences from the cellular to the organismal level are reviewed.

2. ORGANIC MERCURY

2.1. The Chemicals

2.1.1. STRUCTURE AND BREAKDOWN PRODUCTS

Mercury (Hg) exists in both inorganic and organic forms. (For a summary of the key forms, see Table 3.1.) As an inorganic element, Hg can exist in one of three forms: zero (Hg), first (Hg1), and second (Hg2) oxidation states. Mercury ions that are bound covalently to at least one carbon atom are referred to as organic Hg compounds or short-chain alkyl mercurials. Monomethylmercury is the most important chemical form in terms of environmental health concern. As discussed throughout this section, MeHg is ubiquitous in the environment, bioaccumulates in fish tissues, and biomagnifies in freshwater and marine aquatic ecosystems. In addition to MeHg, other organic Hg compounds that have received some

Table 3.1
Selected properties of key mercurial compounds. Information derived from ATSDR (1999)

Characteristic	Elemental Mercury	Mercurous Chloride	Mercuric Chloride	Methylmercury Chloride
Synonyms	mercury, quicksilver, colloidal mercury, metallic mercury	calomel, mercuric (I) chloride	corrosive sublimate, mercury(II) chloride	monomethylmercury, methylmercury(II) chloride
CAS Number	7439-97-6	10112-91-1	7487-94-7	22967-92-6
Linear Formula	Hg	Hg_2Cl_2	$HgCl_2$	CH_3HgCl
Molecular Weight (g/mol)	200.6	472.1	271.5	251.1
Melting Point (°C)	−38.9	525	277	170
Boiling Point (°C)	356.7	383	304	no data
Density (g/cm³)	13.5	7.2	5.4	4.1
Aqueous Solubility (mg/L)	0.056	2	74	1

attention include ethylHg and phenylHg (ATSDR, 1999), though these are generally not of environmental relevance (Hintelmann, 2010).

As discussed above, the total amount of Hg (both organic and inorganic forms) in any given sample is only of limited use, and therefore resolving its chemical speciation is critical. The chemical form of Hg will dictate how it cycles in the environment, enters organisms, and causes harm. Nearly all the Hg in air, water, soil, and sediment exists in the inorganic form, whereas nearly all the Hg that bioaccumulates in aquatic organisms and biomagnifies through the food chain is in the organic form (Wiener et al., 2003).

One critical feature of the Hg cycle is the biomethylation of inorganic Hg into MeHg. This discovery was made by the Swedish researchers Jensen and Jernelov (1967) following a series of simple and convincing aquarium-based studies. Their work, and subsequently that of others, has shown that sulfate-reducing bacteria in aquatic sediments facilitate the formation of MeHg. In addition to biomethylation in sediment, more recent studies indicate that Hg biomethylation may also occur in biofilms, fungi, algal mats, and the water column (St. Louis et al., 2004; Huguet et al., 2010; Thayer, 2010).

2.1.2. SOURCES AND USES

Mercury, like most other elements, has both natural and anthropogenic sources. In terms of natural sources, processes such as weathering, volcanic activities, and wildfires mobilize Hg, and this accounts for approximately one-third of the Hg found in our biosphere. The remainder released into the environment originates from human activities (Pacyna et al., 2010). Much of these emissions come from coal combustion, which is estimated to contribute about 60% of the anthropogenic Hg emissions (Swain et al., 2007). Current Hg releases are significantly greater than those during preindustrial times, based on analyses of museum-preserved bird feathers (Head et al., 2011) and polar bear fur (Dietz et al., 2011), as well as dated sediment cores (Drevnick et al., 2012). As part of a joint Arctic Monitoring and Assessment Programme (AMAP) and United Nations Environment Program (UNEP) report, Pacyna et al. (2010) assess the global inventory of anthropogenic Hg emissions and break it down into key sectors (Figure 3.1).

2.1.3. ECONOMIC IMPORTANCE

Mercury has great importance in a number of commercial and industrial settings as outlined in the previous section, but Hg pollution can have comparable economic impacts. MeHg-contaminated fish can have adverse economic consequences. For example, Sundseth et al. (2010) estimated that MeHg-associated IQ loss resulting from consumption of contaminated fish would amount to about $3.7 billion worldwide in the year 2020.

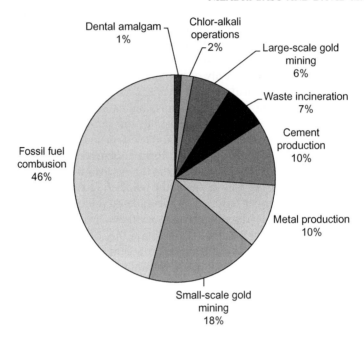

Figure 3.1. Percentage of global anthropogenic emissions of Hg to air in 2005 from various sectors. From AMAP/UNEP (2008) and Pacyna et al. (2010).

Another area of economic concern is related to fish consumption advisories and impacts on environmental quality. In the United States and Canada, it has been estimated that 80% and 97%, respectively, of fish consumption advisories are due to MeHg-contaminated fish (U.S. EPA, 2001, 2009). Advisories affect a multibillion dollar commercial and sportfishing industry. In the Great Lakes region of North America, for example, more than 9.2 million anglers spend a total of 149 million days fishing, and this activity supports nearly 200,000 jobs, generating more than $20 billion annually in economic activity (Allen and Southwick, 2008). Fish consumption advisories also place a disproportionate burden on certain susceptible populations (e.g., indigenous peoples, new immigrants, urban poor) who rely on locally caught fish as an important source of sustenance and culture (Nriagu et al., 2012).

2.1.4. REGULATORY STATUS

Although the adverse public health effects of Hg pollution have been well established, regulating the chemical has been challenging owing to its diverse uses (intentionally and as a by-product) and its movement across local, regional, and international boundaries. Success stories, however, do exist. In the 1970s, actions were taken to prevent the direct discharge of Hg into international waterways by, for example, certain industries and water vessels (Selin and Selin, 2006). In the 1990s, steps were taken to limit the transboundary movement of Hg. For example, within the 1979 UN Convention on Long-Range Transboundary Air Pollution (UNECE CLRTRAP), the 1998 Aarhus Protocol on Heavy Metals identified Hg as a focal contaminant. The 1989 Basel Convention on the Control of Transboundary Movements of Hazardous Wastes and their Disposal (i.e., the Basel Convention) is an international treaty that helps limit the movement of hazardous wastes between countries. Given that Hg is globally used and dispersed, internationally binding treaties are needed to limit exposures. In 2002 the UNEP conducted a Global Hg Assessment, which was followed in 2007 by the establishment of the Global Hg Partnership. This Partnership aims to help countries take action on Hg pollution on a global scale.

2.1.5. WATER QUALITY CRITERIA

Most water quality criteria for Hg exist for inorganic or total Hg (Table 3.2), with fewer criteria available for MeHg. Recent assessments have proposed tissue- and dietary-based guidelines for the protection of fish. In an earlier report using data available up until the mid-1990s, Wiener and Spry (1996) postulated that total Hg levels ranging from 5 to 20 μg/g (wet weight, ww) in the axial muscle and from 5 to 10 μg/g (ww) in the whole body were likely associated with sublethal and lethal effects in fish. Since the mid-1990s, great progress has been made in both the laboratory and the field concerning the aquatic toxicology of Hg. In a recent thoughtful review, Sandheinrich and Wiener (2011) conclude that subclinical changes such as gene-expression alterations, oxidative stress, and effects on reproductive hormones and behavior occur in fish, with total Hg axial muscle concentrations ranging from 0.5 to 1.2 μg/g (ww) and from 0.3 to 0.7 μg/g (ww) in the

Table 3.2
Water quality guidelines for the protection of aquatic life in terms of Hg (THg or Hg2+) exposure. Concentrations are in μg/L

	Country	Value	Reference
Freshwater	USA	1.4 (acute), 0.77 (chronic)	USEPA 2009
	Canada	0.026	CCME 2003
Marine	USA	1.8 (acute), 0.94 (chronic)	USEPA 2009
	Canada	0.016	CCME 2003

whole body. Another assessment was conducted by Beckvar et al. (2005). Based on their review, they recommended a whole-body total Hg concentration of 0.2 µg/g (ww) in adults and juveniles, and for eggs, larvae, and fry they recommended a whole-body residue value of 0.02 µg/g (ww). These values put forth by Sandheinrich and Wiener (2011) and Beckvar et al. (2005) are ~10–20 fold lower than the earlier assessment by Wiener and Spry (1996) and are reflective of newer evidence. The data also demonstrate that concentrations commonly observed in wild fish populations are potentially damaging, thus leading many in the scientific community to conclude that MeHg does indeed pose risks to fish populations.

Depew et al. (2012) argue that gauging risk based on dietary exposures is complementary to the aforementioned tissue residue-level approaches. Following a systematic review of the available literature, Depew et al. (2012) found adverse effects in several species related to growth when dietary MeHg levels exceeded 2.5 µg/g (ww). Dietary concentrations above 0.5 µg/g (ww) were associated with a range of adverse behavioral changes, and concentrations above 0.2 µg/g (ww) were associated with subclinical changes (e.g., biochemical and molecular changes) and reproductive effects.

When the approaches we have detailed (water quality guidelines, tissue residue values, dietary exposure values) are considered and compared to real-world values, it is clear that Hg levels in many areas of the world are sufficiently high to cause toxicity to fish. However, to our knowledge, declines in fish populations solely due to Hg pollution have not yet been documented.

2.2. Environmental Situations of Concern

Reports of Hg poisoning in humans (Grandjean et al., 2010) and wildlife (Basu and Head, 2010) in the mid-20th century accelerated research, and this eventually led to breakthroughs in our understanding of Hg's exposure-disease model so that harm preventative measures could be taken. The greatest environmental situation of concern in an aquatic environment is one that receives significant Hg input and has conditions favorable for Hg methylation and biomagnification. Sites that favor biomethylation are those with, for example, low pH, low alkalinity, and high dissolved organic carbon content. These properties, among others, favor the microbial methylation of inorganic Hg into MeHg. As an example, in newly flooded reservoirs, the concentrations of MeHg in fish increase rapidly after flooding because of increased biomethylation rates (Hall and St. Louis, 2004; Bodaly and Fudge, 1999; Bodaly et al., 2004). After such flooding events, the levels of Hg may remain elevated in resident fish populations for several decades (Bodaly et al., 2007).

Other areas of concern are fish in close proximity to point sources. There are, for example, well-documented cases of elevated Hg levels in fish surrounding a chlor-alkali plant in the English-Wabigoon river system in Ontario (Fimreite and Reynolds, 1973; Kinghorn et al., 2007), in the region of Oak Ridge National Laboratory's Y-12 facility in Tennessee (Brooks and Southworth, 2011), and in a number of Brazilian sites in which small-scale gold mining occurs (Berzas Nevado et al., 2010). The contribution of such point sources (perhaps, except for small-scale mining) is not as great a concern today inasmuch as many facilities (at least in North America and Europe) have been shut down in recent decades. Nonetheless, the amount of Hg released from such operations can be significant (e.g., 350,000 kg at Oak Ridge National Laboratory's Y-12 facility; Brooks and Southworth, 2011), and it has been shown that Hg levels in fish residing near such historic point sources can remain elevated for decades following the closing of facilities (Turner and Southworth, 1999; Kinghorn et al., 2007).

Because Hg is also a globally dispersed contaminant, regions far removed from point sources may be subject to contamination. The Arctic is one such region. Evans et al. (2005) reviewed Hg levels in freshwater fish from a number of sites across the Canadian Arctic and Subarctic, and Dietz et al. (2013) address the toxicological significance of these exposures in resident fish. In general, Hg is present in all fish sampled from the Arctic region, but concentrations may not be sufficiently high in most, but not all, to pose immediate health risks to fish populations.

2.3. Toxicokinetics

2.3.1. UPTAKE

2.3.1.1. Gills. In general, gill uptake of MeHg in fish is minor ($<10\%$) (Rodgers and Beamish, 1981; Hall et al., 1997), although a recent study by Hrenchuk et al. (2012) in yellow perch (*Perca flavescens*) demonstrated waterborne uptakes may exceed 10%. The mechanisms of gill uptake consist of passive and/or active transport (Pickhardt et al., 2006).

2.3.1.2. Gut. The major source of MeHg uptake in fish is the diet (Hall et al., 1997), a fact that underscores the importance of the gut. Once ingested, bioenergetics modeling has estimated that 65% to 90% of the MeHg in prey is assimilated from the gut into the circulatory system (Rodgers, 1994; Leaner and Mason, 2004; Pickhardt et al., 2006). In comparison, in human consumers of MeHg-contaminated fish, about 95% of the ingested Hg is absorbed (Clarkson and Magos, 2006). The digested MeHg is assimilated in the intestine via both passive and active mechanisms

(Leaner and Mason, 2004). Precise details concerning nonspecific transport are unclear (Kidd and Batchelar, 2011), and active transport likely involves uptake through neutral L-enantiomer amino acid carriers that facilitate the uptake of MeHg in the L-cysteine conjugated form (Clarkson and Magos, 2006).

2.3.2. TRANSPORT

Like other vertebrates (Clarkson and Magos, 2006), MeHg in fish is transferred to the blood from the intestine and is then distributed throughout the body in a matter of days (Leaner and Mason, 2004). In blood, about 90% of the MeHg is bound to red blood cells (Oliveira Ribeiro et al., 1999), mainly to hemoglobin.

The relative ease by which MeHg moves throughout the body is due to the interaction of MeHg with the amino acid L-cysteine. The resulting MeHg-L-cysteine complex structurally resembles that of methionine, which is a large neutral amino acid. As an essential amino acid, methionine needs to be taken up by cells, and this process is facilitated by a neutral amino acid carrier. It is this neutral amino acid carrier that is "tricked" to take up MeHg-L-cysteine in place of methionine, resulting in a well-documented form of molecular mimicry (Clarkson and Magos, 2006). However, a recent study by Hoffmeyer et al. (2006) using sophisticated spectroscopic-based methods suggests that such mimicry may be an oversimplification and that more work is needed in this area.

2.3.3. BIOTRANSFORMATION

The conversion of MeHg to inorganic Hg (demethylation) is an important biotransformation step. In mammals, this step is facilitated by intestinal bacteria and specific liver enzymes (Clarkson and Magos, 2006). It is not clear if a similar process occurs in fish, though there is some support for demethylation (Wiener and Spry, 1996; Sandheinrich and Wiener, 2011). Dietary exposure studies involving zebrafish (*Danio rerio*) (Gonzalez et al., 2005) and Arctic charr (*Salvelinus alpinus*) (Oliveira Ribeiro et al., 1999) suggest that over time MeHg demethylation may occur in the liver. Recent research by Drevnick et al. (2008), Barst et al. (2011), and Chumchal et al. (2011) suggests that demethylation may occur in melanomacrophage centers of fish (i.e., spleen, kidney, liver).

There is also limited evidence that fish can methylate Hg. Baatrup et al. (1986) documented increased MeHg in the kidney and liver of fish that were exposed to mercuric chloride, thus suggestive of *in vivo* methylation. The site of methylation is likely in the liver, but not the intestine, based on earlier research on tuna (*Thunnus albacores, Thunnus alaunga*) by Imura et al. (1972) and Pan-Hou and Imura (1981).

2.3.4. SUBCELLULAR PARTITIONING

Owing to its high affinity for protein thiols, MeHg can bind to a number of subcellular components. In rainbow trout (*Oncorhynchus mykiss*) fingerlings exposed to [203]Hg-labeled MeHg via the diet for 2 and 7 weeks, Baatrup and Danscher (1987) used cytochemical approaches to characterize subcellular partitioning in the liver and kidney. In the liver, MeHg was found mainly in the lysosomes, while in the kidney it was associated with a number of organelles and the cellular membrane.

2.3.5. ACCUMULATION IN SPECIFIC ORGANS

A number of studies have reported on MeHg distribution across fish tissues. In general, Hg levels are highest in the blood as well as the liver, kidney, and spleen (Table 3.3). Eventually, MeHg will accumulate in skeletal muscle bound tightly to cysteine (Harris et al., 2003). The half-life of MeHg in muscle can be on the order of a few years, and sequestration in muscle may afford protection against toxicity in other organs. The MeHg found in muscle also affords a long-term assessment of Hg exposure in that fish, unlike blood and liver which have higher turnover rates. When Hg levels are compared across tissues, the ratios are not consistent across species. For example, as shown in Table 3.3, the mean muscle-to-liver ratio of Hg is 1.8 but ranges from 0.6 to 4.5, and the mean muscle-to-brain ratio is 4.6 and ranges from 0.4 to 11. Such variation emphasizes the need to obtain tissue residue values and not simply assess doses based on levels found in muscle as commonly performed.

2.3.6. EXCRETION

As outlined by Kidd and Batchelar (2011), MeHg excretion via gills and kidney is largely unstudied and unknown. MeHg conjugated to glutathione can undergo enterohepatic cycling and eventually be excreted via bile. In general, the excretion of MeHg from fish occurs in two phases. On the one hand, there is a relatively rapid phase of excretion from tissues such as liver and kidney that can occur on the order of 1–2 months (Van Wallenghem et al., 2007). On the other hand, excretion from muscle is much slower and occurs on the order of years (Trudel and Rasmussen, 1997; Amlund et al., 2007).

2.4. Potential for Bioconcentration or Biomagnification

The assimilation of MeHg into food webs is driven by many factors, including inorganic Hg inputs and influential landscape characteristics (Paterson et al., 1998; Wiener et al., 2003; Chasar et al., 2009). As detailed

Table 3.3

Selected field studies to highlight inter-tissue differences in total Hg levels. All values are ng/g

	Muscle	Liver	Kidney	Gill	Gonad	Blood	Brain	Location	Reference
Striped bass	309	531			136	36		Lake Mead, USA	Cizdziel et al. (2003)
Largemouth bass	179	112			47	8.8			
Catfish	175	292			16	6.8			
Bluegill	95	60			15	3.8			
Tilapia	8	14			7.7	1.2			
Spotted seatrout	560	630	1120		60		240	South Florida, USA	Adams et al. (2010)
Bluefish	320	380	570				90	New Jersey, USA	Burger et al. (2012)
European catfish	340	250	140					Po River, Italy	Squadrone et al. (2013)
Scorpion fish	202	285		360			560	Kumamoto, Japan	Watanabe et al. (2012)
Seabream	116	117		24			28		
Jack mackerel	44	53		21			16		
Pacific saury	75	37		8			12		
Japanese whiting	131	59					20		
Japanese flying fish	43	36					10		
Halfbeak	77	43					7		
Carp	31	11	21	96	9			Hutovo Blato, Bosnia and Herzegovina	Has-Schön et al. (2008)
Tench	125	28	40	79	8				
Pumpkinseed	159	36	50	71	10				
Prussian carp	147	33	47	65	9				
Hasselquist	51	67	58	98	5				

previously, fish take up MeHg relatively quickly (within a few months) and eliminate it rather slowly (order of years), thus resulting in the potential for tremendous bioaccumulation. MeHg biomagnifies in aquatic ecosystems, and the concentrations measured in fish tissues can be 10^6–10^7 times greater than levels measured in surface waters (Wiener et al., 2003). Biomagnification factors between piscivorous predators and prey can be 5- to 15-fold (Wiener et al., 2003).

2.5. Mechanism(s) of Toxicity

Methylmercury has a high affinity for thiol groups, and it can readily transfer from one thiol group to another (Clarkson and Magos, 2006). Not surprisingly, existing within a cell are many cellular targets and thus several mechanisms of toxicity. Oxidative stress is a primary mechanism by which MeHg exerts its toxicity that has been reported across several taxa; common measurement outcomes are the production of reactive oxygen species and alterations in enzymes that underlie protective redox systems. Such biochemical changes may result in tissue damage and a myriad of higher biological impacts. In Atlantic salmon (*Salmo salar*) fed MeHg, Berntssen et al. (2003) reported decreased superoxide dismutase and glutathione peroxidase activities as well as increased thiobarbituric acid-reactive substances (TBARS). In field studies, Larose et al. (2008) documented changes in glutathione enzymes (e.g., glutathione S-transferase and glutathione peroxidase) in association with MeHg exposure. At the transcriptional level, Gonzalez et al. (2005) documented a number of genes (e.g., mitochondrial metabolism and oxidative stress pathways) that were affected in zebrafish following MeHg exposure, but interestingly these were largely restricted to the liver and muscle with minimal changes seen in the brain.

Newly proposed sites of MeHg toxic action are melanomacrophage centers. These are macrophage aggregates usually found in the spleen, kidney, and liver of fish, and they may develop or grow in association with exposure to environmental stressors (Agius and Roberts, 2003) including MeHg. In studies of northern pike (*Esox lucius*) from Isle Royale (Lake Superior), Drevnick et al. (2008) documented lipid peroxidation and liver damage in association with Hg exposure. They also measured lipofuscin, a product of lipid peroxidation often found in melanomacrophage centers, to be associated with Hg. A similar finding was made by Raldúa et al. (2007) in fish sampled downstream of a chlor-alkali plant when compared to upstream controls, and also by Schwindt et al. (2008) in a study of several fish species sampled from 14 lakes in the United States. In a study of spotted gar (*Lepisosteus oculatus*) from Caddo

Lake in Texas/Louisiana, Barst et al. (2011) found increased numbers of melanomacrophage centers and also documented higher Hg content in these macrophage centers.

Methylmercury can disrupt blood chemistry. In wolf fish (*Holplias malabaricus*) injected with MeHg, changes were seen in the counts of erythrocytes, leukocytes, neutrophils, and monocytes, as well as increased hematocrit and hemoglobin content (Oliveira Ribeiro et al., 2006). In a field study of seatrout (*Cynoscion nebulosus*) in Florida, 18 of 22 studied blood chemistry parameters were lower in fish sampled from a contaminated site in South Florida when compared to a reference site (Adams et al., 2010). Such changes in blood chemistry may affect an organism's metabolic buffering capacity, especially when challenged with subsequent stressors.

Methylmercury is a proven neurotoxicant. As in other vertebrates, MeHg can cross the blood–brain barrier in fish and exert damage in the central nervous system. In Atlantic salmon fed MeHg, Berntssen et al. (2003) documented neuronal degeneration and vacuolation with effects prominent in the cerebellum, cerebrum, and brain stem. These pathological changes were accompanied by changes in oxidative stress biomarkers and neurochemistry. Altered neurochemistry has been associated with MeHg exposure in mammals (Basu et al., 2010) and birds (Rutkiewicz et al., 2011), and there is also some evidence in fish. The above-mentioned study by Berntssen et al. (2003) documented decreased monoamine oxidase activity in MeHg-fed fish. Monoaminergic changes (e.g., brain serotonin, dopamine) were also observed in catfish (*Clarias batrachus*) exposed to waterborne MeHg (Kirubagaran and Joy, 1990).

2.6. Organism Impacts

There is ample evidence of MeHg toxicity in humans, laboratory models, and wildlife (Scheuhammer et al., 2007; Clarkson and Magos, 2006; Karagas et al., 2012). Typically, fish have been viewed as the source of MeHg to humans and piscivorous wildlife, although now it is recognized that fish are also susceptible to MeHg intoxication. In recent years a growing number of studies have shown the adverse effects in fish at the whole organismal level, and several excellent and comprehensive reviews concerning this matter have been published. The review by Sandheinrich and Wiener (2011) is particularly noteworthy as they concluded that subclinical changes (e.g., biochemical alterations, cellular and tissue damage) and impaired reproduction can occur in fish with muscle concentrations between 0.5 and 1.2 µg/g (ww) or whole-body concentrations between 0.3 and 0.7 µg/g (ww). These concentrations are routinely found in fish from many regions worldwide. For example, in an assessment of Great

Lakes fish, Sandheinrich et al. (2011) used > 43,000 measurements of Hg levels in fish and determined that fish at 3% to 18% of the sites were at increased risk of reproductive impairment or reduced survivorship.

Adverse effects on fish growth and survival have been found in association with relatively high MeHg levels (6 to 20 μg/g ww in the muscle) (Wiener and Spry, 1996). Ecologically relevant exposures may also impair growth as inverse correlations between body condition factor and Hg burdens have been reported by many (Cizdziel et al., 2003; Drevnick et al., 2008; Webb et al., 2006).

A number of studies have shown MeHg to cause neurobehavioral effects in fish. One study that exemplifies the importance of developmental exposure and latencies was conducted by Fjeld et al. (1998). In their study, Arctic grayling (*Thymallus arcticus*) eggs were exposed to waterborne MeHg for 10 days and were then allowed to grow for three years without any further exposures. Following the three-year period, the fish were subject to food (i.e., *Daphnia magna*) acquisition tests showing that those fish from the high-exposure groups fared worse in terms of feeding efficiency and competitive abilities related to controls. In another study, Webber and Haines (2003) documented that golden shiners (*Notemigonus crysoleucas*) exposed to MeHg were hyperactive and exhibited altered shoaling behavior. Impaired swimming activity has been documented in Atlantic salmon (Berntssen et al., 2003) and fathead minnow (*Pimephales promelas*) (Sandheinrich and Miller, 2006).

In terms of reproductive impacts, a number of meaningful studies have been conducted in recent years (reviewed by Crump and Trudeau, 2009). In two different studies (Hammerschmidt et al., 2002; Drevnick and Sandheinrich, 2003), fathead minnows were fed MeHg from the juvenile stage until sexual maturity, after which male–female pairs were allowed to reproduce. Several reproductive end-points were found to be affected. The study by Hammerschmidt et al. (2002) reported impaired reproductive behavior, delayed spawning, and reduced gonad size. Drevnick and Sandheinrich (2003) reported reduced reproduction along with reduced levels of plasma testosterone and estradiol. At the molecular level, Klaper et al. (2006) documented a number of reproductive genes to be affected by MeHg exposure, including vitellogenin and genes related to egg fertilization and development.

2.7. Interactions with Other Toxic Agents

There is much interest in the relationship between Hg and Se. It has long been thought that the demethylation of MeHg and subsequent binding of inorganic Hg with Se in certain organisms represents a detoxification scheme

(Scheuhammer et al., 2012). It has been postulated that the resulting Hg-Se complex thus renders Hg unavailable for toxic actions or, alternatively, that it renders Se unavailable for essential biological functions (Ralston and Raymond, 2010). Regarding the second theory, Mulder et al. (2012), in a study of brown trout (*Salmo trutta*), demonstrated that when Hg is present in a molar excess to Se circulating triiodothyronine (T3) is altered, and this effect may be due to reduced Se bioavailability and subsequent alterations of Se-dependent enzymes that underlie thyroid metabolism (Branco et al., 2012). Laboratory studies have shown that fish exposed to both MeHg and Se have altered tissue residue levels, thus pointing to interactive effects (Branco et al., 2012; Huang et al., 2012).

2.8. Knowledge Gaps and Future Directions

Over the past 50 years tremendous progress has been made in our understanding of how MeHg may affect fish health, although several knowledge gaps remain. First, biomarkers at the molecular and cellular levels need to be linked to adverse outcomes at the organismal level and ultimately to the population level. A number of studies suggest that MeHg at relevant levels may affect fish populations, but this possibility has yet to be established. Second, focused research over the past decade has linked MeHg exposure with reproductive impacts at multiple points along the hypothalamus-pituitary-gonadal (HPG) axis, but physiological impacts on other organ systems have not been well established in fish. Despite being a proven neurotoxicant, the number of published behavioral studies seems limited. In humans, for example, there now exists some evidence suggestive of MeHg-associated changes in the cardiovascular and immune systems (Karagas et al., 2012), but these changes have not been deeply studied in fish. Third, whether or not fish may indeed demethylate MeHg, and if so, the underlying mechanism(s) involved, warrants more research. Some promising results have been obtained (e.g., Drevnick et al., 2008; Barst et al., 2011), though additional work may help us better understand how fish handle MeHg and whether this is the basis for interspecies differences in sensitivity. The inclusion of Hg stable isotope studies may increase our understanding of Hg cycling both in the environment and in the fish body (Kwon et al., 2012). Fourth, continued research is needed to understand the relationship between MeHg and Se with two complementary viewpoints: (1) Se renders Hg unavailable for toxic actions, or (2) Hg renders Se unavailable for essential biological functions, providing a valuable starting point.

3. ORGANOSELENIUM

3.1. The Chemicals

3.1.1. STRUCTURES AND BREAKDOWN PRODUCTS

Selenium (Se) is one of the most hazardous trace elements for fish, following Hg (Luoma and Presser, 2009; Janz et al., 2010). Although inorganic forms of Se are the dominant forms present in the water column, the toxicity of Se to fish is governed by the bioavailability of organoselenium compounds obtained from dietary sources in the food web (Stewart et al., 2010). Thus, the ecotoxicology of Se more closely resembles persistent organic chemicals rather than most trace elements. Although much research on the toxicity of organoselenium compounds to fish has been conducted in recent years, many gaps remain in our understanding of Se as a priority aquatic pollutant (Janz, 2011).

Selenium exists in four oxidation states (VI, IV, 0, –II) and is present in a diverse array of inorganic and organic chemical forms (species) in aquatic environments. Organoselenium compounds comprise the greatest diversity of Se species, with over 50 distinct compounds identified to date (Maher et al., 2010; Wallschläger and Feldmann, 2010). Indeed, Se has one of the most complex and diverse organic chemistries of any environmentally relevant trace element (Luoma and Presser, 2009; Wallschläger and Feldmann, 2010). This feature of the environmental chemistry of Se is related to both its high reactivity and the fact that it is an essential ultratrace micronutrient in most biota, including fish (Schrauzer, 2000; Janz, 2011).

Selenium speciation in abiotic and biotic compartments of aquatic ecosystems is governed both by biogeochemical processes occurring in the water column (oxic processes) and sediments (oxic and anoxic processes) and by biotransformation processes occurring in plants and animals (Maher et al., 2010). Although the dominant Se species found dissolved in the water column are the inorganic oxyanions selenate (SeO_4^{2-} or Se[VI]) and selenite (SeO_3^{2-} or Se[IV]), these forms rarely occur at concentrations resulting in acute or chronic toxicity to fish (Janz et al., 2010; Janz, 2011). The greatest toxicological hazard to fish arises following dietary exposure to certain organoselenium compounds, in particular Se incorporated into proteins in the form of selenomethionine. A detailed discussion of organoselenium biotransformation and its toxicological significance is provided in Section 3.3.3.

Organoselenium compounds can be broadly categorized into methylated and volatile forms, biochemical intermediates, and Se-containing amino acids and proteins (Figure 3.2). It is beyond the scope of this chapter to

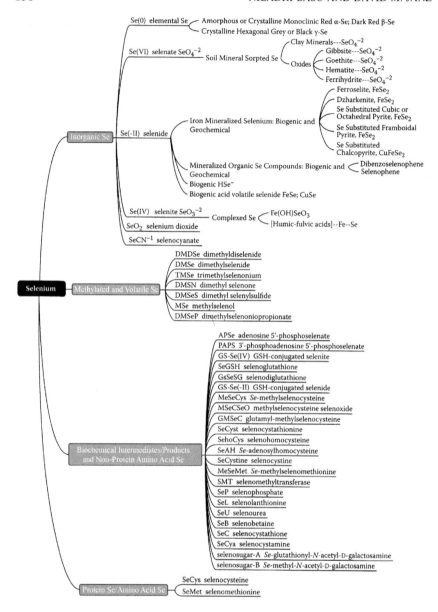

Figure 3.2. Major Se species found in aquatic environments and biota. *Source*: Maher et al. (2010).

provide a thorough discussion of the metabolic and environmental pathways involved in organoselenium speciation, but further details can be found in recent reviews (Maher et al., 2010; Wallschläger and Feldmann, 2010). Although methylselenides (CH_3SeH, $(CH_3)_2Se$, and $(CH_3)_2Se_2$) are known to be produced in Se-contaminated aquatic systems, these forms are volatile and are not believed to be toxicologically significant to fish (Maher et al., 2010).

The majority of Se is present in animals as either selenocysteine or selenomethionine residues incorporated into proteins (Mayland, 1994; Schrauzer, 2000; Suzuki and Ogra, 2002). Selenocysteine, the 21st amino acid, is synthesized in a highly regulated manner via a selenocysteinyl-tRNA directed by a UGA codon reprogrammed by a selenocysteine insertion sequence (Zinoni et al., 1987; Bock et al., 1991). Selenocysteine is then incorporated into essential selenoproteins, which are defined as proteins containing at least one selenocysteine residue (Reeves and Hoffmann, 2009; Gladyshev and Hatfield, 2010). The best characterized selenoproteins in fish are glutathione peroxidases, thioredoxin reductases, iodothyronine deiodinases, and selenophosphate synthetase-2 (Janz, 2011). Fish appear to have the largest selenoproteomes among animals, with genes coding for 32–37 selenoproteins in the species studied to date, compared to 25 selenoproteins in humans (Lobanov et al., 2008, 2009; Reeves and Hoffmann, 2009). In contrast to selenocysteine, selenomethionine is incorporated into proteins in an unregulated, dose-dependent manner since animals cannot distinguish between methionine and selenomethionine during protein synthesis. Thus, with increasing exposure to Se, greater amounts of selenomethionine are incorporated into proteins in the place of methionine, which provides a key aspect of Se toxicology in fish, as discussed further in Section 3.5.

3.1.2. SOURCES

A description of the sources of organoselenium compounds in aquatic systems must begin with inorganic Se, since it is these forms that initially contribute to Se loading. Globally, Se content is variable in soils and sediments but is particularly enriched in black shales, phosphorites, and coal (Presser et al., 2004; Maher et al., 2010). Natural processes such as weathering of rocks, volcanic activity, wildfires, and volatilization from water bodies can influence the loading of inorganic Se into aquatic systems (Nriagu, 1989). However, ecotoxicologically significant inputs of inorganic Se to aquatic ecosystems arise due to anthropogenic activities that involve association of the natural Se sources described above with water (Young et al., 2010a). Industrial activities such as coal, phosphate, uranium, base metal, and precious metal-mining operations; subsequent milling, smelting, and/or combustion processes; and oil refining can greatly increase the loading of Se into aquatic ecosystems (Lemly, 2004; Maher et al., 2010). Agricultural activities such as irrigation and drainage of seleniferous soils, and field application of fertilizers, manure, and biosolids, have also been reported to cause significant Se contamination of receiving water bodies (Bouwer, 1989; Ohlendorf, 2003). In situations where Se poses a toxicological hazard to fish, it is these initial inorganic inputs from anthropogenic activities that provide the "raw materials" for subsequent conversion to organoselenium compounds.

Similar to organomercury compounds, the major sources of organo-selenium compounds in aquatic ecosystems are not through "traditional" point and nonpoint inputs, but arise indirectly via biotransformation of inorganic to organic forms of Se at the base of food webs. Algae, bacteria, fungi, and higher plants (macrophytes) efficiently assimilate and biotrans-form inorganic Se (primarily selenate and selenite) to organoselenium compounds (Fan et al., 2002; Stewart et al., 2010). This is the most important process involved in the bioaccumulation and trophic transfer of organoselenium compounds to fish, since aqueous concentrations of inorganic Se are enriched (bioconcentrated) up to 10^6-fold in such primary and secondary producers, which then convert them to organic, and more toxic, forms (Stewart et al., 2010). Organoselenium compounds subse-quently bioaccumulate via dietary pathways at higher trophic levels in aquatic food webs, including fish. As discussed in Section 3.4, although organoselenium compounds have been shown to biomagnify through successive trophic levels in aquatic ecosystems, the extent of biomagnifica-tion (biomagnification factors generally less than 6) is not as great as that observed with organomercury compounds.

3.1.3. USES AND ECONOMIC IMPORTANCE

Although inorganic Se has a variety of uses (e.g., as a pigment for glass, ceramics, plastics, and paints, as a photovoltaic substance in solar panels and photometers, as an accelerant and vulcanizing agent in the rubber industry, and more recently as a component of the inner core of semiconductor nanocrystals [quantum dots]), the only significant use of organoselenium compounds is for dietary supplements in domestic livestock and humans. As an essential micronutrient that may be limited in certain areas with low soil Se, diets augmented with Se are commonly used for feeding livestock, and occasionally humans. Since greater than 80% of Se in wheat, corn, and soybeans is L(+)-selenomethionine, this is the most appropriate form used for dietary supplements, although other organic and inorganic forms of Se are also used (Schrauzer, 2001). Recently, the increased awareness of anticancer properties related to Se intake has led to increased consumption of Se supplements in the human population.

However, economic costs are associated with aquatic Se contamination. A recent analysis of the economic costs related to wildlife poisonings in the United States arising from just coal-mining wastes has been estimated at greater than $2.3 billion to date (Lemly and Skorupa, 2012).

3.1.4. REGULATORY STATUS

Perhaps more than any aquatic pollutant in recent years, the regulatory status of Se for protecting fish and other aquatic biota has stimulated much debate and discussion. This is especially true in the United States and Canada, where water quality criteria/guidelines for the protection of aquatic life, which are based on aqueous Se concentrations (see Section 3.1.5), are currently being reevaluated. The lack of harmonization of regulatory guidelines is generally due to a failure to link Se biogeochemistry with physiological and ecological processes that determine the bioavailability and toxicity of organoselenium compounds, which are obtained primarily through dietary exposures in fish and other aquatic consumers (Luoma and Presser, 2009; Presser and Luoma, 2010). In short, regulatory guidelines based on aqueous concentrations of (primarily inorganic) Se create great uncertainty since they are unreliable predictors of ecotoxicological risk to fish and other aquatic species. This is because differences in physical (hydrology and geochemistry), ecological (food web dynamics and community structure), and physiological (interspecific differences in uptake, biotransformation, and excretion) processes among aquatic ecosystems dictate the speciation, bioavailability, and toxicity of Se to fish (Luoma and Presser, 2009; Stewart et al., 2010). Thus, perhaps more than any other aquatic pollutant, Se will require some form of site-specific monitoring to reduce uncertainty in assessing the ecological risk to fish and other aquatic species (Presser and Luoma, 2010).

A tissue-based regulatory criterion/guideline for Se has been proposed for some time (DeForest et al., 1999; Hamilton, 2002, 2003). As a result, the United States Environmental Protection Agency (U.S. EPA) has proposed a tissue-based Se criterion for the protection of aquatic life of 7.91 μg Se/g dry mass of fish (whole body), which is currently under revision (U.S. EPA, 2004). Both the United States and Canada will likely set tissue-based aquatic life chronic Se criteria/guidelines for the protection of aquatic life in the near future.

3.1.5. WATER QUALITY CRITERIA

Ambient water quality criteria for the protection of aquatic life from chronic exposure to Se have been developed in the United States, Canada, Australia/New Zealand, and South Africa. The U.S. EPA currently has chronic water quality criteria for the protection of aquatic organisms set at

5 µg Se/L for freshwater and 71 µg Se/L for saltwater (U.S. EPA, 1987). The Canadian Council of Ministers of the Environment (CCME) guideline for protection of aquatic life is 1 µg Se/L for freshwater, with no saltwater guideline (CCME, 2003). The Australian and New Zealand Environment and Conservation Council (ANZECC) criteria (termed "trigger" values) are based on protection of a percentage of aquatic species and is 5 µg Se/L for protection of 99% of freshwater aquatic species, increasing to 34 µg Se/L for protection of 80% of species, with no saltwater guideline (ANZECC, 2000). South Africa has a freshwater guideline of 5 µg Se/L (based on dissolved Se), with no saltwater guideline (Department of Water Affairs and Forestry, 1996). In Canada, certain provincial governments have developed their own guidelines, which vary greatly. In Ontario, the ambient provincial water quality objective (PWQO) for the protection of aquatic life is 100 µg Se/L for freshwater (Ontario Ministry of Environment and Energy, 1999). In contrast, the British Columbia water quality guideline (WQG) is 2 µg Se/L for both freshwater and saltwater (Nagpal, 2001). Clearly, Se regulatory guidelines aimed at protecting fish population sustainability are a subject of significant debate and will likely continue to evolve for many years to come.

3.2. Environmental Situations of Concern

Selenium contamination of freshwater and saltwater ecosystems due to anthropogenic activities is recognized as an ecotoxicological issue of global concern (Lemly, 2004; Young et al., 2010a, b). A wide variety of mining and petroleum industry processes liberate Se from raw materials during ore milling, waste rock disposal, combustion, and/or refining. Specifically, coal combustion wastes from power plants (fly ash), coal mining waste rock disposal, and petroleum refinery wastewater can cause significant Se loading into aquatic ecosystems. Mining operations that extract base metals (e.g., copper, nickel), precious metals (e.g., gold, silver), uranium, and phosphate from ores with marine sedimentary origin can increase Se loading via ore milling wastewater effluents, runoff from waste rock and tailings, and smelting. Agricultural irrigation in regions with seleniferous soils can also result in significant loading of Se to aquatic ecosystems in runoff. It is beyond the scope of this chapter to provide specific details of the many instances of aquatic Se contamination that have been shown to negatively impact fish, but several publications document these situations (Skorupa, 1998; Lemly, 2004; Young et al., 2010a, b; Janz, 2011). In most cases, fish have been shown to be among the most sensitive and susceptible aquatic species impacted by elevated Se arising from these industrial and agricultural activities. In aquatic ecotoxicology, these cases provide some of the clearest cause–effect relationships between exposure to a toxic substance and adverse effects on fish population dynamics (Lemly, 1985; Skorupa, 1998; Janz et al., 2010).

3.3. Toxicokinetics

3.3.1. UPTAKE

3.3.1.1. Gills. Although organoselenium compounds such as selenomethionine and certain methylated forms have been detected in freshwater and saltwater, and may be bioavailable to fish through gill uptake, even in the most Se-contaminated aquatic systems their concentrations are inadequate to cause significant bioaccumulation (Stewart et al., 2010).

3.3.1.2. Gut. As discussed previously, dietary uptake of organoselenium compounds in fish is the predominant exposure pathway, and thus gut uptake is of particular significance for both nutritional requirements and toxicity. L-Selenomethionine is the major ($>80\%$) form of Se found in organisms at all relevant trophic levels (Fan et al., 2002; Maher et al., 2010; Phibbs et al., 2011b). Gastrointestinal absorption of selenomethionine occurs mainly by the epithelial Na^+-dependent neutral amino acid transporter in mammals (Schrauzer, 2000). The symport system B^0 and exchanger (antiport) system b^{0+} of neutral amino acid transporters apparently dominate selenomethionine absorption at the apical membrane, with similar kinetics between methionine and selenomethionine (Nickel et al., 2009). In green sturgeon (*Acipenser medirostris*), selenomethionine and methionine absorption are mediated by the same apical membrane transporter in the gut, although the identity and number of transporters involved are unknown (Bakke et al., 2010). Interestingly, selenomethionine and methionine absorption rates in green sturgeon follow an increasing gradient from proximal to distal portions of the gastrointestinal tract (Bakke et al., 2010). A small portion of gut uptake in fish may also arise from ingesting dissolved organoselenium compounds, which may be greater in marine teleosts; however, this is likely insignificant compared to dietary exposure pathways.

3.3.2. TRANSPORT

Little is known about organoselenium transport in the blood of fish, although insights can be gained from mammals. Selenoprotein P (SelP) is the dominant selenoprotein in rat and human plasma, and its major role involves systemic Se transport and distribution (Motsenbocker and Tappel, 1982; Burk and Hill, 2009). Zebrafish possess two isoforms of SelP, which contain a greater number of selenocysteine residues than mammalian SelP (Kryukov and Gladyshev, 2000; Tujebajeva et al., 2000). This suggests that SelP plays an important role in Se transport in fish, although this remains to be confirmed. In rats, approximately 65% of circulating Se occurs as selenocysteine residues in SelP, and SelP is taken up into organs such as brain and testis by apolipoprotein E receptor-2 (apoER2) and into kidney proximal convoluted tubule cells by megalin (Burk and Hill, 2009). Another selenoprotein found in the mammalian bloodstream is extracellular glutathione peroxidase (GPx3), which has not been identified to date in fish (Kryukov and Gladyshev, 2000). Another possible transport mechanism in fish involves the nonspecific substitution of methionine with seleno-methionine in albumin and other plasma proteins, which has been reported in mammals (Burk et al., 2001).

3.3.3. Biotransformation

Organoselenium compounds undergo complex biotransformation reactions *in vivo* in mammals involved in both nutritional roles and, at greater exposures, toxicity (Mayland, 1994; Schrauzer, 2000; Suzuki and Ogra, 2002). Although some recent work has investigated the biotransformation of organoselenium in fish, this area of study requires much more research to elucidate potentially important mechanisms of toxicity. As mentioned previously, fish obtain the majority of Se from selenomethionine and selenocysteine incorporated into proteins of prey organisms. The biotransformation of selenomethionine appears to be closely related to toxicity in fish, as discussed below.

Selenomethionine absorbed from the gut is transported to the liver via the hepatic portal venous system, where it can undergo two processes. First, a portion of selenomethionine (and all selenocysteine) is biotransformed to the common metabolic intermediate, selenide (Se^{2-}), before being specifically incorporated into essential selenoproteins as selenocysteine (Suzuki and Ogra, 2002). At normal dietary levels of Se, selenomethionine thus acts as an unregulated pool of Se for eventual selenocysteine synthesis (Suzuki and Ogra, 2002). A second portion of intact selenomethionine can be incorporated into proteins in the place of methionine, since animals cannot distinguish between methionine and selenomethionine during protein synthesis. Thus, as dietary selenomethionine exposure increases in aquatic ecosystems contaminated with Se, proportionally greater amounts of selenomethionine will enter the unregulated Se pool via biotransformation or be incorporated into proteins in the place of methionine. Such supraphysiological uptake of selenomethionine is closely linked to toxicity, as discussed in Section 3.5.

Recent field studies have investigated Se speciation in fish inhabiting lakes receiving uranium milling effluent (Phibbs et al., 2011a, 2011b). Through use of synchrotron-based X-ray absorption spectroscopy (XAS), the dominant Se species observed in lake chub (*Couesius plumbeus*) and spottail shiner (*Notropis hudsonius*) collected from an uncontaminated reference lake was found to be selenocystine (a dimer of the essential amino acid, selenocysteine), with detectable but low quantities of selenomethionine. In contrast, selenomethionine was the dominant species observed in fish collected from Se-contaminated lakes. In addition, a significant positive relationship was observed between whole-body total Se concentration and the fraction of Se present as selenomethionine in lake chub and spottail shiner (Phibbs et al., 2011b). This finding suggested that selenomethionine may act as a marker of bioavailable dietary Se in contaminated food webs.

3.3.4. SUBCELLULAR PARTITIONING

Selenium is incorporated into essential selenoproteins (as selenocysteine) or methionine-containing proteins (where selenomethionine is substituted), and thus these proteins will largely dictate subcellular Se partitioning. However, little is known in this regard, and it remains an exciting area for further study.

3.3.5. ACCUMULATION IN SPECIFIC ORGANS

The primary organ for Se accumulation in mammals is the liver (Sato et al., 1980) because it is the dominant site for biotransformation of ingested organoselenium and for selenoprotein synthesis (Burk and Hill, 2009). However, all organs will accumulate Se to a varying extent due to its presence as selenomethionine and selenocysteine in proteins. In fish, organ- and tissue-specific Se accumulation has been determined in adult northern pike, where liver and kidney displayed the greatest Se concentrations, followed by ovary (egg), muscle, and bone (Muscatello et al., 2006). Incorporation of Se into otoliths, calcified structures in the inner ear of fish commonly used for age determination, has been determined using laser ablation inductively coupled plasma mass spectrometry (LA-ICP-MS) and was shown to indicate a chronological record of fish migration into and out of Se-contaminated aquatic ecosystems (Palace et al., 2007).

Relative organoselenium accumulation in organs and tissues of fish is of considerable interest because of the current focus on developing tissue-based regulatory guidelines to protect fish populations (see Section 3.1.5). The ovary (i.e., ovarian follicles) has the greatest ecotoxicological relevance for Se accumulation, since yolk proteins provide the ultimate dose of organoselenium that leads to larval deformities (DeForest and Adams, 2010; Janz et al., 2010; Janz 2011; see Section 3.5). Since the yolk protein precursor vitellogenin differs among fish species in amino acid number and sequence, it has been hypothesized that the relative number of methionine residues in yolk proteins (which are replaced with selenomethionine in a dose-dependent manner) may predict species sensitivity (Janz et al., 2010).

3.3.6. EXCRETION

3.3.6.1. Gills. It is not known whether excretion of organoselenium compounds via gills occurs in fish, although given the efficient absorption of organoselenium and incorporation of Se into proteins, this seems unlikely to be a significant excretion route. It is possible that volatile organic forms of Se (methylselenides) that are produced during organoselenium biotransformation can be excreted via gills, but no studies in fish have been conducted.

3.3.6.2. Kidney. Although information is lacking for fish, urinary excretion of Se is the dominant route of elimination in mammals (Kobayashi et al., 2002). Selenosugars (methylseleno-N-acetylgalactosamine, methylseleno-N-acetylglucosamine, and methylselenogalactosamine) are the major urinary metabolites in mammals at nutritional and low toxic ranges (Gammelgaard et al., 2008). At higher Se exposures, methylated metabolites such as trimethylselenonium appear in mammalian urine and might serve as urinary biomarkers of excess Se intake (Suzuki et al., 2005; Ohta and Suzuki, 2008).

3.3.6.3. Bile. The importance of hepatobiliary Se excretion in fish has not been investigated, but it is a minor route in mammals (Gregus and Klaassen, 1986) involving excretion of Se diglutathione (GS-Se-SG) mediated by the canalicular GSH transporter (Gyurasics et al., 1998).

3.3.6.4. Gut. Although it is likely that a portion of dietary organoselenium is not absorbed and is excreted in feces, this is unlikely to be significant since selenomethionine, representing $>80\%$ of dietary organoselenium, is efficiently absorbed from the GI tract.

3.4. Potential for Bioconcentration or Biomagnification

As discussed in Section 3.1.2, inorganic Se is efficiently bioconcentrated (up to 10^6-fold) at the base of food webs by primary and secondary producers, where it is biotransformed to organoselenium compounds. These dietary sources of organoselenium (predominantly selenomethionine) are then bioaccumulated in successive trophic levels, including fish, which occupy diverse niches and may be primary, secondary, or tertiary consumers. Although Se has been shown to biomagnify through food webs, it does so to a lesser extent than organomercurials or persistent organochlorine contaminants, with biomagnification factors for Se in fish generally ranging from 1.5 to 6 (Lemly, 1985; Muscatello et al., 2008; Muscatello and Janz, 2009; Stewart et al., 2010). Despite relatively low biomagnification factors, the significant bioconcentration or enrichment at the base of food webs can deliver toxicologically significant dietary exposures of organoselenium to fish. In addition, organisms at lower trophic levels such as algae, zooplankton, and benthic invertebrates, which are major prey items for fish, are relatively insensitive to Se toxicity and thus act as vectors, delivering organoselenium compounds to fish (Janz et al., 2010; Stewart et al., 2010).

3.5. Mechanism(s) of Toxicity

Dietary exposure to organoselenium compounds is not considered acutely toxic to fish in even the most Se-contaminated aquatic ecosystems (Cleveland et al., 1993). The most sensitive ecotoxicological effect appears to occur during larval development. Deformities have been observed in larvae as a result of chronic organoselenium exposure of adult female fish in contaminated aquatic systems (Janz et al., 2010; Janz, 2011). During the vitellogenic phase of oogenesis, female fish incorporate excess dietary selenomethionine, in a dose-dependent manner, into the egg yolk precursor protein (vitellogenin) in place of methionine. Following reproduction, the selenomethionine-enriched yolk is resorbed in offspring between hatch and swim-up as their initial energy source, which results in a characteristic and diagnostic suite of larval deformities (see Section 3.6). It has been hypothesized that during yolk resorption, selenomethionine is biotransformed to metabolites such as methylselenol (CH_3SeH) and dimethylselenide (CH_3SeCH_3) that react with glutathione (GSH) to produce reactive oxygen species (ROS) (Spallholz, 1994; Palace et al., 2004; Janz et al., 2010; Misra and Niyogi, 2009). At sufficiently high Se exposures, ROS production can overwhelm antioxidant defenses and cause oxidative damage to proteins, lipids, and DNA, which is postulated to cause larval deformities in fish (Palace et al., 2004), in an etiology similar to certain organic toxicants (Bauder et al., 2005; Wells et al., 2009).

3.6. Organism Impacts

Numerous laboratory studies in fish, employing a multitude of exposure scenarios with varying combinations of inorganic and/or organic Se, and aqueous and/or dietary exposure routes, have reported reduced growth and in certain cases increased mortality (reviewed in Jarvinen and Ankley, 1999; Janz et al., 2010). However, the most ecologically relevant organismal effects following organoselenium exposure appear to be related to reproductive impairment as described above. Skeletal deformities include spinal curvatures of lateral (scoliosis) and dorsoventral (kyphosis and lordosis) aspects. Craniofacial deformities include improper development of the mandible and operculum, shortened facial (oral-nasal) region (due mainly to improper development of the maxilla), reduced forebrain, and microphthalmia (i.e., abnormally small eyes). Fin deformities include shortened or missing fins. In addition, various forms of edema (pericardial, yolk sac, and cranial) are commonly associated with skeletal, craniofacial, and fin deformities. Representative examples of such abnormalities in larval fish are shown in Figure 3.3. These malformations are remarkably consistent

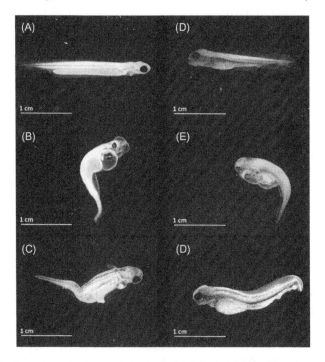

Figure 3.3. Examples of characteristic Se-induced larval deformities in white sucker (*Catostomus commersoni*; panels A–C) and northern pike (panels D–E). Panels A and D are larvae originating from spawning female fish collected at reference sites showing normal development following yolk sac resorption. Panels B, C, E, and F are larvae originating from spawning female fish collected at field sites contaminated with elevated Se. Panel B shows craniofacial and pericardial edema, and a lateral spinal curvature (scoliosis). Panel C shows a shortened pelvic fin and scoliosis. Panel E shows a craniofacial deformity, microphthalmia, pericardial edema, yolk sac edema, and a dorsoventral spinal curvature (kyphosis). Panel F shows a craniofacial deformity and a dorsoventral spinal curvature (lordosis). Reprinted with permission from Muscatello (2009).

among fish species and are a useful diagnostic tool to identify Se toxicity in field studies where fish are exposed to complex mixtures of contaminants and other stressors (Lemly, 1997a; Janz et al., 2010). Importantly, an increasing frequency and severity of these larval deformities can impact juvenile recruitment in fish populations exposed to elevated Se, and an index has been proposed that relates the frequency of deformities in fry to negative impacts on fish population dynamics (Lemly, 1997).

In addition to larval deformities that arise following maternal transfer of organoselenium to offspring, there have been reports of direct effects of elevated dietary organoselenium exposure on swimming physiology and behavior in juvenile and adult fish. Swimming physiology has been shown to be a sensitive and ecologically relevant sublethal response to toxicant exposure in fish (Little and Finger, 1990). An earlier study reported no effects of 90-day organoselenium exposure (up to 25 μg Se g^{-1} diet wet mass [as L-selenomethionine]) on swim duration, frequency of activity, swim speed, and swim distance in juvenile bluegill sunfish (*Lepomis macrochirus*) (Cleveland et al., 1993). In contrast, a recent study investigated the effects of 90-day dietary L-selenomethionine exposure (1.3, 3.7, 9.6, and 26.6 μg Se/g dry weight) on swim performance in adult zebrafish and in their offspring exposed to Se via maternal transfer (Thomas and Janz, 2011). Critical swimming speed (U_{crit}) decreased significantly in adults and in the F_1 generation at 120 days post hatch, suggesting ecologically relevant effects of dietary Se exposure on swim performance (Thomas and Janz, 2011). Surprisingly, the dose-dependent reductions in U_{crit} in adult zebrafish were associated with elevated whole-body concentrations of energy stores (triacylglycerols and glycogen), suggesting a possible effect of dietary Se exposure on energy homeostasis.

In a subsequent study (Thomas et al., 2013), greater whole-body triacylglycerols in adult zebrafish exposed to elevated dietary seleno-methionine coincided with significant downregulation of hepatic mRNA abundance of β-hydroxyacyl coenzyme A dehydrogenase (HOAD) and methionine adenosyltransferase 1α (MAT 1A), enzymes involved in β-oxidation of fatty acids and triacylglycerol transport, respectively. These results suggest that elevated dietary selenomethionine exposure may impair key enzymes involved in energy mobilization in fish, although other mechanisms such as altered glucocorticoid homeostasis may also be involved (Thomas and Janz, 2011; Wiseman et al., 2011). Since the observed effects on swimming performance and energy homeostasis were observed at dietary Se exposures within the same range as those reported to cause increased frequencies of larval deformities, further research is needed to determine the ecophysiological significance of such effects.

3.7. Interactions with Other Toxic Agents

The best characterized interactions of Se with other toxic agents are antagonistic interactions of inorganic Se with a variety of other trace elements in mammals, including As, Ag, Au, Cd, Cu, Co, Cr, Hg, Mo, Ni, Pb, and Zn (Schrauzer, 2009), some of which have also been observed in fish (Janz et al., 2010; Janz, 2011). Less is known about the interactions of organoselenium compounds with other toxicants, particularly in fish. However, MeHg has been reported to inactivate essential selenoproteins in mammals by binding covalently to selenocysteine at the active site of selenoenzymes such as glutathione peroxidases (Ralston et al., 2008).

3.8. Knowledge Gaps and Future Directions

From a basic biological perspective, much remains to be discovered regarding the physiological roles of the majority of selenoproteins in fish, which are believed to possess the largest selenoproteome among animals. To date, work has been limited to genomic studies in only a few fish species. Tissue-specific mRNA expression has only been determined for the majority of selenoproteins in zebrafish (Kryukov and Gladyshev, 2000; Thisse et al., 2003). Further studies investigating physiological functions of selenoproteins will undoubtedly provide fascinating insight into the evolution of selenoprotein families.

From a toxicological perspective, the major data gaps involve better understanding the underlying mechanisms of toxicity in fish. Although it has been hypothesized that biotransformation of organoselenium results in generation of ROS that cause larval deformities in fish, definitive studies involving molecular and pharmacological approaches have not been conducted. In addition, mechanisms of the direct effects of dietary organoselenium exposure on swimming physiology and associated disruption of energy homeostasis have not been elucidated.

From an ecotoxicological and risk assessment/management perspective, the effects of impaired recruitment resulting from the elevated frequencies and severities of larval deformities are not clear. Further studies are warranted to more fully understand the potential impacts of organoselenium exposure on the sustainability of fish populations and communities.

4. ORGANOARSENICALS

4.1. The Chemicals

Arsenic (As) exists in four oxidation states (V, III, O, −III) and is present in a variety of inorganic and organic forms in aquatic ecosystems (Table 3.4).

Table 3.4
Inorganic and organic arsenic compounds found in aquatic ecosystems

Name	Abbreviation	Chemical Formula
Arsenate (arsenic acid)	As[V]	$AsO(OH)_3$
Arsenite (arsenous acid)	As[III]	$As(OH)_3$
Monomethylarsonic acid	MMA[V]	$CH_3AsO(OH)_2$
Monomethylarsonous acid	MMA[III]	$CH_3As(OH)_2$
Dimethylarsenic acid	DMA[V]	$(CH_3)_2AsO(OH)$
Dimethylarsenous acid	DMA[III]	$(CH_3)_2As(OH)$
Dimethylarsinoyl ethanol	DMAE	$(CH_3)_2AsOCH_2CH_2OH$
Dimethylarsinoylacetic acid	DMAA	$(CH_3)_2AsOCH_2COOH$
Trimethylarsine	TMA	$(CH_3)_3As$
Trimethylarsine oxide	TMAO	$(CH_3)_3AsO$
Tetramethylarsonium ion	TETRA	$(CH_3)_4As^+$
Arsenobetaine	AsB	$(CH_3)_3As^+CH_2COO^-$
Arsenocholine	AsC	$(CH_3)_3As^+CH_2CH_2OH$
Arsenic containing ribosides	Arsenosugars	
Arsenic containing neutral lipids	Arsenolipids	

Organoarsenicals include arsenobetaine, arsenocholine, several methylated compounds in the oxidation states of V and III, and a variety of arsenosugars and organolipids (Sharma and Sohn, 2009; Dopp et al., 2010; Reimer et al., 2010). A more detailed inventory of organoarsenical structures and chemical formulas has recently been published (Reimer et al., 2010).

The majority of inorganic As in aquatic environments is of natural origin, although anthropogenic activities such as mining, fossil fuel combustion, and pesticide application can result in local contamination (Sharma and Sohn, 2009). Gold mining is particularly relevant as a source of contamination in surface waters (McIntyre and Linton, 2012). In freshwater, the majority of As is inorganic and present as the oxyanions arsenite (As[III]) and arsenate (As[V]). However, the methylarsenicals MMA[V] and DMA[V] have been known for some time to be present at low concentrations in freshwater (Cullen and Reimer, 1989), which can become greater at warmer water temperatures and in eutrophic lakes due to enhanced biomethylation (Hasegawa et al., 2010).

The majority of toxicity testing for As in fish and other aquatic species has been conducted using arsenite and arsenate, and current guidelines aimed at protecting aquatic biota such as fish are based on total aqueous As. Recent evidence that the trivalent methylated species, MMA[III] and DMA [III], exhibit high toxicity in mammalian systems may result in a reevaluation of As regulations for aquatic biota (see Section 4.5). However, we are not aware of any detailed toxicological studies investigating the acute or chronic toxicity of these methylated forms of As in fish.

Acute and chronic criteria for the protection of aquatic life (based on total aqueous As) for both freshwater and saltwater have been developed in several jurisdictions, including the United States, Canada, the United Kingdom, Australia/New Zealand, South Africa, and the Netherlands (see McIntyre and Linton, 2012 for details). A variety of approaches have been used in deriving these criteria, which has resulted in a wide range of values. Acute aquatic life criteria among jurisdictions range from 8 to 340 $\mu g\,L^{-1}$ and 1.1 to 69 $\mu g\,L^{-1}$ for freshwater and saltwater, respectively. Chronic aquatic life criteria range from 0.5 to 150 $\mu g\,L^{-1}$ and 0.5 to 36 $\mu g\,L^{-1}$ for freshwater and saltwater, respectively (McIntyre and Linton, 2012).

4.2. Environmental Situations of Concern

Naturally elevated As concentrations in groundwater is the most toxicologically significant concern worldwide due to human exposure via drinking water and the known carcinogenicity of As (Mandal and Suzuki, 2002). Aquatic ecotoxicological concerns are uncommon and are associated with anthropogenic activities such as gold mining (Wagemann et al., 1978; de Rosemond et al., 2008), although the role of organoarsenicals is uncertain.

4.3. Toxicokinetics

Suhendrayatna et al. (2002) exposed juvenile tilapia (*Tilapia mossambica*) for 7 days to sublethal aqueous concentrations (1–50 mg As L^{-1}) of MMA[V] or DMA[V] and reported a concentration-dependent whole-body bioaccumulation of total As. Although this study did not directly determine gill uptake, a recent study has identified five members of the aquaglycero-porin subfamily of aquaporins in zebrafish (Hamdi et al., 2009). Aquaglyceroporins (AQPs) are membrane proteins that mediate the osmotically driven flux of glycerol in diverse taxa, including fish (Cutler et al., 2007). Hamdi et al. (2009) found that uptake of arsenite and MMA [III] was facilitated by all AQPs *in vitro* and that mRNA expression of AQP3 was present in zebrafish gill, indicating the potential role of this transport protein in As uptake. Jung et al. (2012) reported that AQP3a, an isoform of AQP3 expressed in killifish (*Fundulus heteroclitus*) gill, did not transport arsenite. It was speculated that this may explain why killifish poorly assimilate As and are tolerant of elevated environmental levels of this metalloid (Jung et al., 2012). However, organoarsenicals are only found at low concentrations in freshwater and saltwater (i.e., $\mu g\,L^{-1}$), suggesting that gill uptake of these organoarsenicals would be minimal in comparison to inorganic As.

Similar to gill uptake, dietary inorganic As accumulates to a greater extent across the gut than organoarsenicals. Rainbow trout exposed to elevated dietary concentrations of DMA[V] and arsanilic acid (p-amino-benzene-arsonic acid; a growth promoter used historically in swine and poultry diets) exhibited concentration-dependent accumulation of total As in whole-body. However, accumulation was much greater for dietary arsenate and arsenite (Cockell and Hilton, 1988). Although no studies have been conducted in fish, a variety of organoarsenicals have been reported to be potentially bioavailable via the gut using an *in vitro* (Caco-2 cells) mammalian model (Laparra et al., 2007).

As with other trace elements discussed in this chapter, biotransformation of As is closely associated with toxicity in vertebrates, including fish (Dopp et al., 2010). Three toxicologically important steps are involved in As biotransformation in mammals: (1) reduction of As[V] to As[III], (2) methylation, and (3) the substitution of hydroxyl by thiol groups (Kumagai and Sumi, 2007; Dopp et al., 2010). Although few studies have investigated the detailed steps of As biotransformation in fish, two potential pathways consistent with that in mammals have been proposed (Kumagai and Sumi, 2007; Dopp et al., 2010). The first is a series of pentavalent to trivalent reductions of As species, and subsequent oxidative methylations catalyzed by arsenic methyltransferase (As3MT, originally named Cyt19) with S-adenosylmethionine (SAM) as the methyl donor. Second, a novel biotransformation pathway has been proposed (Hayakawa et al., 2005) that involves glutathione (GSH) conjugation to As[III] to form arsenite triglutathione, which then undergoes As3MT-catalyzed methylation. Although these pathways of As biotransformation in mammals remain under debate (Dopp et al., 2010), they have not been investigated in fish, and this represents an exciting area of future investigation. Recently, an As3MT homologue was identified in zebrafish, which displays a similar function to mammalian As3MT (Hamdi et al., 2012).

Most of our knowledge regarding As biotransformation in fish comes from studies measuring As speciation in fish tissues. Studies in both freshwater and saltwater fish species report that organoarsenicals comprise the majority of total As in whole-body or muscle (reviewed in McIntyre and Linton, 2012), although the fraction of total As as organoarsenicals varies to a greater extent in freshwater fish. An earlier study (Oladimeji et al., 1979) in rainbow trout exposed to dietary arsenate found that biotransformation to organic form(s) occurred rapidly and extensively throughout the body, with greater than 90% of the dietary inorganic dose converted to organoarsenicals after 24 hours. In a more recent study determining As speciation in muscle of four freshwater fish species (Ciardullo et al., 2010), organoarsenicals represented >98% of total As. Arsenobetaine was the dominant organoarsenical quantified, with arsenocholine, MMA[V], MMA[III], TMAO, TETRA, and several arsenosugars also detected in varying amounts (Ciardullo et al., 2010). In mullet (*Mugil cephalus*) collected from a metal-contaminated lake in Australia, arsenobetaine, DMA[V], TMAO, and arsenocholine were the dominant organoarsenicals detected in tissues and blood, although there was significant variation among tissues (Maher et al., 1999). These findings are largely consistent with the mammalian literature, where inorganic As is efficiently methylated in liver, and to a lesser extent in other tissues, to form DMA[V] as the major product, in addition to other organoarsenicals such as MMA[V], MMA[III], DMA[III], and TMAO (Styblo et al., 2002). One major difference is that mammals do not appear to accumulate significant quantities of arsenobetaine due to rapid excretion of this organoarsenical obtained from dietary sources (Dopp et al., 2010).

Organoarsenicals are widely distributed throughout the body of fish (Oladimeji et al., 1979; Maher et al., 1999; Kumar and Banerjee, 2012). Due to the efficient methylation of inorganic As, the majority of As accumulated in organs and tissues is in organic forms (Maher et al., 1999; McIntyre and Linton, 2012). The liver, as the primary organ responsible for As biotransformation, not surprisingly, accumulates the greatest As concentration, as reported in mullet collected in the field (Maher et al., 1999) and catfish (*Clarias batrachus*) following a 60-day aqueous exposure to arsenate (Kumar and Banerjee, 2012). Arsenic speciation in mullet organs was variable, with arsenobetaine (35%–100% of total As) and DMA[V] (0%–62%) the dominant forms (Maher et al., 1999). This finding is consistent with the pattern of As speciation determined in liver, muscle, and intestine of five freshwater fish species collected from an As-contaminated lake in northern Canada, where arsenobetaine and DMA[V] were the dominant forms (de Rosemond et al., 2008).

Oladimeji et al. (1979) observed that 65% of a single ingested dose of arsenate was excreted within 48 hours in rainbow trout, but that urinary and biliary excretion was very low. These results suggested that excretion of inorganic and/or organic As via gills may be an important route of elimination, which was partially supported in a subsequent study (Oladimeji et al., 1984). Recent work has identified aquaglyceroporin-9 (AQP9) as an efflux transporter for MMA[III] from hepatocytes to blood in mammals (Kumagai and Sumi, 2007). This finding raises the possibility that AQP9, which has been identified in zebrafish (Hamdi et al., 2009), may play a role in organoarsenical excretion in the teleost gill.

In mammals, methylated metabolites such as MMA[V] and DMA[V] are readily excreted in urine (Kumagai and Sumi, 2007), but the role of renal excretion in fish is uncertain. Upregulated expression and activity of multidrug resistant protein-2 (MRP-2), which mediates the efflux of As-glutathione conjugates in liver and kidney of mammals (Dopp et al., 2010), was reported in proximal kidney tubule cells of killifish, indicating that tubular secretion may play a role in As excretion in fish (Miller et al., 2007). Collectively, these findings suggest that the recently proposed glutathione conjugation pathway for As biotransformation (see Section 2) may occur in fish. Further research is needed to determine the importance of renal organoarsenical excretion in fish.

4.4. Potential for Bioconcentration or Biomagnification

Organoarsenicals bioaccumulate in fish from both aqueous and dietary exposure routes, but there is significant variation among fish species and exposure environment (Erickson et al., 2011; McIntyre and Linton, 2012). Saltwater fish generally bioaccumulate greater quantities of organoarsenicals than freshwater species, even though they are usually exposed to smaller aqueous As concentrations. This difference may be due to greater amounts of arsenobetaine in saltwater fish prey, since arsenobetaine has been hypothesized to act as an osmolyte (similar to glycine betaine) in saltwater invertebrates. Freshwater fish also generally accumulate much more variable quantities of organoarsenicals, which may be due to interspecific dietary preferences in freshwater food webs (McIntyre and Linton, 2012). Interestingly, an inverse relationship exists between aqueous As concentration and bioaccumulation factors in fish (Williams et al., 2006; McIntyre and Linton, 2012). This phenomenon was first discussed for other trace elements by McGeer et al. (2003) and is believed to be due to regulatory mechanisms controlling trace element homeostasis in fish. Unlike certain other organometals, biomagnification does not occur with organoarsenicals, and several studies have reported decreasing As concentrations with increasing trophic level (reviewed in McIntyre and Linton, 2012).

4.5. Mechanism(s) of Toxicity and Organism Impacts

Biomethylation of inorganic As[III] (arsenite) was long considered a detoxification process, but the trivalent methylated species, MMA[III] and DMA[III], are now known to be more toxic than inorganic As to mammals (Styblo et al., 2002). Although this has resulted in an increased focus on organoarsenical toxicology in mammalian systems, very little research has been conducted in fish. In mammals, MMA[III] and DMA[III] are potent inhibitors of glutathione reductase, glutathione peroxidase, and thioredoxin reductase, which are key antioxidant enzymes involved in maintaining cellular redox homeostasis (Styblo et al., 2002; Kumagai and Sumi, 2007). A recent hypothesis is that the genotoxicity of trivalent organoarsenicals is due to ROS-induced damage to DNA (Kumagai and Sumi, 2007; Dopp et al., 2010). The relationship between elevated As and increased production of ROS leading to oxidative stress and toxicity in fish represents an exciting area of future research into mechanisms of organoarsenical toxicity.

The majority of past studies investigating organismal As toxicity in fish have been conducted using aqueous and/or dietary exposures to inorganic As, either in controlled laboratory experiments or field studies, as reviewed extensively in McIntyre and Linton (2012). Relatively few studies have examined the direct toxic effects of organoarsenicals on fish physiology. However, given the renewed focus on the toxicity of certain trivalent organoarsenicals, this issue should be an area of future investigation. Many previous studies attributing toxicity to inorganic As may need to be reevaluated, since observed organismal effects may in fact be due to metabolic intermediates such as MMA[III] and DMA[III]. It is beyond the scope of this chapter to provide a detailed review of organismal effects of As in fish, but an excellent recent review of this literature is available (McIntyre and Linton, 2012).

4.6. Knowledge Gaps and Future Directions

The vast majority of toxicological research has focused on exposure to, and the effects of, inorganic As in fish, and much remains to be discovered with respect to organoarsenicals. There are many gaps in our knowledge of uptake, transport, biotransformation, and excretion of organoarsenicals in fish. Since certain methylated As species are now known to be more toxic than inorganic As in mammals, future research should focus on comparative aspects of As biotransformation in fish, which may provide novel insights into the mechanisms of As toxicity.

5. ORGANOTIN COMPOUNDS

5.1. The Chemicals

Organotins are a physically and chemically diverse class of compounds. Their industrial and agricultural use dates back to the 1940s. Currently, more than 800 organotin compounds have been documented, with annual production being about 50,000 tons (Gadaja and Janscó, 2010). Mono and di-organotin compounds are used as stabilizers for polyvinyl chloride (PVC), and these represent drinking water sources of organotins. Disposal of organotin-containing PVC products into landfills may also represent an important source of aquatic pollution. Organotins are also used in preserving agents for wood, paper, textiles, leather, and electrical equipment.

Tributyltin (TBT) is of greatest concern to aquatic ecosystems (Hoch, 2001). Tributyltin (largely as an oxide or methacrylate) has been used globally since the 1970s as a cost-effective, antifouling biocide agent in coats of ocean-going vessels. The solubility of TBT is influenced by factors such as pH, temperature, and dissolved organic matter. For example, regarding pH, the solubility of TBT oxide in water is 750 µg/L at pH 6.6 and 31,000 µg/L at pH 8.1 (Maguire et al. 1983). The main degradation products of TBT, dibutyltin and monobutyltin, are often found in aquatic systems.

Owing to the persistence and toxicity of TBT, in recent years there have been calls to have the compound banned. In the United States, the Organotin Antifouling Paint Control Act (33 U.S.C. 2401) was approved in 1988, and this law bans the use of TBT on most vessels less than 25 meters in length. Through the International Marine Organization of the United Nations Marine Environment Protection Committee, the International Convention on the Control of Harmful Antifouling Systems on Ships went into force in September 2008 and calls for a worldwide ban of TBT and other organotins in antifouling paints.

To protect aquatic life, the U.S. Environmental Protection Agency (EPA) (2003) uses a chronic freshwater criterion of 0.072 µg/L and an acute criterion of 0.46 µg/L. For marine systems, the U.S. EPA chronic criterion is 0.0074 µg/L and the acute criterion is 0.42 µg/L. In Canada, the CCME has set a freshwater guideline of 0.008 µg/L and a saltwater guideline of 0.001 µg/L for the protection of aquatic life.

5.2. Environmental Situations of Concern

The greatest environmental concern caused by TBT involves the waterways associated with ship traffic, such as harbors and marinas. The compound can enter the water column, readily sorb to suspended solids, and eventually settle into sediments. Owing to decades of use and slow decomposition rates, concentrations of TBT in sediments can be significant, thus acting as long-term sources to surrounding water. For example, tin values ranging from 1000 to 2000 ng/L have been reported in many locations (Hoch, 2001). This is of concern considering that imposex in snails is estimated to occur at around 1 ng/L (Hoch, 2001) and similar levels have been associated with shell malformations in oysters (Alzieu et al., 1989). Significantly, in harbors in the United Kingdom and Canada where TBT has been banned the health of mollusk populations eventually did improve (Evans et al., 1996; Minchin et al., 1996; Tester et al., 1996).

5.3. Toxicokinetics

A series of toxicokinetic studies by Martin et al. (1989) revealed important information concerning the uptake, distribution, and elimination of TBT in fish tissues. In one study, rainbow trout were exposed to 0.5 μg/L TBT for 15 days, and tin levels were measured in a range of tissues (concentrations as μg/g ww in brackets): peritoneal fat (9.8), liver (4.8), kidney (3.7), gall bladder and bile (3.1), gill (1.4), gut (1.3), blood (0.9), and muscle (0.5). In addition to measuring TBT in all tissues, the authors also measured dibutyltin, monobutyltin, and inorganic tin. In doing so, they showed that the ratio of metabolites varied across tissues and that the liver is the main site of dealkylation. In another study by these same authors, some of the exposed fish was allowed to depurate for 15 days. An analysis of tissues from these fish showed that tin levels decreased in all tissues but that levels remained elevated in the liver (6.0 μg/g ww) and gall bladder and bile (5.4 μg/g ww). A study by Tsuda et al. (1988) exposed carp to 2.1 μg/L TBT for upwards of 14 days. By sampling carp on exposure days 1, 3, 7, 10, and 14, the authors showed that steady-state levels were achieved after 3 days in the gall bladder, 7 days in the liver, 10 days for muscle, and 14 days for kidney. Bioconcentration factors (BCF) were highest in the kidneys (log BCF = 3.5), followed by gall bladder (3.1), liver (2.8), and muscle (2.7). Tsuda et al. (1988) also exposed fish to dibutyltin and monobutyltin, and showed that these organotins bioconcentrated to a lesser degree than TBT and also had different tissue distributions. These two laboratory exposure studies show that: (a) fish bioconcentrate TBT in a matter of days; (b) tissue distribution patterns exist; (c) dealkylation of TBT to dibutyltin and monobutyltin occurs, largely in the liver; (d) elimination is possible, but slow; and (e) the degree of tin alkylation affects toxicokinetics.

5.4. Potential for Bioconcentration or Biomagnification

Organotins can be taken up by aquatic organisms either directly from water or via the diet. Tributyltin bioconcentration factors can range from < 1 to more than 150,000 (Rüdel, 2003). For example, exposure of rainbow trout to 0.5 μg/L TBT for 64 days resulted in a bioconcentration factor of 406 (Martin et al., 1989). There is evidence that TBT and other organotins can biomagnify (Strand and Jacobsen, 2005; Murai et al., 2008).

5.5. Mechanism(s) of Toxicity and Organism Impacts

Organotins exert toxic effects on a range of organ systems in fish, which is not surprising given their biocidal properties. Tributyltin is a proven endocrine disruptor to aquatic organisms (Matthiessen and Gibbs, 1998). This chemical has been shown to elevate testosterone in female gastropods, thus promoting male sexual characteristics in females. There is growing evidence that TBT can masculinize fish populations (McAllister and Kime, 2004; McGinnis and Crivello, 2011). The underlying mechanisms purported include competitive inhibition of cytochrome P450-dependent aromatase activity, interference of testosterone metabolism, and impaired spermatogenesis. Additional mechanisms may include alteration of genes important in the hypothalamic-pituitary-gonadal (HPG) axis (Sun et al., 2011) and inhibition of peroxisome proliferator-activated receptor (PPAR) activation (Colliar et al., 2011). In addition to endocrine disrupting effects, studies in rockfish showed that TBT can damage DNA and affect a number of genes related to DNA repair (Zuo et al., 2012), affect a number of hepatic epigenetic markers (Wang et al., 2009), and modulate brain glutamatergic signaling (Zuo et al., 2009).

5.6. Knowledge Gaps and Future Directions

Important studies document the adverse health effects associated with organotin exposure in aquatic organisms, with a primary focus on bivalves; much less is known in fish. Furthermore, most studies have focused on TBT, and much remains to be discovered with respect to other organotin compounds. Laboratory-based studies have enabled a mechanistic understanding of toxicity. However, there is a need to corroborate laboratory experiments with field observations. Recent studies demonstrating the epigenetic effects of TBT (Wang et al., 2009) in fish warrants additional research given the potential for such an effect to have long-term (and potentially transgenerational) impacts on fish physiology.

6. OTHER ORGANOMETAL(LOID)S

Certain alkyl derivatives of other elements belonging to group 14 (lead, Pb, and germanium, Ge), group 15 (antimony, Sb), and group 16 (tellurium, Te) of the periodic table have been identified in surface waters. However, limited studies to date suggest that concentrations of these organometal(loid)s in contaminated aquatic ecosystems are not likely to cause toxicities in fish. Thus, only brief mention of these compounds is provided here.

6.1. Organolead Compounds

Organoleads (e.g., tetraethyllead, tetramethyllead) are synthetic compounds used as agents in fuels and engines. Within aquatic ecosystems, these compounds may be degraded into inorganic lead through biological and chemical processes in both the environment (Abadin and Pohl, 2010) and fish tissues (Wong et al., 1981). Organoleads are lipophilic, and within fish they are well absorbed, with liver being a primary site of accumulation and brain being a site of toxic action. In general, organolead compounds are more toxic to fish than inorganic forms (Eisler, 1988), with lethal outcomes resulting from acute (<2 week) exposure of fish to low ppb (µg/L) concentrations. Within organolead compounds, the ethyl derivatives are more toxic than methyl derivatives, and the toxicity increases with higher alkylation (Chau et al., 1980). With prohibition of organolead compounds in most regions, environmental levels have decreased along with ecotoxicological research. However, these compounds still remain used in certain sectors (e.g., general aviation aircraft, auto racing, and recreational marine crafts) and regions (e.g., Middle East).

6.2. Organotellurium Compounds

Similar to Se, Te exists in four oxidation states (VI, IV, 0, –II), but in contrast to the diverse speciation of organoselenium compounds, the only organotellurium compound that has been identified in environmental samples to date is dimethyltelluride (Wallschläger and Feldmann, 2010). Total Te concentrations in uncontaminated ambient waters are at least 10 times lower than Se, and Te has a much greater affinity for particulate matter compared to Se (Harada and Takahashi, 2008).

6.3. Organoantimony Compounds

Antimony (Sb) belongs to group 15 of the periodic table, which includes arsenic and similarly exists in the oxidation states of –III, 0, III, and V. Relevant organoantimony compounds in the environment include the biomethylated species monomethylantimony, dimethylantimony, and trimethylantimony (Filella, 2010). All three species have been detected at low (ng L^{-1}) concentrations in freshwater and saltwater (open ocean). Monomethylantimony and dimethylantimony have been measured at low concentrations in mine effluent (<0.1 µg L^{-1}); these species were also detected for the first time in an aquatic animal (the snail, *Stagnicola* sp.) living downstream of mine effluent discharge (Koch et al., 2000). To our knowledge, there are no studies of organoantimony bioaccumulation or toxicity in fish. Limited studies in other organisms indicate low toxicity of these compounds (Filella, 2010). Thus, it appears that the potential toxicity of organoantimony compounds to fish inhabiting contaminated aquatic ecosystems is unlikely.

6.4. Organogermanium Compounds

Monomethylgermanium and dimethylgermanium have been identified in surface waters (Lewis and Mayer, 1993). Unlike organic forms of lead and tin (other members of group 14), methylgermanium species are nonreactive and resemble silicon in their biogeochemical behavior. To our knowledge no studies have investigated the potential toxicity of methylgermanium compounds in fish.

7. CONCLUSIONS

Certain elements belonging to groups 12 (Hg), 14 (Pb, Sn, and Ge), 15 (As and Sb), and 16 (Se and Te) of the periodic table can be biotransformed to diverse organic species in aquatic ecosystems. The extent of such biotransformation is commonly dependent on the ecological and hydrological characteristics of a specific aquatic system. Several organometal(loid) species can bioaccumulate and biomagnify in aquatic food webs, often behaving more like traditional persistent organic pollutants (e.g., organohalogen compounds). Of the organometal(loid)s, Hg and Se represent the greatest toxicological hazard because of their ubiquity and propensity to bioccumulate in contaminated aquatic ecosystems to levels that cause toxicities in fish. Since current aquatic life criteria/guidelines aimed at protecting fish populations are based on total aqueous concentrations of such elements, these regulatory limits are likely inadequate and create great uncertainty in ecological risk assessments. However, the complex chemistries of these organometal(loid)s cause difficulties in fully understanding their toxicokinetics and toxicodynamics in fish, which represents perhaps the greatest research gap. Tissue-based criteria/guidelines, at least for Hg and Se, should be implemented for the protection of fish and other aquatic biota.

REFERENCES

Abadin, H. G. and Pohl, H. R. (2010). Alkyllead compounds and their environmental toxicology. *Met. Ions Life Sci.* **7**, 153–164.

Adams, D. H., Sonne, C., Basu, N., Nam, D.-H., Leifsson, P. S. and Jensen, A. L. (2010). Mercury contamination in spotted seatrout *Cynoscion nebulosus*: An assessment of liver, kidney, blood and nervous system health. *Sci. Tot. Environ.* **408**, 5808–5816.

Agius, C. and Roberts, R. J. (2003). Melano-macrophage centres and their role in fish pathology. *J. Fish Dis.* **26**, 499–509.

Allen, T. and Southwick, R. (2008). *Sportfishing in America: An Economic Engine and Conservation Powerhouse*. Alexandria, VA: American Sportfishing Association (pp. 12).

Alzieu, C., Sanjuan, J., Michel, P., Borel, M. and Dreno, J. P. (1989). Monitoring and assessment of butyltins in Atlantic Coastal waters. *Mar. Pollut. Bull.* **20**, 22–26.

AMAP/UNEP, Technical Background Report to the Global Atmospheric Mercury Assessment. Arctic Monitoring and Assessment Programme/UNEP Chemicals Branch, 2008, 159 pp.

Amlund, H., Lundebye, A. K. and Berntssen, M. H. (2007). Accumulation and elimination of methylmercury in Atlantic cod (*Gadus morhua* L.) following dietary exposure. *Aquat. Toxicol.* **83**, 323–330.

ANZECC (Australian and New Zealand Environment and Conservation Council) (2000). *Australian and New Zealand Guidelines for Fresh and Marine Water Quality. Volume 1: The Guidelines*. Artarmon (NSW, AUS): Australian Water Association.

ATSDR (1999). *Toxicological Profile for Mercury. Agency for Toxic Substances and Disease Registry*. Atlanta: U.S. Department of Health and Human Services, 617 pp.

Baatrup, E. and Danscher, G. (1987). Cytochemical demonstration of mercury deposits in trout liver and kidney following methylmercury intoxication—differentiation of 2 mercury pools by selenium. *Ecotoxicol. Environ. Saf.* **14**, 129–141.

Baatrup, E., Nielsen, M. G. and Danscher, G. (1986). Histochemical demonstration of 2 mercury pools in trout tissues—mercury in kidney and liver after mercuric-chloride exposure. *Ecotoxicol. Environ. Saf.* **12**, 267–282.

Bakke, A. M., Tashjian, D. H., Wang, C. F., Lee, S. H., Bai, S. C. and Hung, S. S. O. (2010). Competition between selenomethionine and methionine in the intestinal tract of green sturgeon (*Acipenser medirostris*). *Aquat. Toxicol.* **96**, 62–69.

Barst, B. D., Gevertz, A. K., Chumchal, M. M., Smith, J. D., Rainwater, T. R., Drevnick, P. E., Hudelson, K. E., Hart, A., Verbeck, G. F. and Roberts, A. P. (2011). Laser ablation ICP-MS co-localization of mercury and immune response in fish. *Environ. Sci. Technol.* **45**, 8982–8988.

Basu, N. and Head, J. (2010). Mammalian wildlife as complementary models in environmental neurotoxicology. *Neurotoxicol. Teratol.* **32**, 114–119.

Basu, N., Scheuhammer, A. M., Rouvinen-Watt, K., Evans, R. D., Trudeau, V. L. and Chan, H. M. (2010). In vitro and whole animal evidence that methylmercury disrupts GABAergic systems in discrete brain regions in captive mink. *Comp. Biochem. Physiol. Pt C.* **151**, 379–385.

Bauder, M. B., Palace, V. P. and Hodson, P. V. (2005). Is oxidative stress the mechanism of blue sac disease in retene-exposed trout larvae? *Environ. Toxicol. Chem.* (24), 694–702.

Beckvar, N., Dillon, T. M. and Read, L. B. (2005). Approaches for linking whole-body fish tissue residues of mercury or DDT to biological effects thresholds. *Environ. Toxicol. Chem.* **24**, 2094–2105.

Bentley, R. and Chasteen, T. G. (2002). Microbial methylation of metalloids: arsenic, antimony, and bismuth. *Microbiol. Mol. Biol. Rev.* **66**, 250–271.

Berntssen, M. H., Aatland, A. and Handy, R. D. (2003). Chronic dietary mercury exposure causes oxidative stress, brain lesions, and altered behaviour in Atlantic salmon (*Salmo salar*) parr. *Aquat. Toxicol.* **65**, 55–72.

Berzas Nevado, J. J., Rodríguez Martín-Doimeadios, R. C., Guzmán Bernardo, F. J., Jiménez Moreno, M., Herculano, A. M., do Nascimento, J. L. and Crespo-López, M. E. (2010). Mercury in the Tapajós River basin, Brazilian Amazon: A review. *Environ. Int.* **36**, 593–608.

Bock, A., Forchhammer, K., Heider, J. and Baron, C. (1991). Selenoprotein synthesis: Expansion of the genetic code. *Trends Biochem. Sci.* **16**, 463–467.

Bodaly, R. A. and Fudge, R. J. P. (1999). Uptake of mercury by fish in an experimental boreal reservoir. *Arch. Environ. Contam. Toxicol.* **37**, 103–109.

Bodaly, R. A., Beaty, K. G., Hendzel, L. H., Majewski, A. R., Paterson, M. J., Rolfhus, K. R., Penn, A. F., St. Louis, V. L., Hall, B. D., Matthews, C. J., Cherewyk, K. A., Mailman, M.,

Hurley, J. J., Schiff, S. L. and Venkiteswaran, J. J. (2004). Experimenting with hydroelectric reservoirs. *Environ. Sci. Technol.* **38**, 346A–352A.

Bodaly, R. A., Jansen, W. A., Majewski, A. R., Fudge, R. J., Strange, N. E., Derksen, A. J. and Green, D. J. (2007). Postimpoundment time course of increased mercury concentrations in fish in hydroelectric reservoirs of northern Manitoba, Canada. *Arch. Environ. Contam. Toxicol.* **53**, 379–389.

Bouwer, H. (1989). Agricultural contamination: Problems and solutions. *Water Environ. Technol.* **1**, 292–297.

Branco, V., Canário, J., Lu, J., Holmgren and Carvalho, C. (2012). Mercury and selenium interaction in vivo: Effects on thioredoxin reductase and glutathione peroxidase. *Free Radic. Biol. Med.* **52**, 781–793.

Brooks, S. C. and Southworth, G. R. (2011). History of mercury use and environmental contamination at the Oak Ridge Y-12 plant. *Environ. Pollut.* **159**, 219–228.

Burger, J., Jeitner, C., Donio, M., Pittfield, T. and Gochfeld, M. (2012). Mercury and selenium levels, and selenium:mercury molar ratios of brain, muscle and other tissues in bluefish (*Pomatomus saltatrix*) from New Jersey, USA. *Sci. Total Environ.* **443C**, 278–286.

Burk, R. F. and Hill, K. E. (2009). Selenoprotein P—Expression, functions and roles in mammals. *Biochim. Biophys. Acta.* **1790**, 1441–1447.

Burk, R. F., Hill, K. E. and Motley, A. K. (2001). Plasma selenium in specific and non-specific forms. *Biofactors* **14**, 107–114.

CCME (2003). *Canadian Water Quality Guidelines for the Protection of Aquatic Life.* Winnipeg: Canadian Council of Ministers of the Environment.

Challenger, F., Higgenbottom, C. and Ellis, L. (1933). The formation of organo-metalloidal compounds by microorganisms. Part I. Trimethylarsine and dimethylarsine. *J. Chem. Soc.* 95–101.

Chasar, L. C., Scudder, B. C., Stewart, A. R., Bell, A. H. and Aiken, G. R. (2009). Mercury cycling in stream ecosystems. 3. Trophic dynamics and methylmercury bioaccumulation. *Environ. Sci. Technol.* **43**, 2733–2739.

Chau, Y. K., Wong, P. T. S., Kramer, O., Bengert, G. A., Cruz, R. B., Kinrade, J. O., Lye, J. and Van Loon, J. C. (1980). Occurrence of tetraalkyllead compounds in the aquatic environment. *Bull. Environ. Contam. Toxicol.* **24**, 265–269.

Chumchal, M. M., Rainwater, T. R., Osborn, S. C., Roberts, A. P., Abel, M. T., Cobb, G. P., Smith, P. N. and Bailey, F. C. (2011). Mercury speciation and biomagnification in the food web of Caddo lake, Texas and Louisiana, USA, a subtropical freshwater ecosystem. *Environ. Toxicol. Chem.* **30**, 1153–1162.

Ciardullo, S., Aureli, F., Raggi, A. and Cubadda, F. (2010). Arsenic speciation in freshwater fish: Focus on extraction and mass balance. *Talanta* **81**, 213–221.

Cizdziel, J., Hinners, T., Cross, C. and Pollard, J. (2003). Distribution of mercury in the tissues of five species of freshwater fish from Lake Mead, USA. *J. Environ. Monit.* **5**, 802–807.

Clarkson, T. W. and Magos, L. (2006). The toxicology of mercury and its chemical compounds. *Crit. Rev. Toxicol.* **36**, 609–662.

Cleveland, L., Little, E. E., Buckler, D. R. and Wiedmeyer, R. H. (1993). Toxicity and bioaccumulation of waterborne and dietary selenium in juvenile bluegill (*Lepomis macrochirus*). *Aquat. Toxicol.* **27**, 265–279.

Cockell, K. A. and Hilton, J. W. (1988). Preliminary investigations on the comparative chronic toxicity of four dietary arsenicals to juvenile rainbow trout (*Salmo gairdneri* R.). *Aquat. Toxicol.* **12**, 73–82.

Colliar, L., Sturm, A. and Leaver, M. J. (2011). Tributyltin is a potent inhibitor of piscine peroxisome proliferator-activated receptor α and β. *Comp. Biochem. Physiol. C.* **153**, 6–173.

Crump, K. L. and Trudeau, V. L. (2009). Mercury-induced reproductive impairment in fish. *Environ. Toxicol. Chem.* **28**, 895–907.

Cullen, W. R. and Reimer, K. J. (1989). Arsenic speciation in the environment. *Chem. Rev.* **89**, 713–764.

Cutler, C. P., Martinez, A.-S. and Cramb, G. (2007). Review: The role of aquaporin 3 in teleost fish. *Comp. Biochem. Physiol. A* **148**, 82–91.

de Rosemond, S., Xie, Q. and Liber, K. (2008). Arsenic concentration and speciation in five freshwater fish species from Back Bay near Yellowknife, NT, Canada. *Environ. Monit. Assess.* **147**, 199–210.

DeForest, D. K. and Adams, W. J. (2011). Selenium accumulation and toxicity in freshwater fishes. In *Environmental contaminants in biota: Interpreting tissue concentrations* (eds. W. N. Beyer and J. P. Meador), 2nd ed., pp. 193–209. Boca Raton, FL: Taylor and Francis.

DeForest, D. K., Brix, K. V. and Adams, W. J. (1999). Critical review of proposed residue-based selenium toxicity thresholds for freshwater fish. *Human Ecol. Risk Assess* **5**, 1187–1228.

Department of Water Affairs and Forestry (1996). *South African Water Quality Guidelines. Volume 7: Aquatic Ecosystems.* Pretoria: Government Printer, 161 pp.

Depew, D., Basu, N., Burgess, N. M., Campbell, L. M., Devlin, E. W., Drevnick, P., Hammerschmidt, C., Murphy, C. A., Sandheinrich, M. B. and Wiener, J. G. (2012). Toxicity of dietary methylmercury to fish: Derivation of ecologically meaningful threshold concentrations. *Environ. Toxiocol. Chem.* **31**, 1536–1547.

Dietz, R., Born, E. W., Riget, F., Sonne, C., Aubail, A., Drimmie, R. and Basu, N. (2011). Temporal trends and future predictions of mercury concentrations in northwest Greenland polar bear (*Ursus maritimus*) Hair. *Environ. Sci. Technol.* **45**, 1458–1465.

Dietz, R., Sonne, C., Basu, N., Braune, B., O'Hara, T., Letcher, R. J., Scheuhammer, T., Andersen, M., Andreasen, C., Andriashek, D., Asmund, G., Aubail, A., Baagoe, H., Born, E. W., Chan, H. M., Derocher, A. E., Grandjean, P., Knott, K., Kirkegaard, M., Krey, A., Lunn, N., Messier, F., Obbard, M., Olsen, M. T., Ostertag, S., Peacock, E., Renzoni, A., Riget, F. F., Skaare, J. U., Stern, G., Stirling, I., Taylor, M., Wiig, O., Wilson, S. and Aars, J. (2013). What are the toxicological effects of mercury in Arctic biota? *Sci. Total Environ.* **443**, 775–790.

Dopp, E., Hartmann, L. M., Florea, A.-M., Rettenmeier, A. W. and Hirner, A. V. (2004). Environmental distribution, analysis, and toxicity of organometal(loid) compounds. *Crit. Rev. Toxicol.* **34**, 301–333.

Dopp, E., Kligerman, A. D. and Diaz-Bone, R. A. (2010). Organoarsenicals: Uptake, metabolism, and toxicity. *Met. Ions Life Sci.* **7**, 231–265.

Drevnick, P. E. and Sandheinrich, M. B. (2003). Effects of dietary methylmercury on reproductive endocrinology of fathead minnows. *Environ. Sci. Technol.* **37**, 4390–4396.

Drevnick, P. E., Roberts, A. P., Otter, R. R., Hammerschmidt, C. R., Klaper, R. and Oris, J. T. (2008). Mercury toxicity in livers of northern pike (*Esox lucius*) from Isle Royal, USA. *Comp. Biochem. Physiol. C.* **147**, 331–338.

Drevnick, P. E., Engstrom, D. R., Driscoll, C. T., Swain, E. B., Balogh, S. J., Kamman, N. C., Long, D. T., Muir, D. G., Parsons, M. J., Rolfhus, K. R. and Rossmann, R. (2012). Spatial and temporal patterns of mercury accumulation in lacustrine sediments across the Laurentian Great Lakes region. *Environ. Pollut.* **161**, 252–260.

Eisler, R. (1988). Lead hazards to fish, wildlife and invertebrates: A synoptic review. Contaminant Hazard Reviews Report No. 14. Patuxent Wildlife Research Center. U.S. Fish and Wildlife Service.

Erickson, R. J., Mount, D. R., Highland, T. L., Russell Hockett, J. and Jenson, C. T. (2011). The relative importance of waterborne and dietborne arsenic exposure on survival and growth of juvenile rainbow trout. *Aquat. Toxicol.* **104**, 108–115.

Evans, S. M., Evans, P. M. and Leksono, T. (1996). Widespread recovery of dogwelks, *Nucella lapillus* (L.), from the tributyltin contamination in the North Sea and Clyde Sea. *Mar. Pollut. Bull.* **32**, 263–269.

Evans, M. S., Muir, D., Lockhart, W. L., Stern, G., Ryan, M. and Roach, P. (2005). Persistent organic pollutants and metals in the freshwater biota of the Canadian Subarctic and Arctic: An overview. *Sci. Total Environ.* **351–352**, 94–147.

Fan, T. W. M., Teh, S. J., Hinton, D. E. and Higashi, R. M. (2002). Selenium biotransformations into proteinaceous forms by foodweb organisms of selenium-laden drainage waters in California. *Aquat. Toxicol.* **57**, 65–84.

Filella, M. (2010). Alkyl derivatives of antimony in the environment. *Met. Ions Life Sci.* **7**, 267–301.

Fimreite, N. and Reynolds, L. M. (1973). Mercury contamination of fish in northwestern Ontario. *J. Wild. Manage.* **37**, 62–68.

Fjeld, E., Haugen, T. O. and Vøllestad, L. A. (1998). Permanent reductions in the foraging efficiency and competitive ability of grayling (*Thymallus thymallus*) exposed to methylmercury during embryogenesis. *Sci. Tot. Environ.* **213**, 247–254.

Gadaja, T. and Janscó, A. (2010). Organotins. Formation, use, speciation and toxicology. *Met. Ions Life Sci.* **7**, 111–151.

Gammelgaard, B., Gabel-Jensen, C., Sturup, S. and Hansen, H. R. (2008). Complementary use of molecular and element-specific spectrometry for identification of selenium compounds related to human selenium metabolism. *Anal. Bioanal. Chem.* **390**, 1691–1706.

Gladyshev, V. M. and Hatfield, D. L. (2010). Selenocysteine biosynthesis, selenoproteins, and selenoproteomes. In *Recoding: Expansion of Decoding Rules Enriches Gene Expression (Nucleic Acids and Molecular Biology Vol. 24)* (eds. J. F. Atkins and R. F. Gesteland), pp. 3–27. New York: Springer.

Gonzalez, P., Dominique, Y., Massabuau, J. C., Boudou, A. and Bourdineaud, J. P. (2005). Comparative effects of dietary methylmercury on gene expression in liver, skeletal muscle, and brain of the zebrafish (*Danio rerio*). *Environ. Sci. Technol.* **39**, 3972–3980.

Grandjean, P., Satoh, H., Murata, K. and Eto, K. (2010). Adverse effects of methylmercury: Environmental health research implications. *Environ. Health Perspect.* **118**, 1137–1145.

Gregus, Z. and Klaassen, C. D. (1986). Disposition of metals in rats: A comparative study of fecal, urinary and biliary excretion and tissue distribution of eighteen metals. *Toxicol. Appl. Pharmacol.* **85**, 24–38.

Gyurasics, A., Perjesi, P. and Gregus, Z. (1998). Role of glutathione and methylation in the biliary excretion of selenium. The paradoxical effect of sulfobromophthalein. *Biochem. Pharmacol.* **56**, 1381–1389.

Hall, B. D., Bodaly, R. A., Fudge, R. J. P., Rudd, J. W. M. and Rosenberg, D. M. (1997). Food as the dominant pathway of methylmercury uptake by fish. *Water Air Soil Pollut.* **100**, 13–24.

Hall, B. D. and St. Louis, V. L. (2004). Methylmercury and total mercury in plant litter decomposing in upland forests and flooded landscapes. *Environ. Sci. Technol.* **38**, 5010–5021.

Hamdi, M., Sanchez, M. A., Beene, L. C., Liu, Q., Landfear, S. M., Rosen, B. P. and Liu, Z. (2009). Arsenic transport by zebrafish aquaglyceroporins. *BMC Mol. Biol.* **10**, 104–115.

Hamdi, M., Yoshinaga, M., Packianathan, C., Qin, J., Hallauer, J., McDermott, J. R., Yang, H.-C., Tsai, K.-J. and Liu, Z. (2012). Identification of an S-adenosylmethionine (SAM) dependent arsenic methyltransferase in *Danio rerio*. *Toxicol. Appl. Pharmacol.* **262**, 185–193.

Hamilton, S. J. (2002). Rationale for a tissue-based selenium criterion for aquatic life. *Aquat. Toxicol.* **57**, 85–100.

Hamilton, S. J. (2003). Review of residue-based selenium toxicity thresholds for freshwater fish. *Ecotoxicol. Environ. Saf.* **56**, 201–210.

Hammerschmidt, C. R., Sandheinrich, M. B., Wiener, J. G. and Rada, R. G. (2002). Effects of dietary methylmercury on reproduction of fathead minnows. *Environ. Sci. Technol.* **36**, 877–883.

Harada, T. and Takahashi, Y. (2008). Origin of the difference in the distribution behavior of tellurium and selenium in a soil-water system. *Geochim. Cosmochim. Acta* **72**, 1281–1294.

Harris, H. H., Pickering, I. J. and George, G. N. (2003). The chemical form of mercury in fish. *Science* **301**, 1203.

Hasegawa, H., Azizur Rahman, M., Kitahara, K., Itaya, Y., Maki, T. and Ueda, K. (2010). Seasonal changes of arsenic speciation in lake waters in relation to eutrophication. *Sci. Total Environ.* **408**, 1684–1690.

Has-Schön, E., Bogut, I., Rajković, V., Bogut, S., Cacić, M. and Horvatić, J. (2008). Heavy metal distribution in tissues of six fish species included in human diet, inhabiting freshwaters of the Nature Park "Hutovo Blato" (Bosnia and Herzegovina). *Arch. Environ. Contam. Toxicol.* **54**, 75–83.

Hayakawa, T., Kobayashi, Y., Cui, X. and Hirano, S. (2005). A new metabolic pathway of arsenite: Arsenic-glutathione complexes are substrates for human arsenic methyltransferase Cyt19. *Arch. Toxicol.* **79**, 183–191.

Head, J., DeBofsky, A., Hinshaw, J. and Basu, N. (2011). Retrospective analysis of mercury content in feathers of birds collected From the State of Michigan (1895–2007). *Ecotoxicology* **20**, 1636–1643.

Hintelmann, H. (2010). Organomercurials. Their formation and pathways in the environment. *Met. Ions Life Sci.* **7**, 365–401.

Hoch, M. (2001). Organotin compounds in the environment—an overview. *Appl. Geochem.* **16**, 719–743.

Hoffmeyer, R. E., Singh, S. P., Doonan, C. J., Ross, A. R., Hughes, R. J., Pickering, I. J. and George, G. N. (2006). Molecular mimicry in mercury toxicology. *Chem. Res. Toxicol.* **19**, 753–759.

Hrenchuk, L. E., Blanchfield, P. J., Paterson, M. J. and Hintelmann, H. H. (2012). Dietary and waterborne mercury accumulation by yellow perch: a field experiment. *Environ. Sci. Technol.* **46**, 509–516.

Huang, S. S., Strathe, A. B., Fadel, J. G., Johnson, M. L., Lin, P., Liu, T. Y. and Hung, S. S. (2013). The interactive effects of selenomethionine and methylmercury on their absorption, disposition, and elimination in juvenile white sturgeon. *Aquat. Toxicol.* **126**, 274–282.

Huguet, L., Castelle, S., Schäfer, J., Blanc, G., Maury-Brachet, R., Reynouard, C. and Jorand, F. (2010). Mercury methylation rates of biofilm and plankton microorganisms from a hydroelectric reservoir in French Guiana. *Sci. Total Environ.* **408**, 1338–1348.

Imura, N., Pan, S. K. and Ukita, T. (1972). Methylation of inorganic mercury with liver homogenates of tuna. *Chemosphere* **1**, 197–201.

Janz, D. M. (2011). Selenium. In *Fish Physiology, Volume 31B: Homeostasis and Toxicology of Non-Essential Metals* (eds. C. M. Wood, A. P. Farrell and C. J. Brauner), pp. 327–374. London: Academic Press.

Janz, D. M., DeForest, D. K., Brooks, M. L., Chapman, P. M., Gilron, G., Hoff, D., Hopkins, W. D., McIntyre, D. O., Mebane, C. A., Palace, V. P., Skorupa, J. P. and Wayland, M. (2010). Selenium toxicity to aquatic organisms. In *Ecological Assessment of Selenium in the Aquatic Environment* (eds. P. M. Chapman, W. J. Adams, M. L. Brooks, C. G. Delos, S. N. Luoma, W. A. Maher, H. M. Ohlendorf, T. S. Presser and D. P. Shaw), pp. 141–231. Boca Raton, FL: CRC Press.

Jarvinen, A. W. and Ankley, G. T. (1999). *Linkage of Effects to Tissue Residues: Development of a Comprehensive Database for Aquatic Organisms Exposed to Inorganic and Organic Chemicals.* Pensacola, FL: SETAC Press, 364 pp.

Jenkins, R. O. (2010). Biomethylation of arsenic, antimony and bismuth. In *Biological Chemistry of Arsenic, Antimony and Bismuth* (ed. H. Sun) Chichester, UK: John Wiley & Sons.

Jensen, S. and Jernelov, A. (1967). Biological methylation of mercury in aquatic organisms. *Nature* **223**, 753–754.

Jung, D., MacIver, B., Jackson, B. P., Barnaby, R., Sato, J. D., Zeidel, M. L., Shaw, J. R. and Stanton, B. A. (2012). A novel aquaporin 3 in killifish (*Fundulus heteroclitus*) is not an arsenic channel. *Toxicol. Sci.* **127**, 101–109.

Karagas, M. R., Choi, A. L., Oken, E., Horvat, M., Schoeny, R., Kamai, E., Cowell, W., Grandjean, P. and Korrick, S. (2012). Evidence on the human health effects of low-level methylmercury exposure. *Environ. Health Perspect.* **120**, 799–806.

Kidd, K. A. and Batchelar, K. (2011). Mercury. In *Homeostasis and Toxicology of Non-Essential* Metals (Fish Physiology Series), Vol. 31b (eds. C. M. Wood, A. P. Farrell and C. J. Brauner), pp. 238–295. Elsevier Press.

Kinghorn, A., Solomon, P. and Chan, H. M. (2007). Temporal and spatial trends of mercury in fish collected in the English-Wabigoon river system in Ontario, Canada. *Sci. Total Environ.* **372**, 615–623.

Kirubagaran, R. and Joy, K. P. (1990). Changes in brain monoamine levels and monoamine oxidase activity in the catfish, *Clarias batrachus*, during chronic treatments with mercurial. *Bull. Environ. Contam. Toxicol.* **45**, 88–93.

Klaper, R., Rees, C. B., Drevnick, P., Weber, D., Sandheinrich, M. and Caravan, M. J. (2006). Gene expression changes related to endocrine function and decline in reproduction in fathead minnow (*Pimephales promelas*) after dietary methylmercury exposure. *Environ. Health Perspect.* **11**, 1337–1343.

Kobayashi, Y., Ogra, Y., Ishiwata, K., Takayama, H., Aimi, N. and Suzuki, K. T. (2002). Selenosugars are key and urinary metabolites for Se excretion within the required to low-toxic range. *Proc. Natl. Acad. Sci. U.S.A.* **99**, 15932–15936.

Koch, I., Wang, L., Feldmann, J., Andrewes, P., Reimer, K. J. and Cullen, W. R. (2000). Antimony species in environmental samples. *Intern. J. Environ. Anal. Chem.* **77**, 111–131.

Kryukov, G. V. and Gladyshev, V. N. (2000). Selenium metabolism in zebrafish: Multiplicity of selenoprotein genes and expression of a protein containing 17 selenocysteine residues. *Genes Cells* **5**, 1049–1060.

Kumagai, Y. and Sumi, D. (2007). Arsenic: Signal transduction, transcription factor, and biotransformation involved in cellular response and toxicity. *Annu. Rev. Pharmacol. Toxicol.* **47**, 243–262.

Kumar, R. and Banerjee, T. K. (2012). Analysis of arsenic bioaccumulation in different organs of the nutritionally important catfish, *Clarias batrachus* (L.) exposed to the trivalent arsenic salt, sodium arsenite. *Bull. Environ. Contam. Toxicol.* **89**, 445–449.

Kwon, S. Y., Blum, J. D., Carvan, M. J., Basu, N., Head, J. A., Madenjian, C. P. and David, S. R. (2012). Absence of fractionation of mercury isotopes during trophic transfer of methylmercury to freshwater fish in captivity. *Environ. Sci. Technol.* **46**, 7527–7534.

Laparra, J. M., Velez, D., Barbera, R., Montoro, R. and Farre, R. (2007). Bioaccessibility and transport by Caco-2 cells of organoarsenical species present in seafood. *J. Agric. Food Chem.* **55**, 5892–5897.

Larose, C., Canuel, R., Lucotte, M. and Di Giulio, R. T. (2008). Toxicological effects of methylmercury on walleye (*Sander vitreus*) and perch (*Perca flavescens*) from lakes of the boreal forest. *Comp. Biochem. Physiol. Part C* **147**, 139–149.

Leaner, J. J. and Mason, R. P. (2004). Methylmercury uptake and kinetics in sheepshead minnows, *Cyprinidon variegatus*, after exposure to CH_3Hg-spiked food. *Environ. Toxicol. Chem.* **23**, 2138–2146.

Lemly, A. D. (1985). Toxicology of selenium in a freshwater reservoir: Implications for environmental hazard evaluation and safety. *Ecotoxicol. Environ. Saf.* **10**, 314–338.

Lemly, A. D. (1997). A teratogenic deformity index for evaluating impacts of selenium on fish populations. *Ecotoxicol. Environ. Saf.* **37**, 259–266.

Lemly, A. D. (2004). Aquatic selenium pollution is a global environmental safety issue. *Ecotoxicol. Environ. Saf.* **59**, 44–56.

Lemly, A. D. and Skorupa, J. P. (2012). Wildlife and the coal waste policy debate: Proposed rules for coal waste disposal ignore lessons from 45 years of wildlife poisoning. *Environ. Sci. Technol.* **46**, 8595–8600.

Lewis, B. L. and Mayer, H. P. (1993). Biogeochemistry of methylgermanium species in natural waters. *Met. Ions Biol. Syst.* **29**, 79–99.

Little, E. E. and Finger, S. E. (1990). Swimming behaviour as an indicator of sublethal toxicity in fish. *Environ. Toxicol. Chem.* **9**, 13–20.

Lobanov, A. V., Hatfield, D. L. and Gladyshev, V. N. (2008). Reduced reliance on the trace element selenium during evolution of mammals. *Genome Biol.* **9**, R62.

Lobanov, A. V., Hatfield, D. L. and Gladyshev, V. N. (2009). Eukaryotic selenoproteins and selenoproteomes. *Biochim. Biophys. Acta* **1790**, 1424–1428.

Luoma, S. N. and Presser, T. S. (2009). Emerging opportunities in management of selenium contamination. *Environ. Sci. Technol.* **43**, 8483–8487.

Maguire, R. J., Carey, J. H. and Hale, E. J. (1983). Degradation of the tri-n-butyltin species in water. *J. Agric. Food Chem.* **31**, 1060–1065.

Maher, W. A., Roach, A., Doblin, M., Fan, T., Foster, S., Garrett, R., Möller, G., Oram, L. and Wallschläger, D. (2010). Environmental sources, speciation, and partitioning of selenium. In *Ecological Assessment of Selenium in the Aquatic Environment* (eds. P. M. Chapman, W. J. Adams, M. L. Brooks, C. G. Delos, S. N. Luoma, W. A. Maher, H. M. Ohlendorf, T. S. Presser and D. P. Shaw), pp. 47–92. Boca Raton, FL: CRC Press.

Maher, W., Goessler, W., Kirby, J. and Raber, B. (1999). Arsenic concentrations and speciation in the tissues and blood of sea mullet (*Mugil cephalus*) from Lake Macquarie NSW, Australia. *Mar. Chem.* **68**, 169–182.

Mandal, B. K. and Suzuki, K. T. (2002). Arsenic round the world: A review. *Talanta* **58**, 201–235.

Martin, R. C., Dixon, D. G., Maguire, R. J., Hodson, P. V. and Tkacz, R. J. (1989). Acute toxicity, uptake, depuration and tissue distribution of tri-n-butyltin in rainbow trout. *Salmo gardneri*. *Aquat. Toxicol.* **15**, 37–52.

Matthiessen, P. and Gibbs, P. E. (1998). Critical appraisal of the evidence for tributyltin-mediated endocrine disruption in mollusks. *Environ. Toxicol. Chem.* **17**, 37–43.

Mayland, H. (1994). Selenium in plant and animal nutrition. In *Selenium in the Environment* (eds. W. T. Frankenberger, Jr. and S. Benson), pp. 29–45. New York: Marcel Dekker.

McAllister, B. G. and Kime, D. E. (2004). Early life exposure to environmental levels of the aromatase inhibitor tributyltin causes masculinisation and irreversible sperm damage in zebrafish (*Danio rerio*). *Aquat. Toxicol.* **65**, 309–316.

McGeer, J. C., Brix, K. V., Skeaff, J. M., DeForest, D. K., Brigham, S. I., Adams, W. J. and Green, A. (2003). Inverse relationship between bioconcentration factor and exposure concentration for metals: Implications for hazard assessment of metals in the aquatic environment. *Environ. Toxicol. Chem.* **22**, 1017–1037.

McGinnis, C. L. and Crivello, J. F. (2011). Elucidating the mechanism of action of tributyltin (TBT) in zebrafish. *Aquat. Toxicol.* **103**, 25–31.

McIntyre, D. O. and Linton, T. K. (2012). Arsenic. In *Fish Physiology, Volume 31B: Homeostasis and Toxicology of Non-Essential Metals* (eds. C. M. Wood, A. P. Farrell and C. J. Brauner), pp. 297–349. London: Academic Press.

Miller, D. S., Shaw, J. R., Stanton, C. R., Barnaby, R., Karlson, K. H., Hamilton, J. W. and Stanton, B. A. (2007). MRP2 and acquired tolerance to inorganic arsenic in the kidney of killifish (*Fundulus heteroclitus*). *Toxicol. Sci.* **97**, 103–110.

Minchin, D., Stroben, E., Oehlmann, J., Bauer, B., Duggan, C. B. and Keatinge, M. (1996). Biological indicators used to map organotin contamination in Cork Harbour, Ireland. *Mar. Pollut. Bull.* **32**, 188–195.

Misra, S. and Niyogi, S. (2009). Selenite causes cytotoxicity in rainbow trout (*Oncorhynchus mykiss*) hepatocytes by inducing oxidative stress. *Toxicol. In Vitro* **23**, 1249–1258.

Motsenbocker, M. A. and Tappel, A. L. (1982). Selenocysteine-containing proteins from rat and monkey plasma. *Biochim. Biophys. Acta* **704**, 253–260.

Mulder, P. J., Lie, E., Eggen, G. S., Ciesielski, T. M., Berg, T., Skaare, J. U., Jenssen, B. M. and Sørmo, E. G. (2012). Mercury in molar excess of selenium interferes with thyroid hormone function in free-ranging freshwater fish. *Environ. Sci. Technol.* **21**, 9027–9037.

Murai, R., Sugimoto, A., Tanabe, S. and Takeuchi, I. (2008). Biomagnification profiles of tributyltin (TBT) and triphenyltin (TPT) in Japanese coastal food webs elucidated by stable nitrogen isotope ratios. *Chemosphere* **73**, 1749–1756.

Muscatello, J.R. (2009). *Selenium accumulation and effects in aquatic organisms downstream of uranium mining and milling operations in northern Saskatchewan*. Ph.D. thesis, Saskatoon: University of Saskatchewan. 255 pp.

Muscatello, J. R., Belknap, A. M. and Janz, D. M. (2008). Accumulation of selenium in aquatic systems downstream of a uranium mining operation in northern Saskatchewan, Canada. *Environ. Pollut.* **156**, 387–393.

Muscatello, J. R., Bennett, P. M., Himbeault, K. T., Belknap, A. M. and Janz, D. M. (2006). Larval deformities associated with selenium accumulation in northern pike (*Esox lucius*) exposed to metal mining effluent. *Environ. Sci. Technol.* **40**, 6506–6512.

Muscatello, J. R. and Janz, D. M. (2009). Selenium accumulation in aquatic biota downstream of a uranium mining and milling operation. *Sci. Tot. Environ.* **407**, 1318–1325.

Nagpal, N. K. (2001). *Ambient Water Quality Guidelines for Selenium*. Victoria: British Columbia Ministry of Water, Land and Air Protection.

Nickel, A., Kottra, G., Schmidt, G., Danier, J., Hofmann, T. and Daniel, H. (2009). Characterization of transport of selenoamino acids by epithelial amino acid transporters. *Chem. Biol. Interact.* **177**, 234–241.

Nriagu, J. O. (1989). Global cycling of selenium. In *Occurrence and Distribution of Selenium* (ed. M. Inhat), pp. 327–339. Boca Raton, FL: CRC Press.

Nriagu, J., Basu, N. and Charles, S. (2012). Chapter 15—Environmental Justice: The Mercury Connection. In *Mercury in the Environment: Pattern and Process* (ed. M. Bank) Berkeley: University of California Press.

Ohlendorf, H. M. (2003). Ecotoxicology of selenium. In *Handbook of Ecotoxicology* (eds. D. J. Hoffman, B. A. Rattner, G. A. Burton, Jr. and J. Cairns, Jr.), 2nd ed., pp. 465–500. Boca Raton, FL: CRC Press.

Ohta, Y. and Suzuki, K. T. (2008). Methylation and demethylation of intermediates selenide and methylselenol in the metabolism of selenium. *Toxicol. Appl. Pharmacol.* **226**, 169–177.

Oladimeji, A. A., Qadri, S. U., Tam, G. K. H. and deFreitas, A. S. W. (1979). Metabolism of inorganic arsenic to organoarsenicals in rainbow trout (*Salmo gairdneri*). *Ecotoxicol. Environ. Saf.* **3**, 394–400.

Oladimeji, A. A., Quadri, S. U. and deFreitas, A. S. W. (1984). Measuring the elimination of arsenic by the gills of rainbow trout (*Salmo gairdneri*) by using a two compartment respirometer. *Bull. Environ. Contam. Toxicol.* **32**, 661–668.

Oliveira Ribeiro, C. A., Rouleau, C., Pelletier, E., Audet, C. and Tjälve, G. (1999). Distribution kinetics of dietary methylmercury in the artic charr (*Salvelinus alpinus*). *Environ. Sci. Technol.* **33**, 902–907.

Ontario Ministry of Environment and Energy (1999). *Provincial Water Quality Objectives*. Toronto: Queen's Printer, 38 pp.

Pacyna, E. G., Pacyna, J. M., Sundseth, K., Munthe, J., Kindbom, K., Wilson, S., Steenhuisen, F. and Maxson, P. (2010). Global emission of mercury to the atmosphere from anthropogenic sources in 2005 and projections to 2020. *Atmos. Environ.* **44**, 2487–2499.

Palace, V. P., Halden, N. M., Yang, P., Evans, R. E. and Sterling, G. L. (2007). Determining residence patterns of rainbow trout using laser ablation inductively coupled plasma mass spectrometry (LA-ICP-MS) analysis of selenium in otoliths. *Environ. Sci. Technol.* **41**, 3679–3683.

Palace, V. P., Spallholz, J. E., Holm, J., Wautier, K., Evans, R. E. and Baron, C. L. (2004). Metabolism of selenomethionine by rainbow trout (*Oncorhynchus mykiss*) embryos can generate oxidative stress. *Ecotoxicol. Environ. Saf.* **58**, 17–21.

Pan-Hou, H. S. and Imura, N. (1981). Biotransformation of mercurials by intestinal microorganisms isolated from yellowfin tuna. *Bull. Environ. Contam. Toxicol.* **26**, 359–363.

Paterson, M. J., Rudd, J. W. M. and St. Louis, V. (1998). Increases in total and methylmercury in zooplankton following flooding of a peatland reservoir. *Environ. Sci. Technol.* **32**, 3868–3874.

Phibbs, J., Franz, E., Hauck, D. W., Gallego, M., Tse, J. J., Pickering, I. J., Liber, K. and Janz, D. M. (2011a). Evaluating the trophic transfer of selenium in aquatic ecosystems using caged fish, X-ray absorption spectroscopy and stable isotope analysis. *Ecotoxicol. Environ. Saf.* **74**, 1855–1863.

Phibbs, J., Wiramanaden, C. I. E., Hauck, D., Pickering, I. J., Liber, K. and Janz, D. M. (2011b). Selenium uptake and speciation in wild and caged fish downstream of a metal mining and milling discharge. *Ecotoxicol. Environ. Saf.* **74**, 1139–1150.

Pickhardt, P. C., Stepanova, M. and Fisher, N. S. (2006). Contrasting uptake routes and tissue distributions of inorganic and methylmercury in mosquitofish (*Gambusia affinis*) and redear sunfish (*Lepomis microlophus*). *Environ. Toxicol. Chem.* **25**, 2132–2142.

Presser, T. S. and Luoma, S. N. (2010). A methodology for ecosystem-scale modeling of selenium. *Integr. Environ. Assess. Manag.* **6**, 685–710.

Presser, T. S., Piper, D. Z., Bird, K. J., Skorupa, J. P., Hamilton, S. J., Detwiler, S. J. and Huebner, M. A. (2004). The Phosphoria Formation: A model for forecasting global selenium sources to the environment. In *Life Cycle of the Phosphoria Formation: From Deposition to Post-Mining Environment* (ed. J. R. Hein), pp. 299–319. New York: Elsevier.

Raldúa, D., Díez, S., Bayona, J. M. and Barceló, D. (2007). Mercury levels and liver pathology in feral fish living in the vicinity of a mercury cell chlor-alkali factory. *Chemosphere* **66**, 1217–1225.

Ralston, N. V. C., Ralston, C. R., Blackwell, J. L. and Raymond, L. J. (2008). Dietary and tissue selenium in relation to methylmercury toxicity. *Neurotoxicology* **29**, 802–811.

Ralston, N. V. and Raymond, L. J. (2010). Dietary selenium's protective effects against methylmercury toxicity. *Toxicology* **278**, 112–123.

Reeves, M. A. and Hoffmann, P. R. (2009). The human selenoproteome: Recent insights into functions and regulation. *Cell. Mol. Life Sci.* **66**, 2457–2478.

Reimer, K. R., Koch, I. and Cullen, W. R. (2010). Organoarsenicals: Distribution and transformation in the environment. *Met. Ions Life Sci.* **7**, 165–229.

Rodgers, D. W. and Beamish, F. W. H. (1981). Uptake of waterborne methylmercury by rainbow trout (*Salmo gairdneri*) in relation to oxygen consumption and methylmercury concentration. *Can. J. Fish. Aquat. Sci.* **38**, 1309–1315.

Rodgers, D. W. (1994). You are what you eat and a little bit more: Bioenergetics-based models of methylmercury accumulation in fish revisited. In *Mercury Pollution: Integration and Synthesis* (eds. C. J. Watras and J. W. Huckabee), pp. 427–439. Boca Raton, FL: Lewis Publishers.

Rüdel, H. (2003). Case study: Bioavailability of tin and tin compounds. *Ecotoxicol. Environ. Saf.* **56**, 180–189.

Rutkiewicz, J., Nam, D.-H., Cooley, T., Padilla, I. B., Neumann, K., Route, W., Strom, S. and Basu, N. (2011). Mercury exposure and neurochemical biomarkers in bald eagles across several Great Lakes States. *Ecotoxicology* **20**, 1669–1676.

Sandheinrich, M. B. and Miller, K. M. (2006). Effects of dietary methylmercury on reproductive behavior of fathead minnows (*Pimephales promelas*). *Environ. Toxicol. Chem.* **25**, 3053–3057.

Sandheinrich, M. B. and Wiener, J. G. (2011). Methylmercury in freshwater fish—recent advances in assessing toxicity of environmentally relevant exposures. In *Environmental Contaminants in Biota: Interpreting Tissue Concentrations* (eds. W. N. Beyer and J. P. Meador), 2nd ed., pp. 169–190. Boca Raton, FL: CRC Press/Taylor and Francis.

Sandheinrich, M. B., Bhavsar, S. P., Bodaly, R. A., Drevnick, P. E. and Paul, E. A. (2011). Ecological risk of methylmercury to piscivorous fish of the Great Lakes region. *Ecotoxicology* **20**, 1577–1587.

Sato, T., Ose, Y. and Sakai, T. (1980). Toxicological effect of selenium on fish. *Environ. Poll.* **21**, 217–224.

Scheuhammer, A. M., Meyer, M. W., Sandheinrich, M. B. and Murray, M. W. (2007). Effects of environmental methylmercury on the health of wild birds, mammals, and fish. *Ambio* **36**, 12–18.

Scheuhammer, A. M., Basu, N., Evers, D. C., Heinz, G. H., Sandheinrich, M. and Bank, M. (2012). Chapter 11-Toxicology of Mercury in Fish and Wildlife: Recent Advances. In *Mercury in the Environment: Pattern and Process* (ed. M. Bank) Berkeley, CA: University of California Press.

Schrauzer, G. N. (2000). Selenomethionine: A review of its nutritional significance, metabolism and toxicity. *J. Nutr.* **130**, 1653–1656.

Schrauzer, G. N. (2001). Nutritional selenium supplements: Product types, quality, and safety. *J. Amer. Coll. Nutr.* **20**, 1–4.

Schrauzer, G. N. (2009). Selenium and selenium-antagonistic elements in nutritional cancer prevention. *Crit. Rev. Biotechnol.* **29**, 10–17.

Schwindt, A. R., Fournie, J. W., Landers, D. H., Schreck, C. B. and Kent, M. L. (2008). Mercury concentrations in salmonids from western us national parks and relationships with age and macrophage aggregates. *Environ. Sci. Technol.* **42**, 1365–1370.

Selin, N. E. and Selin, H. (2006). Global politics of mercury pollution: the need for multi-scale governance. *Rev. Euro. Commun. Internat. Environ. Law.* **15**, 258–269.

Sharma, V. K. and Sohn, M. (2009). Aquatic arsenic: Toxicity, speciation, transformations, and remediation. *Environ. Int.* **35**, 743–759.

Skorupa, J. P. (1998). Selenium poisoning of fish and wildlife in nature: Lessons from twelve real-world experiences. In *Environmental Chemistry of Selenium* (eds. W. T. Frankenberger, Jr. and R. A. Engberg), pp. 315–354. New York: Marcel Dekker.

Spallholz, J. E. (1994). On the nature of selenium toxicity and carcinostatic activity. *Free Radic. Biol. Med.* **17**, 45–64.

Squadrone, S., Prearo, M., Brizio, P., Gavinelli, S., Pellegrino, M., Scanzio, T., Guarise, S., Benedetto, A. and Abete, M. C. (2013). Heavy metals distribution in muscle, liver, kidney and gill of European catfish (*Silurus glanis*) from Italian Rivers. *Chemosphere* **90**, 358–365.

St. Louis, V. L., Rudd, J. W., Kelly, C. A., Bodaly, R. A., Paterson, M. J., Beaty, K. G., Hesslein, R. H., Heyes, A. and Majewski, A. R. (2004). The rise and fall of mercury methylation in an experimental reservoir. *Environ. Sci. Technol.* **38**, 1348–1358.

Stewart, R., Grosell, M., Buchwalter, D., Fisher, N., Luoma, S. N., Mathews, T., Orr, P. and Wang, X. W. (2010). Bioaccumulation and trophic transfer of selenium. In *Ecological Assessment of Selenium in the Aquatic Environment* (eds. P. M. Chapman, W. J. Adams, M. L. Brooks, C. G. Delos, S. N. Luoma, W. A. Maher, H. M. Ohlendorf, T. S. Presser and D. P. Shaw), pp. 93–139. Boca Raton, FL: CRC Press.

Strand, J. and Jacobsen, J. A. (2005). Accumulation and trophic transfer of organotins in a marine food web from the Danish coastal waters. *Sci. Total Environ.* **350**, 72–85.

Styblo, M., Drobna, Z., Jaspers, I., Lin, S. and Thomas, D. J. (2002). The role of biomethylation in toxicity and carcinogenicity of arsenic: a research update. *Environ. Health Persp.* **110** (Suppl. 5), 767–771.

Suhendrayatna, O. A., Nakajima, T. and Maeda, S. (2002). Studies on the accumulation and transformation of arsenic in fresh organisms: II. Accumulation and transformation of arsenic compounds by *Tilapia mossambica*. *Chemosphere* **46**, 325–331.

Sun, L., Zhang, J., Zuo, Z., Chen, Y., Wang, X., Huang, X. and Wang, C. (2011). Influence of triphenyltin exposure on the hypothalamus-pituitary-gonad axis in male *Sebastiscus marmoratus*. *Aquat. Toxicol.* **104**, 263–269.

Sundseth, K., Pacyna, J. M., Pacyna, E. G., Munthe, J., Belhaj, M. and Astrom, S. (2010). Economic benefits from decreased mercury emissions: Projections for 2020. *J. Clean. Prod.* **18**, 386–394.

Suzuki, K. T. and Ogra, Y. (2002). Metabolic pathway for selenium in the body: Speciation by HPLC-ICP MS with enriched Se. *Food Addit. Contam.* **19**, 974–983.

Suzuki, K. T., Kurasaki, K., Okazaki, N. and Ogra, Y. (2005). Selenosugar and trimethylselenonium among urinary Se metabolites: Dose-and age-related changes. *Toxicol. Appl. Pharmacol.* **206**, 1–8.

Swain, E. B., Jakus, P. M., Rice, G., Lupi, F., Maxson, P. A., Pacyna, J. M., Penn, A., Spiegel, S. J. and Veiga, M. M. (2007). Socioeconomic consequences of mercury use and pollution. *Ambio* **36**, 45–61.

Tester, M., Ellis, D. V. and Thompson, J. A. J. (1996). Neogastropod imposex for monitoring recovery from marine TBT contamination. *Environ. Toxicol. Chem.* **15**, 560–567.

Thayer, J. S. (2010). Roles of organometal(loid) compounds in environmental cycles. *Met. Ions Life Sci.* **7**, 1–32.

Thisse, C., Degrave, A., Kryukov, V. N., Obrecht-Pflumio, S., Krol, A., Thisse, B. and Lescure, A. (2003). Spatial and temporal expression patterns of selenoprotein genes during embryogenesis in zebrafish. *Gene Expr. Patterns* **3**, 525–532.

Thomas, J. K. and Janz, D. M. (2011). Dietary selenomethionine exposure in adult zebrafish alters swimming performance, energetics and the physiological stress response. *Aquat. Toxicol.* **102**, 79–86.

Thomas, J. K., Wiseman, S., Giesy, J. P. and Janz, D. M. (2013). Effects of chronic dietary selenomethionine exposure on repeat swimming performance, aerobic metabolism and methionine catabolism in adult zebrafish (*Danio rerio*). *Aquat. Toxicol.* **130–131**, 112–122.

Trudel, M. and Rasmussen, J. B. (1997). Modeling the elimination of mercury by fish. *Environ. Sci. Technol.* **31**, 1716–1722.

Tsuda, T., Nakanishi, H., Aoki, S. and Takebayashi, J. (1988). Bioconcentration and metabolism of butyltin compounds in carp. *Wat. Res.* **22**, 647–651.

Tujebajeva, R., Ransom, D. G., Harney, J. W. and Berry, M. J. (2000). Expression and characterisation of nonmammalian selenoprotein P in the zebrafish. *Danio rerio. Genes Cells* **5**, 897–903.

Turner, R. R. and Southworth., G. R. (1999). Mercury-contaminated industrial and mining sites in North America: An overview with selected case studies. In *Mercury Contaminated Sites* (eds. R. Ebinghaus, R. R. Turner, L. D. de Lacerda, O. Vasiliev and W. Salomons), Berlin: Springer-Verlag.

U.S. EPA (1987). Ambient water quality criteria for selenium. EPA-440/5-87-006. Washington, DC: Office of Water, Office of Science and Technology.

U.S. EPA (2001). Water quality for the protection of human health: methylmercury. Office of Water, EPA 823-R-01-001.

U.S. EPA (2004). Draft aquatic life water quality criteria for selenium—2004. EPA-822-D-04-001. Washington, DC: Office of Water, Office of Science and Technology. 334 pp.

U.S. EPA (2009). EPA National Listing of Fish Advisories, Office of Water, Washington, DC: (2007). EPA-823-F-09-007.

Van Walleghem, J. L., Blanchfield, P. J. and Hintelmann, H. (2007). Elimination of mercury by yellow perch in the wild. *Environ. Sci. Technol.* **41**, 5895–5901.

Wagemann, R., Snow, N. B., Rosenberg, D. M. and Lutz, A. (1978). Arsenic in sediments, water, and aquatic biota from lakes in the vicinity of Yellowknife, Northwest Territories, Canada. *Arch. Environ. Contam. Toxicol.* **7**, 169–191.

Wallschläger, D. and Feldmann, J. (2010). Formation, occurrence, significance, and analysis of organoselenium and organotellurium compounds in the environment. *Met. Ions Life Sci.* **7**, 319–364.

Wang, Y., Wang, C., Zhang, J., Chen, Y. and Zuo, Z. (2009). DNA hypomethylation induced by tributyltin, triphenyltin, and a mixture of these in *Sebastiscus marmoratus* liver. *Aquat. Toxicol.* **95**, 93–98.

Watanabe, N., Tayama, M., Inouye, M. and Yasutake, A. (2012). Distribution and chemical form of mercury in commercial fish tissues. *J. Toxicol. Sci.* **37**, 853–861.

Webb, M. A. H., Feist, G. W., Fitzpatrick, M. S., Foster, E. P., Schreck, C. B., Plumlee, M., Wong, C. and Gundersen, D. T. (2006). Mercury concentrations in gonad, liver, and muscle of white sturgeon *Acipenser transmontanus* in the lower Columbia River. *Arch. Environ. Contam. Toxicol.* **50**, 443–451.

Webber, H. M. and Haines, T. A. (2003). Mercury effects on predator avoidance behaviour of a forage fish, golden shiner (*Notemigonus crysoleucas*). *Environ. Toxicol. Chem.* **22**, 1556–1561.

Wells, P. G., McCallum, G. P., Chen, C. S., Henderson, J. T., Lee, C. J. J. and Perstin, J. (2009). Oxidative stress in developmental origins of disease: Teratogenesis, neurodevelopmental deficits, and cancer. *Toxicol. Sci.* **108**, 4–18.

Wiener, J. G. and Spry, D. J. (1996). Toxicological significance of mercury in freshwater fish. In *Environmental contaminants in wildlife: Interpreting tissue concentrations* (eds. W. N. Beyer, G. H. Heinz and A. W. Redmon-Norwood), pp. 297–339. Boca Raton, FL: CRC Press.

Wiener, J. G., Krabbenhoft, D. P., Heinz, G. H. and Scheuhammer, A. M. (2003). Ecotoxicology of mercury. In *Handbook of ecotoxicology* (eds. D. J. Hoffman, B. A. Rattner, G. A. Burton, Jr. and J. Cairns, Jr.), 2nd edition, pp. 409–463. Boca Raton, FL: CRC Press.

Williams, L., Schoof, R. A., Yeager, J. W. and Goodrich-Mahoney, J. W. (2006). Arsenic bioaccumulation in freshwater fishes. *Hum. Ecol. Risk Assess.* **12**, 904–923.

Wiseman, S., Thomas, J. K., McPhee, L., Hursky, O., Raine, J. C., Pietrock, M., Giesy, J. P., Hecker, M. and Janz, D. M. (2011). Attenuation of the cortisol response to stress in female rainbow trout chronically exposed to dietary selenomethionine. *Aquat. Toxicol.* **105**, 643–651.

Wong, P. T. S., Chau, Y. K., Kramar, O. and Bengert, G. A. (1981). Accumulation and depuration of tetramethyllead by rainbow trout. *Water Res.* **15**, 621–625.

Young, T. F., Finley, K., Adams, W. J., Besser, J., Hopkins, W. D., Jolley, D., McNaughton, E., Presser, T. S., Shaw, D. P. and Unrine, J. (2010a). What you need to know about selenium. In *Ecological Assessment of Selenium in the Aquatic Environment* (eds. P. M. Chapman, W. J. Adams, M. L. Brooks, C. G. Delos, S. N. Luoma, W. A. Maher, H. M. Ohlendorf, T. S. Presser and D. P. Shaw), pp. 7–45. Boca Raton, FL: CRC Press.

Young, T. F., Finley, K., Adams, W. J., Besser, J., Hopkins, W. D., Jolley, D., McNaughton, E., Presser, T. S., Shaw, D. P. and Unrine, J. (2010b). Appendix A: Selected case studies of ecosystem contamination by Se. In *Ecological Assessment of Selenium in the Aquatic Environment* (eds. P. M. Chapman, W. J. Adams, M. L. Brooks, C. G. Delos, S. N. Luoma, W. A. Maher, H. M. Ohlendorf, T. S. Presser and D. P. Shaw), pp. 257–292. Boca Raton, FL: CRC Press.

Zinoni, F., Birkman, A., Leinflder, W. and Bocj, A. (1987). Cotranslational insertion of selenocysteine into formate dehydrogenase from *Escherichia coli* directed by a UGA codon. *Proc. Natl. Acad. Sci USA* **84**, 3156–3160.

Zuo, Z., Cai, J., Wang, X., Li, B., Wang, C. and Chen, Y. (2009). Acute administration of tributyltin and trimethyltin modulate glutamate and N-methyl-D-aspartate receptor signaling pathway in *Sebastiscus marmoratus*. *Aquat. Toxicol.* **92**, 44–49.

Zuo, Z., Wang, C., Wu, M., Wang, Y. and Chen, Y. (2012). Exposure to tributyltin and triphenyltin induces DNA damage and alters nucleotide excision repair gene transcription in *Sebastiscus marmoratus* liver. *Aquat. Toxicol.* **122–123**, 106–112.

4

EFFECTS ON FISH OF POLYCYCLIC AROMATIC HYDROCARBONS (PAHS) AND NAPHTHENIC ACID EXPOSURES

TRACY K. COLLIER

BERNADITA F. ANULACION

MARY R. ARKOOSH

JOSEPH P. DIETRICH

JOHN P. INCARDONA

LYNDAL L. JOHNSON

GINA M. YLITALO

MARK S. MYERS

Organic Chemical Toxicology of Fishes: Volume 33
FISH PHYSIOLOGY

1. INTRODUCTION

Polycyclic aromatic hydrocarbons (PAHs) are a class of organic pollutants that contain two or more fused aromatic rings composed of hydrogen and carbon. PAHs can also have alkyl groups, such as methyl and ethyl groups, substituted for one or more of their hydrogen atoms, and are generally considered to include heterocyclic aromatic compounds, where sulfur, oxygen, or nitrogen can commonly be substituted for one or more of the carbon atoms in their aromatic rings. PAHs make up a substantial portion of many fossil fuels, including crude oil, oil shales, and tar sands. The petroleum-derived, or petrogenic, PAHs usually have two to three fused rings, are highly alkylated, and are also known as low molecular weight PAHs. When petroleum products and other organic materials such as wood are subject to incomplete combustion, higher molecular weight PAHs (generally thought of as having four to seven fused rings) can be formed as a result of condensation processes. These compounds are generally unsubstituted and are commonly referred to as pyrogenic PAHs. The simplest PAH is naphthalene, consisting of two aromatic benzene rings. Phenanthrene and anthracene are three-ringed compounds, and as the number of rings increases, the number of possible structures multiplies (Figure 4.1). Thus, as a class, PAHs are composed of literally thousands of structures, making their classification, nomenclature, and analysis, as well as studies of their environmental fate and effects, extremely complex.

PAHs are derived from both natural and anthropogenic sources and are released from a wide range of industries and everyday activities. Petrogenic PAHs are associated with petroleum exploration, extraction, transport, and refining, as well as with oil seeps, and can also be formed biogenically. Petrogenic PAHs are made up of 85% or more alkyl congeners, based on analyses of a range of fresh and weathered crude oils and refined petroleum products (Wang et al., 2003). Pyrogenic PAHs are released into the atmosphere and surface waters as a result of fossil fuel and wood combustion, as well as natural forest and grassland fires, and from volcanoes. Unlike many other organic chemical contaminants that are

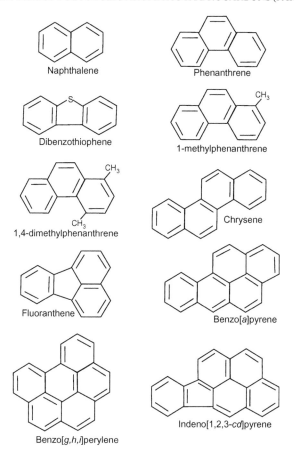

Figure 4.1. Chemical structures of polycyclic aromatic hydrocarbons commonly detected in environmental samples.

manufactured and regulated, PAHs continue to be released on a global scale. Given the multitude of sources, and especially our continued global reliance on petroleum as an energy source, PAHs will continue to be chemical contaminants of concern for decades to come.

In this chapter, we briefly review transformation of PAHs in the aquatic environment, highlighting their efficient metabolism in fish. We also review the evidence that links PAH exposure to a wide range of biological dysfunctions in fish, including neoplasia, reduced reproductive success and other types of endocrine disruption, immunotoxicity, postlarval growth and somatic condition, transgenerational impacts, and finally, recent findings showing that embryonic development of fish is severely affected by extremely low concentrations of PAH exposure. We also include a brief review of the effects of naphthenic acids on fish because these compounds are increasingly recognized as major factors in the toxicity of process waters from a variety of petroleum sources, most notably the immense oil sands deposits found in Alberta, Canada.

2. TRANSFORMATIONS OF PAHs IN THE AQUATIC ENVIRONMENT

2.1. Physical and Chemical Degradation

PAHs are persistent chemical contaminants that are distributed in environmental matrices, including sediments, soils, water, and air. Lower molecular weight PAHs (those containing two to three fused benzene rings) can be found in all of these matrices, whereas higher molecular weight compounds (those containing four to seven fused benzene rings) are more tightly associated (physically and chemically) with sediments/soils and particles than the other abiotic sample types. PAHs in air can be altered via chemical oxidation and photochemical processes (Lima et al., 2005), whereas in sediments/soils and the uppermost portion of the water column, degradation of PAHs, particularly lower molecular weight PAHs, occurs via photooxidation (McElroy et al., 1989; Bertilsson and Widenfalk, 2002). In addition to parent PAHs, oxygenated PAH metabolites formed during these degradation processes can remain associated with sediments up to six months after initial addition to the water column and thus can remain in the environment for extended periods of time (Hinga and Pilson, 1987).

In sediment and water samples, certain microorganisms (e.g., bacteria, fungi) have been shown to mineralize PAHs under aerobic conditions, particularly those compounds that contain two- and three-fused rings (e.g., naphthalene, fluorene), to their basic elements or to biodegrade these compounds to more polar degradation products (Meador et al., 1995; Lima et al., 2005). Information on PAH microbial degradation pathways and identification of degradation products is beyond the scope of this chapter but details on these processes and products can be found in Cerniglia and Heitkamp (1989), Juhasz and Naidu (2000), and Bamforth and Singleton (2005). Some research studies have demonstrated that pyrene, a four-ring PAH, can be mineralized by certain strains of bacteria (e.g., *Mycobacterium*) under optimum growing conditions in the laboratory, but it is unclear if this occurs in the natural environment (Heitkamp et al., 1988; Heitkamp and Cerniglia, 1989). In contrast, other higher molecular weight PAHs (e.g., five- and six-ring compounds) are not readily degraded by microbes and thus are more likely to accumulate in these environmental media (particularly in fine-grained sediments with high organic carbon content) (Meador et al., 1995; Juhasz and Naidu, 2000). Under anoxic conditions, PAHs persist in sediments, particularly in organic sediments (Heitkamp and Cerniglia, 1989).

2.2. Biotransformation in the Aquatic Food Web

In aquatic organisms, exposure to PAHs can occur through dermal exposure, respiration, or consumption of contaminated prey (e.g., annelids, mollusks) or sediment (Varanasi et al., 1989). Biotransformation of PAHs in aquatic organisms occurs to varying degrees depending on a number of factors, including the rate of uptake, metabolic capability, physical condition, feeding strategy, and age (James, 1989; Varanasi et al., 1989). Invertebrates are capable of PAH uptake from their environment and have been shown to have varying levels of PAH-metabolizing capability (James, 1989). Mollusks generally have lower PAH-metabolizing capability compared to certain species of polychaetes, crustaceans, and fish (Meador et al., 1995). In contrast to persistent organic pollutants, PAHs are readily metabolized by teleost fish, with the initial oxidation occurring through the action of cytochrome P4501A, creating epoxides (see Chapter 1 for a complete review of biotransformation). PAH epoxides are short-lived reactive compounds that can hydrolyze nonenzymatically to form phenols or be enzymatically converted by epoxide hydrolase to form dihydrodiols. PAH epoxides, through the action of glutathione-S-transferases, can also be conjugated to glutathione and eventually converted to mercapturic acids. The more stable diols and phenols are also further metabolized in fish through the actions of UDP-glucuronosyl transferases or sulfotransferases, creating glucuronide and sulfate conjugates, respectively. These now highly polar conjugated metabolites are then excreted into urine or secreted into the bile for rapid elimination through the gastrointestinal tract (Roubal et al., 1977; Krahn et al., 1984; Varanasi et al., 1989; Collier and Varanasi, 1991). As a result of this rapid metabolism and elimination in fish, concentrations of parent PAHs are negligible in muscle and other tissues. Thus, bile and urine are good matrices for determining recent exposure to PAHs in teleost fish, with bile being preferred because of its ease in collection. Interestingly, differences in the metabolism of the four-ring PAH, benzo[a]pyrene (BaP), including differences in the types and proportions of metabolites formed, have been demonstrated between two species of freshwater fish (Sikka et al., 1991; Steward et al., 1991). These differences could contribute to variations in susceptibility to these carcinogenic compounds among fish species. Differences in glutathione-S-transferases may also help explain differential susceptibility to chemically induced carcinogenesis among fish species (Collier et al., 1992).

A number of analytical methods have been developed to measure PAH metabolites in fish bile and are reviewed in Beyer et al. (2010). Oil spills, including the *Deepwater Horizon* (*DWH*) spill in the Gulf of Mexico in 2010, have led to a need for additional methods (including rapid analytical methods that measure PAHs and PAH metabolites) to determine PAH exposure in seafood or protected species (e.g., marine mammals, sea turtles) where animals cannot be lethally sampled. For example, a rapid, sensitive HPLC-fluorescence method that measures 14 carcinogenic and noncarcinogenic PAHs in seafood tissues was developed by the U.S. Food and Drug Administration (Gratz et al., 2010) during the *DWH* spill and was used by federal and Gulf state analytical laboratories as part of the seafood safety response (Ylitalo et al., 2012). Development of new analytical methods such as this can provide important information on PAH exposure in aquatic organisms.

3. HEPATIC NEOPLASIA AND RELATED LESIONS IN WILD FISH

A number of case studies in wild fish have strongly linked the occurrence of hepatic neoplasms and neoplasia-related toxicopathic liver lesions to PAH exposure, as measured by ΣPAHs in sediments, PAH metabolites or fluorescent aromatic compounds (FACs) in fish bile, PAH-DNA adducts in liver, or components of the natural diet of these species. The only convincing case studies linking PAH exposure to neoplasia in wild fish are those involving liver neoplasms and a spectrum of hepatic lesions involved in the histogenesis of liver neoplasia. The following sections provide summaries of these interesting and illuminating case studies.

3.1. Mummichog/Killifish, *Fundulus heteroclitus* (Elizabeth River, Virginia, Chesapeake Bay; Delaware Estuary)

Studies in this region on the highly territorial, nonmigratory estuarine mummichog (also referred to as killifish) began with the discovery of an epizootic of a suite of toxicopathic hepatic lesions, including hepatocellular, cholangiocellular, and pancreatic neoplasms, preneoplastic focal lesions (commonly referred to as foci of cellular alteration or altered hepatocellular foci (FCA/AHF)), and lesions manifesting cytotoxicity in fish inhabiting a site in the Elizabeth River, Virginia, that was grossly PAH-contaminated by a nearby facility that pressure-treated wood with creosote (Vogelbein et al., 1990; reviewed in Vogelbein and Unger, 2006). The prevalence of some of these lesions at this site exceeded 70%, and less severely PAH-contaminated sites in the Elizabeth River and James River estuary showed intermediate prevalences of these toxicopathic lesions; they were not detected in fish from relatively uncontaminated areas. Overall, field studies incorporating samplings at 12 sites in the region have shown a strong association between occurrence of these lesions and PAH exposure, with prevalences tracking well along a PAH exposure gradient. Recent studies using stepwise logistic regression have confirmed these trends statistically, whereby both total sediment PAH concentrations and TOC-normalized PAH concentrations exhibited a significantly positive relationship with the risk of occurrence of toxicopathic liver lesions, especially for neoplasms and preneoplastic focal lesions (Vogelbein et al., 2008). Moreover, as was also shown in English sole, immunohistochemical studies on cytochrome P4501A (CYP1A) and glutathione-S-transferase localization in the preneoplastic focal lesions and hepatic neoplasms (Van Veld et al., 1991, 1992) indicate that the hepatocytes that make up these lesions have undergone an adaptive response consistent with the selective resistance to cytotoxicity or resistant hepatocyte model of hepatocarcinogenesis (Solt et al., 1977).

An abundant literature surrounds the issue of toxicity resistance in this population of mummichog/killifish from the grossly creosote-contaminated Atlantic Wood site in the Elizabeth River (reviewed in Van Veld and Nacci, 2008; see that review for more specific references). This collective body of research includes the demonstration of resistance or enhanced tolerance to acute toxicity (i.e., skin and fin erosion, and death) associated with exposure to Atlantic Wood sediment in adults from the Elizabeth River as compared to those from reference sites (the York River), dramatically reduced hepatic CYP1A inducibility in adult Atlantic Wood mummichog as compared to mummichog from uncontaminated reference sites either by interparenteral (i.p.) injections of 3-methylcholanthrene or by exposure to two dilutions of sediment (10%, 30%) from the Atlantic Wood site (Van Veld and Westbrook, 1995), and resistance to teratogenicity in the form of cardiac abnormalities in the offspring of Elizabeth River fish (Ownby et al., 2002). This resistance in the Elizabeth River mummichog is at least partially heritable in laboratory-reared offspring from reproductively mature adults from the Elizabeth River, with resistance to toxicity declining from the F1 to subsequent generations, suggesting that both physiological acclimation and genetic adaptation are components of this resistance (Meyer and Di Giulio, 2002). The mechanism for this resistance to PAH-associated toxicity in both adults and offspring of Elizabeth River mummichog likely involves alterations in the aryl hydrocarbon receptor (AhR)-mediated pathways, specifically those related to CYP1A induction. CYP1A expression and inducibility is depressed in Elizabeth River mummichog, and this inducibility persisted in F1 embryos and larvae reared in the laboratory. However, CYP1A inducibility rose to normal levels in the mature adults developing from these Elizabeth River F1 embryos and in subsequent generations.

Recent studies relate more specifically to differential susceptibility and potential resistance to hepatic carcinogenesis in offspring of this population of Atlantic Wood/Elizabeth River mummichog that are tolerant to PAH toxicity, as compared to those from a reference site, Kings Creek (Wills et al., 2010a). Larvae from Elizabeth River and Kings Creek parents were exposed to BaP for two 24-hr periods, seven days apart, at doses of 50, 100, 200, and 400 µg/L. Fish were sampled for histological examination at three and nine months after the second exposure. Briefly, results showed no toxicopathic, neoplasia-related liver lesions at any dose in either Elizabeth River or Kings Creek juveniles at three months post-exposure, while Kings Creek fish at nine months post-exposure, in the 200 µg/L dose, showed an incidence of 8% (2/25) eosinophilic FCA, compared to 0% in Elizabeth River fish. At the highest dose in the Kings Creek juveniles at nine months, the incidence of eosinophilic FCA was 30%, with two of these fish showing hepatic adenomas or carcinomas (6.7%); in Elizabeth River fish, 2/30 (6.7%) were affected with eosinophilic FCA, and none developed hepatic neoplasms. Even though the sample sizes in these experiments were relatively small, there was a significantly higher ($p = 0.018$) incidence of FCA in Kings Creek juveniles at the highest dose and at nine months (30%), as compared to Elizabeth River juveniles (6.7%) at the same dose and sampling time. Although these results are not definitive, this lower incidence may be related to the reduced inducibility of hepatic CYP1 enzymes in Atlantic Wood/Elizabeth River mummichog.

However, the most convincing evidence supporting a causal link between exposure to sediment-associated PAHs dominated by high molecular weight species and these neoplasms and neoplasia-related lesions was the experimental induction, in a four-year sediment and dietary exposure regimen in mummichog, of high prevalences (up to 40%) of altered hepatocellular foci and moderate prevalences (8.7%) of hepatic neoplasms. This experimental evidence is perhaps the strongest in existence for a direct cause-and-effect relationship between chronic PAH exposure and hepatic neoplasia in fish (Vogelbein and Unger, 2006).

Separate field studies targeting mummichog from five sites in the Delaware Estuary watershed (Pinkney and Harshbarger, 2006) also detected FCA/AHF and hepatic neoplasms at dramatically higher prevalences from sites with elevated concentrations of ΣPAHs in sediments compared to fish from sites with low ΣPAHs. Although this smaller study did not demonstrate a direct statistical correlation between lesion prevalences and PAH exposure, the overall findings support the evidence from the above studies and in a different geographical location.

3.2. Winter Flounder, *Pleuronectes americanus* (Northeast Atlantic Coast)

Multiyear, multisite studies investigating toxicopathic hepatic lesions, including neoplasms, in winter flounder from urban marine and estuarine sites in the northeastern United States have shown significant associations with PAH exposure, in addition to DDT, its metabolites, and chlordanes, but not PCBs (Johnson et al., 1993). Specifically, stepwise logistic regression techniques indicated that significant risk factors for the occurrence of neoplasms were low-molecular-weight PAHs (ΣLAHs) and chlordanes in stomach contents, with relatively low proportions of the variation in risk of lesion occurrence attributed to PAHs (7%) compared to chlordanes (22%). However, substantial proportions (14–29%) of the intersite variation in prevalence of the unique lesion termed hydropic vacuolation of biliary epithelial cells and hepatocytes (HydVac) (Moore et al., 1997) were accounted for by exposure to both ΣLAHs and high- (ΣHAHs) molecular-weight PAHs. HydVac is considered a lesion manifesting cytotoxicity but is clearly also involved in the early stages of the histogenesis of liver neoplasms. However, significant associations between risk of HydVac occurrence and exposure to DDT, its metabolites, and chlordanes, were also shown, and in the case of chlordane explained an even higher proportion (16–43%) of the variation in lesion risk (Johnson et al., 1993). Statistical analysis (by factor and canonical correlation analyses) of an earlier, more limited dataset in winter flounder from similar sites in the northeastern U.S. Atlantic Coast essentially confirmed these associations (Chang et al., 1998). Prevalences of HydVac and hepatic neoplasms dropped considerably and significantly between 1987 and 1999 at the Deer Island flats sewage outfall site in Boston Harbor, presumably resulting from reductions in the chemical content of sludge, source reduction programs, and transfer of the sewage effluent release to an offshore site (Moore et al., 1996; personal communication to Mark Myers). These findings strongly support a causal role for PAHs and certain pesticides in the etiology of these lesions. The current view is that PAHs probably play a role in initiating hepatic neoplasia in winter flounder, but exposure to chlorinated pesticides may play a stronger promotional, cytotoxic, and overall role in the genesis of the toxicopathic lesions and hepatic neoplasms in winter flounder (Moore and Stegeman, 1994; Moore et al., 1997).

3.3. Brown Bullhead Catfish, *Ameiurus nebulosus* (Tributaries of the Southern Great Lakes Region, Chesapeake Bay)

Studies on this benthic species, especially those investigating the Black and Cuyahoga rivers in Ohio, both tributaries entering Lake Erie, have also implicated high-molecular-weight PAHs in sediments as the probable etiologic agents for the hepatic tumors observed (reviewed in Baumann et al., 1991, 1996). Bullhead from areas with high neoplasm prevalences also had significantly elevated hepatic aryl hydrocarbon hydroxylase activity and hepatic DNA adduct concentrations (Balch et al., 1995). Geographically broader studies using multiple histopathological and biochemical biomarkers (Arcand-Hoy and Metcalfe, 1999) found significant prevalences of hepatic neoplasms in the PAH-polluted sites of the Detroit River in Michigan (15%), Black River in Ohio (9.5%), and Hamilton Harbor, Ontario, Canada (6%), with no tumors found in fish from relatively clean reference sites. The liver lesion data were consistent with results of biochemical biomarkers such as bile FACs and hepatic ethoxyresorufin-O-deethylase activity. Other field and caging studies in the Detroit River in Michigan (Maccubbin and Ersing, 1991; Leadley et al., 1998, 1999) found higher hepatic neoplasm prevalences at sites with elevated concentrations of sediment PAHs and FACS in bile (Leadley et al., 1999). Although evidence in the brown bullhead does not reflect the same degree of statistical rigor as the case studies in other species, it collectively supports a strong association between toxicopathic hepatic lesions, especially neoplasms and FCA/AHF, and environmental PAH exposure. Additional supportive field evidence is seen in the clear pattern of initial reduction in age-specific liver neoplasm prevalence after the closing of a coking plant on the Black River with subsequent reduction in PAH exposure (Baumann and Harshbarger, 1995), followed by a brief increase in tumor prevalence three years after dredging activities and resuspension of PAHs in 1990, and then a decrease in age-specific tumor prevalence in 1994 and 1998 (Baumann and Harshbarger, 1998). In this case, reduction in the hypothesized causal variable (PAH exposure) preceded the reduction in the effect, satisfying the required temporal order for cause and effect in epidemiological studies (Susser, 1986).

Separate and more recent field studies with this species in tributaries of the Chesapeake Bay (Pinkney et al., 2009) have shown high levels of PAH exposure, as measured by sediment ΣPAHs, bile FACs (PAH-like metabolites), and hepatic cytochrome P450 activity and PAH-DNA adducts. In addition, higher prevalences of toxicopathic hepatic lesions (including FCA/AHF and neoplasms) have been evidenced in fish from urban sites in the Anacostia River near Washington, DC, as compared to reference sites in the Tuckahoe River. In a truncated dataset including only sites from the Anacostia River and the Tuckahoe River reference site where bile FACs data were available (reviewed in Pinkney et al., 2009), PAH-like metabolites in bile were a significant predictor variable for the occurrence of liver neoplasms in brown bullhead. However, the overall statistical relationship between PAH exposure and the occurrence of hepatic neoplasms and FCA/AHF in the full dataset of sampling sites for this species in this geographic region is not particularly strong or consistent.

3.4. European Flounder, *Platichthys flesus* (North Sea)

Field studies with this species in the Dutch (Vethaak and Wester, 1996) and German (Koehler, 2004) coastal waters of the North Sea, and complementary mesocosm studies in coastal Dutch waters (Vethaak et al., 1996), have documented toxicopathic hepatic lesions involved in the histogenesis of hepatic neoplasia, including FCA/AHF, neoplasms, and HydVac that have been generally associated (with a lesser degree of statistical rigor and certainty) with exposure to environmental contaminants such as PAHs and PCBs. However, a specific statistical association with PAH exposure has not been demonstrated, and the spatial pattern of liver neoplasm prevalence has not suggested a straightforward relationship to PAH exposure. Moreover, there has been a long-term, significant decline in liver neoplasms and related lesions in this species in the North Sea (verified in a linear logistic model) since the early 1990s (Vethaak et al., 2009); these changes have been attributed to improved water and sediment quality over the last 15 to 20 years in Dutch coastal waters (Vethaak et al., 2009). The current view held by European research groups on this issue is that exposures of wild European flounder to tumor promoters such as PCBs and other polyhalogenated hydrocarbons (PHHs) are probably more important risk factors in hepatocarcinogenesis in this species than exposures to genotoxic PAHs (Koehler, 2004; Vethaak et al., 2009).

3.5. English Sole, *Parophrys vetulus* (Puget Sound, Washington, and U.S. Pacific Coast)

Over 35 years of research exists on the occurrence of hepatocellular and biliary neoplasms and a spectrum of other toxicopathic lesions involved in the stepwise histogenesis of liver neoplasia in English sole from Puget Sound (Myers et al., 1987) and other Pacific Coast estuaries (Myers et al., 1994). This histogenesis parallels the experimental histogenesis of liver neoplasia in rats (Farber and Sarma, 1987) and fish (Bailey et al., 1996; Boorman et al., 1997). Within the context of the sequential, stepwise progression of hepatic neoplasia, these hepatic lesions represent the phases of cytotoxicity, compensatory regeneration and proliferation, preneoplastic foci of cellular alteration (FCA) or altered hepatocellular foci (AHF), and hepatocellular and biliary neoplasms. Their geographical patterns of occurrence, combined with chemical contaminant exposure data in English sole at their sites of capture, provide especially strong evidence for a causal relationship with exposure to environmental PAHs and to a lesser degree, certain chlorinated hydrocarbons (CHs) such as PCBs, DDT and its metabolites, and chlordanes (see reviews/articles by Myers and Fournie, 2002; Myers et al., 2003; Johnson et al., 2008, 2013). These studies are based on the initial hypothesis and subsequently derived evidence that these liver lesions in English sole are the result of exposure to PAHs (as hepatoxicants and initiators of carcinogenesis) and, of secondary importance, exposure to chlorinated compounds such as PCBs, DDT and its metabolites, and chlordanes (as hepatotoxicants and nongenotoxic promoters of carcinogenesis) (see specific references in articles by Myers et al., 1987, 2003). To date, evidence from 12 field studies and somewhat fewer laboratory studies that have validated the field findings fulfill all of the classic criteria for causation in epidemiological studies of cancer and other diseases (Colton and Greenberg, 1983; Mausner and Bahn, 1974) as reviewed in Myers et al. (2003).

These criteria, with a brief description of the evidence within them (see Myers et al., 2003 and Collier, 2003 for specific references and more detail) for English sole include the following.

1. **Strength of the association:** Toxicopathic liver lesions were initially correlated ($p < 0.001$) with sediment ΣPAHs in a 32-site Puget Sound study, followed by an 11-site Puget Sound study showing correlations with concentrations of fluorescent aromatic compounds (FACs) in bile ($0.001 < p < 0.01$, depending on lesion type). Subsequent multisite, multiyear field studies in Puget Sound attributed varying proportions of the toxicopathic liver lesion risk to PAHs (12–54%) and far less to PCB (1–8.5%) exposure (see pp. 69–75 of Myers et al., 2003).

2. **Consistency of the association:** The association of the occurrence of these lesions with PAH exposure (as assessed by PAHs in sediments, stomach contents, bile FACs, and as hepatic PAH-DNA adducts) by multivariate analytical techniques such as stepwise logistic regression has held true for over 12 separate multisite, multiyear field studies in Puget Sound, the Pacific Coast from Southern California to Alaska conducted by the Northwest Fisheries Science Center of NOAA Fisheries, or in conjunction with Washington Department of Fish and Wildlife, and Environment Canada, in both adult and subadult English sole (see pp. 75–78 of Myers et al., 2003).

3. **Toxicological and biological plausibility:** The PAHs sole are exposed to in these studies include known higher molecular weight hepatocarcinogens (e.g., BaP) in mammals (see review by Myers et al., 1987 for multiple references) and fish (Metcalfe et al., 1988; Hawkins et al., 1995; Hendricks et al., 1985; Reddy et al., 1999).The spectrum of toxicopathic liver lesions and their complex patterns of co-occurrence closely parallel

those involved in the histogenesis of liver neoplasia in fish (Hendricks et al., 1984; pp. 79–82 in review of Myers et al., 2003) and mammals (Farber and Sarma, 1987). Moreover, laboratory studies in English sole on the uptake and disposition of PAHs such as BaP show that such PAHs can be metabolized/activated by CYP1A to proximate carcinogens that covalently bind to and, in a dose-dependent fashion, form persistent hydrophobic adducts of PAH metabolites with hepatic DNA (reviewed in French et al., 1996). These PAH-DNA adducts have the potential to represent the molecular lesions involved in initiation of hepatocarcinogenesis, and their concentrations have been shown to be significant risk factors for the occurrence of lesions early in the sequential histogenesis of hepatic neoplasia in sole (reviewed in Reichert et al., 1998). Further findings in English sole from PAH-contaminated environments (Myers et al., 1998), showing significantly reduced CYPIA immunostaining in hepatic neoplasms and FCA/AHF, and reduced PAH-DNA adduct concentrations in hepatic neoplasms, as compared to the surrounding normal liver, indicate that hepatic neoplasia in this species follows the selective resistance to cytotoxicity paradigm of hepatocarcinogenesis (Solt et al., 1977). Due to their reduced capacity for CYP1A-mediated activation of PAHs to their cytotoxic and mutagenic intermediates, with a consequent reduction in formation of covalent PAH-DNA adducts from these reactive intermediates, pre-neoplastic and neoplastic hepatocytes and neoplastic biliary epithelial cells in English sole possess an adaptive phenotype that is resistant to the cytotoxicity of agents such as PAHs that require metabolic activation to exert their toxicity (see Myers et al., 1987, 2003; Solt et al., 1977; Farber and Sarma, 1987; Roomi et al., 1985).

In summary, in the resistant hepatocyte model of chemically induced carcinogenesis, upon exposure to chemical carcinogens such as PAHs, uncommon hepatocytes are produced that have become altered through carcinogen-induced mutations (e.g., formation of PAH-DNA adducts), followed by cell proliferation to "fix" the mutation(s), resulting in fully initiated hepatocytes (Solt and Farber, 1976; Columbano et al., 1981; Farber, 1990). These initiated hepatocytes possess adaptive and resistant phenotype possessing mechanisms (e.g., reduced CYP1A expression and decreased activation of toxicants to cytotoxic and mutagenic intermediates) that allow for their survival and proliferation in an environment of toxicant exposure that inhibits the proliferation or survival of surrounding normal cells. These initiated, resistant hepatocytes are protected from the deleterious effects of acute exposure to cytotoxicants, possess a growth advantage in the liver, and therefore form foci of altered hepatocytes (FCA/AHF) and eventually neoplasms.

4. **Temporal sequence (i.e., exposure precedes disease):** Lesions manifesting cytotoxicity and occurring earlier in the histogenetic sequence of neoplasia are first observed and are the most common lesion type in younger fish, whereas the preneoplastic and neoplastic lesions first occur in fish >4 years of age, and their prevalence is strongly associated with increasing fish age (see pp. 82–83 in Myers et al., 2003). Moreover, a unique case study involving site remediation measures and English sole monitoring undertaken in previously creosote-contaminated Eagle Harbor, Washington (Myers et al., 2008), clearly shows a significantly decreasing trend from 1993 to 2004 in PAH exposure (measured by bile FACs and hepatic DNA adduct concentrations) that is paralleled by a significant and decreasing trend in risk of toxicopathic hepatic lesion occurrence. The prevalence of toxicopathic liver lesions in sole from Eagle Harbor has remained stable and at near-reference relative risks from 1997 to 2011. Other studies (1989–2009) targeting English sole from the Seattle Waterfront in Elliott Bay of Puget Sound, Washington, as part of the Puget Sound Assessment and Monitoring Program (PSAMP; now referred to as the Puget Sound Ecosystem Monitoring Program), conducted in conjunction with Washington Department of Fish and Wildlife, have further corroborated this parallel reduction in toxicopathic liver lesion prevalence, with reduced PAH exposure in English sole from a different geographic area in Puget Sound. Results from the PSAMP provide convincing evidence that the incidence of toxicopathic liver disease has declined sharply at the Seattle Waterfront site from a peak of nearly 40% in 1996 to 5% in 2009. This trend has been accompanied by reductions in PAH exposure as measured by FACs in the bile of the same English sole (Puget Sound Partnership, 2010; Collier et al., 2013) (Figure 4.2, Table 4.1). This improvement in a biomarker of fish health likely reflects the efforts to achieve source control and clean up PAH-contaminated sediments in this area.

5. **Dose–response/biological gradient:** Previously discussed correlations and associations in multiple field studies of lesion prevalence with PAH exposure parameters (sediment ΣPAHs, bile FACs, hepatic PAH-DNA adducts) fulfill this criterion. Also, data from three separate field studies show that risk of occurrence of the early hepatic lesions rises significantly with hepatic DNA adduct concentrations, as a dosimeter of chronic PAH exposure. Finally, segmented "hockey-stick" regression analyses of hepatic lesion prevalence in multiple studies have shown relatively consistent thresholds of sediment PAH exposure at which certain hepatic lesion types are first observed (Table 4.1) (see pp. 83–86 in review by Myers et al., 2003; also review by Johnson et al., 2008).

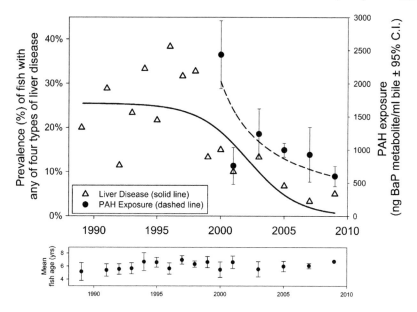

Figure 4.2. Twenty-year trend (1989–2009) in liver disease, and partially overlapping 10-year trend (2000–2009) in PAH exposure in English sole from Elliott Bay, Puget Sound, Washington State. The prevalence of liver disease was calculated using the number of 60 fish per year exhibiting any of four toxicopathic lesion types; preneoplastic foci of cellular alteration, neoplasms, hepatocellular nuclear pleomorphism/megalocytosis, or non-neoplastic proliferative lesions (hepatocellular/biliary regeneration). PAH exposure was estimated as the concentration of benzo[a]pyrene (BaP) equivalents in fish bile. Mean age of fish is shown because increasing age is positively associated with increasing prevalence of liver disease. Adapted from Puget Sound Partnership, 2010, and personal communication from J. E. West and S. M. O'Neill (Washington Department of Fish and Wildlife).

6. **Specificity of the association:** Because of the significant covariance of sediment concentrations of PAHs, PCBs, and certain pesticides in our Pacific Coast-wide English sole studies, it is not possible to attribute the toxicopathic hepatic lesions strictly to PAH exposure (see pp. 86–89 in Myers et al., 2003). However, over multiple studies, the PAHs remain the strongest (explaining the highest proportion of variation in lesion prevalence among all of the chemical exposure variables) and most consistent risk factor for occurrence of these lesions. The PCBs and pesticides contribute less to the overall risk, but are significant and plausible risk factors, as nongenotoxic promoters of liver neoplasia. Additional supportive data for the specificity of the PAH-hepatic lesion association come from results of field studies on sole conducted in Kitimat, British Columbia, in the vicinity of an

Table 4.1

Estimated effects levels associated with increasing sediment PAH concentration for selected liver lesions, indicators of reproductive dysfunction, and DNA damage in English sole. Adapted from Johnson et al. (2002).

PAHs (ppb dry wt)	Any Toxicopathic Liver Lesion (%)	Liver Neoplasms (%)	Inhibited Gonadal Development (%)	Inhibition of Spawning (%)	Infertile Eggs, % of Eggs Spawned	DNA Damage (nmol adducts/ mol bases)
50	0	0	15	12	38	5
100	0	0	15	12	38	5
1000	9	0	15	17	42	25
2000	18	0	15	25	48	36
3000	24	1	15	30	51	43
5000	31	3	18	35	55	51
10,000	40	6	27	43	61	63
100,000	71	16	58	69	80	100

aluminum smelter (Johnson et al., 2009). In this area, prevalences of the same liver lesions were associated with exposure to sediment PAHs, in the relative absence of other confounding chemicals such as PCBs. However, results also showed lower lesion prevalences than might be expected (based on ΣPAHs in sediments) at sites closest to the smelter, in addition to evidence of reduced actual uptake of PAHs in lower-than-expected bile FACs and hepatic PAH-DNA adduct concentrations. This finding suggests the limited bioavailability to English sole of smelter-derived, soot-based PAHs. At sites more distant from the smelter, hepatic lesion prevalences were comparable to those at sites with sediment PAHs from more typical, urban, industrial sources.

7. **Supportive experimental evidence:** The combined data on uptake, metabolism/activation, and binding of PAHs like BaP to DNA with dose-dependent adduct formation, with the potential to represent molecular lesions involved in initiating hepatocarcinogenesis (discussed in (3) above) also fit well into this criterion (see pp. 89–90 in the review by Myers et al., 2003). However, the most important experimental evidence in support of a causal relationship is the laboratory induction of early toxicopathic hepatic lesions (hepatocellular nuclear pleomorphism/ hepatic megalocytosis, hepatocellular regeneration, and preneoplastic focal lesions) in sole by 12 monthly injections into the dorsal sinus of either BaP or a PAH-rich extract from Eagle Harbor, followed by a six-month grow-out (Schiewe et al., 1991).

In addition to the above studies, case studies in the following species and locales were less extensively documented and showed less specific and less proven relationships between similar neoplasia-related lesions and PAH or combined chlorinated hydrocarbon (e.g., PCBs) exposure: white croaker (*Genyonemus lineatus*) and starry flounder (*Platichthys stellatus*), U.S. Pacific Coast (Myers et al., 1994; Stehr et al., 1998; McCain et al., 1992); dab (*Limanda limanda*) from the North Sea and British coastal waters (Kranz and Dethlefsen, 1990; Kohler et al., 1992; Bucke and Feist, 1993; Feist et al., 2004; Stentiford et al., 2003, 2009, 2010); and Atlantic tomcod (*Microgadus tomcod*) from the Hudson River estuary (Smith et al., 1979; Dey et al., 1993).

4. EFFECTS ON REPRODUCTION

There is substantial evidence that PAHs can cause reproductive impairment in a variety of organisms, including fish, through their ability to disrupt endocrine function, and their cytotoxic and mutagenic effects on germ cells (Tuvikene, 1995; Nicolas, 1999; Logan, 2007; Meador, 2008).

4.1. Fecundity of Females

Studies conducted over the past 30 years have shown that PAH exposure is associated with decreased plasma estradiol concentrations, reduced ovarian estradiol production, suppression of vitellogenesis, reduced ovarian growth, ovarian atresia, and reduced fecundity in female fish (e.g., Johnson et al., 1998, 2008; Anderson et al., 1996; Ridgway et al., 1999). More recent studies have confirmed a relationship between PAH exposure and reproductive impairment in female fish. A number of studies in both the field and the laboratory have documented the fact that exposure to PAHs, in mixtures or as single compounds (e.g., naphthalene and BaP), can depress plasma 17-β estradiol in female fish (Monteiro et al., 2000a; Sol et al., 2000; Tintos et al., 2006, 2007; Pollino et al., 2009), as well as suppress the *in vitro* synthesis of reproductive steroids by ovarian tissue (Monteiro et al., 2000b). Such alterations in reproductive hormones are often associated with suppressed vitellogenin production and poor ovarian growth, as reviewed by Nicolas (1999) and Logan (2007). Recent examples of these effects in fish include a study by Pait and Nelson (2009), who found significant negative correlations between sediment PAH concentrations and both gonadosomatic index (GSI) and plasma vitellogenin concentrations in killifish from the Chesapeake Bay in the United States. Similarly, Bugel et al. (2010) found that female killifish from Newark Bay, with elevated concentrations of PAH metabolites in bile, had decreased gonad weight, inhibited gonadal development, and decreased hepatic vitellogenin production. In a laboratory study, Khan (2013) reported reduced GSI and inhibited oocyte development in female cod exposed to a water-soluble fraction of crude oil.

Reduced egg production and declines in GSI have also been documented in zebrafish (*Danio rerio*) exposed to BaP in the water column at concentrations as low as 3 μg/L (Hoffmann and Oris 2006). These authors also found that transcription of several genes related to reproduction, including genes for the steroid metabolizing enzymes, 20 β-hydroxysteroid dehydrogenase and aromatase CYP19A2, and vitellogenin, were affected at BaP exposure concentrations in the 1.5–3.0 μg/L range.

4.2. Fertility of Males

While the reproductive toxicity of PAHs in female fish is well established, less information is available on the effects of PAHs on reproductive function in male fish. In laboratory rodents and humans, exposure to PAHs has been linked to reduced testosterone production, poor sperm quality, and infertility (Ramesh and Archibong, 2011). There is some evidence that these effects occur in fish as well, but results are contradictory. In several studies conducted prior to 2000, PAH exposure was associated with declines in plasma androgen concentrations and/or reduced testicular development in male fish (Burton et al., 1985; McMaster et al., 1996; Truscott et al., 1983; Kiceniuk and Khan, 1987). Sundt and Björkblom (2011) observed inhibited testicular development and a reduction in the quantity of mature sperm in prespawning male Atlantic cod (*Gadus morhua*) exposed to PAHs in produced water for 12 weeks, at concentrations comparable to those found in the environment. Testicular development was altered, showing a rise in the number of spermatogonia and primary spermatocytes and a reduction in the quantity of mature sperm in the produced water-exposed fish compared to control. Khan (2013) reported reduced GSI and disrupted testicular development and sperm production in Atlantic cod exposed to the water-soluble fraction of crude oil. Holth et al. (2008) observed changes in the expression of genes associated with the reproductive system in zebrafish exposed to produced water containing PAHs, at concentrations of 0.54 to 5.4 ppb for 1 to 13 weeks, but found no effects on reproduction or recruitment.

Researchers report mixed effects of PAHs on reproductive steroid concentrations and metabolism in male fish. Martin-Skilton et al. (2006, 2008) found that exposure of juvenile Atlantic cod and turbot (*Scophthalmus maximus*) to North Sea crude oil in water or Prestige fuel oil in the diet increased the metabolism of testosterone (precursor to the biologically active androgen, 11-ketotestosterone, in male fish) and/or reduced circulating concentrations of testosterone in the plasma. However, increased plasma testosterone concentrations and increased *in vitro* gonadotropic hormone (GTH)-stimulated testosterone production have been reported in several fish species (rainbow trout, *Oncorhynchus mykiss*; goldfish, *Carassius auratus*; rainbowfish, *Melanotaenia fluvatilis*; and grouper, *Epinephelus areolatus*) exposed to a variety of PAHs, including naphthalene, β-naphthoflavone, BaP, and retene (Evanson and Van Der Kraak, 2001; Wu et al., 2003; Pollino et al., 2009). A study by Sun et al. (2011) may shed some light on these conflicting findings. These researchers examined the effects of waterborne phenanthrene at doses of 0.06, 0.6, and 6 μg/L on testicular development in a species of rockfish, *Sebastiscus marmoratus*. After 50 days' exposure, the gonadosomatic indices and percentage of sperm produced, concentrations of salmon-type gonadotropin-releasing hormone, follicle-stimulating hormone, luteinizing hormone mRNA, 17β-estradiol, and γ-glutamyl transpeptidase activity were all depressed at the 0.06 and 0.6 μg/L doses, but were comparable to the control animals at the 6 μg/L dose. The researchers also found that these effects were related to phenanthrene accumulation in the brain, which increased with exposure to the two moderate doses of phenenathrene, but declined again at the highest exposure concentration. The decline in brain phenanthrene at the highest dose was related to changes in glutathione-S-transferase activity. Sun et al. (2011) hypothesized that the U-shaped dose–response curve for the effects of phenanthrene on testicular function was mediated via alteration of biotransformation enzyme activity in the brain. Although there is little additional information on the effects of phenanthrene on brain GST activity in fish, other studies have documented upregulation of hepatic GST activity in fish exposed to phenanthrene (Shallaja and D'Silva, 2003; Correia et al., 2007; Oliveira et al., 2008; Pathiratne and Hemachandra, 2010). However, results are somewhat inconsistent and dose dependent; for example, some researchers report GST induction at low doses of phenanthrene but little effect or suppression at higher doses (Oliveira et al., 2008; Pathiratne and Hemachandra, 2010).

Results of field studies of male reproductive function at PAH-contaminated sites are mixed. Bugel et al. (2010) evaluated the reproductive health of male Atlantic killifish inhabiting the heavily industrialized Newark Bay, which is contaminated with PAHs as well as other contaminants. The Newark Bay males had decreased gonad weight, altered testis development, and decreased gonadal aromatase mRNA expression. Similarly, Pait and Nelson (2005) observed a significant negative correlation between GSI of male killifish and PAH concentrations in sediments in Chesapeake Bay. However, in both of these studies, killifish were exposed to other contaminants in combination with PAHs, which may also have contributed to the observed effects. On the other hand, Sol et al. (2008) found little correlation between exposure to chemical contaminants, including PAHs, and either 11-ketotestosterone concentrations or gonad weight in male English sole from Puget Sound. Similarly, male California halibut (*Paralichthys californicus*) exposed to sediments from natural oil seeps did not show a dose-responsive decrease in testosterone concentrations or GSI (Seruto et al., 2005). Differences in the timing, duration, and level of exposure, the specific PAHs to which fish were exposed, and other contaminants present at these study sites, could all contribute to the conflicting responses observed in these studies.

PAHs are genotoxic compounds known to reduce sperm quantity and quality in humans and laboratory rodents (Xia et al., 2009; Ramesh and Archibong, 2011) and have strong cytotoxic effects on the testis in marine mammals (Godard et al., 2006). However, less information is available on their effects on sperm quality in fish. Nagler and Cyr (1997) observed reduced sperm quality in male plaice exposed to sediments with high concentrations of PAHs. White et al. (1999) found reduced egg hatchability and larval survival in offspring of fathead minnow (*Pimephales promelas*) exposed to BaP, but whether this finding was associated with genotoxic effects on sperm was not determined. Bohne-Kjersem et al. (2009) observed changes in some proteins and enzymes (Pzp-(pregnancy zone protein)-resembling protein and alpha-2-antiplasmin) involved in mammalian spermatogenesis, sperm capacitation, and fertilization in juvenile Atlantic cod exposed to crude oil at a concentration of 0.25 mg/L. These studies suggest an effect of PAHs on sperm quality in male fish.

4.3. Mechanisms of Endocrine Disruption

PAHs exert their endocrine disrupting effects through various mechanisms. Like dioxin, they can influence steroid production and metabolism through their interaction with the aryl hydrocarbon receptor and other dioxin-responsive elements (Williams et al., 1998; Navas and Segner, 2001; Villeneuve et al., 2002; Logan, 2007) (see Chapter 1), and they have also been shown to interact with other receptors and responsive elements that play a role in regulation of endocrine processes, including cAMP responsive elements, estrogen-responsive elements (EREs), and elements related to other nuclear receptors, including peroxisome proliferator-activated receptor, retinoid X receptor, and retinoic acid receptor (Cheshenko et al., 2008; Bilbao et al., 2010).

Several studies have demonstrated that PAHs can alter the expression and activity of the P450 enzyme aromatase, which is involved in the conversion of androgens into estrogens, generally suppressing activity and reducing estrogen synthesis (see reviews by Cheshenko et al., 2008; Le Page et al., 2011). In fish, there are two isoforms of aromatase—a predominantly ovarian form, CYP19A1, and a brain form, CYP19A2. The effects of BaP and other PAHs on aromatase activity have been investigated in multiple fish species including mummichog, gray mullet (*Chelon labrosus*), Atlantic cod, turbot (*Schophthalmus maximus*), and zebrafish (Kazeto et al., 2004; Hoffmann and Oris, 2006; Patel et al., 2006; Martin-Skilton, 2006, 2008; Dong et al., 2008; Bilbao et al., 2010). The degree to which the expression of these enzymes is altered appears to vary with dose, length of exposure, and life stage, although of the two forms, CYP19A2 appears to be the most consistently sensitive to PAH exposure.

PAHs may also affect endocrine processes through their influence on brain monoaminergic neurotransmitters, such as dopamine, serotonin, and noradrenaline. In a series of studies, Gesto and colleagues (Gesto et al., 2006, 2008, 2009) established that exposure of immature rainbow trout to the naphthalene altered levels of these three neurotransmitters in several brain regions, including the hypothalamus, preoptic region, pituitary, and brain stem. This could be significant, as serotonin is known to play a role in the modulation of diverse physiological processes, including reproduction, in fish (Rahman et al., 2011).

Exposure to PAHs may also alter estrogen and testosterone metabolism, although the effects are not necessarily consistent among studies. For example, while earlier studies suggested that PAH exposure might inhibit estrogen metabolism in fish (Snowberger and Stegeman, 1987; Stein et al., 1988), Butala et al. (2004) reported that exposure to BaP increased formation of the estradiol metabolites, 2- and 4-hydoxyestradiol, in liver, gill, and gonad microsomes of channel catfish (*Ictalurus punctatus*). Martin-Skilman et al. (2006, 2008) observed increased UDP-glucuronosyltransfer-ase activity in juvenile turbot exposed to fuel oil, and increased glucuronidation and clearance of testosterone in turbot exposed to crude oil.

5. EFFECTS ON EMBRYONIC AND LARVAL DEVELOPMENT

Of the many aspects of fish physiology potentially affected by PAHs, a wealth of laboratory and field studies indicate that embryonic development is the most sensitive process identified to date. In particular, developing fish embryos are especially sensitive to low concentrations of waterborne PAHs derived from spilled crude oil or its refined products. Largely because of the 1989 *Exxon Valdez* oil spill in Prince William Sound, AK, the impacts of crude oil exposure on fish development have been intensively studied for over two decades. This line of research has led to the identification of two major toxic pathways for injury to fish early life history stages. The first is the recognition of cardiac-specific toxicity associated primarily with the tricyclic families of PAHs that are most abundant in petroleum products, that is, fluorenes, dibenzothiophenes, and phenanthrenes. This form of toxicity is specific to the organogenesis phase of development during which the initial embryonic heart tube begins to function and undergoes looping and morphogenesis of atrial and ventricular chambers to form the larval heart. A second major pathway is light-induced toxicity (phototoxicity). Phototoxicity is not necessarily specific to a particular stage of development, but occurs in fish species (and other organisms) that have translucent early life history stages with little pigment. Some PAHs produce phototoxic reactions in cells and tissues, through activation by ultraviolet (UV) radiation and generation of reactive oxygen species, which leads to membrane damage and cell death (Arfsten et al., 1996; Yu, 2002). In laboratory studies with diverse aquatic animals, this UV-enhanced phototoxicity of PAHs occurs at much lower tissue concentrations that produce acute toxicity in the absence of UV light (Diamond, 2003). Although the precise compounds responsible for phototoxicity of petroleum products are currently unidentified, they are presumed to be polyaromatic compounds due to the necessity of this chemical structure for activity (Veith et al., 1995).

The elucidation of tricyclic PAH cardiotoxicity followed directly from field studies that assessed the impacts of the *Exxon Valdez* spill. Initial efforts focused on the two most important commercial fish species in Prince William Sound at the time of the spill: pink salmon (*Oncorhynchus gorbuscha*) and Pacific herring (*Clupea pallasi*). Due to the spawning habits of both species, they in essence put their embryos in harm's way by depositing fertilized eggs adjacent to (herring) or in stream deltas that crossed (salmon) oiled shorelines. Samples collected from pink salmon redds showed elevated rates of embryonic lethality up to four years following the spill (Rice et al., 2001), while herring larvae sampled in plankton tows along oiled shorelines showed high rates of morphological abnormalities, the most frequent of which was fluid accumulation (edema) around the heart (Marty et al., 1997a). Laboratory studies showed that exposure to Alaska North Slope crude oil (ANSCO), representative of oil from the hold of the *Exxon Valdez*, caused the same types of abnormalities in salmon and herring embryos observed in the field in a dose-dependent manner (Carls et al., 1999; Heintz et al., 1999; Marty et al., 1997a, 1997b). Since that time, multiple studies have demonstrated a common syndrome of embryonic oil exposure marked by fluid accumulation (edema or ascites) and variable degrees of craniofacial and body axis malformation (Figure 4.3). These effects have been documented for geologically distinct crude oils (ANSCO,

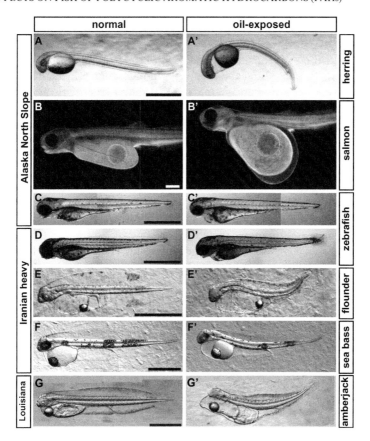

Figure 4.3. Crude oil exposure during embryogenesis causes a common fluid accumulation syndrome across multiple species of fish, geological sources of oil, and exposure methods. Left panels show control embryos with the egg shell dissected or newly hatched larvae; right panels are corresponding oil-exposed clutch mates. All exposures occurred from shortly after fertilization to the hatching stage. (A) Pacific herring and (B) pink salmon exposed to ANSCO using oiled gravel columns; (C) zebrafish exposed to a mechanically dispersed water-accommodated fraction of ANSCO; (D) zebrafish exposed to Iranian heavy crude oil using an oiled gravel column; (E) olive flounder (*Paralichthys olivaceus*) and (F) Japanese sea bass (*Lateolabrax japonicas*) exposed to Iranian heavy crude oil using an oiled gravel column; (G) Yellowtail amberjack (*Seriola lalandi*) exposed to a mechanically dispersed water-accommodated fraction of a Louisiana crude oil. (A) Incardona et al., 2009; (B) Carls and Incardona, unpublished; (C) and (D) Jung et al., 2013; (E) and (F) Jung and Incardona, unpublished; (G) Incardona et al., unpublished. See color plate at the back of the book.

Iranian heavy, Bass Strait, and Medium South American (Couillard, 2002; Incardona et al., 2005; Jung et al., 2013; McIntosh et al., 2010; Pollino and Holdway, 2002) and in a diversity of fish species beyond pink salmon and Pacific herring, including mummichog (Couillard, 2002), zebrafish (Carls et al., 2008; Incardona et al., 2005), rainbowfish (Pollino and Holdway, 2002), and Atlantic herring (*Clupea harengus*) (McIntosh et al., 2010). Importantly, the syndrome associated with crude oil exposure is independent of the method used to expose embryos and is identical whether aqueous PAHs are derived by passage of water though oiled gravel generator columns, physical dispersion of oil droplets, or chemical dispersion with detergents (Carls et al., 2008; Incardona et al., 2005; Jung et al., 2013) (Figures 4.3 and 4.4).

A series of studies using the zebrafish model determined that heart failure and the corresponding loss of circulatory function underlie the majority of the morphological defects associated with "canonical" crude oil toxicity described above (Incardona et al., 2011a). The characteristic accumulation of fluid in edematous embryos is a secondary or downstream consequence of severe cardiac malfunction, including heart malformation. Comparison between the effects of single PAHs and petrogenic PAH mixtures from crude oil in zebrafish embryos established that individual PAHs produce different forms of toxicity via distinct mechanistic pathways (Incardona et al., 2004, 2005, 2006, 2011b; Scott et al., 2011). A major finding was that the tricyclic compounds fluorene, dibenzothiophene, and phenanthrene drive most of the toxicity in petrogenic PAH mixtures (Figure 4.4). Importantly, as spilled oiled is weathered, the fractions of these tricyclic PAH classes within the total dissolved PAHs measured in water increases relative to the smaller two-ring compounds over time. The alkylated derivatives in particular become the dominant compounds during prolonged periods of weathering (Carls et al., 1999; Heintz et al., 1999; Short and Heintz, 1997). These three-ring compounds cause cardiac dysfunction relatively early in the organogenesis phase of development, when the heart is both taking shape and starting to circulate blood to other still-forming organs. The developing heart is unable to effectively pump blood, which leads to gross injury in the form of heart malformation, edema, craniofacial defects, and body axis defects (Incardona et al., 2004, 2005).

The primary effects of the tricyclic PAHs on cardiac function are most likely a consequence of ion channel blockade (Incardona et al., 2004, 2005, 2009). The observed abnormalities in heart rhythm (slowed or irregular heart rate) and contractility (pumping ability) are consistent with tricyclics acting as pharmacological agents to block the sodium, potassium, or calcium channels that together regulate the electrical activity of the heart. Consistent with this, the physiological effects of crude oil on the fish heart are indistinguishable from the side effects of some human drugs such as the antihistamine terfenadine, which were removed from the market after they were found to induce fatal arrhythmias (De Bruin et al., 2005; Mitcheson et al., 2000). In zebrafish, PAHs (Incardona et al., 2004, 2005) and drugs

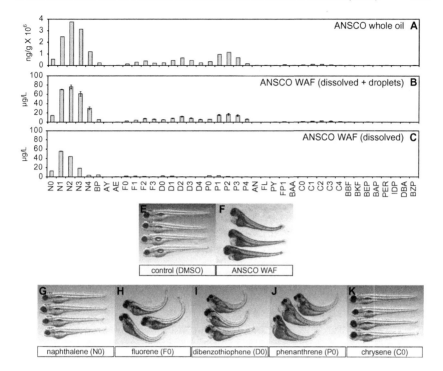

Figure 4.4. Individual tricyclic compounds in crude oil are sufficient to produce cardiotoxicity in zebrafish embryos similar to the entire petrogenic PAH mixture. (A) PAHs measured in ANSCO are shown for the whole oil after artificially weathering to remove monoaromatic compounds; (B) a mechanically dispersed water-accommodated fraction containing both aqueous (dissolved) PAHs and whole oil droplets, and (C) the same water-accommodated fraction after filtration to remove droplets, leaving behind only the aqueous fraction. Although the mixtures are dominated by naphthalenes, the next most abundant classes are the tricyclic compounds. (E–K) Hatching stage zebrafish larvae exposed to (E) solvent control (dimethylsulfoxide); (F) mechanically dispersed water-accommodated fraction of ANSCO containing 1.5 mg/L ΣPAHs; (G) 10 mg/L naphthalene; (H) 10 mg/L fluorene; (I) 10 mg/L dibenzothiophene; (J) 10 mg/L phenanthrene; and (K) 2 mg/L chrysene. Embryos are representative of studies described in Incardona et al. (2004 and 2005). Exposures were performed in small-volume (2.5 mL) multiwall polystyrene plates with high nominal concentrations of PAHs above the solubility limit of each compound. Actual aqueous concentrations were not measured but had to be at or below the solubility limit for each compound (e.g., about 7 μM for phenanthrene).

that have ion channel blocking side effects (Langheinrich et al., 2003; Milan et al., 2003) cause the same arrhythmias. These insights from the zebrafish model were extended to Pacific herring in controlled laboratory exposures using ANSCO (Incardona et al., 2009). Moreover, reduced heart rate and pericardial edema were documented in both caged and naturally spawned herring embryos collected from shorelines in San Francisco Bay up to two years after they were oiled in the aftermath of the 2007 *Cosco Busan* bunker oil spill (Incardona et al., 2012a).

The *Cosco Busan* field studies completed a full circle in which a gross syndrome was identified in field samples collected after the *Exxon Valdez* spill; the specific target organ and toxic mechanisms leading to that syndrome were identified in laboratory studies with a model fish species (zebrafish); and then the same specific defects observed in laboratory studies were confirmed in embryos collected from the impacted zone of another oil spill 18 years later. At the same time, this ground-truthing of petrogenic PAH cardiotoxicity in the field was overshadowed by a more dramatic toxicity observed in herring spawn at oiled shorelines that was subsequently determined to be a result of phototoxicity (Incardona et al., 2012a, 2012b). During the initial assessment three months following the *Cosco Busan* spill, differential toxicity was observed in herring embryos that incubated either in the shallow subtidal zone, beneath at least a meter of turbid San Francisco Bay water, or in the potentially sun-lit intertidal zone (also called the photic zone). While sublethal cardiac impacts were observed in subtidal embryos, embryos deposited in the intertidal zone of oiled sites developed to the hatching stage (e.g., fully pigmented eyes) before succumbing to lethality that involved extensive tissue deterioration (Incardona et al., 2012a).

A previous study in Pacific herring demonstrated how canonical crude oil sublethal toxicity could be converted to a more potent and acutely lethal toxicity by an interaction with UV radiation (Barron et al., 2003). As was the case for canonical petrogenic PAH toxicity, the zebrafish model provided a useful system for assessing the phototoxic potential of bunker oils typically used to power large ships like the *Cosco Busan*. A comparison of two different bunker oils with ANSCO in a simple zebrafish embryo phototoxicity assay demonstrated that the bunker oils contained an even more potent phototoxic activity than the unrefined crude oil (Hatlen et al., 2010). Embryos exposed to bunker oil in a dark incubator showed all the signs of canonical petrogenic PAH cardiotoxicity. When transferred to clean water and exposed briefly to sunlight, however, they responded with a rapid cytolytic response in transparent tissues, leading to massive tissue disruption and mortality (Hatlen et al., 2010). A broadscan GC/MS analysis of bunker oil compounds in water-accommodated fractions showed that a wide range of heterocyclic and potentially polar aromatic compounds are present at much higher concentrations in the bunker oils relative to unrefined crude oil (Hatlen et al., 2010). Assays in zebrafish embryos exposed to single PAH compounds showed that the PAHs most abundant in both crude and bunker oils did not produce strong phototoxic reactions, nor did the most abundant three-ringed heterocycles such as carbazole (Incardona et al., 2012b). The precise compounds causing bunker (or crude) oil phototoxicity remain unclear.

Regardless of the identity of the phototoxic compounds, assays in Pacific herring embryos did show that bunker oil remained potently phototoxic after at least two months of weathering (Incardona et al., 2012b). Moreover, exposure to effluent from gravel columns coated with *Cosco Busan* bunker oil containing ΣPAH concentrations less than 2 µg/L resulted in a tissue disruption pattern that was essentially identical both to that observed in zebrafish assays using water-accommodated fractions (WAFs) from mechanically dispersed oil (Hatlen et al., 2010) and to dead embryos collected from intertidal zones at oiled sites three months after the *Cosco Busan* spill (Incardona et al., 2012a, 2012b) (Figure 4.5).

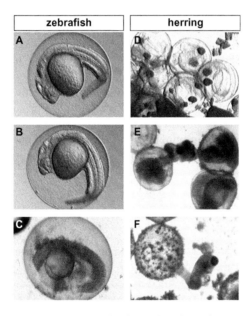

Figure 4.5. Phototoxicity of bunker oil in zebrafish and herring embryos. Left column, zebrafish embryos at 24 hours of development. (A) No-oil control following a 2-h exposure to ambient sunlight. (B) Embryo exposed to bunker oil for 24 hours followed by a 2-h outdoor exposure protected from light with aluminum foil. (C) Embryo exposed to bunker oil for 24 hours followed by a 2-h exposure to ambient sunlight outdoors. Right column, herring embryos. (D) Healthy hatching stage herring embryos. (E) Dead hatching stage embryos collected from the intertidal zone of the *Cosco Busan* spill site. (F) Dead hatching stage embryo following laboratory combined exposure to *Cosco Busan* bunker oil (expo 1.8 µg/L) and sunlight. (A–C) Hatlen et al., 2010; (D–F) Incardona et al., 2012b.

These recent findings of both major pathways of early life stage toxicity for PAHs (or other aromatics from petroleum) have important implications for the future of research on this topic. The recognition of cardiotoxicity as the driver for embryonic malformation syndromes has identified important hypotheses for the nature of delayed mortality that can occur in fish that survive very low-concentration PAH exposure during embryogenesis. Mark and recapture studies in pink salmon repeatedly demonstrated a reduction in marine survival for externally normal juvenile fish that were exposed to ANSCO as embryos (Heintz, 2007; Heintz et al., 2000). A parallel study using zebrafish showed that embryos exposed to ANSCO (ΣPAH 20–30 µg/L) grew into adults with no external abnormalities, but with abnormal heart shape that was associated with reduced aerobic performance as measure by critical swimming speed (Hicken et al., 2011). It is therefore possible that the observed reduced marine survival in the pink salmon studies was due to cardiovascular impairment resulting from mild heart defects downstream of cardiotoxicity during organogenesis. Future studies are likely to focus on cardiac anatomy and cardiovascular performance in oil-exposed salmon and herring.

The *Cosco Busan* studies highlight how poorly characterized crude oil and its products like bunker oil remain, with respect to both chemical complexity and biological effects. Identification of the primary phototoxic compounds will require a major effort using novel approaches in analytical chemistry and a focus on the "unresolved complex mixture," which is typically a large fraction in bunker oils. These chemical approaches will also need to be integrated into a program assessing biological effects.

6. TRANSGENERATIONAL EFFECTS

The transgenerational effects of chemical contaminants, mediated through epigenetic mechanisms, are an emerging area of research (Skinner et al., 2011). Epigenetics is defined as the study of mitotically or meiotically heritable changes in gene function that occur without a change in the DNA sequence (Vandegehuchte and Janssen, 2011). Epigenetic regulation of gene expression typically occurs through DNA methylation or histone modification (Head et al., 2012). Environmental influences, including toxic chemicals, can alter DNA methylation and epigenetic programming in both germ and somatic cells (Skinner et al., 2011). When an epigenetic mutation occurs in a somatic cell, it may lead to disease in the affected individual, but it will not be transmitted to the next generation. However, if the germ line is modified, it can affect not only the offspring of the individual with the mutation, but also subsequent generations (Skinner et al., 2011). Research on humans and rodents shows that exposure to PAHs can cause epigenetic modifications to somatic- and germ-cell lines, potentially leading to transgenerational effects. Exposure to PAHs in cigarette smoke can cause changes in DNA methylation in lung tissue that are correlated with the development of lesions associated with tumorigenesis (Phillips and Goodman, 2009; Liu et al., 2009, Tommasi et al., 2010). There is also some evidence that maternal smoking may be associated with altered DNA methylation patterns in the fetus that can influence nervous system and brain development (Stein, 2012), or even increase the risk of adult-onset diseases such as asthma, reduced fecundity, and obesity (Suter et al., 2010). In another recent study, Manikkam et al. (2012) found that exposure to jet fuel promoted early-onset female puberty and reduced fertility in both male and female rats not only in the exposed generation and its offspring, but in the unexposed F3 generation. These conditions appeared to be mediated through changes in DNA methylation observable in the sperm of all exposure lineage males.

Exposure to PAHs may also lead to changes in DNA methylation in fish (Fang et al., 2010), including changes associated with carcinogenesis (Fang et al., 2010; Mirbahai et al., 2011a, 2011b). Changes in DNA methylation in liver neoplasms have been observed in both laboratory-reared zebrafish, in which tumors were induced with a model carcinogen (Mirbahai et al., 2011a), and wild dab that developed neoplasms as a result of environmental exposure to PAHs (Mirbahai et al., 2011b). These studies indicate that epigenetic mechanisms occur in fish, that they are affected by exposure to PAHs, and that they may play a role in disease occurrence. There are also reports of multigenerational effects of PAH exposure. For example, when White et al. (1999) exposed fathead minnow to BaP, hatch frequency of eggs produced by F1 females was reduced and the survival of the F2 larvae was poor. Several studies also indicate that fish populations from contaminated areas have developed resistance to the effects of PAHs that may be heritable (Wirgin and Waldman, 2004). For example, Elizabeth River and other U.S. Atlantic urban coastal populations of killifish have adapted to high concentrations of PAHs by having reduced responsiveness to these compounds in the aryl-hydrocarbon receptor (AHR) pathway (Ownby et al., 2002; Meyer et al., 2002; Wills et al., 2010b). This alteration is at least partially heritable through the F2 generation and is thought to play a role in mediating their resistance to teratogenicity and lethality caused by PAHs (Meyer and Di Giulio, 2002, 2003; Meyer et al., 2002). However, the mechanisms underlying the development of PAH resistance in killifish are not completely understood. Although the genetic basis for resistance has not been fully elucidated (Mulvey et al., 2002, 2003; Hahn et al., 2004;

Meyer et al., 2005; Oleksiak et al., 2011), there is little evidence for epigenetic changes associated with modified AHR pathway functioning. Resistant and sensitive populations display no consistent differences in either AHR or AHR promoter DNA methylation patterns (Arzuaga et al., 2004; Timme-Laragy et al., 2005; Aluru et al., 2011), and there is no evidence that the resistant phenotype is passed on to future generations through epigenetically modified germ cells. Thus, as yet, there appear to be no reports in fish of true transgenerational effects of PAHs occurring as a result of epigenetic modifications to fish germ lines.

7. OTHER EFFECTS WITH CONSEQUENCES FOR SURVIVAL OF POSTLARVAL STAGES

7.1. Effects on Growth and Condition

Normal growth and adequate energy reserves are critical to the survival of juvenile fish and have a strong influence on the reproductive potential of adults. Several studies conducted prior to 2000 indicated that exposure to PAHs in crude oil or from urban and industrial sources could lead to reduced growth rates and reduced condition indices in fish (Al-Yakoob et al., 1996; Carls et al., 1996; Kubin, 1997; Moles and Norcross, 1998; Ridgway et al., 1999). In recent years these findings have been confirmed and expanded to include the effects of PAHs on lipid profiles and energy metabolism.

Reduced growth and/or condition factors have been documented in several studies conducted since 2000 in which fish were exposed to PAH mixtures comparable to those present in urban sediments. Rice et al. (2000) found significantly reduced weight in juvenile English sole fed polychaete worms reared on sediments containing 3000–4000 µg/kg dry wt of PAHs, after an exposure period of only 12 days. The growth rate was markedly lower (0.05–0.1% per day) in exposed fish than in control fish (1.1–1.2% per day). Meador et al. (2006) reported reduced weight in juvenile Chinook (*Oncorhynchus tshawytscha*) salmon fed a diet containing PAHs at a dose of 18 µg/g fish/day, as well as increased variation in weight distribution in Chinook salmon at a dietary dose of 0.7 µg/PAHs/ g fish/day Saborido-Rey et al. (2007) also showed reductions in the growth and condition factor in turbot exposed to dietary PAHs at concentrations from 25 to 260 µg/g dry wt.

Several other studies document reduced growth and condition indices in fish exposed to PAHs in the form of crude oil and related petroleum products. Heintz et al. (2000) report reduced growth of juvenile pink salmon exposed as embryos to water contaminated with PAHs from gravel coated with weathered crude oil at doses as low as 18 µg/L PAH. Kerambrun et al. (2012) exposed juvenile sea bass (*Dicentrarchus labrax*) to PAHs in the form of Arabian light crude oil for 48 and 96 hours, at concentrations ranging from 11 to 49 µg/L, and fish were then held for 26–28 days in clean seawater. They found that the 48-hour exposure led to a significant decrease in specific growth rate (length) and in RNA:DNA ratios, which are an indicator of growth rate. Significant decreases in the specific growth rates for both length and weight, recent growth rate estimated from otolith analysis, RNA:DNA ratio, and the Fulton's K condition index were observed in sea bass 26 d after the 96-h exposure. Holth et al. (2008) reported reduced condition factor in zebrafish exposed to PAHs in produced water at concentrations as low as 0.54 µg/L. Claireaux et al. (2004) reported reduced condition factor in common sole (*Solea solea*) exposed to fuel oil for 24 h and then transferred to clean water and monitored for two months. Similar reductions in condition factor were found in sole caught from sites exposed to the Erika oil spill (Gilliers et al., 2006). Amara et al. (2007) found that juvenile common sole from contaminated estuaries along the French coast of the Southern Bight of the North Sea and the Eastern English Channel had lower growth rates (estimated from otolith microstructure) and lower condition index than sole from less contaminated estuaries in the same region, but other contaminants as well as PAHs were present at the impacted sites.

Several investigators report changes in energy metabolism, particularly lipid metabolism, in fish exposed to PAHs. Meador et al. (2006) observed reductions in total body lipid and triacylglycerol (TAG) in juvenile Chinook salmon fed a PAH-contaminated diet at a dose of 18 µg/g fish/day and a reduction in plasma triglyceride and lipase concentrations at a dose of 2.3 µg/g fish/day. Bohne-Kjersem et al. (2009) found evidence from a proteonomic study of an effect of crude oil exposure on lipid metabolism. In juvenile cod, both prepro-apolipoprotein and apolipoprotein B, which are important regulators of triglyceride and cholesterol metabolism, were differentially expressed after 24 days of exposure to crude oil at concentrations from 0.06 to 1 mg/L. Holth et al. (2008) also observed effects on genes associated with lipid metabolism in zebrafish exposed to PAHs in produced water at concentrations ranging from 0.54 to 5.4 µg/L for 1, 7, and 13 weeks. In a mesocosm study, Claireaux et al. (2004) showed that the ratio of TAG to structural lipids (TAG:ST ratio) declined by 75% relative to controls in juvenile sole exposed for 24 hr to PAHs in petroleum fuel (0.1 and 0.5% fuel in water), then held for three months in clean seawater. Kerambrun et al. (2012) found a similar trend in TAG:ST ratios in juvenile Atlantic cod exposed to light crude oil (total PAH concentrations of 10–50 µg/L), but these differences were not statistically significant, perhaps because feeding with commercial pellets minimized these nutritional deficits. Amara et al. (2007) report lower TAG:ST ratios in juvenile common sole from PAH-contaminated estuaries along the French coast of the Southern Bight of the North Sea and the Eastern English Channel compared to sole from less affected estuaries in the region; however, other contaminants in addition to PAHs were also present at the affected sites and may have contributed to their results.

Other researchers report changes in plasma glucose concentrations and other measures associated with energy metabolism in fish exposed to PAHs. Tintos et al. (2007) found that exposure of immature female rainbow trout to naphthalene (intraperitoneally injected or in slow-release coconut oil implants at doses of 10 and 50 mg/kg body wt) led to increased glycogenolysis in the liver, use of exogenous glucose, and glycolysis and decreased gluconeogenesis. They hypothesized that the increased energy production in liver might be related to increased detoxification activity in liver after PAH exposure. Gesto et al. (2008) observed depressed plasma glucose concentrations in rainbow trout exposed to naphthalene, BaP, or β-naphthoflavone and subjected to prolonged stress (high stock density for 72 hr). Meador et al. (2006) showed decreased plasma glucose concentrations in juvenile Chinook salmon fed a diet containing 18 μg/g fish/day of PAHs. Several additional studies report short-term increases in plasma glucose concentrations in fish exposed to PAHs, often associated with increased concentrations of plasma cortisol and so likely a response to the stress of toxicant exposure (Tintos et al., 2008; Oliveira et al., 2007). In a field study, Oliveira et al. (2011) observed high glucose and lactate concentrations in plasma of the gray golden mullet (*Liza aurata*), from a PAH-contaminated site in comparison to fish from an uncontaminated site.

Contaminants, including PAHs, may also affect growth, condition, and energy metabolism through their effects on thyroid function. Brown et al. (2004) reviewed the effect of chemical contaminants on fish thyroid function and identified several studies in which exposure to model PAHs or water-soluble fractions of crude oil led to changes in plasma concentrations of the thyroid hormones triiodothyronine (T3) and thyroxine (T4). However, the specific effects found in these studies were not consistent, with PAH exposure sometimes leading to increases and in other cases to decreases in T3 and T4 concentrations. More recent studies are similar in their outcome. For example, Wu et al. (2003) found that exposure to dietary BaP at concentrations of 0.25 and 12.5 μg/g body wt led to increases in free T3 concentrations in plasma of juvenile grouper (*Epinephelus areolatus*). However, in a field study, Oliveira et al. (2011) found low plasma T3 concentrations in golden gray mullet from a PAH-contaminated site in Portugal. Similarly, He et al. (2012) found that exposure of embryonic rockfish (*Sebasticus marmoratus*) to pyrene reduced concentrations of T3 in whole embryos and downregulated thyroid receptor gene expression in rockfish embryos, but no effect was found on T4 concentrations. In adult eel (*Anguilla anguilla*), exposure to β-naphthoflavone significantly decreased plasma T4 concentrations, whereas thyroid-stimulating hormone and T3 remained constant (Teles et al., 2005). Other studies suggest that separate classes of contaminants, such as PCBs, are more strongly correlated than PAHs with alterations in thyroid function in fish (Brar et al., 2010).

7.2. Immunotoxicity

PAHs have been found to modulate both the innate and adaptive immune response in fish (reviewed in Rice and Arkoosh, 2002; Burnett, 2005; Reynaud and Deschaux, 2006). The immune system protects against internal and external insults. A properly functioning immune system is an important fitness trait that is critical for survival of the individual and the population (Segner et al., 2012a). An altered immune system due to contaminant exposure, either in conjunction with other stressors (Jacobson et al., 2003) or alone (Arkoosh et al., 2000, 2001, 2010), can result in increased disease susceptibility. This increase in disease susceptibility can lead to potential population-level effects (Arkoosh et al., 1998; Loge et al., 2005; Spromberg and Meador, 2005). Extrapolating changes due to contaminants from the individual or molecular level to alterations in the population or community is a key concept in ecotoxicology (Newman and Clements, 2008).

The mechanisms of PAH toxicity on the immune system in mammals have been well studied, but few studies exist in fish. However, the mechanisms of PAH toxicity are thought to be conserved from fish to mammals (Reynaud and Deschaux, 2006). Two potential mechanisms are: (1) metabolic activation (P450-mediated biotransformation) of PAH is required for immunotoxicity (Reynaud et al., 2001; Carlson et al., 2004a, b); and (2) PAHs can modulate intracellular calcium levels, suggesting that inhibition of the immune response is calcium dependent (Reynaud et al., 2001, 2003).

Innate or nonspecific immunity is the first line of defense against pathogen invasion (Magnadóttir, 2006; Whyte, 2007). Innate immunity prevents pathogens from attacking, invading, and multiplying on or in the host. Fish have integumental, humoral, and cellular innate defenses against pathogens (Bols et al., 2001; Ellis, 2001). Most studies examining the effects of PAHs on innate immunity have focused on humoral and cellular components, and not on the integumental component.

Studies examining the effect of PAHs on the humoral component of the innate immune system have investigated lysozyme and complement activity (Hutchinson et al., 2003; Bado-Nilles et al., 2009). *In vitro* exposure of European sea bass (*Dicentrarchus labrax* Linné) to a number of individual PAHs produced both an enhancement and a suppression of alternative complement activity and lysozyme concentration, depending on the concentration and the PAH examined (Bado-Nilles et al., 2009). However, dab exposed to sediments spiked with PAHs did not have an altered lysozyme concentration (Hutchinson et al., 2003). Therefore, it is apparent that the effects of PAH exposure on the humoral component of the immune system depend not only on the concentration of the PAH but also on the method of exposure.

Studies examining the effect of PAHs on the cellular component of the innate immune system have focused mainly on phagocytes (Reynaud et al., 2001, 2004; Carlson et al., 2002). Kidney phagocytes from Japanese medaka (*Oryzias latipes*) injected with BaP had a reduced amount of reactive oxygen species (ROS) activity (Carlson et al., 2002). Also, dab exposed to sediments contaminated with PAHs had a reduced amount of ROS activity in the kidney (Hutchinson et al., 2003). However, the kidney phagocytes of common carp (*Cyprinus carpio* L.) exposed to 3-methylcholanthrene had an enhanced production of ROS (Reynaud et al., 2001). Likewise, ROS production increased in kidney phagocytes of English sole exposed to a mixture of PAHs (Clemons et al., 1999). Phagocytes are critical for the optimal functioning of the innate immune system, as well as the adaptive immune system. Humoral and cellular pattern recognition receptors (PRRs) recognize pattern recognition proteins (PRPs) on microbes and malignant or "nonself" host tissue. The PRP on macrophage cells triggers phagocytosis and eventually leads to activation of adaptive or specific immunity (Janeway et al., 2005; Magnadóttir, 2010).

Adaptive or specific immunity takes longer to develop than innate immunity and consists of both cellular and humoral components (reviewed in Burnett, 2005). Cellular components of adaptive immunity in fish have been examined with mitogen assays. The mitogen assay examines the ability of critical immune cells, for example, B and T cells, to proliferate in response to selective plant lectins and pharmacologic agents. Studies examining the immunotoxicological effects of PAHs on lymphocyte proliferation have exposed fish either through their diets (Connelly and Means, 2010), in cell culture (Reynaud et al., 2003), or by injection (Reynaud and Deschaux, 2005).

Depending on the PAH used and the route of exposure, both immunosuppression and immunostimulatory effects of the mitogen response have been observed. For example, bluegills (*Lepomis macrochirus*) exposed individually to either two- (2-methylnaphthalene) or three-ringed (2-aminoanthracene and 9,10-dimethylanthracene) PAHs in their diets exhibit either increased or suppressed mitogenesis, depending on the PAH. However, a mixture of these same PAHs suppressed the response (Connelly and Means, 2010). Suppression to B- and T-cell mitogens was also observed in common carp exposed to 3-methylcholanthrene either intraperitoneally or in culture after the cells were isolated (Reynaud et al., 2003; Reynaud and Deschaux, 2005). Japanese medaka injected with BaP also had a suppressed mitogen response to both T- and B-cell mitogens (Carlson et al., 2002).

The humoral or soluble components of adaptive immunity include the production of immunoglobulins (antibodies) and cytokines (Segner et al., 2012b). Few studies have examined the effect of PAHs on the humoral components of adaptive immunity. Rainbow trout intraperitoneally injected with retene, an alkyl PAH, had an increased antibody titer to *Aeromonas salmonicida* relative to fish that were not exposed to the PAH (Hogan et al., 2010). The plaque-forming cell (PFC) assay determines the numbers of B-cells producing antibodies to a hapten. Juvenile Chinook salmon exposed to 7,12,-dimethylbenz[a]anthracene (DMBA) had suppressed primary and secondary PFC responses (Arkoosh et al., 1994). Japanese medaka were also found to have a suppressed primary PFC response after intraperitoneal injection with BaP.

Cellular assays examining various components of the innate or adaptive immune system are important in examining the immunotoxicity of PAHs. However, they are not measures of whole fish toxicity resulting from PAH exposure (Segner et al., 2012a). A more integrated and holistic approach for assessing the immunotoxicity of a chemical examines the host's susceptibility to a pathogen (Arkoosh and Collier, 2002; Arkoosh et al., 2005). With the disease challenge assay, immunotoxicity is examined with an activated rather than a resting immune system (Köllner et al., 2002). Few studies report the toxicity of environmentally relevant PAH concentrations on disease susceptibility in fish. Juvenile rainbow trout fed a diet containing a mixture of 10 high-molecular-weight PAHs were found to be more susceptible to the pathogen *Aeromonas salmonicida* (Bravo et al., 2011). Japanese medaka injected with BaP and juvenile Chinook salmon injected with a mixture of PAHs or DMBA were also more susceptible to the pathogens *Yersinia ruckeri* (Carlson et al., 2002) and *Listonella anguillarum* (Arkoosh et al., 2001), respectively. Interestingly, zebrafish exposed simultaneously to phenanthrene and the bacteria *Mycobacterium marinum* demonstrated a greater survival than those exposed to phenanthrene alone (Prosser et al., 2011). The authors hypothesized that the bacteria-induced inflammatory response interfered with the metabolism of phenanthrene to more toxic compounds, suggesting an antagonistic interaction between the stressors. This hypothesis is supported by previous studies demonstrating that proinflammatory cytokines can inhibit CYP1A biotransformation of the parent PAH compound to more potentially toxic metabolites (reviewed in Morgan, 2001). However, the exact mechanisms that result in the suppression of the CYP1A biotransformation pathway are not known (Prosser et al., 2011).

Studies with fish also exist that examine the immunotoxic effects of petroleum products such as heavy oil (Nakayama et al., 2008; Song et al., 2012a, 2012b), oil-produced water (Perez-Casanova et al., 2010, 2012), the water-soluble fraction of oil (Kennedy and Farrell, 2008; Danion et al., 2011a, 2011b), and creosote (Karrow et al., 1999, 2001). The immunotoxic effects of these products are thought to be correlated to the PAH component of the petroleum products (Reynaud and Deschaux, 2006). Functional immunological end-points and gene expression have been found to be sensitive to these products, and the effects are species and product specific (Barron, 2012).

In summary, the observed response of the teleost immune system to PAH exposure has been variable. The majority of studies investigating the effects on teleost immune function have focused mainly on individual PAHs, for example, BaP, 3-methylcholanthrene, and the synthetic DMBA (Hogan et al., 2010). A number of variables influence a host's susceptibility to these PAHs and how the contaminant ultimately influences the function of the immune system. These variables include the host, the PAH or PAH mixture used, the PAH concentration, and the route of PAH exposure (e.g., *in vivo*, *in vitro*, and *in situ*) (Tryphonas et al., 2005). A uniform approach to PAH exposure assessment on immune function would facilitate comparisons of the effects of individual and mixed PAHs on a single species as well as across species.

8. NAPHTHENIC ACIDS AS A NEW CONCERN

8.1. Background

Naphthenic acids are complex mixtures of organic compounds that are natural (and corrosive) constituents of petroleum. Naphthenic acids are present in produced waters from certain petroleum deposits, most notably the immense oil sands deposits in northern Alberta, Canada. Naphthenic acids are also found in waters discharged from oil production platforms in the North Sea (Røe Utvik, 1999). Because they can make up as much as 3–4% by weight of petroleum and are generally more water soluble than PAHs, the environmental toxicology of naphthenic acids is receiving increasing attention. This is especially true in areas such as Alberta, where the rate of extraction of bitumen is increasing, with concomitant production of millions of gallons of produced water contaminated with both naphthenic acids and PAHs. It has been estimated that 100 tonnes of naphthenic acids are being released from the daily processing of 500,000 tonnes of oil sands in Alberta (Clemente and Fedorak, 2005); large retention ponds have been built to handle the discharges of produced waters.

In terms of their chemistry, naphthenic acids are alkyl-substituted saturated cyclic and noncyclic carboxylic acids with the general chemical formula $C_nH_{2n+z}O_2$, where n indicates the number of carbon atoms and z is zero or a negative even integer that specifies the hydrogen deficiency arising from ring structures. When z is divided by -2, the result is the number of rings in the compound (see Figure 4.6). The acyclic portions of naphthenic acids can be highly branched, contributing to the complexity of these mixtures.

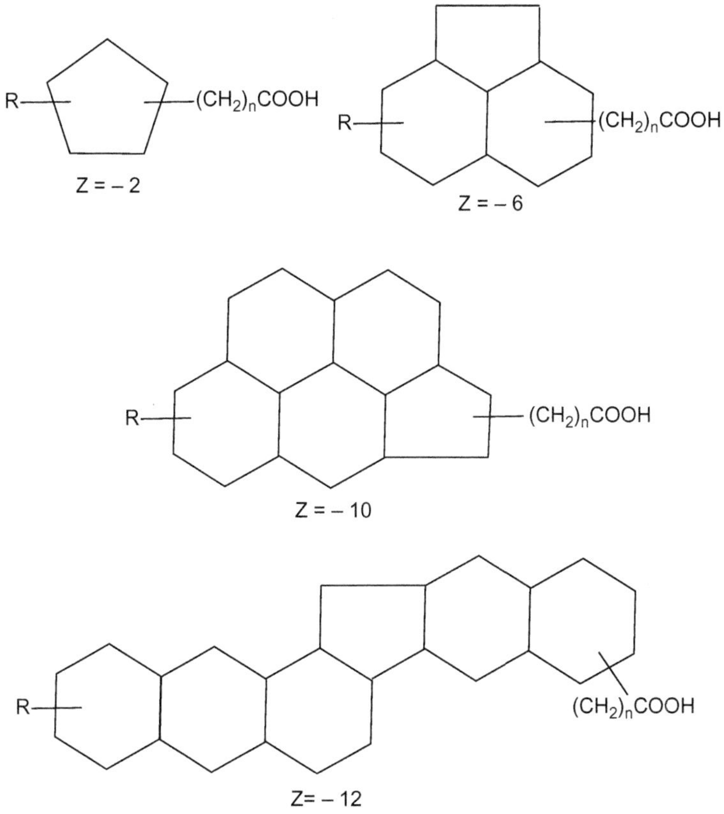

Figure 4.6. Chemical structure of four types of naphthenic acids.

8.2. Analytical Methods and Tissue Residues

The standard methods for measuring naphthenic acids in water involve extracting the samples and analyzing for the carboxylic acid moiety using Fourier transform infrared spectroscopy (Jivraj et al., 1995). When this method is used on biological samples, however, it will also detect naturally occurring fatty acids (which do fit the naphthenic acid formula given above, when $z = 0$). In order to ascertain the uptake of naphthenic acids in fish, Young et al. (2007) developed a method to qualitatively determine the presence of petroleum-derived naphthenic acids in fish tissues, involving anion exchange chromatography, followed by derivatization and GC/MS. Their results demonstrated that naphthenic acids were taken up by rainbow trout exposed via feeding or via waterborne routes, and that fish exposed to produced waters from oil sands extraction also accumulated naphthenic acids in their bodies. The method was subsequently modified to use GC/MS in single ion-monitoring mode, allowing for estimation of concentrations down to 0.1 mg/kg wet weight tissue (Young et al., 2008). Application of this method to fish captured near oil sands deposits or downstream of oil sands operations showed that several individuals of three species of fish (walleye (*Stizostedion vitreum*), northern pike (*Esox lucius*), and lake whitefish (*Coregonus clupeaformis*)) had concentrations of free naphthenic acids as high as 2.8 mg/kg wet weight. However, a subsequent sampling of several fish from the same area did not show any fish having detectable concentrations (> 0.1 mg/kg) of free naphthenic acids in their muscle tissue. However, values were reported for apparent naphthenic acids in liver tissue, which the authors contend were not due to naphthenic acids because the chromatogram did not show an unresolved hump characteristic of naphthenic acids (Young et al., 2011). This same paper reported analysis of several tissues from rainbow trout that had been exposed to waterborne naphthenic acids for 10 days at a target concentration of 3 mg/L. Data from analyses of heart, liver, gills, kidney, and muscle tissue of four individual fish showed that liver had the highest concentrations (88 ± 54 mg/kg wet wt), followed by kidney and gill, with heart and muscle tissue lowest. Naphthenic acids are thought to rapidly depurate from fish after exposure (Young et al., 2008). A comparatively small number of studies have investigated either the toxicity of commercially available naphthenic acid mixtures to fish or the toxicity of environmentally relevant naphthenic acids, extracted from oil sands process waters or oil production platform process waters. Studies of process waters themselves are not reviewed here because of the confounding effects of other contaminants, notably the PAHs covered in the rest of this chapter. Determining the methods for bioremediation of naphthenic acids in oil sands process waters is an area of active research, reviewed in Quagraine et al. (2005).

8.3. Toxicology of Naphthenic Acids

Naphthenic acids are acutely toxic in bioassays using *Daphnia magna* as test organisms, and LC_{50}s were in the low (< 1 to 10) millimolar range. Toxicity generally increased with increases in molecular weight (Frank et al., 2009), though the addition of a second carboxylic acid moiety resulted in reduced acute toxicity. Juvenile (2-month-old) salmonids exposed to naphthenic acids for 10 days showed 50% mortality at a water concentration of 25 mg/L (Dokholyan and Magomedov, 1984). Exposure of fish (wild-caught yellow perch *Perca flavescens* and laboratory-exposed Japanese medaka) to commercially available naphthenic acid mixtures has demonstrated that these compounds cause increased embryo deformities and reduced length at hatch (Peters et al., 2007). Exposure of these same species to oil sands process waters containing naphthenic acids resulted in less toxicity than would have been predicted based on the concentrations of naphthenic acids alone. The authors speculated that either differences in naphthenic acid compositions or the presence of highly toxic impurities in the commercial mixture may have accounted for this difference.

Petrogenic naphthenic acids (extracted from North Sea oil platform process waters) have recently been shown to act as estrogen receptor agonists and androgen receptor antagonists *in vitro*, using yeast cell assays (Thomas et al., 2009). Evidence arguing against naphthenic acids being reproductive toxins in fish comes from the work of Lister et al. (2008). In that study, when goldfish were exposed to oil sands process waters, gonads of both males and females showed reduced steroidogenesis, but when naphthenic acids from those same process waters were extracted and administered to goldfish for 7 days, the reductions in steroidogenesis were not observed, suggesting that compounds other than naphthenic acids were responsible for the effects of the whole process water. Conversely, a recent study in fathead minnows showed that naphthenic acids extracted from oil sands process waters did cause reproductive dysfunction, reducing female fecundity, as well as concentrations of reproductive hormones and secondary sexual characteristics in males (Kavanagh et al., 2012). These authors also showed that HCO_3^- ions, but not Na^+, Cl^-, or SO_4^- ions, could ameliorate the toxicity of naphthenic acids extracted from oil sands process waters. This is important because process waters often contain substantial amounts of salts.

To see if exposure to naphthenic acids caused histopathological changes in fish, one study using yellow perch as the test organism compared the effects of commercially available naphthenic acids to naphthenic acids extracted from oil sands process water (Nero et al., 2006). The authors reported 100% mortality at low ppm (i.e., between 1 and 10) concentrations of both mixtures, within 96 hours, and in fish exposed to lower concentrations of both mixtures for three weeks, there were increases in cellular proliferation in the gills (epithelial, mucous, and chloride cells). No histological changes were observed in the liver of exposed fish compared to control fish.

One recent study examined the immunotoxic effects of naphthenic acids in fish, using naphthenic acids that were extracted from oil sands process waters and then injected intraperitoneally into rainbow trout (MacDonald et al., 2013). The authors demonstrated that there were no appreciable concentrations of PAHs present in the extracted material; they did so by testing for induction of CYP1A and fluorescent aromatic compounds in the bile, described earlier in this chapter as indicators of PAH exposure in vertebrate species. While there were indications of immunotoxicity in the naphthenic acid-exposed fish, other fish in the same study, injected with BaP, manifested different and more severe symptoms. The authors suggested that the extracted naphthenic acids "act via a generally toxic mechanism rather than by specific toxic effects on immune cells." (p. 95) In another recent study, Hagen et al. (2012) investigated the effects of waterborne mixtures of commercial naphthenic acids on immune responses in goldfish and concluded that acute exposure (1 week) of these compounds caused the upregulation of several pro-inflammatory cytokines and increased resistance to parasitic infection, whereas longer term exposures (8 weeks) resulted in opposite effects on both end-points.

As this discussion makes clear, there is a paucity of information on the toxicology of naphthenic acids in fish. Considerably more information is needed about the occurrence and biological exposures of naphthenic acids in aquatic systems, and about the ecotoxicological relevance of those exposures. Recent advances in techniques for extraction and analysis of naphthenic acids should lead to more investigations utilizing environmentally realistic exposures, and greater clarity concerning the environmental effects of these petroleum-derived substances.

9. FUTURE DIRECTIONS

While substantial progress has been made in understanding the effects of PAHs and related compounds on fish, a number of questions remain to be addressed. The findings highlighted in this chapter showing the exquisite sensitivity of fish early life stages to petrogenic compounds both through cardiotoxicity and phototoxicity, as well as the carcinogenicity of PAHs and their impacts on the metabolic, immune, and reproductive systems, together with continued reliance on fossil fuels as a global energy source, are compelling reasons to increase research on the toxicology of PAHs and other petrochemicals. Future research needs to include work on analytical chemistry, the mechanisms of PAH toxicity, the effects of mixtures and multiple stressors, the monitoring of remediation, and improved risk assessment.

9.1. Better Analytical Chemistry

Better analytical chemistry will help determine what specific compounds are responsible for the variably toxic effects of PAHs. Some specific needs include better characterization of the PAH content of petroleum, particularly the components of fresh and weathered oil that have not previously been well studied (e.g., Aeppli et al., 2012) and the relationship of different classes of PAHs to toxicity.

9.2. Understanding the Mechanisms of Toxicant Action

The mechanisms underlying many toxic effects of PAHs, including certain embryo deformities, immunotoxicity, endocrine disruption, carcino-genicity, and phototoxicity, are still not completely understood, particularly for the wide array of alkyl PAH and alkyl monoaromatics typical of oil. There is also a need for better assessment of the additive, antagonistic, and synergistic effects of mixtures of PAHs, as well as PAHs in combination with other contaminants. The potential for PAHs to cause epigenetic transgenerational effects in fish is also as yet uncertain. Increased application of genomic techniques should increase our understanding of the underlying mechanisms responsible for the phenotypic effects demonstrated in earlier studies of impaired immunocompetence, reproductive impairment, metabolic disturbance, liver disease, and other impacts. Proteomics in concert with genomics can help determine whether the effects of PAH exposure are at the level of transcription or translation (Pruett et al., 2007).

9.3. Continued Use of Field Study Approaches

Research approaches in future field studies that integrate several sublethal end-points strongly associated with PAH exposure, such as toxicopathic lesions involved in hepatic neoplasia, reproductive effects (e.g., inhibited gonadal development, inhibition of spawning, percentage infertile eggs), with PAH exposure indicators such as sediment ΣPAH concentrations, FACs in bile, and hepatic PAH-DNA adducts in resident fish (e.g.,

English sole; see Table 4.1) are strongly encouraged. Such epizootiological studies have been instrumental in establishing PAHs as a risk factor for neoplasia and related lesions, and could serve a similar purpose for other toxic effects. We also recommend increased use of these above end-points in field studies that evaluate the recovery of fish health at sites that have undergone significant remediation, in order to show the effectiveness of management actions. Examples demonstrating that remediation can lead to improved ecological conditions and improved fish health include English sole from Eagle Harbor (Myers et al., 2008) and Elliott Bay, Washington (State of the Sound, 2009; Figure 4.2), brown bullhead in the Black River, Ohio (Baumann and Harshbarger, 1998), and winter flounder in Boston Harbor, Massachusetts (Moore et al., 1996).

9.4. Cumulative Effects with Other Stressors

We have little information on the cumulative and interactive effects of PAHs and nonchemical stressors, particularly as they relate to the delayed effects of PAHs exposure. Development of methods to test the capacity of PAH-exposed larvae, fry, and other sensitive life stages to handle natural environmental stressors (e.g., temperature flux, salinity stress, predator–prey interactions) would be very useful for estimating risks to recruitment and population abundance.

9.5. Modeling of Population Impacts

Future studies should use modeling approaches to link end-points, such as disease susceptibility, growth and metabolic alterations, endocrine disruption, neoplasia-related processes in liver, and early life history impacts, to demographic traits (e.g., survival and reproduction) that can lead to population-level effects (Spromberg and Meador, 2005, 2006).

9.6. Improved Framework for Regulation of PAHs to Protect Fish

Finally, there is a need for improved regulation of PAHs for the protection of fish. Currently, sediment quality standards set by various regulatory agencies are usually based on toxicity to benthic invertebrates (Long et al., 1998), which may avoid impacts on the fish prey base but is not necessarily protective of direct toxicity to fish. At best, standards are based on critical body residues in fish in combination with biota-sediment bioaccumulation factors (Meador et al., 2002a, 2002b). However, in the case of PAHs, which are readily metabolized in fish, this approach is not applicable. We advocate regulations based on studies targeting resident fish in their natural habitat and incorporating a suite of integrated, sublethal effects of PAHs, and PAH exposure thresholds for these effects in these fish (Johnson et al., 2002). Use of injury thresholds based on dietary PAH exposure or levels of PAH metabolites in fish bile are two additional approaches for regulating PAH exposure in fish (e.g., see Meador et al., 2008).

ACKNOWLEDGMENT

The authors wish to acknowledge the constructive and helpful review comments provided by Dr. Peter Hodson, Professor, School of Environmental Studies, Queen's University, Kingston, ON, Canada and Dr. Jonny Beyer, senior researcher at the Norwegian Institute for Water Research, Oslo, Norway, and Associate Professor, University of Stavanger, Stavanger, Norway.

REFERENCES

Aeppli, C., Carmichael, C. A., Nelson, R. K., Lemkau, K. L., Graham, W. M., Redmond, M. C., Valentine, D. L. and Reddy, C. M. (2012). Oil weathering after the *Deepwater Horizon* disaster led to the formation of oxygenated residues. *Environmental Science and Technology* **46**, 8799–8807.

Al-Yakoob, S. N., Gundersen, D. and Curtis, L. (1996). Effects of the water-soluble fraction of partially combusted crude oil from Kuwait's oil fires (from Desert Storm) on survival and growth of the marine fish *Menidia beryllina*. *Ecotoxicology and Environmental Safety* **35**, 142–149.

Aluru, N., Karchner, S. I. and Hahn, M. E. (2011). Role of DNA methylation of AHR1 and AHR2 promoters in differential sensitivity to PCBs in Atlantic killifish, *Fundulus heteroclitus*. *Aquatic Toxicology* **101**, 288–294.

Amara, R., Meziane, T., Gilliers, C., Hermel, G. and Laffargue, P. (2007). Growth and condition indices in juvenile sole *Solea solea* measured to assess the quality of essential fish habitat. *Marine Ecology Progress Series* **351**, 201–208.

Anderson, M. J., Miller, M. R. and Hinton, D. E. (1996). In vitro modulation of 17-β-estradiol-induced vitellogenin synthesis: effects of cytochrome P4501A1 inducing compounds on rainbow trout (*Oncorhynchus mykiss*) liver cells. *Aquatic Toxicology* **34**, 327–350.

Arcand-Hoy, L. D. and Metcalfe, C. D. (1999). Biomarkers of exposure of brown bullheads (*Ameiurus nebulosus*) to contaminants in the lower Great Lakes, North America. *Environmental Toxicology and Chemistry* **18**, 740–749.

Arfsten, D. P., Schaeffer, D. J. and Mulveny, D. C. (1996). The effects of near ultraviolet radiation on the toxic effects of polycyclic aromatic hydrocarbons in animals and plants: a review. *Ecotoxicology and Environmental Safety* **33**, 1–24.

Arkoosh, M. R., Clemons, E., Myers, M. and Casillas, E. (1994). Suppression of B-cell mediated immunity in juvenile Chinook salmon (*Oncorhynchus tshawytscha*) after exposure to either a polycyclic aromatic hydrocarbon or to polychlorinated biphenyls. *Immunopharmacology and Immunotoxicology* **16**, 293–314.

Arkoosh, M. R., Casillas, E., Clemons, E., Kagley, A. N., Olson, R., Reno, P. and Stein, J. E. (1998). Effect of pollution on fish diseases: potential impacts on salmonid populations. *Journal of Aquatic Animal Health* **10**, 182–190.

Arkoosh, M. R., Casillas, E., Clemons, E., Huffman, P., Kagley, A., Collier, T. and Stein, J. E. (2000). Increased susceptibility of juvenile Chinook salmon to infectious disease after exposure to chlorinated and aromatic compounds found in contaminated urban estuaries. *Marine Environmental Research* **50**, 470–471.

Arkoosh, M. R., Clemons, E., Huffman, P., Kagley, A. N., Casillas, E., Adams, N., Sanborn, H. R., Collier, T. K. and Stein, J. E. (2001). Increased susceptibility of juvenile Chinook salmon to vibriosis after exposure to chlorinated and aromatic compounds found in contaminated urban estuaries. *Journal of Aquatic Animal Health* **13**, 257–268.

Arkoosh, M. R. and Collier, T. K. (2002). Ecological risk assessment paradigm for salmon: analyzing immune function to evaluate risk. *Human and Ecological Risk Assessment* **8**, 265–276.

Arkoosh, M. R., Boylen, D., Stafford, C. L., Johnson, L. L. and Collier, T. K. (2005). Use of disease challenge assay to assess immunotoxicity of xenobiotics. In *Techniques in Aquatic Toxicology* (ed. G. K. Ostrander), pp. 19–38. New York: Taylor and Francis Group.

Arkoosh, M. R., Boylen, D., Dietrich, J. P., Anulacion, B. F., Ylitalo, G. M., Bravo, C. F., Johnson, L. L., Loge, F. J. and Collier, T. K. (2010). Disease susceptibility of salmon exposed to polybrominated diphenyl ethers (PBDEs). *Aquatic Toxicology* **98**, 51–59.

Arzuaga, X., Calcaño, W. and Elskus, A. (2004). The DNA de-methylating agent 5-azacytidine does not restore CYP1A induction in PCB resistant Newark Bay killifish (*Fundulus heteroclitus*). *Marine Environmental Research* **58**, 517–520.

Bado-Nilles, A., Quentel, C., Thomas-Guyon, H. and Le Floch, S. (2009). Effects of two oils and 16 pure polycyclic aromatic hydrocarbons on plasmatic immune parameters in the European sea bass, *Dicentrarchus labrax* (Linne). *Toxicology in Vitro* **23**, 235–241.

Bailey, G., Williams, D. and Hendricks, J. (1996). Fish models for environmental carcinogenesis: the rainbow trout. *Environmental Health Perspectives* **104**, 5–21.

Balch, G. C., Metcalfe, C. D., Reichert, W. L. and Stein, J. E. (1995). Biomarkers of exposure of brown bullheads to contaminants in Hamilton Harbor, Ontario. In *Biomonitors and Biomarkers as Indicators of Environmental Change: A Handbook* (eds. F. M. Butterworth, L. D. Corkum and J. Guzman-Rincon), pp. 249–273. New York: Plenum Press.

Bamforth, S. M. and Singleton, I. (2005). Bioremediation of polycyclic aromatic hydrocarbons: current knowledge and future directions. *Journal of Chemical Technology and Biotechnology* **80**, 723–736.

Barron, M. G., Carls, M. G., Short, J. W. and Rice, S. D. (2003). Photoenhanced toxicity of aqueous phase and chemically dispersed weathered Alaska North Slope crude oil to Pacific herring eggs and larvae. *Environmental Toxicology and Chemistry* **22**, 650–660.

Barron, M. G. (2012). Ecological impacts of the *Deepwater Horizon* oil spill: implications for immunotoxicity. *Toxicologic Pathology* **40**, 315–320.

Baumann, P. C., Mac, M. J., Smith, S. B. and Harshbarger, J. C. (1991). Tumor frequencies in walleye (*Stizostedion vitreum*) and brown bullhead (*Ictalurus nebulosus*) and sediment contaminants in tributaries of the Laurentian Great Lakes. *Canadian Journal of Fisheries and Aquatic Sciences* **48**, 1804–1810.

Baumann, P. C. and Harshbarger, J. C. (1995). Decline in liver neoplasms in wild brown bullhead catfish aftercoking plant closes and environmental PAHs plummet. *Environmental Health Perspectives* **103**, 168–170.

Baumann, P. C., Smith, I. R. and Metcalfe, C. D. (1996). Linkages between chemical contaminants and tumors in benthic Great Lakes fish. *Journal of Great Lakes Research* **22**, 131–152.

Baumann, P. C. and Harshbarger, J. C. (1998). Long term trends in liver neoplasm epizootics of brown bullhead in the Black River, Ohio. *Environmental Monitoring and Assessment* **53**, 213–223.

Bertilsson, S. and Widenfalk, A. (2002). Photochemical degradation of PAHs in freshwaters and their impact on bacterial growth—influence of water chemistry. *Hydrobiologia* **469**, 23–32.

Beyer, J., Jonsson, G., Porte, C., Krahn, M. M. and Ariese, F. (2010). Analytical methods for determining metabolites of polycyclic aromatic hydrocarbon (PAH) pollutants in fish bile: a review. *Environmental Toxicology and Pharmacology* **30**, 224–244.

Bilbao, E., Raingeard, D., de Cerio, O. D., Ortiz-Zarragoitia, M., Ruiz, P., Izagirre, U., Orbea, A., Marigómez, I., Cajaraville, M. P. and Cancio, I. (2010). Effects of exposure to Prestige-like heavy fuel oil and to perfluorooctane sulfonate on conventional biomarkers and target gene transcription in the thicklip grey mullet *Chelon labrosus*. *Aquatic Toxicology* **98**, 282–296.

Bohne-Kjersem, A., Skadsheim, A., Goksøyr, A. and Grøsvik, B. E. (2009). Candidate biomarker discovery in plasma of juvenile cod (*Gadus morhua*) exposed to crude North Sea oil, alkyl phenols and polycyclic aromatic hydrocarbons (PAHs). *Marine Environmental Research* **68**, 268–277.

Bols, N. C., Brubacher, J. L., Ganassin, R. C. and Lee, L. E. J. (2001). Ecotoxicology and innate immunity in fish. *Developmental and Comparative Immunology* **25**, 853–873.

Boorman, G. A., Botts, S., Bunton, T. E., Fournie, J. W., Harshbarger, J. C., Hawkins, W. E., Hinton, D. E., Jokinen, M. P., Okihiro, M. S. and Wolfe, M. J. (1997). Diagnostic criteria for degenerative, inflammatory, proliferative nonneoplastic and neoplastic liver lesions in medaka (*Oryzias latipes*): consensus of a National Toxicology Program Pathology Working Group. *Toxicologic Pathology* **25**, 202–210.

Brar, N. K., Waggoner, C., Reyes, J. A., Fairey, R. and Kelley, K. M. (2010). Evidence for thyroid endocrine disruption in wild fish in San Francisco Bay, California, USA. Relationships to contaminant exposures. *Aquatic Toxicology* **96**, 203–215.

Bravo, C. F., Curtis, L. R., Myers, M., Meador, J. P., Johnson, L. L., Buzitis, J., Collier, T. K., Morrow, J. D., Laetz, C. A., Loge, F. J. and Arkoosh, M. R. (2011). Biomarker response and disease susceptibility in juvenile rainbow trout *Oncorhynchus mykiss* fed a high molecular weight PAH mixture. *Environmental Toxicology and Chemistry* **30**, 1–11.

Brown, S. B., Adams, B. A., Cyr, D. G. and Eales, J. G. (2004). Contaminant effects on the teleost fish thyroid. *Environmental Toxicology and Chemistry* **23**, 1680–1701.

Bucke, D. and Feist, S. W. (1993). Histopathological changes in the livers of dab, *Limanda limanda (L.)*. *Journal of Fish Diseases* **16**, 281–296.

Bugel, S. M., White, L. A. and Cooper, K. R. (2010). Impaired reproductive health of killifish (*Fundulus heteroclitus*) inhabiting Newark Bay, NJ, a chronically contaminated estuary. *Aquatic Toxicology* **96**, 182–193.

Burnett, K. G. (2005). Impact of enviromental toxicants and natural variables on the immune system of fishes. In *Biochemistry and Molecular Biology of Fishes. Vol. VI, Environmental Toxicology* (eds. T. P. Mommsen and T. W. Moon), pp. 231–253. Amsterdam: Elsevier.

Burton, D., Burton, M. P., Truscott, B. and Idler, D. R. (1985). Epidermal cellular proliferation and differentiation in sexually mature male *Salmo salar* with androgen levels depressed by oil. *Proceedings of the Royal Society of London Series B Biological Sciences* **225**, 121–128.

Butala, H., Metzger, C., Rimoldi, J. and Willett, K. L. (2004). Microsomal estrogen metabolism in channel catfish. *Marine Environmental Research* **58**, 489–494.

Carls, M. G., Holland, L., Larsen, M., Lum, J. L., Mortensen, D. G., Wang, S. Y. and Wertheimer, A. C. (1996). Growth, feeding, and survival of pink salmon fry exposed to food contaminated with crude oil. *American Fisheries Society Symposium* **18**, 608–618.

Carls, M. G., Rice, S. D. and Hose, J. E. (1999). Sensitivity of fish embryos to weathered crude oil: part I. Low-level exposure during incubation causes malformations, genetic damage, and mortality in larval Pacific herring (*Clupea pallasi*). *Environmental Toxicology and Chemistry* **18**, 481–493.

Carls, M. G., Holland, L., Larsen, M., Collier, T. K., Scholz, N. L. and Incardona, J. P. (2008). Fish embryos are damaged by dissolved PAHs, not oil particles. *Aquatic Toxicology* **88**, 121–127.

Carlson, E., Li, Y. and Zelikoff, J. (2002). Exposure of Japanese medaka (*Oryzias latipes*) to benzo[a]pyrene suppresses immune function and host resistance against bacterial challenge. *Aquatic Toxicology* **56**, 289–301.

Carlson, E. A., Li, Y. and Zelikoff, J. T. (2004a). Suppressive effects of benzo[a]pyrene upon fish immune function: evolutionarily conserved cellular mechanisms of immunotoxicity. *Marine Environmental Research* **58**, 731–734.

Carlson, E. A., Li, Y. and Zelikoff, J. T. (2004b). Benzo[a]pyrene-induced immunotoxicity in Japanese medaka (*Oryzias latipes*): relationship between lymphoid CYP1A activity and humoral immune suppression. *Toxicology and Applied Pharmacology* **201**, 40–52.

Cerniglia, C. E. and Heitkamp, M. A. (1989). Microbial degradation of polycyclic aromatic hydrocarbons (PAH) in the aquatic environment. In *Metabolism of Polycyclic Aromatic Hydrocarbons in the Aquatic Environment* (ed. U. Varanasi), pp. 41–68. Boca Raton, FL: CRC Press.

Chang, S., Zdanowicz, V. S. and Murchelano, R. A. (1998). Associations between liver lesions in winter flounder (*Pleuronectes americanus*) and sediment chemical contaminants from north-east United States estuaries. *ICES Journal of Marine Science: Journal du Conseil* **55**, 954–969.

Cheshenko, K., Pakdel, F., Segner, H., Kah, O. and Eggen, R. I. L. (2008). Interference of endocrine disrupting chemicals with aromatase CYP19 expression or activity, and consequences for reproduction of teleost fish. *General and Comparative Endocrinology* **155**, 31–62.

Claireaux, G., Désaunay, Y., Akcha, F., Aupérin, B., Bocquené, G., Budzinski, H., Cravedi, J.-P., Davoodi, F., Galois, R., Gilliers, C., Goanvec, C., Guérault, D., Imbert, N., Mazéas, O., Nonnotte, G., Nonnotte, L., Prunet, P., Sébert, P. and Vettier, A. (2004). Influence of oil exposure on the physiology and ecology of the common sole *Solea solea*: experimental and field approaches. *Aquatic Living Resources* **17**, 335–351.

Clemente, J. S. and Fedorak, P. M. (2005). A review of the occurrence, analyses, toxicity and biodegradation of napthenic acids. *Chemosphere* **60**, 585–600.

Clemons, E., Arkoosh, M. R. and Casillas, E. (1999). Enhanced superoxide anion production in activated peritoneal macrophages from English sole (*Pleuronectes vetulus*) exposed to polycyclic aromatic compounds. *Marine Environmental Research* **47**, 71–87.

Collier, T. and Varanasi, U. (1991). Hepatic activities of xenobiotic metabolizing enzymes and biliary levels of xenobiotics in English sole (*Parophrys vetulus*) exposed to environmental contaminants. *Archives of Environmental Contamination and Toxicology* **20**, 462–473.

Collier, T. K., Singh, S. V., Awasthi, Y. C. and Varanasi, U. (1992). Hepatic xenobiotic metabolizing enzymes in two species of benthic fish showing different prevalences of contaminant-associated liver neoplasms. *Toxicology and Applied Pharmacology* **113**, 319–324.

Collier, T. K. (2003). Forensic ecotoxicology: establishing causality between contaminants and biological effects in field studies. *Human and Ecological Risk Assessment* **9**, 259–266.

Collier, T. K., Chiang, M. W. L., Au, D. W. T. and Rainbow, P. S. (2013). Biomarkers currently used in environmental monitoring. In *Ecological Biomarkers: Indicators of Ecotoxicological Effects* (eds. C. Amiard-Triquet, J.-C. Amiard and P. S. Rainbow), pp. 385–409. Boca Raton, FL: CRC Press Taylor and Francis Group.

Colton, T. and Greenberg, E. R. (1983). Cancer epidemiology. In *Statistics in Medical Research: Methods and Issues, with Applications in Cancer Research* (eds. V. Mike and K. E. Stanley), pp. 23–70. Wiley Subscription Services, Inc., A Wiley Company.

Columbano, A., Rajalakshmi, S. and Sarma, D. S. (1981). Requirement for cell proliferation for the initiation of liver carcinogenesis as assayed by three different procedures. *Cancer Research* **41**, 2079–2083.

Connelly, H. and Means, J. C. (2010). Immunomodulatory effects of dietary exposure to selected polycyclic aromatic hydrocarbons in the bluegill (*Lepomis macrochirus*). *International Journal of Toxicology* **29**, 532–545.

Correia, A. D., Goncalves, R., Scholze, M., Ferreira, M. and Reis-Henriques, M. A. (2007). Biochemical and behavioral responses in gilthead seabream (*Sparus aurata*) to phenanthrene. *Journal of Experimental Marine Biology and Ecology* **347**, 109–122.

Couillard, C. M. (2002). A microscale test to measure petroleum oil toxicity to mummichog embryos. *Environmental Toxicology* **17**, 195–202.

Danion, M., Le Floch, S., Lamour, F., Guyomarch, J. and Quentel, C. (2011a). Bioconcentration and immunotoxicity of an experimental oil spill in European sea bass (*Dicentrarchus labrax* L.). *Ecotoxicology and Environmental Safety* **74**, 2167–2174.

Danion, M., Le Floch, S., Kanan, R., Lamour, F. and Quentel, C. (2011b). Effects of in vivo chronic hydrocarbons pollution on sanitary status and immune system in sea bass (*Dicentrarchus labrax* L.). *Aquatic Toxicology* **105**, 300–311.

De Bruin, M. L., Pettersson, M., Meyboom, R. H. B., Hoes, A. W. and Leufkens, H. G. M. (2005). Anti-HERG activity and the risk of drug-induced arrhythmias and sudden death. *European Heart Journal* **26**, 590–597.

Dey, W. P., Peck, T. H., Smith, C. E. and Kreamer, G.-L. (1993). Epizoology of hepatic neoplasia in Atlantic tomcod (*Microgadus tomcod*) from the Hudson River estuary. *Canadian Journal of Fisheries and Aquatic Sciences* **50**, 1897–1907.

Diamond, S. A. (2003). Photoactivated toxicity in aquatic environments. In *UV Effects in Aquatic Organisms and Ecosystems* (eds. D. Hader and G. Joir), pp. 219–250. New York: John Wiley & Sons.

Dokholyan, B. K. and Magomedov, A. K. (1984). Effect of sodium naphthenate on survival and some physiological–biochemical parameters of some fishes. *Journal of Icthyology* **23**, 125–132.

Dong, W., Wang, L., Thornton, C., Scheffler, B. E. and Willett, K. L. (2008). Benzo(a)pyrene decreases brain and ovarian aromatase mRNA expression in *Fundulus heteroclitus*. *Aquatic Toxicology* **88**, 289–300.

Ellis, A. E. (2001). Innate host defense mechanisms of fish against viruses and bacteria. *Developmental and Comparative Immunology* **25**, 827–839.

Evanson, M. and Van Der Kraak, G. J. (2001). Stimulatory effects of selected PAHs on testosterone production in goldfish and rainbow trout and possible mechanisms of action. *Comparative Biochemistry and Physiology Part C: Toxicology and Pharmacology* **130**, 249–258.

Fang, X., Dong, W., Thornton, C., Scheffler, B. and Willett, K. L. (2010). Benzo(a)pyrene induced glycine N-methyltransferase messenger RNA expression in *Fundulus heteroclitus* embryos. *Marine Environmental Research* **69** (Suppl. 1), S74–S76.

Farber, E. and Sarma, D. S. R. (1987). Biology of disease—hepatocarcinogenesis: a dynamic cellular perspective. *Laboratory Investigations* **56**, 4–22.

Farber, E. (1990). Clonal adaptation during carcinogenesis. *Biochemical Pharmacology* **39**, 1837–1846.

Feist, S., Lang, T., Stentiford, G. and Koehler, A. (2004). Use of liver pathology of the European flatfish dab (*Limanda limanda* L.) and flounder (*Platichthys flesus* L.) for monitoring. *ICES Techniques in Marine Environmental Sciences*, 38. Copenhagen: ICES.

Frank, R. A., Fischer, K., Burnison, B. K., Arsenault, G., Headley, J. V., Peru, K. M., Van Der Kraak, G. and Solomon, K. R. (2009). Effect of carboxylic acid content on the acute toxicity of oil dands naphthenic acids. *Environmental Science and Technology* **43**, 266–271.

French, B. L., Reichert, W. L., Hom, T., Nishimoto, M., Sanborn, H. R. and Stein, J. E. (1996). Accumulation and dose-response of hepatic DNA adducts in English sole (Pleuronectes vetulus) exposed to a gradient of contaminated sediments. *Aquatic Toxicology* **36**, 1–16.

Gesto, M., Tintos, A., Soengas, J. L. and Míguez, J. M. (2006). Effects of acute and prolonged naphthalene exposure on brain monoaminergic neurotransmitters in rainbow trout (*Oncorhynchus mykiss*). *Comparative Biochemistry and Physiology Part C: Toxicology and Pharmacology* **144**, 173–183.

Gesto, M., Soengas, J. L. and Míguez, J. M. (2008). Acute and prolonged stress responses of brain monoaminergic activity and plasma cortisol levels in rainbow trout are modified by PAHs (naphthalene, β-naphthoflavone and benzo(a)pyrene) treatment. *Aquatic Toxicology* **86**, 341–351.

Gesto, M., Tintos, A., Soengas, J. L. and Míguez, J. M. (2009). β-Naphthoflavone and benzo(a) pyrene alter dopaminergic, noradrenergic, and serotonergic systems in brain and pituitary of rainbow trout (*Oncorhynchus mykiss*). *Ecotoxicology and Environmental Safety* **72**, 191–198.

Gilliers, C., Le Pape, O., Désaunay, Y., Bergeron, J. P., Schreiber, N., Guerault, D. and Amara, R. (2006). Growth and condition of juvenile sole (*Solea solea L.*) as indicators of habitat quality in coastal and estuarine nurseries in the Bay of Biscay with a focus on sites exposed to the Erika oil spill. *Scienta Marina* **70S1**, 183–192.

Godard, C. A. J., Wise, S. S., Kelly, R. S., Goodale, B., Kraus, S., Romano, T., O'Hara, T. and Wise, J. P., Sr (2006). Benzo[a]pyrene cytotoxicity in right whale (*Eubalaena glacialis*) skin, testis and lung cell lines. *Marine Environmental Research* **62** (Suppl. 1), S20–S24.

Gratz, S., Mohrhaus, A., Gamble, B., Gracie, J., Jackson, D., Roetting, J., et al. (2010). *Screen for the presence of polycyclic aromatic hydrocarbons in select seafoods using LC-Fluorescence. Laboratory Information Bulletin No. 4475.* Silver Spring, MD: U.S. Food and Drug Administration.

Hagen, M. O., Garcia-Garcia, E., Oladiran, A., Karpman, M., Mitchell, S., El-Din, M. G., Martin, J. W. and Belosevic, M. (2012). The acute and sub-chronic exposures of goldfish to naphthenic acids induce different host defense responses. *Aquatic Toxicology* **109**, 143–149.

Hahn, M. E., Karchner, S. I., Franks, D. G. and Merson, R. R. (2004). Aryl hydrocarbon receptor polymorphisms and dioxin resistance in Atlantic killifish (*Fundulus heteroclitus*). *Pharmacogenetics and Genomics* **14**, 131–143.

Hatlen, K., Sloan, C. A., Burrows, D. G., Collier, T. K., Scholz, N. L. and Incardona, J. P. (2010). Natural sunlight and residual fuel oils are an acutely lethal combination for fish embryos. *Aquatic Toxicology* **99**, 56–64.

Hawkins, W. E., Walker, W. W. and Overstreet, R. M. (1995). Carcinogenicity tests using aquarium fish. *Toxicology Mechanisms and Methods* **5**, 225–263.

He, C., Zuo, Z., Shi, X., Sun, L. and Wang, C. (2012). Pyrene exposure influences the thyroid development of *Sebastiscus marmoratus* embryos. *Aquatic Toxicology* **124–125**, 28–33.

Head, J. A., Dolinoy, D. C. and Basu, N. (2012). Epigenetics for ecotoxicologists. *Environmental Toxicology and Chemistry* **31**, 221–227.

Heintz, R. A., Short, J. W. and Rice, S. D. (1999). Sensitivity of fish embryos to weathered crude oil: part II. Increased mortality of pink salmon (*Oncorhynchus gorbuscha*) embryos incubating downstream from weathered *Exxon Valdez* crude oil. *Environmental Toxicology and Chemistry* **18**, 494–503.

Heintz, R. A., Rice, S. D., Wertheimer, A. C., Bradshaw, R. F., Thrower, F. P., Joyce, J. E. and Short, J. W. (2000). Delayed effects on growth and marine survival of pink salmon *Oncorhynchus gorbuscha* after exposure to crude oil during embryonic development. *Marine Ecology Progress Series* **208**, 205–216.

Heintz, R. A. (2007). Chronic exposure to polynuclear aromatic hydrocarbons in natal habitats leads to decreased equilibrium size, growth, and stability of pink salmon populations. *Integrated Environmental Assessment and Management* **3**, 351–363.

Heitkamp, M. A., Franklin, W. and Cerniglia, C. E. (1988). Microbial metabolism of polycyclic aromatic hydrocarbons: isolation and characterization of a pyrene-degrading bacterium. *Applied and Environmental Microbiology* **54**, 2549–2555.

Heitkamp, M. A. and Cerniglia, C. E. (1989). Polycyclic aromatic hydrocarbon degradation by a *Mycobacterium sp.* in microcosms containing sediment and water from a pristine ecosystem. *Applied and Environmental Microbiology* **55**, 1968–1973.

Hendricks, J., Meyers, T. and Shelton, D. (1984). Histological progression of hepatic neoplasia in rainbow trout (*Salmo gairdneri*). *National Cancer Institute Monograph Series* **65**, 321–336.

Hendricks, J. D., Meyers, T. R., Shelton, D. W., Casteel, J. L. and Bailey, G. S. (1985). Hepatocarcinogenicity of benzo[a]pyrene to rainbow trout by dietary exposure and intraperitoneal injection. *Journal of the National Cancer Institute* **74**, 839–851.

Hicken, C. E., Linbo, T. L., Baldwin, D. H., Myers, M. S., Holland, L., Larsen, M., Scholz, N. L., Collier, T. K., Rice, G. S., Stekoll, M. S. and Incardona, J. P. (2011). Sub-lethal exposure to crude oil during embryonic development alters cardiac morphology and reduces aerobic capacity in adult fish. *Proceedings of the National Academy of Sciences U S A* **108**, 7086–7090.

Hinga, K. R. and Pilson, M. E. Q. (1987). Persistence of benz[a]anthracene degradation products in an enclosed marine ecosystem. *Environmental Science and Technology* **21**, 648–653.

Hoffmann, J. L. and Oris, J. T. (2006). Altered gene expression: a mechanism for reproductive toxicity in zebrafish exposed to benzo[a]pyrene. *Aquatic Toxicology* **78**, 332–340.

Hogan, N. S., Lee, K. S., Kollner, B. and van den Heuvel, M. R. (2010). The effects of the alkyl polycyclic aromatic hydrocarbon retene on rainbow trout (*Oncorhynchus mykiss*) immune response. *Aquatic Toxicology* **100**, 246–254.

Holth, T. F., Nourizadeh-Lillabadi, R., Blaesbjerg, M., Grung, M., Holbech, H., Petersen, G. I., Alestrom, P. and Hylland, K. (2008). Differential gene expression and biomarkers in zebrafish (*Danio rerio*) following exposure to produced water components. *Aquatic Toxicology* **90**, 277–291.

Hutchinson, T. H., Field, M. D. R. and Manning, M. J. (2003). Evaluation of non-specific immune functions in dab, *Limanda limanda* L., following short-term exposure to sediments contaminated with polyaromatic hydrocarbons and/or polychlorinated biphenyls. *Marine Environmental Research* **55**, 193–202.

Incardona, J. P., Collier, T. K. and Scholz, N. L. (2004). Defects in cardiac function precede morphological abnormalities in fish embryos exposed to polycyclic aromatic hydrocarbons. *Toxicology and Applied Pharmacology* **196**, 191–205.

Incardona, J. P., Carls, M. G., Teraoka, H., Sloan, C. A., Collier, T. K. and Scholz, N. L. (2005). Aryl hydrocarbon receptor-independent toxicity of weathered crude oil during fish development. *Environmental Health Perspectives* **113**, 1755–1762.

Incardona, J. P., Day, H. L., Collier, T. K. and Scholz, N. L. (2006). Developmental toxicity of 4-ring polycyclic aromatic hydrocarbons in zebrafish is differentially dependent on AH receptor isoforms and hepatic cytochrome P450 1A metabolism. *Toxicology and Applied Pharmacology* **217**, 308–321.

Incardona, J. P., Carls, M. G., Day, H. L., Sloan, C. A., Bolton, J. L., Collier, T. K. and Scholz, N. L. (2009). Cardiac arrhythmia is the primary response of embryonic Pacific herring (*Clupea pallasi*) exposed to crude oil during weathering. *Environmental Science and Technology* **43**, 201–207.

Incardona, J. P., Collier, T. K. and Scholz, N. L. (2011a). Oil spills and fish health: exposing the heart of the matter. *Journal of Exposure Science and Environmental Epidemiology* **21**, 3–4.

Incardona, J. P., Linbo, T. L. and Scholz, N. L. (2011b). Cardiac toxicity of 5-ring polycyclic aromatic hydrocarbons is differentially dependent on the aryl hydrocarbon receptor 2 isoform during zebrafish development. *Toxicology and Applied Pharmacology* **257**, 242–249.

Incardona, J. P., Vines, C. A., Anulacion, B. F., Baldwin, D. H., Day, H. L., French, B. L., Labenia, J. S., Linbo, T. L., Myers, M. S., Olson, O. P., Sloan, C. A., Sol, S., Griffin, F. J., Menard, K., Morgan, S. G., West, J. E., Collier, T. K., Ylitalo, G. M., Cherr, G. N. and Scholz, N. L. (2012a). Unexpectedly high mortality in Pacific herring embryos exposed to the 2007 *Cosco Busan* oil spill in San Francisco Bay. *Proceedings of the National Academy of Sciences U S A* **109**, E51–58.

Incardona, J. P., Vines, C. A., Linbo, T. L., Myers, M. S., Sloan, C. A., Anulacion, B. F., Boyd, D., Collier, T. K., Morgan, S., Cherr, G. N. and Scholz, N. L. (2012b). Potent phototoxicity of marine bunker oil to translucent herring embryos after prolonged weathering. *PLoS One* **7**, e30116.

Jacobson, K. C., Arkoosh, M. R., Kagley, A. N., Clemons, E. R., Collier, T. K. and Casillas, E. (2003). Cumulative effects of natural and anthropogenic stress on immune function and disease resistance in juvenile Chinook salmon. *Journal of Aquatic Animal Health* **15**, 1–12.

James, M. O. (1989). Biotransformation and deposition of PAH in aquatic invertebrates. In *Metabolism of Polycyclic Aromatic Hydrocarbons in the Aquatic Environment* (ed. U. Varanasi), pp. 69–91. Boca Raton, FL: CRC Press.

Janeway, C., Travers, P., Walport, M. and Shlomchik, M. (2005). Innate immunity. In *Immunobiology* (eds. C. Janeway, P. Travers, M. Walport and M. Shlomchik), 6th ed., pp. 37–100. New York: Garland Science Publishing.

Jivraj, M. N., MacKinnon, M. and Fung, B. (1995). *Naphthenic acid extraction and quantitative analysis with FT-IR spectroscopy. Syncrude Analytical Manuals,* 4th ed. Edmonton, AB: Syncrude Canada Ltd.

Johnson, L. L., Stehr, C. M., Olson, O. P., Myers, M. S., Pierce, S. M., Wigren, C. A., McCain, B. B. and Varanasi, U. (1993). Chemical contaminants and hepatic lesions in winter flounder (*Pleuronectes americanus*) from the northeast coast of the United States. *Environmental Science and Technology* **27**, 2759–2771.

Johnson, L. L., Landahl, J. T., Kubin, L. A., Horness, B. H., Myers, M. S., Collier, T. K. and Stein, J. E. (1998). Assessing the effects of anthropogenic stressors on Puget Sound flatfish populations. *Journal of Sea Research* **39**, 125–137.

Johnson, L. L., Collier, T. K. and Stein, J. E. (2002). An analysis in support of sediment quality thresholds for polycyclic aromatic hydrocarbons (PAHs) to protect estuarine fish. *Aquatic Conservation: Marine and Freshwater Ecosystems* **12**, 517–538.

Johnson, L. L., Arkoosh, M. R., Bravo, C. F., Collier, T. K., Krahn, M. M., Meador, J. P., Myers, M. S., Reichert, W. L. and Stein, J. E. (2008). The effects of polycyclic aromatic hydrocarbons in fish from Puget Sound Washington. In *The Toxicology of Fishes* (eds. R. T. DiGiulio and D. E. Hinton), pp. 874–919. Boca Raton, FL: CRC Press.

Johnson, L. L., Ylitalo, G. M., Myers, M. S., Anulacion, B. F., Buzitis, J., Reichert, W. L. and Collier, T. K. (2009). Polycyclic Aromatic Hydrocarbons and Fish Health Indicators in the Marine Ecosystem in Kitimat, British Columbia. U.S. Department of Commerce, NOAA Tech. Memo., NMFS-NWFSC-98. p. 123.

Johnson, L. L., Anulacion, B. F., Arkoosh, M. R., Burrows, D. G., da Silva, D. A. M., Dietrich, J. D., Myers, M. S., Spromberg, J. A. and Ylitalo, G. M. (2013). Effects of legacy persistent organic pollutants (POPs) in fish—current and future challenges. In *Fish Physiology in Review* (ed. K. B. Tierney), Elsevier.

Juhasz, A. L. and Naidu, R. (2000). Bioremediation of high molecular weight polycyclic aromatic hydrocarbons: a review of the microbial degradation of benzo[a]pyrene. *International Biodeterioration and Biodegradation* **45**, 57–88.

Jung, J.-H., Hicken, C. E., Boyd, D., Anulacion, B. F., Carls, M. G., Shim, W. J. and Incardona, J. P. (2013). Geologically distinct crude oils cause a common cardiotoxicity syndrome in developing zebrafish. *Chemosphere* **91**, 1146–1155.

Karrow, N. A., Boermans, H. J., Dixon, D. G., Hontella, A., Solomon, K. R., Whyte, J. J. and Bols, N. C. (1999). Characterizing the immunotoxicity of creosote to rainbow trout (Oncorhynchus mykiss): a microcosm study. *Aquatic Toxicology* **45**, 223–239.

Karrow, N. A., Bols, N. C., Wyte, J. J., Solomon, K. R., Dixon, D. G. and Boermans, H. J. (2001). Effects of creosote exposure on rainbow trout pronephros phagocyte activity and the percentage of lymphoid B cells. *Journal of Toxicology and Environmental Health Part A* **63**, 363–381.

Kavanagh, R. J., Frank, R. A., Burnison, B. K., Young, R. F., Fedorak, P. M., Solomon, K. R. and Van Der Kraak, G. (2012). Fathead minnow (*Pimephales promelas*) reproduction is impaired when exposed to a naphthenic acid extract. *Aquatic Toxicology* **116-117**, 34–42.

Kazeto, Y., Place, A. R. and Trant, J. M. (2004). Effects of endocrine disrupting chemicals on the expression of CYP19 genes in zebrafish (*Danio rerio*) juveniles. *Aquatic Toxicology* **69**, 25–34.

Kennedy, C. J. and Farrell, A. P. (2008). Immunological alternations in juvenile Pacific herring, *Clupea pallasi,* exposed to aqueous hydrocarbons derived from crude oil. *Environmental Pollution* **153**, 638–648.

Kerambrun, E., Le Floch, S., Sanchez, W., Thomas Guyon, H., Meziane, T., Henry, F. and Amara, R. (2012). Responses of juvenile sea bass, *Dicentrarchus labrax*, exposed to acute concentrations of crude oil, as assessed by molecular and physiological biomarkers. *Chemosphere* **87**, 692–702.

Khan, R. A. (2013). Effects of polycyclic aromatic hydrocarbons on sexual maturity of Atlantic cod, *Gadus morhua*, following chronic exposure. *Environment and Pollution* **2**, 1–10.

Kiceniuk, J. W. and Khan, R. A. (1987). Effect of petroleum hydrocarbons on Atlantic cod, *Gadus morhua*, following chronic exposure. *Canadian Journal of Zoology* **65**, 490–494.

Koehler, A. (2004). The gender-specific risk to liver toxicity and cancer of flounder (*Platichthys flesus (L.)*) at the German Wadden Sea coast. *Aquatic Toxicology* **70**, 257–276.

Kohler, A., Deisemann, H. and Lauritzen, B. (1992). Histological and cytochemical indices of toxic injury in the liver of dab *Limanda limanda*. *Marine Ecology Progress Series* **91**, 141–153.

Köllner, B., Kotterba, G. and Fisher, U. (2002). Evaluation of immune functions of rainbow trout—how can environmental influences be detected? *Toxicology Letters* **131**, 83–95.

Krahn, M. M., Myers, M. S., Burrows, D. G. and Malins, D. C. (1984). Determination of metabolites of xenobiotics in the bile of fish from polluted waterways. *Xenobiotica* **14**, 633–646.

Kranz, H. and Dethlefsen, V. (1990). Liver anomalies in dab *Limanda limanda* from the southern North Sea with special consideration given to neoplastic lesions. *Diseases of Aquatic Organisms* **9**, 171–185.

Kubin, L. A. (1997). A study of growth in juvenile English sole exposed to sediments amended with aromatic compounds [Master's Thesis]. Bellingham, WA: Western Washington University.

Langheinrich, U., Vacun, G. and Wagner, T. (2003). Zebrafish embryos express an orthologue of HERG and are sensitive toward a range of QT-prolonging drugs inducing severe arrhythmia. *Toxicology and Applied Pharmacology* **193**, 370–382.

Le Page, Y., Vosges, M., Servili, A., Brion, F. and Kah, O. (2011). Neuroendocrine effects of endocrine disruptors in teleost fish. *Journal of Toxicology and Environmental Health, Part B* **14**, 370–386.

Leadley, T. A., Balch, G., Metcalfe, C. D., Lazar, R., Mazak, E., Habowsky, J. and Haffner, G. D. (1998). Chemical accumulation and toxicological stress in three brown bullhead (*Ameiurus nebulosus*) populations of the Detroit River, Michigan, USA. *Environmental Toxicology and Chemistry* **17**, 1756–1766.

Leadley, T. A., Arcand-Hoy, L. D., Haffner, G. D. and Metcalfe, C. D. (1999). Fluorescent aromatic hydrocarbons in bile as a biomarker of exposure of brown bullheads (*Ameiurus nebulosus*) to contaminated sediments. *Environmental Toxicology and Chemistry* **18**, 750–755.

Lima, A. L. C., Farrington, J. W. and Reddy, C. M. (2005). Combustion-derived polycyclic aromatic hydrocarbons in the environment—a review. *Environmental Forensics* **6**, 109–131.

Lister, A., Nero, V., Farwell, A., Dixon, D. G. and Van Der Kraak, G. (2008). Reproductive and stress hormone levels in goldfish (*Carassius auratus*) exposed to oil sands process-affected water. *Aquatic Toxicology* **87**, 170–177.

Liu, W.-b., Liu, J.-y., Ao, L., Zhou, Z.-y., Zhou, Y.-h., Cui, Z.-h., Yang, H. and Cao, J. (2009). Dynamic changes in DNA methylation during multistep rat lung carcinogenesis induced by 3-methylcholanthrene and diethylnitrosamine. *Toxicology Letters* **189**, 5–13.

Logan, D. T. (2007). Perspective on ecotoxicology of PAHs to fish. *Human and Ecological Risk Assessment: An International Journal* **13**, 302–316.

Loge, F., Arkoosh, M. R., Ginn, T. R., Johnson, L. L. and Collier, T. K. (2005). Impact of environmental stressors on the dynamics of disease transmission. *Environmental Science and Technology* **39**, 7329–7336.

Long, E. R., Field, L. J. and MacDonald, D. D. (1998). Predicting toxicity in marine sediments with numerical sediment quality guidelines. *Environmental Toxicology and Chemistry* **17**, 714–727.

Maccubbin, A. and Ersing, N. (1991). Tumors in fish from the Detroit River. *Hydrobiologia* **219**, 301–306.

MacDonald, G. Z., Hogan, N. S., Kollner, B., Thorpe, K. L., Phalen, L. J., Wagner, B. D. and van den Heuvel, M. R. (2013). Immunotoxic effects of oil sands-derived naphthenic acids to rainbow trout. *Aquatic Toxicology* **126**, 95–103.

Magnadóttir, B. (2006). Innate immunity of fish (overview). *Fish and Shellfish Immunology* **20**, 137–151.

Magnadóttir, B. (2010). Immunological control of fish diseases. *Marine Biotechnology* **12**, 361–379.

Manikkam, M., Guerrero-Bosagna, C., Tracey, R., Haque, M. M. and Skinner, M. K. (2012). Transgenerational actions of environmental compounds on reproductive disease and identification of epigenetic biomarkers of ancestral exposures. *PLoS One* **7**, e31901.

Martin-Skilton, R., Thibaut, R. and Porte, C. (2006). Endocrine alteration in juvenile cod and turbot exposed to dispersed crude oil and alkylphenols. *Aquatic Toxicology* **78** (Suppl.), S57–S64.

Martin-Skilton, R., Saborido-Rey, F. and Porte, C. (2008). Endocrine alteration and other biochemical responses in juvenile turbot exposed to the *Prestige* fuel oil. *Science of The Total Environment* **404**, 68–76.

Marty, G. D., Hose, J. E., McGurk, M. D., Brown, E. D. and Hinton, D. E. (1997a). Histopathology and cytogenetic evaluation of Pacific herring larvae exposed to petroleum hydrocarbons in the laboratory or in Prince William Sound, Alaska, after the *Exxon Valdez* oil spill. *Canadian Journal of Fisheries and Aquatic Sciences* **54**, 1846–1857.

Marty, G. D., Short, J. W., Dambach, D. M., Willits, N. H., Heintz, R. A., Rice, S. D., Stegeman, J. J. and Hinton, D. E. (1997b). Ascites, premature emergence, increased gonadal cell apoptosis, and cytochrome P4501A induction in pink salmon larvae continuously exposed to oil-contaminated gravel during development. *Canadian Journal of Zoology-Revue Canadienne De Zoologie* **75**, 989–1007.

Mausner, J. S. and Bahn, A. K. (1974). *Epidemiology: An Introductory Text.* Philadelphia: Saunders.

McCain, B. B., Chan, S. L., Krahn, M. M., Brown, D. W., Myers, M. S., Landahl, J. T., Pierce, S., Clark, R. C. and Varanasi, U. (1992). Chemical contamination and associated fish diseases in San Diego Bay. *Environmental Science and Technology* **26**, 725–733.

McElroy, A. E., Farrington, J. W. and Teal, J. M. (1989). Bioavailability of polycyclic aromatic hydrocrabons in aquatic environment. In *Metabolism of Polycyclic Aromatic Hydrocarbons in the Aquatic Environment* (ed. U. Varanasi), pp. 1–39. Boca Raton, FL: CRC Press.

McIntosh, S., King, T., Wu, D. and Hodson, P. V. (2010). Toxicity of dispersed weathered crude oil to early life stages of Atlantic herring (*Clupea harengus*). *Environmental Toxicology and Chemistry* **29**, 1160–1167.

McMaster, M. E., Van Der Kraak, G. J. and Munkittrick, K. R. (1996). An epidemiological evaluation of the biochemical basis for xteroid hormonal cepressions in fish exposed to industrial wastes. *Journal of Great Lakes Research* **22**, 153–171.

Meador, J. P., Stein, J. E., Reichert, W. L. and Varanasi, U. (1995). Bioaccumulation of polycyclic Aromatic hydrocarbons by marine organisms. In *Reviews of Environmental Contamination and Toxicology* (ed. G. Ware), pp. 79–165. New York: Springer.

Meador, J. P., Collier, T. K. and Stein, J. E. (2002a). *Use of Tissue and Sediment-based Threshold Concentrations of Polychlorinated Biphenyls (PCBs) to Protect Juvenile Salmonids Listed under the U.S. Endangered Species Act. Aquatic Conservation: Marine and Freshwater Ecosystems.* Hoboken, NJ: John Wiley & Sons, pp. 493–516.

Meador, J. P., Collier, T. K. and Stein, J. E. (2002b). Determination of a tissue and sediment threshold for tributyltin to protect prey species of juvenile salmonids listed under the U.S. Endangered Species Act. *Aquatic Conservation: Marine and Freshwater Ecosystems* **12**, 539–551.

Meador, J. P., Sommers, F. C., Ylitalo, G. M. and Sloan, C. A. (2006). Altered growth and related physiological responses in juvenile Chinook salmon (*Oncorhynchus tshawytscha*) from dietary exposure to polycyclic aromatic hydrocarbons (PAHs). *Canadian Journal of Fisheries and Aquatic Sciences* **63**, 2364–2376.

Meador, J. P. (2008). Polycyclic aromatic hydrocarbons. In *Encyclopedia of Ecology* (eds. S. E. Jorgensen and B. Fath), pp. 2881–2891. Oxford: Academic Press.

Metcalfe, C. D., Cairns, V. W. and Fitzsimons, J. D. (1988). Experimental induction of liver tumours in rainbow trout (*Salmo gairdneri*) by contaminated sediment from Hamilton Harbour, Ontario. *Canadian Journal of Fisheries and Aquatic Sciences* **45**, 2161–2167.

Meyer, J. and Di Giulio, R. T. (2002). Patterns of heritability of decreased EROD activity and resistance to PCB 126-induced teratogenesis in laboratory-reared offspring of killifish (*Fundulus heteroclitus*) from a creosote-contaminated site in the Elizabeth River, VA, USA. *Marine Environmental Research* **54**, 621–626.

Meyer, J. N., Nacci, D. E. and Di Giulio, R. T. (2002). Cytochrome P4501A (CYP1A) in killifish (*Fundulus heteroclitus*): heritability of altered expression and relationship to survival in contaminated sediments. *Toxicololigical Science* **68**, 69–81.

Meyer, J. N. and Di Giulio, R. T. (2003). Heritable adaptation and fitness costs in killifish (*Fundulus beteroclitus*) inhabiting a polluted estuary. *Ecological Applications* **13**, 490–503.

Meyer, J. N., Volz, D. C., Freedman, J. H. and Di Giulio, R. T. (2005). Differential display of hepatic mRNA from killifish (*Fundulus heteroclitus*) inhabiting a Superfund estuary. *Aquatic Toxicology* **73**, 327–341.

Milan, D. J., Peterson, T. A., Ruskin, J. N., Peterson, R. T. and MacRae, C. A. (2003). Drugs that induce repolarization abnormalities cause bradycardia in zebrafish. *Circulation* **107**, 1355–1358.

Mirbahai, L., Williams, T. D., Zhan, H., Gong, Z. and Chipman, J. K. (2011a). Comprehensive profiling of zebrafish hepatic proximal promoter CpG island methylation and its modification during chemical carcinogenesis. *BMC Genomics* **12**, 3.

Mirbahai, L., Yin, G., Bignell, J. P., Li, N., Williams, T. D. and Chipman, J. K. (2011b). DNA methylation in liver tumorigenesis in fish from the environment. *Epigenetics* **6**, 1319-1333.

Mitcheson, J. S., Chen, J., Lin, M., Culberson, C. and Sanguinetti, M. C. (2000). A structural basis for drug-induced long QT syndrome. *Proceedings of the National Academy of Sciences U S A* **97**, 12329-12333.

Moles, A. and Norcross, B. L. (1998). Effects of oil-laden sediments on growth and health of juvenile flatfishes. *Canadian Journal of Fisheries and Aquatic Sciences* **55**, 605-610.

Monteiro, P. R. R., Reis-Henriques, M. A. and Coimbra, J. (2000a). Plasma steroid levels in female flounder (*Platichthys flesus*) after chronic dietary exposure to single polycyclic aromatic hydrocarbons. *Marine Environmental Research* **49**, 453-467.

Monteiro, P. R. R., Reis-Henriques, M. A. and Coimbra, J. (2000b). Polycyclic aromatic hydrocarbons inhibit in vitro ovarian steroidogenesis in the flounder (*Platichthys flesus L.*). *Aquatic Toxicology* **48**, 549-559.

Moore, M. J. and Stegeman, J. J. (1994). Hepatic neoplasms in winter flounder *Pleuronectes americanus* from Boston Harbor, Massachusetts, USA. *Diseases of Aquatic Organisms* **20**, 33-48.

Moore, M. J., Shea, D., Hillman, R. E. and Stegeman, J. J. (1996). Trends in hepatic tumours and hydropic vacuolation, fin erosion, organic chemicals and stable isotope ratios in winter flounder from Massachusetts, USA. *Marine Pollution Bulletin* **32**, 458-470.

Moore, M. J., Smolowitz, R. M. and Stegeman, J. J. (1997). Stages of hydropic vacuolation in the liver of winter flounder *Pleuronectes americanus* from a chemically contaminated site. *Diseases of Aquatic Organisms* **31**, 19-28.

Morgan, E. (2001). Regulation of cytrochrome P450 by inflammation mediators: why and how? *Drug Metabolism and Disposition* **29**, 207-212.

Mulvey, M., Newman, M. C., Vogelbein, W. and Unger, M. A. (2002). Genetic structure of *Fundulus heteroclitus* from PAH-contaminated and neighboring sites in the Elizabeth and York rivers. *Aquatic Toxicology* **61**, 195-209.

Mulvey, M., Newman, M. C., Vogelbein, W. K., Unger, M. A. and Ownby, D. R. (2003). Genetic structure and mtDNA diversity of *Fundulus heteroclitus* populations from polycyclic aromatic hydrocarbon-contaminated sites. *Environmental Toxicology and Chemistry* **22**, 671-677.

Myers, M. S., Rhodes, L. D. and McCain, B. B. (1987). Pathologic anatomy and patterns of occurrence of hepatic neoplasms, putative preneoplastic lesions, and other idiopathic hepatic conditions in English sole (*Parophrys vetulus*) from Puget Sound, Washington. *Journal of the National Cancer Institute* **78**, 333-363.

Myers, M. S., Stehr, C. M., Olson, O. P., Johnson, L. L., McCain, B. B., Chan, S.-L. and Varanasi, U. (1994). Relationships between toxicopathic hepatic lesions and exposure to chemical contaminants in English sole (*Pleuronectes vetulus*), starry flounder (*Platichthys stellatus*), and white croaker (*Genyonemus lineatus*) from selected marine sites on the Pacific Coast,USA. *Environmental Health Perspectives* **102**, 200-215.

Myers, M. S., French, B. L., Reichert, W. L., Willis, M. L., Anulacion, B. F., Collier, T. K. and Stein, J. E. (1998). Reductions in CYP1A expression and hydrophobic DNA adducts in liver neoplasms of English sole (*Pleuronectes vetulus*): further support for the "resistant hepatocyte" model of hepatocarcinogenesis. *Marine Environmental Research* **46**, 197-202.

Myers, M. S. and Fournie, J. W. (2002). Histopathological biomarkers as integrators of anthropogenic and environmental stressors. In *Biological Indicators of Aquatic Ecosystem Stress* (ed. S. M. Adams), pp. 221-287. Bethesda, MD: American Fisheries Society.

Myers, M. S., Johnson, L. L. and Collier, T. K. (2003). Establishing the causal relationship between polycyclic aromatic hydrocarbon (PAH) exposure and hepatic neoplasms and neoplasia-related liver lesions in English sole (*Pleuronectes vetulus*). *Human and Ecological Risk Assessment: An International Journal* **9**, 67–94.

Myers, M. S., Anulacion, B. F., French, B. L., Reichert, W. L., Laetz, C. A., Buzitis, J., Olson, O. P., Sol, S. and Collier, T. K. (2008). Improved flatfish health following remediation of a PAH-contaminated site in Eagle Harbor, Washington. *Aquatic Toxicology* **88**, 277–288.

Nagler, J. J. and Cyr, D. G. (1997). Exposure of male american plaice (*Hippoglossoides platessoides*) to contaminated marine sediments decreases the hatching success of their progeny. *Environmental Toxicology and Chemistry* **16**, 1733–1738.

Nakayama, K., Kitamura, S. I., Murakami, Y., Song, J. Y., Jung, S. J., Oh, M. J., Iwata, H. and Tanabe, S. (2008). Toxicogenomic analysis of immune system-related genes in Japanese flounder (*Paralichthys olivaceus*) exposed to heavy oil. *Marine Pollution Bulletin* **57**, 445–452.

Navas, J. M. and Segner, H. (2001). Estrogen-mediated suppression of cytochrome P4501A (CYP1A) expression in rainbow trout hepatocytes: role of estrogen receptor. *Chemico-Biological Interactions* **138**, 285–298.

Nero, V., Farwell, A., Lee, L. E., Van Meer, T., MacKinnon, M. D. and Dixon, D. G. (2006). The effects of salinity on naphthenic acid toxicity to yellow perch: gill and liver histopathology. *Ecotoxicology and Environmental Safety* **65**, 252–264.

Newman, M. C. and Clements, W. H. (2008). *Ecotoxicology: A Comprehensive Treatment*. Boca Raton, FL: CRC Press, Taylor and Francis Group.

Nicolas, J.-M. (1999). Vitellogenesis in fish and the effects of polycyclic aromatic hydrocarbon contaminants. *Aquatic Toxicology* **45**, 77–90.

Oleksiak, M. F., Karchner, S. I., Jenny, M. J., Mark Welch, D. B. and Hahn, M. E. (2011). Transcriptomic assessment of resistance to effects of an aryl hydrocarbon receptor (AHR) agonist in embryos of Atlantic killifish (*Fundulus heteroclitus*) from a marine Superfund site. *BMC Genomics* **12**, 263–280.

Oliveira, M., Pacheco, M. and Santos, M. A. (2007). Cytochrome P4501A, genotoxic and stress responses in golden grey mullet (*Liza aurata*) following short-term exposure to phenanthrene. *Chemosphere* **66**, 1284–1291.

Oliveira, M., Pacheco, M. and Santos, M. A. (2008). Organ specific antioxidant responses in golden grey mullet (*Liza aurata*) following a short-term exposure to phenanthrene. *Science of the Total Environment* **396**, 70–78.

Oliveira, M., Pacheco, M. and Santos, M. A. (2011). Fish thyroidal and stress responses in contamination monitoring—An integrated biomarker approach. *Ecotoxicology and Environmental Safety* **74**, 1265–1270.

Ownby, D. R., Newman, M. C., Mulvey, M, Vogelbein, W. K., Unger, M. A. and Arzayus, L. F. (2002). Fish (*Fundulus heteroclitus*) populations with different exposure histories differ in tolerance of creosote contaminated sediments. *Environmental Toxicology and Chemistry* **21**, 1897–1902.

Pait, A. S. and Nelson, J. O. (2009). A survey of indicators for reproductive endocrine disruption in *Fundulus heteroclitus* (killifish) at selected sites in the Chesapeake Bay. *Marine Environmental Research* **68**, 170–177.

Patel, M. R., Scheffler, B. E., Wang, L. and Willett, K. L. (2006). Effects of benzo(a)pyrene exposure on killifish (*Fundulus heteroclitus*) aromatase activities and mRNA. *Aquatic Toxicology* **77**, 267–278.

Pathiratne, A. and Hemachandra, C. K. (2010). Modulation of ethoxyresorufin O-deethylase and glutathione S-transferase activities in Nile tilapia (*Oreochromis niloticus*) by polycyclic aromatic hydrocarbons containing two to four rings: implications in biomonitoring aquatic pollution. *Ecotoxicology* **19**, 1012–1018.

Perez-Casanova, J. C., Hamoutene, D., Samuelson, S., Burt, K., King, T. L. and Lee, K. (2010). The immune response of juvenile Atlantic cod (*Gadus morhua* L.) to chronic exposure to produced water. *Marine Environmental Research* **70**, 26–34.

Perez-Casanova, J. C., Hamoutene, D., Hobbs, K. and Lee, K. (2012). Effects of chronic exposure to the aqueous fraction of produced water on growth, detoxification, and immune factors of Atlantic cod. *Ecotoxicology and Environmental Safety* **86**, 239–249.

Peters, L. E., MacKinnon, M., Van Meer, T., van den Heuvel, M. R. and Dixon, D. G. (2007). Effects of oil sands process-affected waters and naphthenic acids on yellow perch (*Perca flavescens*) and Japanese medaka (*Oryzias latipes*) embryonic development. *Chemosphere* **67**, 2177–2183.

Phillips, J. M. and Goodman, J. I. (2009). Inhalation of cigarette smoke induces regions of altered DNA methylation (RAMs) in SENCAR mouse lung. *Toxicology* **260**, 7–15.

Pinkney, A. E. and Harshbarger, J. C. (2006). Tumor prevalence in mummichogs from the Delaware Estuary watershed. *Journal of Aquatic Animal Health* **18**, 244–251.

Pinkney, A. E., Harshbarger, J. C. and Rutter, M. A. (2009). Tumors in brown bullheads in the Chesapeake Bay watershed: analysis of survey data from 1992 through 2006. *Journal of Aquatic Animal Health* **21**, 71–81.

Pollino, C. A. and Holdway, D. A. (2002). Toxicity testing of crude oil and related compounds using early life stages of the crimson-spotted rainbowfish (*Melanotaenia fluviatilis*). *Ecotoxicology and Environmental Safety* **52**, 180–189.

Pollino, C. A., Georgiades, E. and Holdway, D. A. (2009). Physiological changes in reproductively active rainbowfish (*Melanotaenia fluviatilis*) following exposure to naphthalene. *Ecotoxicology and Environmental Safety* **72**, 1265–1270.

Prosser, M. P., Unger, M. A. and Vogelbein, W. K. (2011). Multistressor interactions in the zebrafish (*Danio rerio*): concurrent phenanthrene exposure and *Mycobacterium marinum* infection. *Aquatic Toxicology* **102**, 177–185.

Pruett, S. B., Holladay, S. D., Prater, M. R., Yucesoy, B. and Luster, M. I. (2007). The promise of genomics and proteomics in immunotoxicology and immunopharmacology. In Target Organ Toxicology Series, *Immunotoxicology and Immunopharmacology* (eds. R. Luebke, R. House and I. Kimber), pp. 79–95. Boca Raton, FL: CRC Press, Taylor and Francis Group.

Puget Sound Partnership (2010). 2009 State of the Sound. Puget Sound Partnership, Olympia, WA. Publication No. PSP09-08.

Quagraine, E., Peterson, H. and Headley, J. (2005). *In situ* bioremediation of naphthenic acids contaminated tailing pond waters in the Athabasca Oil Sands Region—demonstrated field studies and plausible options: a review. *Journal of Environmental Science and Health, Part A: Toxic/Hazardous Substances and Environmental Engineering* **40**, 685–722.

Rahman, M. S., Khan, I. A. and Thomas, P. (2011). Tryptophan hydroxylase: a target for neuroendocrine disruption. *Journal of Toxicology and Environmental Health, Part B* **14**, 473–494.

Ramesh, A. and Archibong, A. E. (2011). Reproductive toxicity of polycyclic aromatic hydrocarbons: occupational relevance. In *Reproductive and Developmental Toxicology* (ed. R. C. Gupta), pp. 577–591. San Diego, CA: Academic Press.

Reddy, A. P., Spitsbergen, J. M., Mathews, C., Hendricks, J. D. and Bailey, G. (1999). Experimental hepatic tumorigenicity by environmental hydrocarbon dibenzo(a)pyrene. *Journal of Environmental Pathology, Toxicology and Oncology* **18**, 261–269.

Reichert, W. L., Myers, M. S., Peck-Miller, K., French, B., Anulacion, B. F., Collier, T. K., Stein, J. E. and Varanasi, U. (1998). Molecular epizootiology of genotoxic events in marine fish: linking contaminant exposure, DNA damage, and tissue-level alterations. *Mutation Research/Reviews in Mutation Research* **411**, 215–225.

Reynaud, S., Duchiron, C. and Deschaux, P. (2001). 3-Methylcholanthrene increases phorbol 12-myristate 13-acetate-induced respiratory burst activity and intracellular calcium levels in common carp (*Cyprinus carpio* L) macrophages. *Toxicology and Applied Pharmacology* **175**, 1–9.

Reynaud, S., Duchiron, C. and Deschaux, P. (2003). 3-Methylcholanthrene inhibits lymphocyte proliferation and increases intracellular calcium levels in common carp (*Cyprinus carpio* L). *Aquatic Toxicology* **63**, 319–331.

Reynaud, S., Duchiron, C. and Deschaux, P. (2004). 3-Methylcholanthrene induces lymphocyte and phagocyte apoptosis in common carp (*Cyprinus carpio* L) in vitro. *Aquatic Toxicology* **66**, 307–318.

Reynaud, S. and Deschaux, P. (2005). The effects of 3-methylcholanthrene on lymphocyte proliferation in the common carp (*Cyprinus carpio* L.). *Toxicology* **211**, 156–164.

Reynaud, S. and Deschaux, P. (2006). The effects of polycyclic aromatic hydrocarbons on the immune system of fish: a review. *Aquatic Toxicology* **77**, 229–238.

Rice, C. A., Myers, M. S., Willis, M. L., French, B. L. and Casillas, E. (2000). From sediment bioassay to fish biomarker—connecting the dots using simple trophic relationships. *Marine Environmental Research* **50**, 527–533.

Rice, C. D. and Arkoosh, M. R. (2002). Immunological indicators of environmental stress and disease susceptibility in fishes. In *Biological Indicators of Stress in Aquatic Ecosystem Stress* (ed. S. M. Adams), pp. 187–220. Bethesda, MD: American Fisheries Society Publication.

Rice, S. D., Thomas, R. E., Carls, M. G., Heintz, R. A., Wertheimer, A. C., Murphy, M. L., Short, J. W. and Moles, A. (2001). Impacts to pink salmon following the *Exxon Valdez* oil spill: persistence, toxicity, sensitivity, and controversy. *Reviews of Fisheries Science* **9**, 165–211.

Ridgway, L. L., Chapleau, F., Comba, M. E. and Backus, S. M. (1999). Population characteristics and contaminant burdens of the white sucker (*Catostomus commersoni*) from the St. Lawrence River near Cornwall, Ontario and Massena, New York. *Journal of Great Lakes Research* **25**, 567–582.

Røe Utvik, T. I. (1999). Chemical characterisation of produced water from four offshore oil production platforms in the North Sea. *Chemosphere* **39**, 2593–2606.

Roomi, M. W., Ho, R. K., Sarma, D. S. R. and Farber, E. (1985). A common biochemical pattern in preneoplastic hepatocyte nodules generated in four different models in the rat. *Cancer Research* **45**, 564–571.

Roubal, W. T., Collier, T. K. and Malins, D. C. (1977). Accumulation and metabolism of carbon-14 labeled benzene, naphthalene, and anthracene by young coho salmon (*Oncorhynchus kisutch*). *Archives of Environmental Contamination and Toxicology* **5**, 513–529.

Saborido-Rey, F., Domínguez-Petit, R., Tomás, J., Morales-Nin, B. and Alonso-Fernandez, A. (2007). Growth of juvenile turbot in response to food pellets contaminated by fuel oil from the tanker "Prestige". *Marine Ecology Progress Series* **345**, 271–279.

Schiewe, M. H., Weber, D. D., Myers, M. S., Jacques, F. J., Reichert, W. L., Krone, C. A., Malins, D. C., McCain, B. B., Chan, S.-L. and Varanasi, U. (1991). Induction of foci of cellular alteration and other hepatic lesions in English sole (*Parophrys vetulus*) exposed to an extract of an urban marine sediment. *Canadian Journal of Fisheries and Aquatic Sciences* **48**, 1750–1760.

Scott, J. A., Incardona, J. P., Pelkki, K., Shepardson, S. and Hodson, P. V. (2011). AhR2-mediated; CYP1A-independent cardiovascular toxicity in zebrafish (*Danio rerio*) embryos exposed to retene. *Aquatic Toxicology* **101**, 165–174.

Segner, H., Moller, A. M., Wenger, M. and Casanova-Nakayama, A. (2012a). Fish immunotoxicology: research at the crossroads of immunology, ecology and toxicology. In *Interdisciplinary Studies on Environmental Chemistry—Environmental Pollution and Ecotoxicology* (eds. M. Kawaguchi, K. Misaki, H. Sato, T. Yokokawa, T. Itai, et al.), pp. 1–12. Tokyo: TERRAPUB.

Segner, H., Wenger, M., Moller, A. M., Kollner, B. and Casanova-Nakayama, A. (2012b). Immunotoxic effects of environmental toxicants in fish—how to assess them? *Environmental Science and Pollution Research* **19**, 2465–2476.

Seruto, C., Sapozhnikova, Y. and Schlenk, D. (2005). Evaluation of the relationships between biochemical endpoints of PAH exposure and physiological endpoints of reproduction in male California Halibut (*Paralichthys californicus*) exposed to sediments from a natural oil seep. *Marine Environmental Research* **60**, 454–465.

Shallaja, M. S. and D'Silva, C. (2003). Evaluation of impact of PAH on a tropical fish, Oreochromis mossambicus using multiple biomarkers. *Chemosphere* **53**, 835–841.

Sikka, H. C, Steward, A. R., Kandaswami, C., Rutkowski, J. P., Zaleski, J., Kumar, S., Earley, K. and Gupta, R. C. (1991). Metabolism of benzo(a)pyrene and persistence of DNA adducts in the brown bullhead (*Ictalurus nebulosus*). *Comparative Biochemical Physiology* **100C** (1-2), 25–28.

Short, J. W. and Heintz, R. A. (1997). Identification of *Exxon Valdez* oil in sediments and tissues from Prince William Sound and the Northwestern Gulf of Alaska based on a PAH weathering model. *Environmental Science and Technology* **31**, 2375–2384.

Skinner, M. K., Manikkam, M. and Guerrero-Bosagna, C. (2011). Epigenetic transgenerational actions of endocrine disruptors. *Reproductive Toxicology* **31**, 337–343.

Smith, C. E., Peck, T. H., Klauda, R. J. and McLaren, J. B. (1979). Hepatomas in Atlantic tomcod *Microgadus tomcod* (Walbaum) collected in the Hudson River estuary in New York. *Journal of Fish Diseases* **2**, 313–319.

Snowberger, E. A. and Stegeman, J. J. (1987). Patterns and regulation of estradiol metabolism by hepatic microsomes from two species of marine teleosts. *General and Comparative Endocrinology* **66**, 256–265.

Sol, S. Y., Johnson, L. L., Horness, B. H. and Collier, T. K. (2000). Relationship between oil exposure and reproductive parameters in fish collected following the *Exxon Valdez* oil spill. *Marine Pollution Bulletin* **40**, 1139–1147.

Sol, S., Johnson, L., Boyd, D., Olson, O. P., Lomax, D. and Collier, T. (2008). Relationships between anthropogenic chemical contaminant exposure and associated changes in reproductive parameters in male English sole (*Parophrys vetulus*) collected from Hylebos Waterway, Puget Sound, Washington. *Archives of Environmental Contamination and Toxicology* **55**, 627–638.

Solt, D. B. and Farber, E. (1976). New principle for analysis of chemical carcinogenesis. *Nature* **263**, 701–703.

Solt, D. B., Medline, A. and Farber, E. (1977). Rapid emergence of carcinogen-induced hyperplastic lesions in a new model for the sequential analysis of liver carcinogenesis. *American Journal of Pathology* **88**, 595–618.

Song, J. Y., Nakayama, K., Kokushi, E., Ito, K., Uno, S., Koyama, J., Rahman, M. H., Murakami, Y. and Kitamura, S. I. (2012a). Effect of heavy oil exposure on antibacterial activity and expression of immune-related genes in Japanese flounder *Paralichthys olivaceus*. *Environmental Toxicology and Chemistry* **31**, 828–835.

Song, J. Y., Ohta, S., Nakayama, K., Murakami, Y. and Kitamura, S. I. (2012b). A time-course study of immune response in Japanese flounder *Paralichthys olivaceus* exposed to heavy oil. *Environmental Science and Pollution Research* **19**, 2300–2304.

Spromberg, J. A. and Meador, J. P. (2005). Population-level effects on Chinook salmon from chronic toxicity test measurement endpoints. *Integrated Environmental Assessment and Management* **1**, 9–21.

Spromberg, J. A. and Meador, J. P. (2006). Relating chronic toxicity responses to population-level effects: a comparison of population-level parameters for three salmon species as a function of low-level toxicity. *Ecological Modelling* **199**, 240–252.

Stehr, C. M., Johnson, L. L. and Myers, M. S. (1998). Hydropic vacuolation in the liver of three species of fish from the U.S. West Coast: lesion description and risk assessment associated with contaminant exposure. *Diseases of Aquatic Organisms* **32**, 119–135.

Stein, J. E., Hom, T., Sanborn, H. and Varanasi, U. (1988). The metabolism and disposition of 17β-estradiol in English sole exposed to a contaminated sediment extract. *Marine Environmental Research* **24**, 252–253.

Stein, R. A. (2012). Epigenetics and environmental exposures. *Journal of Epidemiology and Community Health* **66**, 8–13.

Stentiford, G. D., Longshaw, M., Lyons, B. P., Jones, G., Green, M. and Feist, S. W. (2003). Histopathological biomarkers in estuarine fish species for the assessment of biological effects of contaminants. *Marine Environmental Research* **55**, 137–159.

Stentiford, G. D., Bignell, J. P., Lyons, B. P. and Feist, S. W. (2009). Site-specific disease profiles in fish and their use in environmental monitoring. *Marine Ecology Progress Series* **381**, 1–15.

Stentiford, G. D., Bignell, J. P., Lyons, B. P., Thain, J. E. and Feist, S. W. (2010). Effect of age on liver pathology and other diseases in flatfish: implications for assessment of marine ecological health status. *Marine Ecology Progress Series* **411**, 215–230.

Steward, A. R, Kandaswami, C., Chidambaram, S., Ziper, C., Rutkowski, J. P. and Sikka, H. C. (1991). Disposition and metabolic fate of benzo-a-pyrene in the common carp. *Aquatic Toxicology* **20** (4), 205–218.

Sun, L., Zuo, Z., Luo, H., Chen, M., Zhong, Y., Chen, Y. and Wang, C. (2011). Chronic exposure to phenanthrene influences the spermatogenesis of male *Sebastiscus marmoratus*: U-shaped effects and the reason for them. *Environmental Science and Technology* **45**, 10212–10218.

Sundt, R. C. and Björkblom, C. (2011). Effects of produced water on reproductive parameters in prespawning Atlantic cod (*Gadus morhua*). *Journal of Toxicology and Environmental Health, Part A* **74**, 543–554.

Susser, M. (1986). Rules of inference in epidemiology. *Regulatory Toxicology and Pharmacology* **6**, 116–128.

Suter, M., Abramovici, A., Showalter, L., Hu, M., Shope, C. D., Varner, M. and Aagaard-Tillery, K. (2010). In utero tobacco exposure epigenetically modifies placental CYP1A1 expression. *Metabolism* **59**, 1481–1490.

Teles, M., Oliveira, M., Pacheco, M. and Santos, M. A. (2005). Endocrine and metabolic changes in *Anguilla anguilla* L. following exposure to beta-naphthoflavone—a microsomal enzyme inducer. *Environment International* **31**, 99–104.

Thomas, K. V., Langford, K., Petersen, K., Smith, A. J. and Tollefsen, K. E. (2009). Effect-directed identification of naphthenic acids as important in vitro xeno-estrogens and anti-androgens in North Sea offshore produced water discharges. *Environmental Science and Technology* **43**, 8066–8071.

Timme-Laragy, A. R., Meyer, J. A., Waterland, R. A. and Di Giulio, R. T. (2005). Analysis of CpG methylation in the killifish CYP1A promoter. *Comparative Biochemistry and Physiology Part C: Toxicology and Pharmacology* **141**, 406–411.

Tintos, A., Gesto, M., Alvarez, R., Míguez, J. M. and Soengas, J. L. (2006). Interactive effects of naphthalene treatment and the onset of vitellogenesis on energy metabolism in liver and gonad, and plasma steroid hormones of rainbow trout *Oncorhynchus mykiss*. *Comparative Biochemistry and Physiology Part C: Toxicology and Pharmacology* **144**, 155–165.

Tintos, A., Gesto, M., Míguez, J. M. and Soengas, J. L. (2007). Naphthalene treatment alters liver intermediary metabolism and levels of steroid hormones in plasma of rainbow trout (*Oncorhynchus mykiss*). *Ecotoxicology and Environmental Safety* **66**, 139–147.

Tintos, A., Gesto, M., Míguez, J. M. and Soengas, J. L. (2008). β-Naphthoflavone and benzo(a)pyrene treatment affect liver intermediary metabolism and plasma cortisol levels in rainbow trout *Oncorhynchus mykiss*. *Ecotoxicology and Environmental Safety* **69**, 180–186.

Tommasi, S., Kim, S. I., Zhong, X., Wu, X., Pfeifer, G. P. and Besaratinia, A. (2010). Investigating the epigenetic effects of a prototype smoke-derived carcinogen in human cells. *PLoS One* **5**, e10594.

Truscott, B., Walsh, J. M., Burton, M. P., Payne, J. F. and Idler, D. R. (1983). Effect of acute exposure to crude petroleum on some reproductive hormones in salmon and flounder. *Comparative Biochemistry and Physiology Part C: Comparative Pharmacology* **75**, 121–130.

Tryphonas, H., Fournier, M., Bakley, E. R., Smits, J. and Brousseau, P. (2005). *Investigative Immunotoxicology*. Boca Raton, FL: CRC Press.

Tuvikene, A. (1995). Responses of fish to aromatic hydrocarbons. *Annales Zoologici Fennici* **32**, 295–309.

Van Veld, P., Ko, U., Vogelbein, W. K. and Westbrook, D. J. (1991). Glutathione S-transferase in intestine, liver and hepatic lesions of mummichog (*Fundulus heteroclitus*) from a creosote-contaminated environment. *Fish Physiology and Biochemistry* **9**, 369–376.

Van Veld, P. A., Vogelbein, W. K., Smolowitz, R., Woodin, B. R. and Stegeman, J. J. (1992). Cytochrome P450IA1 in hepatic lesions of a teleost fish (*Fundulus heteroclitus*) collected from a polycyclic aromatic hydrocarbon-contaminated site. *Carcinogenesis* **13**, 505–507.

Van Veld, P. A. and Westbrook, D. J. (1995). Evidence for depression of cytochrome P4501A in a population of chemically resistant mummichog (*Fundulus heteroclitus*). *Environmental Science* **3**, 221–234.

Van Veld, P. A. and Nacci, D. E. (2008). Toxicity resistance. In *The Toxicology of Fishes* (eds. R. T. Di Guilio and D. E. Hinton), pp. 597–641. Boca Raton, FL: CRC Press.

Vandegehuchte, M. and Janssen, C. (2011). Epigenetics and its implications for ecotoxicology. *Ecotoxicology* **20**, 607–624.

Varanasi, U., Stein, J. E. and Nishimoto, M. (1989). Biotransformation and disposition of polycyclic aromatic hydrocarbons in fish. In *Metabolism of Polycyclic Aromatic Hydrocarbons in the Aquatic Environment* (ed. U. Varanasi), pp. 93–149. Boca Raton, FL: CRC Press.

Veith, G. D., Mekenyan, O. G., Ankley, G. T. and Call, D. J. (1995). A QSAR analysis of substituent effects on the photoinduced acute toxicity of PAHs. *Chemosphere* **30**, 2129–2142.

Vethaak, A., Jol, J., Meijboom, A., Eggens, M., Rheinallt, T., Wester, P., van de Zande, T., Bergman, A., Dankers, N., Ariese, F., Baan, R., Everts, J., Opperhuizen, A. and Marquenie, J. (1996). Skin and liver diseases induced in flounder (*Platichthys flesus*) after longterm exposure to contaminated sediments in large-scale mesocosms. *Environmental Health Perspectives* **104**, 1218–1229.

Vethaak, A. and Wester, P. (1996). Diseases of flounder *Platichthys flesus* in Dutch coastal and estuarine waters, with particular reference to environmental stress factors. II. Liver histopathology. *Diseases of Aquatic Organisms* **26**, 99–116.

Vethaak, A. D., Jol, J. G. and Pieters, J. P. F. (2009). Long-term trends in the prevalence of cancer and other major diseases among flatfish in the southeastern North Sea as indicators of changing ecosystem health. *Environmental Science and Technology* **43**, 2151–2158.

Villeneuve, D. L., Khim, J. S., Kannan, K. and Giesy, J. P. (2002). Relative potencies of individual polycyclic aromatic hydrocarbons to induce dioxinlike and estrogenic responses in three cell lines. *Environmental Toxicology* **17**, 128–137.

Vogelbein, W. K., Fournie, J. W., Van Veld, P. A. and Huggett, R. J. (1990). Hepatic neoplasms in the mummichog *Fundulus heteroclitus* from a creosote-contaminated site. *Cancer Research* **50**, 5978–5986.

Vogelbein, W. K. and Unger, M. A. (2006). Liver carcinogenesis in a non-migratory fish: The association with polycyclic aromatic hydrocarbon exposure. *Bulletin of European Association of Fish Pathologists* **26**, 11–20.

Vogelbein, W.K., Unger, M.A., Gauthier, D.T., 2008. The Elizabeth River Monitoring Program 2006–2007: Association between mummichog liver histopathology and sediment chemical contamination. A Final Report. Virginia Dept. of Environmental Quality. p. 34.

Wang, Z., Hollebone, B. P., Fingas, M., Fieldhouse, B., Sigouin, L., Landriault, M., Smith, P., Noonan, J. and Thouin, G. (2003). Characteristics of Spilled Oils, Fuels, and Petroleum Products: 1. Composition and Properties of Selected Oils. EPA/600/R-03/072. Research Triangle Park, NC: U.S. Environmental Protection Agency, 280 pp.

White, P. A., Robitaille, S. and Rasmussen, J. B. (1999). Heritable reproductive effects of benzo[a]pyrene on the fathead minnow (*Pimephales promelas*). *Environmental Toxicology and Chemistry* **18**, 1843–1847.

Whyte, K. S. (2007). The innate immune response of finfish—A review of current knowledge. *Fish and Shellfish Immunology* **23**, 1127–1151.

Williams, D. E., Lech, J. J. and Buhler, D. R. (1998). Xenobiotics and xenoestrogens in fish: modulation of cytochrome P450 and carcinogenesis. *Mutation Research/Fundamental and Molecular Mechanisms of Mutagenesis* **399**, 179–192.

Wills, L. P., Jung, D., Koehrn, K., Zhu, S., Willett, K. L., Hinton, D. E. and Di Giulio, R. T. (2010a). Comparative chronic liver toxicity of benzo[a]pyrene in two populations of the Atlantic killifish with different exposure histories. *Environmental Health Perspectives* **118**, 1376–1381.

Wills, L. P., Matson, C. W., Landon, C. D. and Di Giulio, R. T. (2010b). Characterization of the recalcitrant CYP1 phenotype found in Atlantic killifish (*Fundulus heteroclitus*) inhabiting a Superfund site on the Elizabeth River, VA. *Aquatic Toxicology* **99**, 33–41.

Wirgin, I. and Waldman, J. R. (2004). Resistance to contaminants in North American fish populations. *Mutation Research* **552**, 73–100.

Wu, R. S. S., Pollino, C. A., Au, D. W. T., Zheng, G. J., Yuen, B. B. H. and Lam, P. K. S. (2003). Evaluation of biomarkers of exposure and effect in juvenile areolated grouper (*Epinephelus areolatus*) on foodborne exposure to benzo[a]pyrene. *Environmental Toxicology and Chemistry* **22**, 1568–1573.

Xia, Y., Zhu, P., Han, Y., Lu, C., Wang, S., Gu, A., Fu, G., Zhao, R., Song, L. and Wang, X. (2009). Urinary metabolites of polycyclic aromatic hydrocarbons in relation to idiopathic male infertility. *Human Reproduction* **24**, 1067–1074.

Ylitalo, G. M., Krahn, M. M., Dickhoff, W. W., Stein, J. E., Walker, C. C., Lassitter, C. L., Garrett, E. S., Desfosse, L. L., Mitchell, K. M., Noble, B. T., Wilson, S., Beck, N. B., Benner, R. A., Koufopoulos, P. N. and Dickey, R. W. (2012). Federal seafood safety response to the *Deepwater Horizon* oil spill. *Proceedings of the National Academy of Sciences* **109**, 20274–20279.

Young, R. F., Orr, E. A., Goss, G. G. and Fedorak, P. M. (2007). Detection of naphthenic acids in fish exposed to commercial naphthenic acids and oil sands process-affected water. *Chemosphere* **68**, 518–527.

Young, R. F., Wismer, W. V. and Fedorak, P. M. (2008). Estimating naphthenic acids concentrations in laboratory-exposed fish and in fish from the wild. *Chemosphere* **73**, 498–505.

Young, R. F., Michel, L. M. and Fedorak, P. M. (2011). Distribution of naphthenic acids in tissues of laboratory-exposed fish and in wild fishes from near the Athabasca oil sands in Alberta, Canada. *Ecotoxicology and Environmental Safety* **74**, 889–896.

Yu, H. (2002). Environmental carcinogenic polycyclic aromatic hydrocarbons: photochemistry and phototoxicity. *Journal of Environmental Science and Health (C) Environmental Carcinogenesis and Ecotoxicological Review* **20**, 149–183.

5

ESTROGENIC ENDOCRINE DISRUPTING CHEMICALS IN FISH

CHRISTOPHER J. KENNEDY
HEATHER L. OSACHOFF
LESLEY K. SHELLEY

1. INTRODUCTION

Communication within and between cells is afforded by both the endocrine and nervous systems. These systems operate in a highly integrated manner to regulate and coordinate physiological processes as diverse as

Organic Chemical Toxicology of Fishes: Volume 33
FISH PHYSIOLOGY

homeostasis, energy availability, growth, development, and reproduction. Endocrine messengers (hormones) are released by cells, tissues, or ductless glands and have actions on cells intracellularly (intracrine action), locally (autocrine and paracrine action), or distant to the point of release usually via the circulatory system (endocrine). Cellular recipients of a hormonal signal may be one of several cell types, possibly residing in different tissues and triggering a range of physiological effects. For example, 17β-estradiol (E2) has autocrine, paracrine, endocrine, and neuromodulatory actions.

Four categories of vertebrate hormones exist based on molecular structure: (1) those derived from single amino acids (e.g., norepinephrine, 3,5,3'-triiodothyronine [T_3]), (2) peptides and proteins (e.g., growth hormone, follicle stimulating hormone), (3) steroid hormones (e.g., estrogens, glucocorticoids, vitamin D), and (4) eicosanoids (e.g., prostaglandins, leukotrienes). Several important endocrine glands have been identified in fish (Figure 5.1).

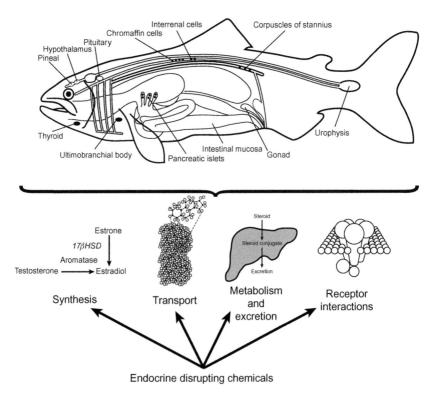

Figure 5.1. The main endocrine tissues in fish and the four main mechanisms by which EDCs can affect the endocrine system.

Although knowledge of the neuroendocrine system in fish is advancing rapidly, the complexity of this system seems to increase with the discovery of new messenger molecules, new messenger-producing cells, and the cells and targets that respond to them, resulting in an ever changing landscape of interconnected signal transduction pathways. In addition, endocrine alterations occur in response to the developmental and life history stage, as well as changing internal and external environments. Equally unknown are the possible subtle physiological changes associated with hormonal alteration, and the subsequent effects on individual biology and potentially population-level events. There has likely never been a more important time to understand fish endocrinology than the present, as there is rising concern that certain environmental contaminants and naturally occurring compounds impact the endocrine systems of animals (Figure 5.1).

Clear evidence exists that some xenobiotics can mimic or obstruct endocrine function in vertebrates and affect normal biological function in a wide-ranging manner (Krimsky, 2000). These chemicals have been termed endocrine-active chemicals (EACs) or endocrine disrupting compounds (EDCs). Knowledge regarding EDC action is rooted in the discovery of the synthetic estrogen diethylstilbestrol (DES) (Dodds et al., 1938; Burlington and Lindeman, 1950) and the weakly estrogenic pesticide 1,1,1-trichloro-2,2-di(4-chlorophenyl)ethane (DDT) with a related structure. As early as 1949, it was reported that crop dusters who frequently handled DDT had reduced sperm counts (Patlak, 1996). The tragic use of the synthetic estrogen DES by pregnant women for nearly 30 years (1940s–1970s) to help prevent miscarriages resulted in infertility and increased rates of vaginal clear-cell adenocarcinomas in daughters (Colborn et al., 1993). In 1962, Rachel Carson made reference to altered steroid metabolism as well as increased xenoestrogen burdens from the environment (e.g., from DDT) as potential causes for concern to humans and wildlife (Carson, 1962). In the mid-1960s, the poor hatching success of herring gulls (Keith, 1966) was followed by evidence that DDT produced characteristically estrogenic responses in the reproductive tracts of rats and birds (Bitman and Cecil, 1970). The concurrent increases in an understanding of the molecular events surrounding hormone receptor interactions, and mounting evidence of developmental and reproductive abnormalities in top predator species suggested an affected endocrine system as a cause (Colborn and Clement, 1992). EDC exposure can seriously affect the development and function of individual organisms and the ability to reproduce, and may even cause transgenerational effects (i.e., effects on offspring but not the parent organisms).

The U.S. Environmental Protection Agency (U.S. EPA) defines the term *EDC* as an "exogenous agent that interferes with the synthesis, secretion, transport, binding, action or elimination of natural hormones in the body that are responsible for the maintenance of homeostasis, reproduction, development, and/or behavior." A definition that includes not only the above primary ways EDCs are thought to interfere with the normal functioning of the endocrine system, but also some of the higher level effects they induce has also been used (Kavlock et al., 1996). A more general definition of an EDC is an exogenous substance that causes adverse health effects in an intact organism, or its progeny, consequent to changes in endocrine function.

In the aquatic environment, an EDC's ability to affect an individual or population depends on a number of factors, including the potency or efficacy of the EDC, its concentration, duration of exposure, bioconcentration potential, presence of other EDCs, life stage exposed, season, and other environmental stressors present (e.g., temperature and salinity). Responses used to detect endocrine active compounds (EACs) typically include cell proliferation, binding to the estrogen receptor, and production of specific proteins in cell cultures or whole animals, gamete counts, malformations, hatch rates, and the like (Kime, 1998).

2. EDC

The number of suspected or confirmed EDCs is ever increasing and includes industrial intermediates and plasticizers, as well as classic contaminants such as the polycyclic aromatic hydrocarbons (PAHs), polychlorinated biphenyls (PCBs), dioxins, certain pesticides, and even a number of trace elements (Knudsen and Pottinger, 1998). Naturally occurring plant compounds also have hormone-like properties and are termed phytoestrogens (Mitksicek, 1995). These become a concern when plant products are released in large quantities to the aquatic environment as a result of human activities such as pulp and paper production.

EDCs are diverse in sources, structures (they tend to be synthetic and mostly have one or more phenyl ring), persistence, and effects, with perhaps little in common with one another, except that they elicit a response that adversely alters the endocrine system. This chapter focuses mainly on those EDCs that affect the estrogen pathway. This is the most widely studied model system for EDCs and highlights the complexity of this environmental issue and its ecological relevance to fish. This section presents the classes of EDC, with some example compounds, and details their sources, uses, regulatory status (if available), and toxic actions.

2.1. Xenoestrogen Classes

2.1.1. NATURAL AND SYNTHETIC STEROLS

Sterols are a family of compounds that have a base structure consisting of steroid rings containing an alcohol group. Many natural plants, fungi, or animal compounds belong to the sterol family, including cholesterol and stigmasterol (neither of which is an EDC). Key sterol hormones in vertebrates include 17β-estradiol (E2; Figure 5.2A), testosterone, and 11-ketotestosterone. The phytosterol β-sitosterol (Figure 5.2B) has been shown to alter the endocrine status of goldfish (*Carassius auratus*; MacLatchy and Van Der Kraak, 1995).

Synthetic sterol compounds of concern are those used in birth control or hormone replacement therapies for women, which include 17α-ethinylestradiol (EE2; Figure 5.2C), mestranol (the pro-drug of EE2), and norethisterone (also known as norethindrone).

Sources of sterol compounds include pulp and paper mill effluents (typically high in phytosterols), sewage effluents, and agricultural runoff (high urine and fecal estrogen/androgen hormones, synthetic hormones, and dietary phytosterols), decaying wood/plants and leaves (releasers of

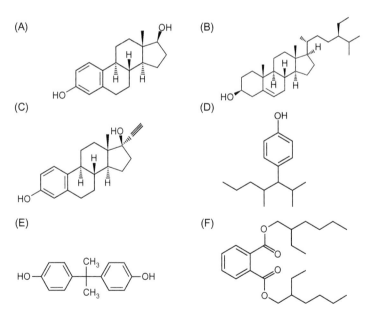

Figure 5.2. Chemical structures of xenoestrogens. (A) 17β-estradiol (E2). (B) β-sitosterol. (C) 17α-ethinylestradiol (EE2). (D) nonylphenol (NP). (E) bisphenol A (BPA). (F) di(2-ethylhexyl) phthalate (DEHP).

phytosterols), and fish spawning grounds (piscine excretion of estrogen/ androgen hormones) (Kolodziej et al., 2004; Hewitt et al., 2006; Furtula et al., 2012).

2.1.2. DETERGENTS/SURFACTANTS

The main class of surfactants investigated in endocrine-disrupting studies is the alkylphenol ethoxylates (APEs). These compounds are used in household and laboratory products, cosmetics, pesticide formulants (agricultural and forestry practices), de-icers, and industrial processes (mostly as emulsifiers, cleaners, or dust-control agents). The structures of some detergents and surfactants, or their breakdown products, are sufficiently close to estrogen hormones such that interactions with ERs occur. A key example of an APE is the by-product of nonylphenol ethoxylate (NPE) manufacture or degradation, which is nonylphenol (NP; Figure 5.2D). This particular compound is found in the aquatic environment worldwide as it is discharged in relatively high concentrations in wastewater effluents.

The United States has aquatic life ambient water quality criteria for NP, a criterion each for freshwater and saltwater, for both acute and chronic exposures. For example, the freshwater acute criterion states that the 1 h average concentration of NP should not exceed 28 µg/L more than once every three years on the average (additional criteria listed by the EPA, 2005). Many countries are evaluating the risk of NP/NPEs and the APEs in general, as they are well documented to be persistent and moderately bioaccumulative in the aquatic environment (EPA, 2005). In the United States, the EPA has initiated a voluntary phaseout of NPEs (during 2012–2014), and the European Union has enacted measures to reduce or eliminate the use of APEs in products as a part of the REACH program (Registration, Evaluation, Authorization, and Restriction of Chemical substances). For the foreseeable future, NP/NPE and the APEs remain an environmental contaminant of concern; however, if replacement compounds are utilized, it is possible this group of contaminants could become a historical issue.

2.1.3. PESTICIDES AND FORMULANTS

A variety of pesticides can act as estrogenic EDCs (e.g., 1,1,1-trichloro-2,2-di(4-chlorophenyl)ethane [DDT] or 1,1-dichloro-2,2-bis(p-dichlorodiphenyl) ethylene [DDE], methoxychlor, and endosulfan). These compounds may interact with ERs and alter estrogen-responsive parameters. DDT is a legacy pesticide, and its use was phased out in most countries decades ago. However, it is still measured in the aquatic environment, likely due to its resistance to degradation, its resuspension of contaminated sediments, and its long-range transport. Methoxychlor and endosulfan, both insecticides, have been banned or phased out (10 years ago for methoxychlor and by 2012 for endosulfan), yet these pesticides may still be detected in the aquatic environment for decades, which will continue to cause concern with respect to long-term endocrine-disrupting activity.

Exposure of Atlantic salmon (*Salmo salar*) to NP, a formulant for forestry-applied pesticides, coincided with a decrease in returns and catches (Fairchild et al., 1999). The pesticide (aminocarb) with NP as a formulant was applied at a time coinciding with the final stages of smolt development; the concentration of NP in the water was consistent with that found in effluents from industrial or municipal sewage sources. Furthermore, exposure to NP during smolt development impaired osmoregulation and smolt development, altered plasma hormone concentrations responsible for regulating smoltification, and led to behavioral changes that affected migration success and timing (Madsen et al., 1997, 2004; McCormick et al., 2005). Atlantic salmon exposed to pulses of NP during smolt development before being transferred to saltwater had decreased growth and the growth hormone/insulin-like growth factor I axis was disrupted, which may underlie the impaired survival in the marine environment (Arsenault et al., 2004).

2.1.4. PLASTIC-RELATED COMPOUNDS

Compounds related to the production, synthesis, or stability of plastics have been reported to be EDCs (Oehlmann et al., 2009). Two key compounds include bisphenol A (BPA; Figure 5.2E) and di(2-ethylhexyl) phthalate (DEHP; Figure 5.2F), which is a member of the phthalate family of compounds. Not all phthalates exhibit endocrine-disrupting activity, but several compounds in the family are known EDCs (DEHP is considered the chemical with the most estrogenic potency). BPA was reported in 1936 as being a full estrogen agonist (Dodds and Lawson, 1936), before it was determined that this hormonally active drug could be polymerized to produce polycarbonate plastic. Phthalates are nearly ubiquitous in modern domestic products. They can be found in personal care products, adhesives, medicinal products, packaging, toys, paints/inks, and many other items. Plastic-related EDCs can be ingested by humans, and therefore, BPA and phthalates have been assessed for risk largely in association with human health concerns; the World Health Organization has developed a drinking water quality guideline level of 0.008 mg/L for DEHP. Following studies reporting BPA as an EDC, several countries (Canada was the first country to declare BPA a toxic substance) took action to ban it from items relating to infants.

BPA and phthalates are most often evaluated by food and drug agencies, as opposed to environmental regulators. However, fish are exposed to BPA and phthalates in waters receiving leachate from landfills and wastewater effluents. Countries included in the European Union, and the United States and Canada, are considering regulating this family of compounds for environmental protection since these chemicals have been found in measurable concentrations in aquatic environments (Kolpin et al., 2002).

2.1.5. PULP AND PAPER MILL DISCHARGES

Pulp and paper mill effluents cause reproductive effects in fish (reviewed by Parrott et al., 2006 and Hewitt et al., 2006). Such effects were possibly masked by the historical direct lethality to fish caused by dioxins and furans; these effects have largely been eliminated over the last decade with improvements to mill treatment processes to reduce or eliminate these toxic compounds. Natural, wood-derived compounds remain that cause reproductive effects in fish (Hewitt et al., 2006), now attributed mainly to phytosterols and resin acids. However, effluents are complex mixtures, which make definitive cause-and-effect relationships difficult to establish. Even though effluent composition varies from mill to mill, similar effects are found in fish regardless of mill process technology or operation, or differences in wood finish (Hewitt et al., 2006). These effects include vitellogenin (VTG) induction, depression of plasma sex steroids concentrations, changes in sexual development, and decreases/delays in the expression of secondary sex characteristics, demasculinization of male fish, masculinization of female fish, and feminization of male fish, among others (Parrott et al., 2006; Hewitt et al., 2006). The focus in this section is on the feminizationof fish.

Pulp and paper mill effluents feminize fish. Parrott et al. (2003, 2004) found increased female to male sex ratios in fathead minnows (*Pimephales promelas*) and showed that juveniles and males produced ovipositors (female-specific egg-laying organs) when exposed to bleached sulfite mill effluent. In a study with bleached Kraft mill effluent-exposed Chinook salmon (*Oncorhynchus tshawytscha*) fry, Afonso et al. (2002) reported partially and completely feminized genetic males, containing phenotypic ovaries, intersex testis, or underdeveloped testis. Whole-body zebrafish (*Danio rerio*) VTG concentrations were approximately 26-fold higher in fish exposed to 50% pulp and paper mill effluent when compared to controls (Orn et al., 2001). In a Chilean study, Orrego et al. (2006) placed caged juvenile rainbow trout (*Oncorhynchus mykiss*) upstream, downstream, and in an area of the river directly impacted by a pulp and paper mill discharge. They found that exposed trout had increased gonadal somatic index (GSI) and plasma VTG levels combined with an induction of gonad maturation (deduced by the presence of vitellogenic oocytes).

A large body of research exists regarding specific candidate compounds that resulted in the effects noted above, one of which is β-sitosterol. It is a phytosterol measured in many studies, at higher levels than other compounds, and has known estrogenic/EDC activity in fish, including the ability to induce VTG production and reduce sex-steroid levels (MacLatchy and Van Der Kraak, 1995; Tremblay and Van Der Kraak, 1999; Hewitt et al., 2006). β-sitosterol was evaluated in a study by Orrego et al. (2010), who injected immature rainbow trout with 5 mg/kg β-sitosterol as well as other Chilean pulp and paper mill-related materials (extracts from primary and secondary treated effluents and dehydroabietic acid [DHAA]) and E2 as a positive control (5 mg/kg). Regardless of fish sex, VTG protein was induced by β-sitosterol, E2, and the extracts, but not in the DHAA treatment. Therefore, β-sitosterol may be one of the compounds contributing to the effects of pulp and paper mill effluent on fish feminization.

2.1.6. SEWAGE TREATMENT PLANT EFFLUENTS

Domestic wastes contain natural estrogen hormones and additional types of xenoestrogens, including detergents, plastics, pesticides, and pharmaceutical products. Domestic waste is treated by sewage treatment plants (STPs), which degrade portions of the xenoestrogens depending on treatment type (Lishman et al., 2006; Filby et al., 2010; reviewed in Oulton et al., 2010). However, some endocrine active compounds still remain in the treated effluent. In the UK, fish exposed to STP effluent were feminized (rainbow trout: Purdom et al., 1994; Harries et al., 1996, roach (*Rutilus rutilus*): Jobling et al., 1998, 2002; Rodgers-Gray et al., 2001). STP effluents also alter other estrogen-responsive parameters (Batty and Lim, 1999; Afonso et al., 2002; Osachoff et al., 2013; Bjerregaard et al., 2006; Hoger et al., 2006; Lavado et al., 2004; Larsson et al., 1999; Folmar et al., 1996; Vajda et al., 2008, 2011).

Current evidence suggests that natural and synthetic estrogen hormones play the largest roles in sewage estrogenicity, even though other xenoestrogens are present, such as alkylphenols and APEs (Johnson and Sumpter, 2001; Aerni et al., 2004; Corcoran, 2010). This effect is due to the potency of estrogen hormones compared to other compounds (Thorpe et al., 2003). The concentrations of estrogen hormones in STP effluents are generally in the ng/L range (Belfroid et al., 1999; Larsson et al., 1999; Ternes et al., 1999; Spengler et al., 2001; Aerni et al., 2004; Johnson et al., 2005; Kim et al., 2007; Kuster et al., 2008; Sousa et al., 2010; Williams et al., 2012). Other compounds such as BPA, phthalates, and NP/APEs are usually found in the μg/L range (Larsson et al., 1999; Aerni et al., 2004; Johnson et al., 2005; Barber et al., 2007; Sousa et al., 2010).

Animal wastes from agricultural operations also contribute xenoestrogens to the environment. Estrone (E1) concentrations of 2–7 ng/L were found in downstream water from a bovine feeding operation (Soto et al.,

2004). Hormone implants contribute to the estrogenic and androgenic activity in effluents or waste released by feedlot operations. Additional nontraditional sources of EDCs include fish hatcheries, dairy farms, and fish spawning grounds (Kolodziej et al., 2004). Concentrations of E1, testosterone, and androstenedione in the ng/L range have been found in fish hatchery effluents and a Chinook spawning ground. At a dairy farm, E1 concentrations in the waste were as high as 650 ng/L. On a watershed level, nonpoint-source estrogen hormones from agricultural sources were detected (Jeffries et al. 2010; Furtula et al. 2012).

2.1.7. Other EDCs

There are other, less potent xenoestrogens (with conflicting reports in the literature of estrogenic, androgenic, or anti-estrogenic properties), including dioxins, furans, polychlorinated biphenyls (PCBs), polybrominated diphenyl ethers (PBDEs), and cadmium or other metals (see Denslow and Sepulveda [2007] for review). These compounds are all regulated and have water quality criteria based on adverse toxic effects that (most often) do not relate to endocrine disruption.

Synthetic compounds used as fragrances in detergents and pharmaceuticals and personal care products (PPCPs) have been determined to have xenoestrogenic properties, in particular, two polycyclic musks: galaxolide (1,3,4,6,7,8-hexahydro-4,6,6,7,8,8-hexamethylcyclopenta(g)-2-benzopyran; HHCB) and tonalide (6-acetyl-1,1,2,4,4,7 hexamethyltetraline; AHTN). There are no regulations or water quality criteria for the synthetic musks at present.

Studies have documented the potency of toxicants compared to E2 (compilation in Table 5.1). There is incomplete consensus in the literature, so the cited values are presented. The approximate order of potency is from high to low: natural or synthetic estrogen hormones > pesticides or NP or BPA > DEHP (Table 5.1).

3. TOXICOKINETICS

As detailed in Chapter 1, the concentration at the site of action is dependent on a chemical's toxicokinetics, that is, its *absorption*, *distribution*, *metabolism*, and *excretion*. General conclusions regarding the toxicokinetics of EDCs are difficult because of their structural diversity. EDCs are usually found in very low concentrations in water and unless accumulation occurs in tissues to relatively higher concentrations, effects are unlikely, since interactions with the estrogen receptor are less than natural E2.

Table 5.1

Compilation of xenoestrogenic potencies based on literature values

Compound	Abbreviation	Source	Functions	Potency (fold, compared to E2)	Endpoint	Citation
17β-estradiol	E2	natural hormone; sewage effluent	reproduction and behavior in vertebrates	1.0		
17α-ethinylestradiol	EE2	natural hormone; sewage effluent	contraceptive	28	FHM VTG induction (EC_{50})	Brian et al. (2005)[a]
				11–27	Rbt VTG induction (EC_{50})	Thorpe et al. (2003)
				31–33	Zebrafish VTG induction (EC_{50}) and ovarian somatic index (OSI)	Van den Belt et al. (2004)
				100	Zebrafish (transgenic assay)	Legler et al. (2002)
				133	Medaka testis-ova formation	Metcalfe et al. (2001)[b]
estrone	E1	natural hormone; sewage effluent	storage form of E2 (ketone or oxidized form)	0.3–0.4	Rbt VTG induction (EC_{50})	Thorpe et al. (2003)
				0.5	Medaka testis-ova formation	Metcalfe et al. (2001)[b]
				0.5–0.8	Zebrafish VTG induction (EC_{50}) and ovarian somatic index (OSI)	Van den Belt et al. (2004)
				~1	FHM reproductive (various)	Dammann et al. (2011)
				1	Zebrafish (transgenic assay)	Legler et al. (2002)

(Continued)

Table 5.1 (continued)

β-sitosterol	–	sewage and pulp and paper mill effluents	phytosterol	~0.6–0.8	Rbt VTG induction	Orrego et al. (2010)[c]
o,p′-DDT	DDT	historical contamination	pesticide	0.02	Zebrafish (transgenic assay)	Legler et al. (2002)
estriol	E3	natural hormone; sewage effluent	vertebrate pregnancy hormone	0.005	Medaka testis-ova formation	Metcalfe et al. (2001)[b]
methoxychlor	–	sewage effluent and run-off	pesticide	0.0025		Thorpe et al. (2001)
nonylphenol	NP	mainly sewage effluent	surfactant (breakdown product)	0.00356	FHM VTG induction (EC$_{50}$)	Brian et al. (2005)[a]
				0.0006–0.0007	Zebrafish VTG induction (EC$_{50}$) and ovarian somatic index (OSI)	Van den Belt et al. (2004)
bisphenol A	BPA	leachate from plastics	synthesis of plastics	0.00016	FHM VTG induction (EC$_{50}$)	Brian et al. (2005)[a]
				0.00068	Medaka testis-ova formation	Metcalfe et al. (2001)[b]
di(2-ethylhexyl) phthalate	DEHP	leachate from plastics	synthesis of plastics	$<8 \times 10^{-7}$	Medaka testis-ova formation	Metcalfe et al. (2001)[b]

[a]Calculated from EC$_{50}$ values reported in Table 2.
[b]Calculated from mean LOEL values reported in Table 7.
[c]Calculate from VTG concentrations shown in Figure 5.

3.1. The Importance of the Octanol-Water Partitioning Coefficient

Two of the most important properties relating to the biological fate of any organic compound are aqueous solubility and the log octanol-water partition coefficient (log K_{ow}) value (see Chapter 1), which can be used to estimate or predict other properties of more immediate environmental and ecotoxicological interest. They are also correlated with the organic carbon-normalized partition coefficient (K_{oc}), bioconcentration, bioaccumulation, and biomagnification factors (BCFs, BAFs, BMFs), and indices of biodegradability.

Both natural hormones released into the environment and synthetic compounds that can act as EDCs span a wide range of log K_{ow} values, from those that are somewhat hydrophobic to those that are considered superhydrophobic. For example, the log K_{ow} values for estrogen hormones range from 2.6 to 4.0 (Khanal et al., 2007). Values reported for BPA fall between 3.1 and 3.3 (Staples et al., 1998) and for NP the range is 4.0 to 4.5 (Barber et al., 2009). Even within a class, log K_{ow} values can range widely, as is seen for the various phthalate compounds: DMP (1.8), DEP (2.8), DnBP (4.4), BBP (4.6), DEHP (7.0), and DnOP (8.18) (Dargnat et al., 2009). Metabolites can vary in log K_{ow} values (e.g., biotranformation to very water-soluble ions), or they can be similar to the parent compound: For p,p′-DDT, p,p′-DDE, and p,p′-DDD, log K_{ow} values are 6.91 (Swann et al., 1981), 6.51, and 6.02 (Howard and Meylan, 1997), respectively. This diversity in log K_{ow} results in disparate fates in the environment, uptake into aquatic organisms, and levels of accumulation.

3.2. EDC Uptake

The main sources of EDC uptake in fish are water and food, or if benthic species consume sediments, these can also be a major source, as many EDCs with high log K_{oc} values would bind tightly to sediments. The major route of EDC uptake, as with all other aspects of toxicokinetics, will be chemical specific; those EDCs with high water solubility (log K_{ow} < 1) will not move into organisms easily due to their lack of lipophilicity. In the range of log K_{ow} 1–3, there will be an increasing likelihood of entering the gills, as the chemical will have sufficient water solubility to be delivered to the gill surface and lipophilic enough to enter. Between log K_{ow} 4 and 6, uptake at the gills will plateau and will be limited by the chemical's ability to dissolve in water, and the major route will likely be the gastrointestinal tract associated with food. The potential for uptake from food versus water increases with higher log K_{ow} values as these chemicals may accumulate to relatively high concentrations in prey organisms rather than be available in water (Gobas and Mackay, 1987; McKim et al., 1985). Unless there exists an active uptake mechanism for a particular water-soluble EDC, EDCs that easily enter a fish and accumulate are hydrophobic.

3.3. Distribution

Once EDCs enter fish, water-soluble hormones or EDCs dissolve in plasma, while hydrophobic/lipophilic compounds are transported in association with plasma proteins such as albumin, lipoproteins, or specific hormone transport proteins. The high-affinity, high-capacity proteins include sex steroid-binding globulin (SSBG) for E2 or androgens, corticosteroid-binding globulin (CBG) for cortisol, and transthyretin for thyroid hormones. SSBGs have been identified in several fish species, including common carp (*Cyprinus carpio*), channel catfish (*Ictalurus punctatus*), Atlantic salmon, and rainbow trout (Kloas et al., 2000; Tollefsen et al., 2002; Gale et al., 2004; Tollefsen, 2007). The protein-bound chemical is released when the affinity of another biomolecule or tissue component is greater than that of the plasma protein. EDCs that do not readily pass through cell membranes or make use of specialized transport mechanisms have a restricted distribution, whereas other xenobiotics that readily pass through cell membranes can become distributed widely throughout the body. When bound, the biotransformation and metabolic clearance rates of hormones or EDCs can be greatly reduced.

3.4. Biotransformation

Numerous mechanisms exist to limit and regulate the accumulation of EDCs and other xenobiotics in cells to avoid adverse physiological effects that reflect the concentrations and persistence of the chemical at a target site and its interaction at that site. Biotransformation is the most widely studied, and this process has substantial bearing on the tissue levels of hormones and EDCs, their half-lives ($t_{1/2}$s) and clearances, as well as metabolite patterns, all of which can affect the severity and duration of an endocrine-related response. In fish, estrogen and other steroid hormones are degraded by the CYP3A and CYP2K subfamily of cytochrome P450 enzymes (Stegeman, 1993), with the major site of hydroxylation at the 6β position (Zimniak and Waxman, 1993; Waxman et al., 1988). It has been suggested that rainbow trout CYPKM2 and CYPLMC2 (LM2) are similar proteins, both belonging to the CYP2K subfamily (Andersson, 1992). Further biotransformation of parent compound or hydroxylated metabolites occurs via glucuronidation and sulfonation and is an important route of clearance of active hormones, generating a pool of hormone-conjugates that may be excreted or re-converted to the active form in peripheral tissues. Natural estrogens, phytoestrogens, and EDCs of a synthetic origin will be biotransformed in a chemical-specific manner, and empirical studies must be undertaken to determine their specific rates of biotransformation. Several EDCs can be metabolized extremely rapidly, with $t_{1/2}$s measured in hours. In juvenile rainbow trout, a plasma half-life for BPA was 3.75 h following injection. Such a rapid metabolism of EDCs would suggest a lower estrogenic sensitivity, since both accumulation and binding to ERs would be low compared to E2 (Lindholst et al., 2001, 2003). In comparison, polychlorinated biphenyls (PCBs) and DDT are persistent organic pollutants (POPs) and EDCs that bioconcentrate and biomagnify in fish. Fish are generally considered to have poor capability to biotransform PCBs (Matthews and Dedrick, 1984; Boon et al., 1997); however, recent work would suggest that fish do have some capacity to biotransform PCBs to hydroxylated metabolites (Wong et al., 2002, 2004). The ability of fish to biotransform PCBs is restricted ($t_{1/2}$s can reach 150 d [Buckman et al., 2006]) compared to birds and mammals, but the mechanisms involved appear to be similar in all species and dependent on chlorine substitution pattern. Calculated $t_{1/2}$s for PCBs vary, and depend on the degree of substitution; PCBs with log K_{ow} values of 6.5 and higher have $t_{1/2}$ of >1000 d reported in large rainbow trout (Niimi and Oliver, 1983). For juvenile trout, $t_{1/2}$s on the order of 100–200 d have been determined in this hydrophobicity range (Paterson et al., 2007). A higher degree of chlorination on the PCB molecule usually results in a more hydrophobic molecule that is recalcitrant to biotransformation, typically resulting in longer $t_{1/2}$s: PCB 84 (log $K_{ow} = 6$), PCB 132 (log $K_{ow} = 6.6$), and PBC 174 (log $K_{ow} = 7.1$) have $t_{1/2}$s of 41, 58, and 77 d, respectively (Konwick et al., 2006).

DDT and its metabolites, DDE and DDD, are all persistent EDCs that highlight how important knowledge of toxicokinetics is in evaluating the potential risks and toxicological implications of EDC exposure. For example, the $t_{1/2}$s of the isomers of DDT (o,p'-DDT and p,p'-DDT [log K_{ow}s = 5.7, 6]) and one of its metabolites DDD (o,p'-DDD and p,p'-DDD [log K_{ow} = 6.1, 5.5]) are different at 37, 27, 41, and 43 d, respectively (Konwick et al., 2006). Furthermore, the toxicity of the DDT metabolites DDE and DDD is also different from the parent compound; DDT and DDD both act as xenobiotic estrogens, but DDE acts as an anti-androgen (Lotufo et al., 2000). The bioaccumulation of DDT in fish and its biomagnification in the food chain have been well documented (Wang 2005). Both metabolites, DDE and DDD, are very readily stored in lipid-rich tissues such as adipose tissue and liver.

Many compounds, including some hormones, must be acted upon by biotransformation enzymes and bioactivated into an active form. This is also true for many "pro-EDCs" such as methoxychlor, indol-3-carbinol, and vinclozolin (Dehal and Kupfer, 1994; Kelce et al., 1994), whose metabolites are the ultimate endocrine active form. The demethylation of methoxychlor in channel catfish and largemouth bass (*Micropterus salmoides*) gives rise to the mono- and bis-demethylated primary metabolites OH-methoxychlor and 2,2-bis(p-hydroxyphenyl)-1,1,1-trichloroethane, which are estrogenic and anti-androgenic, respectively (Nyagode, 2007). Interestingly, biotransformation may also alter the interaction with a receptor; for example, benzo[a]pyrene is both estrogenic and anti-estrogenic, while its metabolite 2,8-dihydroxybenzo[a]pyrene is estrogenic (Gould et al., 1998). The pharmacological action of a parent molecule or a metabolite may also be very different between cells or between species. Tamoxifen is an estrogen agonist/antagonist in the rat uterus (Branham, 1993) but is a full agonist in mammary tissue (Gottardis et al., 1988). Recently, hydroxylated PCBs (OH-PCBs) were found in a number of fish species from the Great Lakes (Campbell et al., 2003), which present new concerns because their toxicity may be greater than their parent compounds (Purkey et al., 2004), particularly with respect to endocrine disruption (Gerpe et al., 2000; Carlson and Williams, 2001). OH-PCBs found in fish may be derived as metabolites from CYP enzyme-mediated Phase I biotransformation of PCBs via an insertion of an OH-group (Yoshimura et al., 1987).

Multidrug resistance (MDR) or multixenobiotic resistance (MXR) mechanisms are mediated by the expression of a variety of transmembrane transport proteins, the most common among them being P-glycoprotein (P-gp) (Bard, 2000) (see Chapter 1). P-gp (ABCB1) is a 170 kDa ATP-dependent efflux pump located in the plasma membrane of most organisms examined to date, including fish. It is capable of transporting a wide range of planar, moderately hydrophobic xenobiotics, and as such comprises a primary line of cellular defense against exogenous compounds. P-gp, in conjunction with other membrane-bound transporter families, has been considered a "Phase III" defense system capable of modulating the toxic effects of compounds by manipulating their bioavailability and disposition in an organism's body. Some hormones are also known substrates of P-gp and include the steroid hormones cortisol, aldosterone, estriol (Sturm and Segner, 2005; Ueda et al., 1997), and dexamethasone (Bard, 2000), as well as the EDCs benzo[a]pyrene (Yeh et al., 1992) and nonlyphenol ethoxylates (Loo and Clarke, 1998). P-gp has been reported in bile canaliculi, gill chondrocytes, pseudobranch, kidney renal tubules, pancreatic exocrine tissue, gas gland, and intestinal epithelium of guppy (*Poecilia reticulate*) (Hemmer et al., 1995) and is highly expressed along the luminal mucosa of the distal intestine in channel catfish (Kleinow et al., 1999). This suggests that P-gp may play a significant role in the excretion of accumulated EDCs and the prevention of their uptake from the diet.

3.5. Excretion

Gills, gut, and kidney excrete EDCs in fish and require water (Chapter 1). Chemicals with a relatively low log K_{ow} and/or log octanol-air partitioning coefficient (K_{OA}) values are quickly eliminated in most organisms and typically do not bioconcentrate or biomagnify even if they are not subject to biotransformation. However, chemicals of high K_{ow} and high K_{OA} are very slowly eliminated, and if not biotransformed easily, they will likely bioaccumulate.

3.6. Potential for Bioconcentration or Biomagnification

Bioconcentration (BCF), bioaccumulation (BAF), and biomagification factors (BMFs) vary with the physicochemical characteristics of each EDC and the biology of the organisms (e.g., presence of transporters, biotransformation ability, etc.) and will range widely (e.g., values for BCFs < 5000 are not considered to be very bioaccumulative) for EDCs from different classes of compound. For example, the BAF in fish for EE2 was determined to be low at 332 (Lai et al., 2002). The BCFs for BPA in fish have been estimated to be between 5 and 68 (Staples et al., 1998). Nonylphenol has been shown to bioaccumulate to values up to 410 in fish (Ahel et al., 1993; Snyder et al., 2001a). BCFs for DDT, DDE, and DDD are 19,600, 4400, and 12,500 in fish, respectively (EPA, 1989). For PCBs, which do not biodegrade easily, in combination with very high partition coefficients and provided that there are the necessary conditions for PCBs to bioaccumulate to extremely high levels, BCFs > 100,000 have been reported in many fish species (Ivanciuc et al., 2006).

4. MECHANISMS OF TOXICITY

Xenoestrogens exert their toxic effects through a multitude of mechanisms. A single chemical may act through multiple pathways depending on life stage, gender, availability of receptors or targets, and concentration of the chemical. EDCs were once thought to interact only with nuclear estrogen receptors (ERs) that are present in most tissues throughout the organism. Currently, it is known that many xenoestrogens also interact with a greater affinity with other nuclear receptors or even cell-surface receptors. Effects can also occur as a result of interactions at higher levels, such as alteration in the hypothalamus-pituitary-gonad (HPG) axis that regulates hormonal activity. In addition, xenoestrogens have been found to alter the transcription or activity of hormone receptors and enzymes involved in biotransformation or steroid biosynthesis pathways. Each of these potential mechanisms will be discussed in more detail in the following section, with specific examples used to highlight key points.

4.1. Interaction with Estrogen Receptors

4.1.1. NUCLEAR ESTROGEN RECEPTORS

Steroid hormones exert their effects through interaction with intracellular nuclear hormone receptors, which act as transcription factors. Circulating steroid hormones, such as E2, diffuse into target cells and interact with the estrogen receptor (ER) present inside the cell. Once bound, the liganded ER dimerizes with another ER, and this complex can associate with estrogen response elements (EREs) and initiate the transcription of specific genes. A multitude of genes are estrogen-responsive such as those related to reproduction (e.g., VTG, vitelline envelope proteins [VEPs]) or the immune system (e.g., cytokines), which explains why E2 has such far-reaching and diverse physiological effects. This signal transduction pathway involving the transcription of new genes is slow in that changes in cellular activity are not apparent for hours or even days.

Different species of fish have different complements of ER subtypes. Rainbow trout, for example, have four different ERs: ERα1, ERα2, ERβ1, and ERβ2 (Nagler et al., 2007). In contrast, many other fish such as fathead minnow (*Pimephales promelas*), goldfish (*Carrasius auratus*), and tilapia (*Oreochromic niloticus*) do have an ERα2 isoform, although it is possible that this missing ER has not yet been identified (Filby and Tyler, 2005; Wang et al., 2005; Nelson and Habibi, 2010). These differences in the complement of ERs present in various fish species may account for some of the interspecies differences noted in sensitivity to xenoestrogen toxicity.

In addition, the transcription or expression of ERs varies between tissues and life stages. Nagler et al. (2007) demonstrated that while transcription of all four forms of ERs occurs in almost every tissue in the body, they are not equally transcribed (ERα1 and ERβ2 mRNA levels are higher in general) and the predominant isoform is different between tissues (e.g., ERα1 is the most predominant form in the liver, while ERβ2 is most predominant in the spleen). Others have demonstrated that the transcription of ERs can change with developmental stage and shifting predominance of one ER form to another, or increased abundance of an ER isoform during ontogeny may play a role in hormone-driven developmental processes (Filby and Tyler, 2005; Boyce-Derricott et al., 2010).

It is also generally believed that, as in mammals, each form of ER in fish has a different physiological role in terms of response to estrogens or xenoestrogens. This hypothesis is supported by a limited number of studies that show differences in binding affinities between the various ERs for ligands; by studies in which specific ER subtype agonists are antagonists used to activate or suppress a specific ER; or by studies that use molecular techniques (such as RNA silencing) to knock down specific ER subtypes. However, one of the difficulties in performing these types of studies is the lack of fish ER-specific agonists and antagonists, since it has been clearly demonstrated that mammalian compounds do not always work the same in fish. Regardless, a number of studies in different fish species have demonstrated that ERα transcription is induced by exposure to E2 and that the induction is associated with increased transcription of VTG. Leaños-Castañeda and van der Kraak (2007), using subtype-specific (mammalian) ER agonists and antagonists, found that ERβ was required for the induction of VTG transcription in rainbow trout hepatocytes, even though it is ERα transcript abundance that is increased in response to E2. A study by Nelson and Habibi (2010) in goldfish shed further light on this apparent contradiction. Studies using agonists/antagonists and RNA silencing suggest that all ER subtypes are required for normal VTG induction (Nelson and Habibi, 2010). Since ERβ1 and ERβ2 subtypes have a greater binding affinity for E2 and are thus more sensitive to lower E2 concentrations, ERβs serve to ensure ERα induction in response to increasing levels of E2 at sexual maturation. This primes the liver and enables a greater overall sensitivity of the liver to E2 so that VTG production driven by all ER subtypes can be quickly increased.

While the natural ligands for the ER are endogenous estrogens (E2, E1, E3), the chemicals typically identified as xenoestrogens also have some level of affinity for binding ER (reviewed in Rempel and Schlenk, 2008; Dang, 2010). Most xenoestrogens have lower affinity for ER than E2, with the exception of EE2 which has substantially higher binding affinity. Interestingly, in channel catfish, it has been demonstrated that synthetic xenoestrogens (EE2, NP, 4-octylphenol, endosulfan) have greater binding affinity to ERα than ERβ, whereas the converse is true for natural estrogens (Gale et al., 2004). Regardless, although the binding affinities of the xenoestrogens are not typically as strong as those for E2, suggesting higher concentrations of these chemicals would be required to elicit effects, they are still capable of activating the ER. In this way, they trigger the transcription of estrogen-responsive genes.

Differential regulation of ER transcription and expression also occur as a result of endogenous estrogen or xenoestrogen exposure. It has been demonstrated that exposure to E2 consistently results in the upregulation of ERα in the liver, gonad, brain, and leukocytes (MacKay et al., 1996; Marlatt et al., 2008; Shelley et al., 2013). Similarly, exposure to EE2 increased ERα1 transcription in liver from rainbow trout and the rare minnow (*Gobiocyrpis rarus*) (Boyce-Derricott et al., 2009; Wang et al., 2011). Following exposure to NP, ERα transcript abundance also increased in Altantic salmon liver and brain, but was decreased in the gill of Atlantic salmon and in liver from *G. rarus*, suggesting that the effects are species and tissue specific (Meucci and Aurkwe, 2006a; Wang et al., 2011). Transcriptional downregulation of ERβ1 has also been reported in liver of some fish following exposure to E2, EE2, or NP (Marlatt et al., 2008; Wang et al., 2011). Generally, the transcription of ERβ2 has been unaffected by exposure to E2 or other xenoestrogens. While the functional significance of the xenoestrogen-induced alterations to ER abundance is uncertain (most of the studies did not consider translational or posttranslational controls that might affect ER expression and temper changes in transcription), the dysregulation of ER transcription does provide some insight into a potential mechanism underlying xenoestrogen effects.

4.1.2. MEMBRANE ESTROGEN RECEPTORS

In addition to slow, genomic responses, estrogens and xenoestrogens are also capable of causing rapid responses that occur in seconds or minutes and are not affected by transcription or translation inhibitors. These are nongenomic, membrane-mediated ER effects. While some debate has arisen over which receptor(s) are responsible, some variants of the nuclear ERα and ERβ that associate with the plasma membrane or endoplasmic reticulum can mediate rapid, nongenomic effects, such as the recently identified ERα36 (Kang et al., 2010). Other researchers have identified a G-protein linked estrogen receptor (GPER, formerly GPR30) as being responsible for rapid signal transduction associated with estrogens (as reviewed by Thomas, 2012). Regardless of the receptor identity, rapid estrogen signal transduction leads to a rapid increase in secondary signaling molecules including intracellular calcium, adenylate cyclase activity and cAMP, and protein kinases (e.g., MAPK), consistent with traditional signaling through G-protein receptors.

In fish, estrogens can also act through nongenomic mechanisms. Studies using cells derived from the gonad or oocytes exposed to E2 conjugated to bovine serum albumin (BSA) have found that alterations in the target cell still occur, even though the E2-BSA conjugate cannot pass through the membrane. For example, Loomis and Thomas (2000) demonstrated that production of 11-ketotestosterone in Atlantic croaker (*Micropogonias undulates*) testicular tissue is sensitive to E2, with the effects mediated by membrane receptors. This study used multiple approaches including the use of E2-BSA conjugates, ER antagonists and agonists, transcription and translation inhibitors, and receptor binding studies using membrane preparations; all of the findings point toward a steroid membrane receptor. Similar findings using multiple approaches have been used to suggest a role for a membrane ER (mER) in both oocyte maturation in zebrafish and regulation of milt quality and quantity in goldfish (Pang and Thomas, 2010; Mangiamele and Thompson, 2012). Recently, Liu et al. (2009) reported on the identification of an mER homologous to human and rodent GPER in testes and brain of zebrafish. While it seems likely that E2 can exert nongenomic effects on target tissues and that mERs exist in fish, currently we have no information on how xenoestrogens might act through fish mERs, as this remains an area of active investigation.

4.1.3. XENOESTROGEN POTENCIES

Most compounds that mimic E2 are not nearly as potent as an estrogen hormone (Table 5.1). This relates to binding affinity for the ERs. Potency is an important topic when we are considering mixtures of xenoestrogens. For example, NP and other APEs are typically present in a sewage effluent in μg/L concentrations, whereas estrogen hormones, such as E2 and EE2, are present at pg/L or ng/L concentrations (Snyder et al., 2001b). Yet, the estrogenicity of the effluent would be related to the estrogen hormones with little contribution from the APEs. This is due to the weak estrogenicity of the APEs and the strong potency of the estrogen hormones. Estrogenic mixtures are also discussed in Section 6.

4.2. Interaction with Other Nuclear or Peptide Receptors

In addition to interaction with ERs, the xenoestrogenic activity of EDCs is also associated with other receptors in the nuclear hormone receptor superfamily and can elicit downstream effects of those receptors.

The estrogen-related receptor (ERR) is considered to be an orphan nuclear receptor since no endogenous ligand has been identified; it does appear, however, to be sensitive to activation by xenoestrogens and by BPA in particular, with BPA having a 100-fold higher affinity for ERRγ than ER (Takayanagi et al., 2006). Although ERR does not bind estrogens, it can be a transcription factor for genes containing either an ERE or an ERR-response element (ERRE). The three main forms of ERR (ERRα, ERRβ, and ERRγ) have different tissue distributions and are differentially expressed during development. Among fish, ERR has been identified and characterized in zebrafish and killifish (*Fundulus heteroclitus*), and it may have a role in developmental processes in the brain and muscle (Bardet et al., 2004; Tarrant et al., 2006). Even though ERR does not bind E2, E2 exposure downregulated transcription of ERRα in female heart tissue (Tarrant et al., 2006). Exposure of zebrafish embryos to BPA for short duration during neurogenesis resulted in hyperactivity in larvae and impaired learning ability of adults, possibly in part a result of ERRγ–BPA interaction (Saili et al., 2012). Without conclusive studies in fish that link xenoestrogen effects to an ERR-mediated pathway, it remains an interesting avenue for future research.

Several xenoestrogens have affinity for the nuclear androgen receptor (AR). Thus, it is possible that some of the estrogenic effects noted for xenoestrogens may be a result of AR antagonism. Indeed, o,p′-DDE, BPA and butyl benzyl phthalate act as antagonists of the AR in an *in vitro* yeast-based system (Sohoni and Sumpter, 1998). Blocking the AR, but not activating it, would interfere with the ability of androgens to interact with the AR, thus decreasing AR-dependent gene transcription. The above study also showed that NP is a weak AR agonist, potentially causing mixed estrogenic and androgenic effects. In a hermaphroditic fish (*Rivulus marmoratus*), BPA exposure downregulated AR in the liver and NP exposure downregulated AR in liver and gonad (Seo et al., 2006). This study also found that secondary males exposed to NP upregulated AR in the liver, suggesting that the effects of NP are dependent on either gender or the developmental stage.

Growth hormone (GH), a peptide hormone produced by the pituitary, interacts with the transmembrane growth hormone receptor (GHR) to stimulate the release of the insulin-like growth factors (IGF, IGF-1, and IGF-2) that affect somatic growth and osmoregulation in fish. There is also an interaction between the HPG axis and GH axis, with GH release stimulated by gonadotropin-releasing hormone (GnRH) and GH acting to increase luteinizing hormone (LH) and follicle-stimulating hormone (FSH) activity in the gonad. GH is involved in spermatogenesis in males and can also increase the activity of CYP19 (aromatase), which converts testosterone to E2 and may play a role in ensuring proper developmental timing for maturation of oocytes. Given the carefully controlled and orchestrated interplay between the two hormone systems, disruption of GH by xenoestrogens could conceivably indirectly alter reproductive end-points in fish. There is some evidence to support this hypothesis. For example, Davis et al. (2008) demonstrated that exposure of male tilapia to E2 led to a downregulation of the GH/IGF axis and was associated with decreased transcription of the genes for GHR1, GHR2, IGF-1, and IGF-2 and upregulation of ERα and VTG, creating a gene-expression profile more consistent with female tilapia. Apart from the effects on reproduction, GH/IGF is also involved in osmoregulation and is important in promoting seawater adaptation in fish. Waterborne exposure to three xenoestrogens (E2, NP, and β-sitosterol) increased plasma chloride concentrations following seawater transfer and suppressed or abolished the normal increase in GH axis gene transcription (GHR1, GHR2, IGF-1, IGF-2, IGF receptor-1A [IGFR1A], and IGFR1B) observed in both liver and gill of untreated control fish (Hanson et al., 2012). Whether this effect was due to negative feedback regulation (i.e., downregulation) of the GH axis directly by the xenoestrogens or through ER-related crosstalk is not known, but the findings underscore the critical role played by the GH/IGF axis in anadromous fish seawater adaptation.

The pregnane X receptor (PXR) is a nuclear receptor that can interact with a wide range of endogenous steroid hormones and xenobiotics, including EDCs. Once activated, the PXR induces transcription of a number of genes involved in xenobiotic metabolism and detoxification, including phase I enzymes such as cytochrome P450 3A (CYP3A), phase II enzymes such as glutathione-S-transferase (GST), and efflux pumps such as multidrug resistance protein 1 (P-gp, MDR1). Similarly, the constitutive androstane receptor (CAR) is also a xenobiotic sensor, albeit for a more limited range of ligands, which can also upregulate the transcription of various xenobiotic metabolizing enzymes, including some of the same genes as PXR plus some unique genes for CYP2B, several phase II enzymes, and multidrug resistance-associated protein (MRP1). These receptors are also responsible for the metabolism of endogenous steroids; thus activation or antagonism of these receptors can result in changes in the transcription of genes that may ultimately alter steroid homeostasis. Kretschmer and Baldwin (2005) summarized the reported CAR and PXR receptivity for xenoestrogens in human and mouse models; they noted that many xenoestrogens have been found to have agonistic activity for one or both of these receptors, which can lead to deleterious consequences in the organism. Much less information is available for fish species. Winter flounder (*Pleuronectes amercanus*) exposed to NP had increased CYP3A expression and metabolism of testosterone by hydroxylases, effects expected with activation of PXR by NP (Baldwin et al., 2005). Similar findings regarding CYP3A transcriptional upregulation and PXR were noted in Atlantic salmon exposed to NP (Meucci and Arukwe, 2006b). Several phthalates also activated PXR-mediated transcription in an *in vitro* mammalian model, but similar studies in fish have not been done to date (Hurst and Waxman, 2004). Thus, PXR and/or CAR-mediated upregulation of biotransformation pathways seem to be an important mechanism of xenoestrogen action in fish that leads to disruption of homeostasis through altered metabolism of both endogenous and exogenous compounds.

4.3. Crosstalk between ER and the Aryl Hydrocarbon Receptor (AhR)

In mammalian systems, the ER is involved in crosstalk with other receptor-mediated signaling systems, including the nuclear aryl hydrocarbon receptor (AhR). The AhR has been characterized in a wide range of fish species and acts as a xenosensor for environmental contaminants such as dioxin, dioxin-like compounds, and PAHs, with ligand binding triggering the transcription of xenobiotic metabolizing genes containing a xenobiotic response element (XRE). One of the upregulated genes is CYP1A1, the activity of which is commonly evaluated in the ethoxyresorufin-O-deethylase (EROD) assay. Increased EROD activity is used as a biomarker of AhR activation since exposure to AhR ligands leads to substantial increases in both CYP1A1 transcription and activity.

In fish, the crosstalk between the AhR and ER signaling pathways is either uni- or bidirectional. For example, co-exposure to β-naphthoflavone (BNF, a potent AhR agonist) and EE2 (ER agonist) in rainbow trout hepatocytes downregulated VTG production by 40% compared to EE2 exposure alone, while CYP1A1 gene transcription and protein levels remained the same as in the BNF alone exposure (Gräns et al., 2010). This is an example of unidirectional crosstalk where AhR is affecting the ER signaling pathway. In hepatocytes exposed to EE2 alone, however, basal CYP1A1 mRNA abundance was decreased, suggesting that a different type of crosstalk occurs in AhR-uninduced cells where ER affects AhR-mediated gene transcription. Pretreatment of zebrafish embryos with dioxin (either 2,3,7,8-TCDD or 1,2,3,7,8-pentachlorodibenzo-*p*-dioxin), followed by EE2 exposure inhibited EE2/ER-induced *vtg* transcription in larvae, which could be at least partially reversed by using an AHR2 morpholino to decrease *ahr2* transcription (Bugel et al., 2013). This finding suggests that the timing of exposure might be important (pretreatment). However, dioxins resist biodegradation and likely persisted in the tissues of the zebrafish larvae during the EE2 exposure. In contrast, co-exposure of goldfish to E2 (ER agonist) and benzo[a]pyrene (BaP, AhR agonist) downregulated genes related to both ER (ERα and VTG) and AhR (AhR2, CYP1A1) activation compared to responses to E2 or BaP alone, suggesting that there is a bidirectional modulation of ER and AhR signaling pathways (Yan et al., 2012). Similar bidirectional ER and AhR crosstalk occurred in Atlantic salmon hepatocytes using NP as the ER agonist and PCB-77 as the AhR agonist (Mortensen and Arukwe, 2007). In addition, as is seen in mammals, ligand-activated AhR can recruit and activate ERs in the absence of ER ligands (known as ER "hijacking") in fish, resulting in the transcription of genes containing an ERE, and this involves complex receptor interactions and protein stability (Gjernes et al., 2012). The study of crosstalk mechanisms in fish between ER and other receptor signaling pathways is in its infancy. This avenue of research offers the potential to understand the mechanistic basis underlying the mixture toxicity of xenoestrogens and other common environmental contaminants such as dioxins and PAHs.

4.4. Nonreceptor Mediated Mechanisms

Xenoestrogens can also disrupt steroid homeostasis through effects at several other levels of the HPG axis. Under normal physiological conditions, hormones are produced in tissues, primarily the gonads, in response to signals initiated by neurons in the hypothalamus, which release GnRH to stimulate pituitary release of LH and FSH; this release in turn stimulates steroidogenesis and/or gametogenesis. Hormones are then synthesized from cholesterol through the steroidogenic pathway and released into the blood, where they associate with plasma sex steroid-binding globulins (SSBGs) and occasionally other plasma proteins such as albumin. At a target cell, the hormone dissociates from the SSBP and is free to interact with receptors in the cell. Ultimately, steroid hormones are metabolized and excreted following biotransformation that occurs primarily in the liver. Production of GnRH can be regulated by signals from other neurons mediated by neurotransmitters such as dopamine or GABA. There are also feedback loops in the HPG axis, so that concentrations of plasma hormones can act at the neuroendocrine level to control their own release. Overall, the HPG axis is a delicately balanced system that tightly controls hormone levels and activity to ensure that reproductive (and other) processes progress in the expected, sequential fashion. Xenoestrogens have been found to cause disruptive effects at most levels of the HPG axis; examples will be highlighted in the following sections.

4.4.1. Xenoestrogen Effects on Steroidogenesis

In fish, steroidogenesis of the sex hormones typically occurs in the gonad, although some enzymes in the pathway, such as CYP19 (aromatase), occur in other tissues like the brain. Here we illustrate how xenoestrogens can alter steroid production, ultimately leading to changes in hormone concentrations.

The rate-limiting step in steroidogenesis involves the transfer of cholesterol across the mitochondrial membrane by steroidogenic acute regulatory (StAR) protein. The second step involves the conversion of cholesterol to pregnenalone involving the CYP450scc (side chain cleavage) enzyme, which is used as a precursor for production of numerous steroid hormones. These steps are potential targets for xenoestrogens, and alterations would have substantial downstream effects (Arukwe, 2008). To this end, EE2 downregulated StAR and P450scc mRNA abundance in male fathead minnow in a concentration-dependent manner; these changes were associated with a decrease in plasma testosterone (Garcia-Reyero et al., 2009). In contrast, while StAR transcription was downregulated in the ovary by E2 exposure via implant in female rainbow trout, P450scc was unaffected (Nakamura et al., 2009). Other studies also showed that StAR and P450scc mRNA abundance can be altered in oocytes by nonylphenol exposure, but the effects varied depending on duration of exposure and concentration (Kortner and Arukwe, 2007).

CYP19 catalyzes the conversion of testosterone to estrogen and plays a key role in controlling estrogen levels to orchestrate normal sexual differentiation, reproduction, and development of reproductive behaviors. In fish the main forms are CYP19A, which is primarily found in the ovaries, and CYP19B, which is primarily found in the brain. CYP19 characterization, functions, and the effects of xenoestrogens on the expression and activity of CYP19 were thoroughly reviewed by Cheshenko et al. (2008). The promoter regions of the CYP19A and CYP19B genes have multiple response elements such as ERE, XRE, and a cAMP response element, indicating that the expression of these genes is modulated by multiple endogenous and environmental factors, including the presence of xenoestrogens. Generally, EE2 exposure is associated with increased transcription of one or both of the CYP19 genes in liver or brain (Filby et al., 2007; Kortner et al., 2009). However, for other xenoestrogen studies, the same xenoestrogen has positive, negative, or no effect on CYP19. For example, CYP19A mRNA and aromatase activity in the brain of Atlantic salmon was decreased following three days of NP exposure (5 and 50 μg/L), while CYP19B mRNA was unaffected (Kortner et al., 2009). In contrast and also in the brain of Atlantic salmon of similar size and weight exposed to the same concentrations of NP for three days, CYP19B and CYP1A were unaffected (trending downward but not significantly); CYP19B was upregulated after seven days of exposure (Meucci and Arukwe, 2006a). These studies illustrate two issues: the occasional difficulty in comparing the effects of xenoestrogens between studies and the importance of considering xenobiotic impacts over time since the effects may not remain the same over the course of continuing exposures.

Other sex steroid production enzymes that are sensitive to xenoestrogen modulation include 3β-hydroxysteroid dehydrogenase (3β-HSD), 11β-hydroxylase (P450 11β), and cytochrome P450 17 (CYP17; 17α-hydroxylase/17,20 lyase) (Filby et al., 2007; Garcia-Reyaro et al., 2009;

Nakamura et al., 2009). Generally, these enzymes were all downregulated by xenoestrogen exposure, but $20\alpha/\beta$-HSD (which promotes the production of maturation-inducing hormone) was upregulated in rainbow trout ovarian microsomes by NP (Thibault and Porte, 2004).

4.4.2. XENOESTROGENS AND BIOTRANSFORMATION/METABOLIC PATHWAYS

Xenoestrogens interact with both PXR and CAR, and ER can crosstalk with AhR signaling to alter transcription of genes related to xenobiotic metabolism. While upregulation of these genes might aid the organism in eliminating xenobiotics and preventing some toxic effects, it can also have the unintended effect of increasing or altering the metabolism of endogenous steroids, leading to perturbation of plasma hormone concentrations and ratios.

The interference of xenoestrogens with the synthesis and clearance of key sex hormones through obstructions in biotransformation reactions may alter the bioavailable amounts of active hormones and be a potential mechanism of endocrine disruption by either prolonging or hastening their retention. For example, in chub (*Leucisus cephauls*), alkylphenols are reported to be inhibitors of sulfotransferase enzymes, which add sulfate to estrone to increase its excretion (Kirk et al., 2003). Exposure to exogenous E2 or NP reduced CYP1A1 activity, and E2 also suppressed GST activity (Vaccaro et al., 2005). As noted previously, EE2 and E2 decreased CYP1A1 mRNA abundance in rainbow trout hepatocytes only in the absence of AhR ligands (Navas and Segner, 2001; Gräns et al., 2010). Juvenile salmon treated with NP experienced concentration-dependent alterations in steroid hydroxylases, cytochrome P450 isozymes, and conjugating enzyme levels that lowered plasma concentrations of E2 (Arukwe et al., 2009). Alkylphenols reduced glucuronidation of p-nitrophenol and testosterone in salmonids (Arukwe et al., 2009; Thibault et al., 2001), whereas polychlorinated biphenyls (PCBs) increased glucuronidation of estradiol (Andersson et al., 1985; Forlin and Haux, 1985). Estrogenic NP and androgenic organotins and fenarimol compounds are active chemicals toward steroid biotransformation, specifically inhibiting the glucuronidation of estradiol (Thibault and Porte, 2004).

Metabolic activation or bioactivation of xenoestrogens to more potent active metabolites can occur during biotransformation reactions. Schlenk et al. (1998) found that monodemethylated methoxychlor, a metabolite of methoxychlor produced by hepatic microsomes, exhibited substantially greater binding affinity for ER and produced greater VTG responses in channel catfish compared to the parent compound. Similarly, the BPA metabolite [4-methyl-2,4-bis(4-hydroxyphenyl)pent-1-ene; MBP] had greater toxicity to medaka embryos than the parent BPA and caused VTG induction at substantially lower concentrations (Ishibashi et al., 2005).

4.4.3. XENOESTROGENS AND SEX STEROID-BINDING PROTEINS (SSBPs)

SSBPs serve as high-affinity carrier proteins for steroid hormones and likely influence the bioavailability of hormones as well as reduce their metabolism and elimination. Xenoestrogens like EE2, NP, and other alkylphenols, BPA, phthalates, and endosulfan, bind to SSBPs in fish blood. EDC concentrations must be high in order to alter the binding of hormones to SSBPs since EDCs have a much lower affinity than endogenous steroids such as E2, testosterone, and progesterone. However, species differences exist in binding affinities of xenoestrogens for SSBPs. For example, EE2 had a relative binding affinity for catfish SSBP that was three times higher than E2, but in Atlantic salmon EE2 had a relative binding affinity about 100 times lower than E2 (Tollefsen et al., 2002; Gale et al., 2004). Also, xenoestrogens could bind to SSBPs in areas where hormone concentrations are lower (e.g., at the gill or gut) and be displaced from the SSBPs in areas where hormone concentrations are higher (e.g., the gonad). Interestingly, *in vivo* exposure to either EE2 or di-*n*-butyl phthalate increased SSBP binding sites in Atlantic salmon. This result suggests that steroid hormone binding, particularly of androgens, would increase and lead to the estrogenic effects seen with several weak xenoestrogens such as the phthalates (Tollefsen et al., 2002). However, the toxicological implications of xenoestrogen binding to SSBPs have not been established.

4.4.4. XENOESTROGENS AND NEUROENDOCRINE EFFECTS

Xenoestrogen effects on neuroendocrine function in fish, specifically on the hypothalamus and pituitary levels of the HPG axis, have been reviewed (Page et al., 2011). The mechanism through which E2 (positive or negative feedback loops) or xenoestrogens exert effects on GnRH release is unclear. However, it appears that while the neurons that produce GnRH do not express estrogen receptors themselves, other neurons (e.g., those expressing KiSS1-derived peptide receptor [GPR54 or the Kisspeptin receptor]) that innervate the GnRH-producing neurons do express ER and may modulate GnRH production in response to elevated E2 (and potentially xenoestrogen) concentrations. Early studies demonstrated that exposure to either E2 or o,p′-DDT in the diet stimulated the release of gonadotropins (specifically LH) in Atlantic croaker (*Micropogonias undulates*; Khan and Thomas, 1998). Both E2 and NP also increased LHβ expression at the transcriptional level in female Atlantic salmon only, which correlated with the hepatic transcription of *vtg* and *zrp* (Yadetie and Male, 2002). Subsequent studies in the zebrafish (*Danio rerio*) have found that exposure of fish to xenoestrogens such as EE2 or NP during early developmental stages can have consequences for GnRH neuron development, an effect that is ER dependent (Vosges et al., 2012). Numerous studies have found that estrogens and xenoestrogens can alter gene expression and activity of CYP19 (aromatase, converts testosterone to estradiol) in the brain of fish, which could cause changes in E2 concentrations locally, although the effects may be dependent on species, gender, and developmental stage (reviewed by Cheshenko et al., 2008).

5. XENOESTROGEN EFFECTS

Xenoestrogens have myriad adverse effects on eggs, larvae, juveniles, and adult laboratory and wild fish, which involve wide-ranging physiological systems such as reproduction, immunity, osmoregulation, metabolism, endocrine development, and growth (Kime, 1998). Therefore, many organismal end-points can be used to evaluate xenoestrogenic effects.

5.1. Reproductive Impacts

Xenoestrogens mimic E2, the main vertebrate reproduction estrogen hormone, resulting in reproductive effects that are well documented at all levels of biological organization (Denslow and Sepulveda, 2007; Goksoyr, 2006).

5.1.1. MOLECULAR AND BIOCHEMICAL EFFECTS

In mature female fish, the liver produces VTG, which is transported in blood to the ovary where it is cleaved into lipovitellin and phosvitin and incorporated into the egg as an embryonic energy source (Mommsen and Walsh, 1988; Arukwe and Goksoyr, 2003). Xenoestrogens induce VTG production in fish (demonstrated in arctic char, carp, fathead minnow, flounder, goldfish, Japanese medaka, plaice, rainbow trout, rare minnow, roach, sand goby (*Pomatoschistus minutus*), sheepshead minnow (*Cyprinodon variegatus*), and zebrafish). Larval, juvenile, and male fish lack ovarian tissue to store VTG, and excess or unstored VTG causes kidney lesions/necrosis (Herman and Kincaid, 1988; Schwaiger et al., 2000; Zha et al., 2007), and male reproductive impairment (Kramer et al., 1998; Thorpe et al., 2007).

Choriogenins, also known as (VEPs) or zona pellucida/radiata proteins (Arukwe and Goksoyr, 2003; Mommsen and Korsgaard, 2008), form the structure of the chorion or egg envelope. Estrogens or xenoestrogens interact with hepatic ERs and initiate estrogen-responsive gene expression producing these egg proteins. Concentration-dependent increases in VTG transcript and protein were seen in sheepshead minnow exposed to E2 and NP (Hemmer et al., 2002). VTG mRNA usually peaks during exposure to EE2 and returns rapidly to baseline levels, while VTG protein peaks days after exposure has stopped and remains elevated for weeks in flounder (*Platichthys flesus*), sand goby (*Pomatoschistus minutus*), and plaice (*Pleuronectes platessa*) (Craft et al., 2004). In carp (*Cyprinus carpio*) exposed to E2, EE2, NP, and BPA, 28 hepatic gene transcripts were altered by exposure to these xenoestrogens; all belong to pathways involved in the physiological stress response, oogenesis, protein biosynthesis, and iron transport/homeostasis (Moens et al., 2007). In flounder, gene-expression alterations by E2 include choriogenin and VTG induction, and transcripts that relate to other biological functions (Williams et al., 2007). This sheds light on the mechanisms of ER-related crosstalk (Filby et al., 2006), due to the presence of ERs in many tissues (Nagler et al., 2007; Shelley et al., 2013).

Liver is a target tissue for xenoestrogens (Mommsen and Walsh, 1988). E2 and NP-treated rainbow trout show decreased liver glycogen, increased proportions of unsaturated lipids and nucleic acid content, and altered protein synthesis (Cakmak et al., 2006). Glycogen decreases can reflect an increase in energy demand (Mommsen and Walsh, 1988), and lipid or carbohydrate effects could be due to VTG (a large phospholipoglycoprotein) assembly and production, since protein concentration increases in plasma can be as high as a million-fold.

Xenoestrogens affect E2 and 11-KT concentrations by altering their synthesis, secretion, and/or function (Jobling et al., 2002; Hoger et al., 2006; Meier et al., 2007). For example, circulating levels of 11-KT are suppressed in fathead minnow males exposed to 50 ng/L E2 (Martinovic et al., 2006). Reductions in plasma 11-KT concentrations in male rainbow trout occur when exposed to 100 ng/L EE2 (Schultz et al., 2003).

5.1.2. TISSUE EFFECTS

Xenoestrogens cause both increases and decreases in gonadal somatic index (GSI) and liver somatic index (LSI). Ovarian and testicular GSI (Jobling et al., 1996; Schultz et al., 2003; Kristensen et al., 2005; Thorpe et al., 2007), and liver composition and size (Cakmak et al., 2006) are affected by increases in estrogen-responsive gene transcription and protein production (Persson et al., 1997; Parrott and Blunt, 2005; Schultz et al., 2003).

Xenoestrogen potency varies (Metcalfe et al., 2001; Thorpe et al., 2003) in adults, but can also alter gonad differentiation and development in eggs, larvae, and juveniles (Metcalfe et al., 2001; Afonso et al., 2002; van Aerle et al., 2002; Nash et al., 2004). Exposure results in partial or complete alterations of testicular tissue by the inclusion of ovarian tissue, oocytes, or egg proteins (Metcalfe et al., 2001; Jobling et al., 2002; van Aerle et al., 2002). Partial gonadal changes also occur and include ovo-testis in "intersex" fish. Some ovo-testis can function, but a delayed progression of spermatogenesis leads to reduced sperm counts (Jobling et al., 1996, 2002; Kristensen et al., 2005). Full testicular alterations also occur and result in genetic males with phenotypic ovaries (Afonso et al., 2002; Kristensen et al., 2005).

5.1.3. ORGANISM EFFECTS

Growth and condition factor (K) are inconsistently altered by xenoestrogens. In Japanese medaka, higher and lower morphometric parameters (wet weights, lengths, condition factors) occurred following exposure to several xenoestrogens (Metcalfe et al., 2001). Other examples include increased weight in the male roach following exposure to postsewage effluent (Rodgers-Gray et al., 2001); accelerated growth in fry and juveniles of EE2-exposed three-spine stickleback (*Gasterosteus aculeatus*) (Bell et al., 2004); decreased condition factor in EE2-treated fathead minnows (Filby et al., 2007); and decreased weight of zebrafish exposed to 100 ng/L E2 (Brion et al., 2004).

Physiological and behavioral performance is also affected by xenoestrogens. Feminized males may have reduced courting, nesting, and mating abilities. Male fathead minnows exposed to E2 were unable to hold nest sites in competition with untreated males (Hyndman et al., 2010). Fathead minnow pairs exposed to E1 have poor reproductive success including a reduced number of spawning events and reduced total egg production (Thorpe et al., 2007). Secondary sexual characteristics were also reduced in fathead minnow males exposed to 0.96 ng/L EE2, which resulted in decreased fertilization success (Parrott and Blunt, 2005).

5.1.4. POPULATION EFFECTS

Partial or fully feminized male fish result in skewed male:female sex ratios (Scholz, 2000; Kristensen et al., 2005; Parrott and Blunt, 2005) through a number of different mechanisms that translate into a lack of recruitment of wild fish. For example, feminized males reproductively underperform compared to untreated male competitors in courtship behavior (Kristensen et al., 2005; Martinovic et al., 2007). Reduced embryonic survival in rainbow trout eggs also results from paternal exposure to EE2 (Brown et al., 2007). Other effects on offspring viability include reduced progress to the eyed egg stage (Schultz et al., 2003), reduced egg survival (Nash et al., 2004), and inhibited or delayed egg production (Dammann et al., 2011). Life-long exposure of fathead minnows to 5 ng/L

EE2 resulted in complete reproductive failure in offspring, which produced no viable eggs, due to a lack of viable sperm from abnormal testis (Nash et al., 2004). In a long-term seven-year study, EE2 application of 5–6 ng/L (environmentally relevant concentration; Kolpin et al., 2002) to experimental lakes resulted in the collapse of a fathead minnow population (Kidd et al., 2007).

5.2. Immunotoxicity

Estrogens can disrupt other, nonreproductive processes including immune function (Iwanowicz and Ottinger, 2009; Milla et al., 2011). Immunotoxicity is defined as either up- or downregulation of immune system components or processes. The most definitive and ecologically relevant measure of immunotoxicity is the host resistance challenge. Cellular-level assays are also commonly used and include leukocyte differential or absolute counts, or measurement of cellular function (e.g., phagocytosis, respiratory burst). Increasingly, molecular-level end-points are also being included in immunotoxicological assessments in combination with functional end-points (e.g., Wenger et al., 2011, 2012; Shelley et al., 2012, 2013).

The endocrine system regulates and modulates immune system responses, and elevated plasma hormone concentrations that occur during sexual maturation have been associated with increased susceptibility to infections (Currie and Woo, 2007). E2-exposed fish also experience increased mortality following subsequent exposure to a pathogen (Wang and Belosevic, 1997; Wenger et al., 2011, 2012). At the cellular level, E2 exposure is associated with alterations in complement, phagocytosis, respiratory burst activity, and B and T cell proliferation, effects that are species and life-stage dependent (Law, 2001; Thilagam et al., 2009; Wenger et al., 2011; Shelley et al., 2013). In addition, microarray studies frequently identify immune system-related genes and pathways as being differentially regulated by estrogen and xenoestrogen exposures (Benninghoff and Williams, 2008; Shelley et al., 2012; Wenger et al., 2012; Osachoff et al., 2013). ERs have been identified in rainbow trout leukocytes at the mRNA level, and ERs have also been detected in the head kidney and spleen in rainbow trout and channel catfish, offering a pathway for estrogen and xenoestrogens to exert effects (Xia et al., 2000; Nagler et al., 2007; Shelley et al., 2013).

Leukocyte differential or absolute cell counts are often used as an indicator of immunotoxic effects. The number or proportion of lymphocytes is decreased following exposure to E2, NP, EE2, and estrogenic sewage effluents (Schwaiger et al., 2000; Liney et al., 2004; Thilagam et al., 2009;

Shelley et al., 2012). Lymphocytes are most often involved in adaptive or humoral immune responses, and a decrease in this cell population can impair the ability of fish to eliminate bacterial pathogens in particular.

Exposure to low concentrations (1 µg/L) of NP for a prolonged period (54 days) is associated with decreased phagocytosis and lymphocyte proliferation in rainbow trout (Hébert et al., 2009). Exposure of rainbow trout to NP (18 µg/L) for four days results in increased mortality in subsequent host resistance challenges with *Listonella anguillarum* (Shelley et al., 2012). Several potential pathways have been identified at the genomic level that could contribute to decreased disease resistance in these fish. These pathways include the transcriptional downregulation of immune-related signaling pathways and several important iron-binding proteins (ferritin and serotransferrin) (Shelley et al., 2012). Similar downregulation of iron binding (transferrin) or regulatory hormones (hepcidin) is seen in fish following exposures to E2 and other xenoestrogens (EE2, DES, and methoxychlor), and could represent a common target for xenoestrogens (Larkin et al., 2003; Robertson et al., 2009).

Legacy organochlorine compounds that exhibit estrogenic activity (lindane, endosulfan, and DDT) are shown to cause developmental immunotoxicity where exposure during early life stages results in effects later in life (Milston et al., 2003). Short-term exposures of Chinook salmon (*Oncorhynchus tshawytscha*) to o,p′-DDE during early-life developmental stages led to long-term impairment of humoral immune responses up to one year later. The long-term immunotoxicological effects may be a result of chemical persistence or of exposure at a critical stage or target in a developing immune system. Exposure to other xenoestrogens, including EE2 early in development, causes changes in immune gene expression and spleen growth and differentiation later in life (Shved et al., 2009).

5.3. Osmoregulatory Toxicity

Yolk-sac Atlantic salmon fry exposed to waterborne E2 (2 µg/L) or NP (10 µg/L) for 21 days and raised for one year in clean freshwater experienced higher mortality than unexposed fish during grow-out in freshwater (Lerner et al., 2007). E2 and NP exposure reduced gill Na/K-ATPase activity, seawater tolerance, and seawater preference, while the latency period to enter seawater was increased. The exposure duration was short, the concentration of NP used was environmentally relevant, and the chemicals used (NP and E2) have low (or no) persistence in tissue. This highlights the consequences of the timing of exposure, where exposures that occur during particularly sensitive stages of development can have implications long after the exposure has ceased.

5.4. Ethotoxicity

The endocrine system has bidirectional communication with the nervous system (i.e., neuroendocrine interaction), and ethnotoxicity? (alterations in behavioral end-points) has been recognized as a consequence of EDC exposure. Focus has been on behaviors involved in reproduction, including mating, courtship, and nesting. Other nonreproductive behaviors that are sensitive to perturbation by xenoestrogen exposure are foraging, antipredator and avoidance responses, and social behavior. For example, EE2-exposed three-spine stickleback (*Gasterosteus aculeatus*) increased activities that enhance the risk of predation (Bell, 2004), while exposure to E2, genistein or equol, led to decreased aggressive displays in fighting fish (*Betta splendens*) (Clotfelter and Rodriguez, 2006).

Shoaling is a behavior that provides protection by confusing predators and that dilutes individual risk (Pavlov and Kasumyan, 2000). Short-term (five-day) exposure of rainbow trout to waterborne NP (40 and 80 µg/L) reduced the tendency to shoal, and decreased the ability to forage for food (Ward et al., 2006). In contrast, zebrafish exposed to EE2 (5 ng/L) for 14 days stayed with a shoal more frequently than control fish, and the exposed fish had elevated anxiety levels (increased freezing and remained at the bottom of the tank) in a novel tank test (Reyhanian et al., 2011). Zebrafish exposed to a higher EE2 concentration (25 ng/L) spent more time near the surface of the tank in the novel tank test, thus increasing their risk of predation, which suggested that the anxiogenic response had a nonmonotonic concentration–response curve. Guppies (*Poecilia reticulata*) exposed to 3 or 10 ng/L EE2 for 21 or 28 days exhibited similar anxiogenic responses, with no changes in shoaling behavior (Hallgren et al., 2011). While increased shoaling behavior might decrease predation risk, increased anxiety responses at lower EE2 concentrations would negatively affect foraging ability and reproductive opportunities. Again, the consequences of exposure to xenoestrogens and the duration of effects may depend on the timing of exposures. Guppies exposed to 2 or 20 ng/L of E2 during early development and raised for six months in clean water exhibited altered behavior (observed as a greater tendency to remain near the bottom of a novel tank) as adults (Volkova et al., 2012).

6. INTERACTIONS WITH OTHER TOXIC AGENTS

Exposure to single chemicals rarely occurs in nature, and fish are commonly exposed to complex mixtures of legacy contaminants and currently used chemicals from a variety of sources. Interactions between chemicals can be additive, antagonistic, potentiating, or synergistic. Estrogenic EDC examples exist for these types of interactions. Additivity of estrogen hormones has been shown for VTG induction in fathead minnow (Brian et al., 2005; Thorpe et al., 2006) and sea bass (*Dicentrarchus labrax*; Correia et al., 2007). Antagonism also occurs: tamoxifen and cortisol reduced E2-induced VTG protein production in Arctic char (Berg et al., 2004), and mixtures of cadmium and E2 reduced estrogen-responsive gene transcripts (Vetillard 2005) and VTG protein production in rainbow trout (Olsson et al., 1995). E2 exposure potentiated the toxicity of injected cadmium (Olsson et al., 1995).

7. KNOWLEDGE GAPS AND FUTURE DIRECTIONS

A significant environmental concern regarding fish now exists; a very diverse group of synthetic and natural chemicals can alter endocrine function, causing a multitude of adverse effects. Although debate about the significance of endocrine disruption in fish continues, regulation of these chemicals in aquatic habitats has begun. One major challenge to regulators is to develop screening methods that will identify EDCs. Several methods have been developed and used successfully for xenoestrogens. The assembly, synthesis, and interpretation of data from these studies are not easily accomplished because the available information comes mainly from standard toxicological tests using a variety of species under varying exposure regimes and relies on standard end-points that are not specifically designed to determine if responses are endocrine-related. Molecular and biochemical approaches to identify estrogen-specific measurements are being used (e.g., VTG production in male fish); however, the relationship between biomarkers at these lower levels of biological organization may be difficult to extrapolate to higher levels. New *in vivo* and *in vitro* toxicity tests specifically designed to test potential endocrine disruptors for estrogen-, androgen-, and thyroid-mediated endocrine effects have been established (OECD, EPA), and are designed for a variety of species including fish. With the diversity of life histories of fish species and differing sensitivities exhibited by teleosts to pollutants, selection of a sentinel species for extrapolation within this group may also prove problematic. Endocrine systems are multi-action, multimechanism systems that pose challenges for researchers. Interpreting data from these studies can be difficult owing to crosstalk between signaling pathways, hormone feedback and regulatory aspects to endocrine function, and redundancy in hormone systems.

Currently, several areas can be identified where research can be focused to more fully understand EDC effects and their significance. The identification of EDCs in the environment is paramount, and screening protocols or tests for xenoestrogens and other types of endocrine disruption need to be developed. A suite of biomarkers for *in situ* measurements in the environment, perhaps with the use of sentinel species and sensitive life stages, should be developed. Most EDCs are found in mixtures, and interactions between mixture components have been seen in a number of studies. Several approaches can be taken to assess mixtures of other xenobiotics and may be useful in addressing EDC toxicity such as the Toxic Equivalency Factor approach. Additional work in this area is required, however, to further develop and refine mixture toxicology models. Underlying any screening tool or biomarker development, end-point selection, and even extrapolation capability is the need to more fully understand the basic physiological functioning of the endocrine systems of organisms such as fish. All of these research gaps need to be incorporated into a comprehensive framework for testing that will allow for the coordinated assessment of the potential for chemicals to significantly disrupt the endocrine systems of fish.

REFERENCES

Aerni, H., Kobler, B., Rutishauser, B., Wettstein, F., Fischer, R., Giger, W., Hungerbuhler, A., Marazuela, M. D., Peter, A., Schonenberger, R., Vogeli, A. C., Suter, M. and Eggen, R. (2004). Combined biological and chemical assessment of estrogenic activities in wastewater treatment plant effluents. *Anal. Bioanal. Chem.* 378, 688–696.

Afonso, L. O., Smith, J. L., Ikonomou, M. G. and Devlin, R. H. (2002). Y-chromosomal DNA markers for discrimination of chemical substance and effluent effects on sexual differentiation in salmon. *Environ. Health Perspect.* 110, 881–887.

Ahel, M., McEvoy, J. and Giger, W. (1993). Bioaccumulation of the lipophilic metabolites of nonionic surfactants in freshwater organisms. *Environ. Pollut.* 79, 243–248.

Andersson, T., Pesonen, M. and Johansson, C. (1985). Differential induction of cytochrome P450-dependent monooxygenase, epoxide hydrolase, glutathione transferase and UDP glucuronosyltransferase activities in the liver of the rainbow trout by B-naphthoflavone or Clophen A50. *Biochem. Pharmacol.* 34, 3309–3314.

Andersson, T. (1992). Purification, characterization, and regulation of a male-specific cytochrome P450 in the rainbow trout kidney. *Mar. Environ. Res.* 34, 109–112.

Arsenault, J. T. M., Fairchild, W. L., MacLatchy, D. L., Burridge, L., Haya, K. and Brown, S. B. (2004). Effects of water-borne 4-nonylphenol and 17β-estradiol exposures during parr-smolt transformation on growth and plasma IGF-I of Atlantic salmon (*Salmo salar* L.). *Aquat. Toxicol.* 66, 255–265.

Arukwe, A. (2008). Steroidogenic acute regulatory (StAR) protein and cholesterol side-chain cleavage (P450scc)-regulated steroidogenesis as an organ-specific molecular and cellular target for endocrine disrupting chemicals in fish. *Cell Biol. Toxicol.* 24, 527–540.

Arukwe, A. and Goksoyr, A. (2003). Eggshell and egg yolk proteins in fish: hepatic proteins for the next generation: oogenetic, population, and evolutionary implications of endocrine disruption. *Comp. Hepatol.* 2, 4.

Arukwe, A., Forlin, L. and Goksøyr, A. (2009). Xenobiotic and steroid biotransformation enzymes in Atlantic salmon (*Salmo salar*) liver treated with an estrogenic compound, 4-nonylphenol. *Environ. Toxicol. Chem.* 16, 2576–2583.

Baldwin, W. S., Rolling, J. A., Peterson, S. and Chapman, L. M. (2005). Effects of nonylphenol on hepatic testosterone metabolism and the expression of acute phase proteins in winter flounder (*Pleuronectes americanus*): Comparison to the effects of Saint John's wort. *Comp. Biochem. Physiol. C* 140, 87–96.

Barber, L. B., Lee, K. E., Swackhamer, D. L. and Schoenfuss, H. L. (2007). Reproductive responses of male fathead minnows exposed to wastewater treatment plant effluent, effluent treated with XAD8 resin, and an environmentally relevant mixture of alkylphenol compounds. *Aquat. Toxicol.* 82, 36–46.

Barber, L. B., Keefe, S. H., Leblanc, D. R., Bradely, P. M., Chapelle, F. H., Meyer, M. T., Loftin, K. A., Kolpin, D. W. and Rubio, F. (2009). Fate of sulfamethoxazole, 4-nonylphenol, and 17B-estradiol in groundwater contaminated by wastewater treatment plant effluent. *Environ. Sci. Technol.* 43, 4843–4850.

Bard, S. M. (2000). Multixenobiotic resistance as a cellular defense mechanism in aquatic organisms. *Aquat. Toxicol.* 48, 357–389.

Bardet, P. L., Obrecht-Pflumio, S., Thisse, C., Laudet, V., Thisse, B. and Vanacker, J. M. (2004). Cloning and developmental expression of five estrogen-receptor related genes in the zebrafish. *Dev. Genes Evol.* 214, 240–249.

Batty, J. and Lim, R. (1999). Morphological and reproductive characteristics of male mosquitofish (*Gambusia affinis holbrooki*) inhabiting sewage-contaminated waters in New South Wales, Australia. *Arch. Environ. Contam. Toxicol.* 36, 301–307.

Belfroid, A. C., Van der Horst, A., Vethaak, A. D., Schäfer, A. J., Rijs, G. B. J., Wegener, J. and Cofino, W. P. (1999). Analysis and occurrence of estrogenic hormones and their glucuronides in surface water and waste water in The Netherlands. *Sci. Total Environ.* 225, 101–108.

Bell, A. M. (2004). An endocrine disrupter increases growth and risk behavior in three-spined stickleback (*Gasterosteus aculeatus*). *Horm. Behav.* 45, 108–114.

Benninghoff, A. D. and Williams, D. E. (2008). Identification of a transcriptional fingerprint of estrogen exposure in rainbow trout liver. *Toxicol. Sci.* 101, 65–80.

Berg, H., Modig, C. and Olsson, P. (2004). 17beta-estradiol induced vitellogenesis is inhibited by cortisol at the post-transcriptional level in Arctic char (*Salvelinus alpinus*). *Reprod. Biol. Endo.* 2, 62.

Bitman, J. and Cecil, H. C. (1970). Estrogenic activity of DDT analogs and polychlorinated biphenyls. *J. Agric. Food Chem.* 18, 1108–1112.

Bjerregaard, L. B., Korsgaard, B. and Bjerregaard, P. (2006). Intersex in wild roach (*Rutilus rutilus*) from Danish sewage effluent-receiving streams. *Ecotoxicol. Environ. Saf.* 64, 321–328.

Boon, J. P., van der Meer, J., Allchin, C. R., Law, R. J., Klungsøyr, J., Leonards, P. E. G., Spliid, H., Storr-Hansen, E., Mckenzie, C. and Wells, D. E. (1997). Concentration-dependent changes of PCB patterns in fish-eating mammals: structural evidence for induction of cytochrome P450. *Arch. Environ. Contam. Toxicol.* 33, 298–311.

Boyce-Derricott, J., Nagler, J. J. and Cloud, J. G. (2009). Regulation of hepatic estrogen receptor isoform mRNA expression in rainbow trout (*Oncorhynchus mykiss*). *Gen. Comp. Endocrinol.* 161, 73–78.

Boyce-Derricott, J., Nagler, J. J. and Cloud, J. G. (2010). The ontogeny of nuclear estrogen receptor isoform expression and the effect of 17β-estradiol in embryonic rainbow trout (*Oncorhynchus mykiss*). *Mol. Cell. Endocrinol.* 315, 277–281.

Branham, W. S., Lyncook, B. D., Andrews, A. and Sheehan, D. M. (1993). Growth of separated and recombined neonatal rat uterine luminal epithelium and stroma on extracellular matrix: effects of *in vivo* tamoxifen exposure. *In Vitro Cell Develop. Biol. Animal.* 29A, 408–414.

Brian, J. V., Harris, C. A., Scholze, M., Backhaus, T., Booy, P., Lamoree, M., Pojana, G., Jonkers, N., Runnalls, T., Bonfà, A., Marcomini, A. and Sumpter, J. P. (2005). Accurate prediction of the response of freshwater fish to a mixture of estrogenic chemicals. *Environ. Health Perspect.* 113, 721–728.

Brion, F., Tyler, C. R., Palazzi, X., Laillet, B., Porcher, J. M., Garric, J. and Flammarion, P. (2004). Impacts of 17β-estradiol, including environmentally relevant concentrations, on reproduction after exposure during embryo-larval-, juvenile- and adult-life stages in zebrafish (*Danio rerio*). *Aquat. Toxicol.* 68, 193–217.

Brown, K. H., Schultz, I. R. and Nagler, J. J. (2007). Reduced embryonic survival in rainbow trout resulting from paternal exposure to the environmental estrogen 17alpha-ethynylestradiol during late sexual maturation. *Reproduction* 134, 659–666.

Buckman, A. H., Wong, C. S., Chow, E. A., Brown, S. B., Soloman, K. R. and Fisk, A. T. (2006). Biotransformation of polychlorinated biphenyls (PCBs) and biotransformation of hydroxylated PCBs in fish. *Aquat. Toxicol.* 78, 176–185.

Bugel, S. M., White, L. A. and Cooper, K. R. (2013). Inhibition of vitellogenin gene induction by 2,3,7,8-tetrachlordibenzo-*p*-dioxin is mediated by aryl hydrocarbon receptor 2 (AhR2) in zebrafish (*Danio rerio*). *Aquat. Toxicol.* 126, 1–8.

Burlington, H. and Lindeman, V. F. (1950). Effect of DDT on testes and secondary sex characters of white Leghorn cockerels. *Proc. Soc. Exp. Biol. Med.* 74, 48–51.

Cakmak, G., Togan, I. and Severcan, F. (2006). 17β-Estradiol induced compositional, structural and functional changes in rainbow trout liver, revealed by FT-IR spectroscopy: A comparative study with nonylphenol. *Aquat. Toxicol.* 77, 53–63.

Campbell, L. M., Muir, D. C. G., Whittle, D. M., Backus, S., Norstrom, R. J. and Fisk, A. T. (2003). Hydroxylated PCBs and other chlorinated phenolic compounds in lake trout (*Salvelinus namaycush*) blood plasma from the Great Lakes region. *Environ. Sci. Technol.* 37, 1720–1725.

Carlson, D. B. and Williams, D. E. (2001). 4-Hydroxy-2′,4′,6′-trichlorobiphenyl and 4-hydroxy-2′,3′,4′,5′-tetrachlorobiphenyl are estrogenic in rainbow trout. *Environ. Toxicol Chem.* 20, 351–358.

Carson, R. (1962). *Silent Spring.* New York: Fawcett Crest.

Cheshenko, K., Pakdel, F., Segner, H., Kah, O. and Eggen, R. I. L. (2008). Interference of endocrine disrupting chemicals with aromatase CYP19 expression or activity, and consequences for reproduction of teleost fish. *Gen. Comp. Endocrinol.* 155, 31–62.

Clotfelter, E. D. and Rodriguez, A. C. (2006). Behavioral changes in fish exposed to phytoestrogens. *Environ. Poll.* 144, 833–839.

Colborn, T. and Clement, C. (1992). Chemically induced alterations in sexual and functional development: The wildlife/human connection. *Adv. Modern Environ. Toxicol. 21.* Princeton, NJ: Princeton Scientific.

Colborn, T., vom Saal, F. S. and Soto, A. M. (1993). Developmental effects of endocrine disrupting chemicals in wildlife and humans. *Environ. Health Persp.* 101, 378–383.

Corcoran, J., Winter, M. J. and Tyler, C. R. (2010). Pharmaceuticals in the aquatic environment: A critical review of the evidence for health effects in fish. *Crit. Rev. Toxicol.* 40, 287–304.

Correia, A., Freitas, S., Scholze, M., Goncalves, J., Booij, P., Lamoree, M., Mananos, E. and Reis-Henriques, M. (2007). Mixtures of estrogenic chemicals enhance vitellogenic response in sea bass. *Environ. Health Perspect.* 115 (Suppl 1), 115–121.

Craft, J. A., Brown, M., Dempsey, K., Francey, J., Kirby, M. F., Scott, A. P., Katsiadaki, I., Robinson, C. D., Davies, I. M., Bradac, P. and Moffat, C. F. (2004). Kinetics of vitellogenin protein and mRNA induction and depuration in fish following laboratory and environmental exposure to oestrogens. *Mar. Environ. Res.* 58, 419–423.

Currie, J. L. and Woo, P. T. K. (2007). Susceptibility of sexually mature rainbow trout, *Oncorhynchus mykiss* to experimental cryptobiosis caused by *Cryptobia salmositica*. *Parisitol. Res* 101, 1057–1067.

Dammann, A. A., Shappell, N. W., Bartell, S. E. and Schoenfuss, H. L. (2011). Comparing biological effects and potencies of estrone and 17β-estradiol in mature fathead minnows, *Pimephales promelas*. *Aquat. Toxicol.* 105, 559–568.

Dang, Z. C. (2010). Comparison of relative binding affinities to fish and mammalian estrogen receptors: The regulatory implications. *Toxicol. Lett.* 192, 298–315.

Dargnat, C., Teil, M., Chevreuil, M. and Blanchard, M. (2009). Phthalate removal throughout waste water treatment plant case study of Marne Aval station (France). *Sci. Tot. Environ.* 407, 1235–1244.

Davis, L. K., Pierce, A. L., Hiramatsu, N., Sullivan, C. V., Hirano, T. and Grau, E. G. (2008). Gender-specific expression of multiple estrogen receptors, growth hormone receptors, insulin-like growth factors and vitellogenins, and effects of 17β-estradiol in the male tilapia (*Oreochromis mossambicus*). *Gen. Comp. Endocrinol.* 156, 544–551.

Dehal, S. S. and Kupfer, D. (1994). Metabolism of the proestrogenic pesticide methoxychlor by hepatic P450 monooxygenases in rats and humans. Dual pathways involving novel ortho ring-hydroxylation by CYP2B. *Drug. Metab. Dispos.* 22, 937–946.

Denslow, N. and Sepulveda, M. (2007). Ecotoxicological effects of endocrine disrupting compounds on fish reproduction. In *The Fish Oocyte* (eds. P. J. Babin, J. Cerda and E. Lubzens), pp. 255–322. Netherlands: Springer.

Dodds, E. C. and Lawson, W. (1936). Synthetic oestrogenic agents without the phenanthrene nucleus. *Nature* 137, 996.

Dodds, E. C., Goldberg, L. and Lawson, W. (1938). Estrogenic activity of certain synthetic compounds. *Nature* 141, 247–248.

Fairchild, W. L., Swansburg, E. O., Arsenault, J. T. and Brown, S. B. (1999). Does an association between pesticide use and subsequent declines in catch of Atlantic salmon (*Salmo salar*) represent a case of endocrine disruption? *Environ. Health Perspect.* 107, 349–358.

Filby, A. L. and Tyler, C. R. (2005). Molecular characterization of estrogen receptors 1, 2a and 2b and their tissue and ontogenic expression profiles in fathead minnow (*Pimephales promelas*). *Biol. Reprod.* 73, 648–662.

Filby, A. L., Thorpe, K. L. and Tyler, C. R. (2006). Multiple molecular effect pathways of an environmental oestrogen in fish. *J. Mol. Endocrinol.* 37, 121–134.

Filby, A. L., Thorpe, K. L., Maack, G. and Tyler, C. R. (2007). Gene expression profiles revealing the mechanisms of anti-androgen- and estrogen-induced feminization in fish. *Aquat. Toxicol.* 81, 219–231.

Filby, A. L., Shears, J. A., Drage, B. E., Churchley, J. H. and Tyler, C. R. (2010). Effects of advanced treatments of wastewater effluents on estrogenic and reproductive health impacts in fish. *Environ. Sci. Technol.* 44, 4348–4354.

Folmar, L. C., Denslow, N. D., Rao, V., Chow, M., Crain, D. A., Enblom, J., Marcino, J. and Guillette, L. J., Jr. (1996). Vitellogenin induction and reduced serum testosterone concentrations in feral male carp (*Cyprinus carpio*) captured near a major metropolitan sewage treatment plant. *Environ. Health Perspect.* 104, 1096–1101.

Forlin, L. and Haux, C. (1985). Increased excretion in the bile of 17B[^{3}H]-estradiol derived radioactivity in the rainbow trout treated with B-naphthoflavone. *Aquat. Toxicol.* 6, 197–208.

Furtula, V., Osachoff, H., Derksen, G., Juahir, H., Colodey, A. and Chambers, P. (2012). Inorganic nitrogen, sterols and bacterial source tracking as tools to characterize water quality and possible contamination sources in surface water. *Water Res.* 46, 1079–1092.

Gale, W. L., Patiño, R. and Maule, A. G. (2004). Interaction of xenobiotics with estrogen receptors α and β and a putative plasma sex hormone-binding globulin from channel catfish (*Ictalurus punctatus*). *Gen. Comp. Endocrinol.* 136, 338–345.

Garcia-Reyero, N., Kroll, K. J., Liu, L., Orlando, E. F., Watanabe, K. H., Sepúlveda, M. S., Villeneuve, D. L., Perkins, E. J., Ankley, G. T. and Denslow, N. D. (2009). Gene expression responses in male fathead minnows exposed to binary mixtures of an estrogen and antiestrogen. *BMC Genom.* 10, 308.

Gerpe, M., Kling, P., Berg, A. J. and Olsson, P. E. (2000). Arctic char (*Salvelinus alpinus*) metallothionein: cDNA sequence, expression, and tissue-specific inhibition of cadmium-mediated metallothionein induction by 17 beta-estradiol, 4-OH-PCB 30, and PCB 104. *Environ. Toxicol. Chem.* 19, 638–645.

Gjernes, M. H., Schlenk, D. and Arukwe, A. (2012). Estrogen receptor-hijacking by dioxin-like 3,3′,4,4′,5-pentachlorobiphenyl (PCB126) in salmon hepatocytes involves both receptor activation and receptor protein stability. *Aquat. Toxicol.* 124–125, 197–208.

Gobas, F. A. P. C. and Mackay, D. (1987). Dynamics of hydrophobic organic chemical bioconcentration in fish. *Environ. Toxicol. Chem.* 6, 495–504.

Goksoyr, A. (2006). Endocrine disruptors in the marine environment: mechanisms of toxicity and their influence on reproductive processes in fish. *J. Toxicol. Environ. Health A* 69, 175–184.

Gottardis, M. M., Robinson, S. P., Satyaswaroop, P. G. and Jordan, V. C. (1988). Contrasting effects of tamoxifen on endometrial and breast tumor growth in the athymic mouse. *Cancer Res.* 48, 812–815.

Gould, J. C., Leonard, L. S., Manesse, S. C., Wagner, B. L., Conner, K., Zacharewski, T., Safe, S., McDonnell, D. P. and Gaido, K. W. (1998). Bisphenol A interacts with the estrogen receptor a in a distinct manner from estradiol. *Mol. Cell. Endocrinol.* 142, 203–214.

Gräns, J., Wassmur, B. and Celander, M. C. (2010). One-way inhibiting cross-talk between arylhydrocarbon receptor (AhR) and estrogen receptor (ER) signaling in primary cultures of rainbow trout hepatocytes. *Aquat. Toxicol.* 100, 263–270.

Hallgren, S., Volkova, K., Reyhanian, N., Olsén, K. H. and Hällström, I. P. (2011). Anxiogenic behaviour induced by 17α-ethynylestradiol in male guppies (*Poecilia reticulata*). *Fish Physiol. Biochem.* 37, 911–918.

Hanson, A. M., Kittilson, J. D., McCormick, S. D. and Sheridan, M. A. (2012). Effects of 17β-estradiol, 4-nonylphenol and β-sitosterol on the growth hormone-insulin-like growth factor system and seawater adaptation of rainbow trout (*Oncorhynchus mykiss*). *Aquaculture* 362–363, 241–247.

Harries, J. E., Sheahan, D. A., Matthiessen, P., Neall, P., Rycroft, R., Tylor, T., Jobling, S., Routledge, E. J. and Sumpter, J. P. (1996). A survey of estrogenic activity in United Kingdom inland waters. *Environ. Toxicol. Chem.* 15, 1993–2002.

Hébert, N., Gagné, F., Cyr, D., Pellerin, J., Blaise, C. and Fournier, M. (2009). Effects of 4-nonylphenol on the immune system of rainbow trout. *Fres. Environ. Bull.* 18, 757–761.

Hemmer, M. J., Courtney, L. A. and Ortego, L. S. (1995). Immunohistochemical detection of P-glycoprotein in teleost tissues using mammalian polyclonal and monoclonal antibodies. *J. Exp. Zool.* 272, 69–77.

Hemmer, M. J., Bowman, C. J., Hemmer, B. L., Friedman, S. D., Marcovich, D., Kroll, K. J. and Denslow, N. D. (2002). Vitellogenin mRNA regulation and plasma clearance in male sheepshead minnows (Cyprinodon variegatus) after cessation of exposure to 17β-estradiol and p-nonylphenol. *Aquat. Toxicol.* 58, 99–112.

Herman, R. L. and Kincaid, H. L. (1988). Pathological effects of orally administered estradiol to rainbow trout. *Aquaculture* 72, 165–172.

Hewitt, M. L., Parrott, J. L. and McMaster, M. E. (2006). A decade of research on the environmental impacts of pulp and paper mill effluents in Canada: sources and characteristics of bioactive substances. *J. Toxicol. Environ. Health, Pt. B* 9, 341–356.

Hoger, B., Taylor, S., Hitzfeld, B., Dietrich, D. R. and van den Heuvel, M. R. (2006). Stimulation of reproductive growth in rainbow trout (*Oncorhynchus mykiss*) following exposure to treated sewage effluent. *Environ. Toxicol. Chem.* 25, 2753–2759.

Howard, P. H. and Meylan, W. M. (1997). *Handbook of Physical Properties of Organic Chemicals*. Boca Raton, FL: Lewis Publishers.

Hurst, C. H. and Waxman, D. J. (2004). Environmental phthalate monoesters activate pregnane X receptor-mediated transcription. *Toxicol. Appl. Pharmacol.* 199, 266–274.

Hyndman, K. M., Biales, A., Bartell, S. E. and Schoenfuss, H. L. (2010). Assessing the effects of exposure timing on biomarker expression using 17β-estradiol. *Aquat. Toxicol.* 96, 264–272.

Ishibashi, H., Watanabe, N., Matsumura, N., Hirano, M., Nagao, Y., Shiratsuchi, H., Kohra, S., Yoshihara, S. and Arizono, K. (2005). Toxicity to early life stages and an estrogenic effect of a bisphenol A metabolite, 4-methyl-2,4-bis(4-hydroyxphenyl)pent-1-ene on the medaka (*Oryzias latipes*). *Life Sci.* 77, 2643–2655.

Ivanciuc, T., Ivanciuc, O. and Klein, D. J. (2006). Modeling the bioconcentration factors and bioaccumulation factors of polychlorinated biphenyls with posetic quantitative super-structure/activity relationships (QSSAR). *Mol. Diversity* 10, 133–145.

Iwanowicz, L. R. and Ottinger, C. A. (2009). Estrogens, estrogen receptors and their role as immunoregulators in fish. In *Fish Defenses Immunology* (eds. G. Zaccone, J. Meseguer, A. Garcia-Ayala and B. G. Kapoor), Vol. 1: pp. 277–322. Enfield: Science Publishers.

Jeffries, K. M., Jackson, L. J., Ikonomou, M. G. and Habibi, H. R. (2010). Presence of natural and anthropogenic organic contaminants and potential fish health impacts along two river gradients in Alberta, Canada. *Environ. Toxicol. Chem.* 29, 2379–2387.

Jobling, S., Sumpter, J. P., Sheahan, D., Osborne, J. A. and Matthiessen, P. (1996). Inhibition of testicular growth in rainbow trout (*Oncorhynchus mykiss*) exposed to estrogenic alkylphenolic chemicals. *Environ. Toxicol. Chem.* 15, 194–202.

Jobling, S., Nolan, M., Tyler, C. R., Brighty, G. and Sumpter, J. P. (1998). Widespread sexual disruption in wild fish. *Environ. Sci. Technol.* 32, 2498–2506.

Jobling, S., Beresford, N., Nolan, M., Rodgers-Gray, T., Brighty, G. C., Sumpter, J. P. and Tyler, C. R. (2002). Altered sexual maturation and gamete production in wild Roach (*Rutilus rutilus*) living in rivers that receive treated sewage effluents. *Biol. Reprod.* 66, 272–281.

Johnson, A. C. and Sumpter, J. P. (2001). Removal of endocrine-disrupting chemicals in activated sludge treatment works. *Environ. Sci. Technol.* 35, 4697.

Johnson, A. C., Aerni, H., Gerritsen, A., Gibert, M., Giger, W., Hylland, K., Jürgens, M., Nakari, T., Pickering, A., Suter, M. J., Svenson, A. and Wettstein, F. E. (2005). Comparing steroid estrogen, and nonylphenol content across a range of European sewage plants with different treatment and management practices. *Water Res.* 39, 47–58.

Kang, L., Zhang, Z., Xie, Y., Tu, Y., Wang, D., Liu, Z. and Wang, Z. Y. (2010). Involvement of estrogen receptor variant ER-α36, not GPR30, in nongenomic estrogen signaling. *Mol. Endocrinol.* 24, 709–721.

Kavlock, R. J., Daston, G. P., DeRosa, C., Fenner-Crisp, P., Gray, L. E., Jr, Kaattari, S., Lucier, G., Luster, M., Mac, M. J., Maczka, C., Miller, R., Moore, J., Rolland, R., Scott, G., Sheehan, D. M., Sinks, T. and Tilson, H. A. (1996). Research needs for the assessment of health and environmental effects of endocrine disruptors: a report of the U.S. EPA-sponsored workshop. *Environ. Health Perspect.* 104, 715–740.

Keith, J. A. (1966). Reproduction in a population of herring gulls (Larus argentatus) contaminated by DDT. *J. Appl. Ecol.* 3, 57–70.

Kelce, W. R., Monosson, E., Gamcsik, M. P., Laws, S. C. and Gray, L. E., Jr (1994). Environmental hormone disruptors: evidence that vinclozolin developmental toxicity is mediated by antiandrogenic metabolites. *Toxicol. Appl. Pharmacol.* 126, 276–285.

Khan, I. A. and Thomas, P. (1998). Estradiol-17β and o,p'-DDT stimulate gonadotropin release in Atlantic croaker. *Mar. Environ. Res.* 46, 149–152.

Khanal, S. K., Xie, B., Thompson, M. L., Sung, S., Ong, S. and van Leeuwen, J. H. (2007). Fate, transport, and biodegradation of natural estrogens in the environment and engineered systems. *Environ. Sci. Technol.* 40, 6537–6546.

Kidd, K. A., Blanchfield, P. J., Mills, K. H., Palace, V. P., Evans, R. E., Lazorchak, J. M. and Flick, R. W. (2007). Collapse of a fish population after exposure to a synthetic estrogen. *Proc. Natl. Acad. Sci. U. S. A.* 104, 8897–8901.

Kim, S. D., Cho, J., Kim, I. S., Vanderford, B. J. and Snyder, S. A. (2007). Occurrence and removal of pharmaceuticals and endocrine disruptors in South Korean surface, drinking, and waste waters. *Water Res.* 41, 1013–1021.

Kime, D. E. (1998). *Endocrine Disruption in Fish.* Norwell, MA: Kluwer Academic Publishers, p. 396.

Kirk, C. J., Bottomley, L., Minican, N., Carpenter, H., Shaw, S., Kohli, N., Winter, M., Taylor, E. W., Waring, R. H., Michelangeli, F. and Harris, R. M. (2003). Environmental endocrine disrupters dysregulate estrogen metabolism and Ca^{2+} homeostasis in fish and mammals via receptor-independent mechanisms. *Comp. Biochem. Physiol. A* 135, 1–8.

Kleinow, K. M., Doi, A. M. and Smith, A. A. (1999). Distribution of C219 detectable P-glycoprotein transporter in the catfish. Eds: A. A. Elskus, W. K. Vogelbein, S. M. McLaughlin and A. S. Kane. In: PRIMO 10 (Pollutant Responses in Marine Organisms), Williamsburg, VA.

Kloas, W., Schrag, B., Ehnes, C. and Segner, H. (2000). Binding of xenobiotics to hepatic estrogen receptor and plasma sex steroid binding protein in the teleost fish, the common carp (*Cyprinus carpio*). *Gen. Comp. Endocrinol.* 199, 287–299.

Knudsen, F. R. and Pottinger, T. G. (1998). Interaction of endocrine disrupting chemicals, singly and in combination, with estrogen-, androgen-, and corticosteroid-binding sites in rainbow trout (*Oncorhynchus mykiss*). *Aquat. Toxicol.* 44, 159–170.

Kolodziej, E. P., Harter, T. and Sedlak, D. L. (2004). Dairy wastewater, aquaculture, and spawning fish as sources of steroid hormones in the aquatic environment. *Environ. Sci. Technol.* 38, 6377–6384.

Kolpin, D. W., Furlong, E. T., Meyer, M. T., Thurman, E. M., Zaugg, S. D., Barber, L. B. and Buxton, H. T. (2002). Pharmaceuticals, hormones, and other organic wastewater contaminants in U.S. Streams, 1999–2000: A national reconnaissance. *Environ. Sci. Technol.* 36, 1202–1211.

Konwick, B. J., Garrison, A. W., Avants, J. K. and Fisk, A. T. (2006). Bioaccumulation and biotransformation of chiral triazole fungicides in rainbow trout (*Oncorhynchus mykiss*). *Aquat. Tox.* 80, 372–381.

Kortner, T. M. and Arukwe, A. (2007). The xenoestrogen, 4-nonylphenol, impaired steroidogenesis in previtellogenic oocyte culture of Atlantic cod (*Gadus morhua*) by targeting the StAR protein and P450scc expressions. *Gen. Comp. Endocrinol.* 150, 419–429.

Kortner, T. M., Mortensen, A. S., Hansen, M. D. and Arukwe, A. (2009). Neural aromatase transcript and protein levels in Atlantic salmon (*Salmo salar*) are modulated by the ubiquitous water pollutant, 4-nonylphenol. *Gen. Comp. Endocrinol.* 164, 91–99.

Kramer, V. J., Miles-Richardson, S., Pierens, S. L. and Giesy, J. P. (1998). Reproductive impairment and induction of alkaline-labile phosphate, a biomarker of estrogen exposure, in fathead minnows (*Pimephales promelas*) exposed to waterborne 17β-estradiol. *Aquat. Toxicol.* 40, 335–360.

Kretschmer, X. C. and Baldwin, W. S. (2005). CAR and PXR: Xenosensors of endocrine disrupters? *Chemico-Biol. Interact.* 155, 111–128.

Krimsky, S. (2000). *Hormonal Chaos. The Scientific and Social Origins of the Environmental Endocrine Hypothesis.* Baltimore, MD: Johns Hopkins University Press, pp. 284.

Kristensen, T., Baatrup, E. and Bayley, M. (2005). 17α-ethinylestradiol reduces the competitive reproductive fitness of the male guppy (*Poecilia reticulata*). *Biol. Reprod.* 72, 150–156.

Kuster, M., López de Alda, M. J., Hernando, M. D., Petrovic, M., Martín-Alonso, J. and Barceló, D. (2008). Analysis and occurrence of pharmaceuticals, estrogens, progestogens and polar pesticides in sewage treatment plant effluents, river water and drinking water in the Llobregat river basin (Barcelona, Spain). *J. Hydrol.* 358, 112–123.

Lai, K. M., Scrimshaw, M. D. and Lester, J. N. (2002). Prediction of the bioaccumulation factors and body burden of natural and synthetic estrogens in aquatic organisms in the river systems. *Sci. Tot. Environ.* 289, 159–168.

Larkin, P., Folmar, L. C., Hemmer, M. J., Poston, A. J. and Denslow, N. D. (2003). Expression profiling of estrogenic compounds using a sheepshead minnow cDNA macroarray. *Environ. Health Perspect.* 111, 839–846.

Larsson, D. G. J., Adolfsson-Erici, M., Parkkonen, J., Pettersson, M., Berg, A. H., Olsson, P. E. and Förlin, L. (1999). Ethinyloestradiol—an undesired fish contraceptive? *Aquat. Toxicol.* 45, 91–97.

Lavado, R., Thibault, R., Raldúa, D., Martín, R. and Porte, C. (2004). First evidence of endocrine disruption in feral carp from the Ebro River. *Toxicol. Appl. Pharmacol.* 196, 247–257.

Law, W. Y., Chen, W. H., Song, Y. L., Dufour, S. and Chang, C. F. (2001). Differential in vitro suppressive effects of steroids on leukocyte phagocytosis in two teleosts, tilapia and common carp. *Gen. Comp. Endocrinol.* 121, 163–172.

Leaños-Castañeda, O. and van der Kraak, G. (2007). Functional characterization of estrogen receptor subtypes, ERα and ERβ, mediating vitellogenin production in the liver of rainbow trout. *Toxicol. Appl. Pharmacol.* 224, 116–125.

Lerner, D. T., Björnsson, B. T. and McCormick, S. D. (2007). Larval exposure to 4-nonylphenol and 17β-estradiol affects physiological and behavioural development of seawater adaptation in Atlantic salmon smolts. *Environ. Sci. Technol.* 41, 4479–4485.

Lindholst, C., Soren, N., Pedersen, S. N. and Bjerregaard, P. (2001). Uptake, metabolism and excretion of bisphenol A in the rainbow trout (*Oncorhynchus mykiss*). *Aquat. Toxicol.* 55, 75–84.

Lindholst, C., Wynne, P. M., Marriott, P., Pedersen, S. N. and Bjerregaard, P. (2003). Metabolism of bisphenol A in zebrafish (*Danio rerio*) and rainbow trout (*Oncorhynchus mykiss*) in relation to estrogenic response. *Comp. Biochem. Physiol. C.* 135, 169–177.

Liney, K. E., Hagger, J. A., Tyler, C. R., Depledge, M. H., Galloway, T. S. and Jobling, S. (2004). Health effects in fish of long-term exposure to effluents from wastewater treatment works. *Environ. Health Perspect.* 114, 81–89.

Lishman, L., Smyth, S. A., Sarafin, K., Kleywegt, S., Toito, J., Peart, T., Lee, B., Servos, M., Beland, M. and Seto, P. (2006). Occurrence and reductions of pharmaceuticals and personal care products and estrogens by municipal wastewater treatment plants in Ontario, Canada. *Sci. Total Environ.* 367, 544–558.

Liu, Z., Zhu, P., Sham, K. W. Y., Yuen, J. M. L., Xie, C., Zhang, Y., Liu, Y., Li, S., Huang, X., Cheng, C. H. K. and Lin, H. (2009). Identification of a membrane estrogen receptor in zebrafish with homology to mammalian GPER and its high expression in early germ cells of the testis. *Biol. Reprod.* 80, 1253–1261.

Loo, T. W. and Clarke, D. M. (1998). Nonylphenol ethoxylates, but not nonyphenol, are substrates of the human multidrug resistance P-glycoprotein. *Biochem. Biophys. Res. Commun.* 247, 478–480.

Loomis, A. K. and Thomas, P. (2000). Effects of estrogens and xenoestrogens on androgen production by Atlantic croaker testes *in vitro*: Evidence for a nongenomic action mediated by an estrogen membrane receptor. *Biol. Reprod.* 62, 995–1004.

Lotufo, G. R., Landrum, P. F., Gedeon, M. L., Tigue, E. A. and Herche, L. R. (2000). Comparative toxicity and toxicokinetics of DDT and its major metabolites in freshwater amphipods. *Environ. Toxicol. Chem.* 19, 368–379.

MacKay, M. E., Raelson, J. and Lazier, C. B. (1996). Up-regulation of estrogen receptor mRNA and estrogen receptor activity by estradiol in the liver of rainbow trout and other teleostean fish. *Comp. Biochem. Physiol. C* 115, 201–209.

MacLatchy, D. L. and Van Der Kraak, G. J. (1995). The phytoestrogen β-sitosterol alters the reproductive endocrine status of goldfish. *Toxicol. Appl. Pharmacol.* 134, 305–312.

Madsen, S. S., Mathiesen, A. B. and Korsgaard, B. (1997). Effects of 17β-estradiol and 4-nonylphenol on smoltification and vitellogenesis in Atlantic salmon (Salmo salar). *Fish Physiol. Biochem.* 17, 302–312.

Madsen, S. S., Skovbølling, S., Nielsen, C. and Korsgaard, B. (2004). 17β-estradiol and 4-nonylphenol delay smolt development and downstream migration in Atlantic salmon, Salmo salar. *Aquat. Toxicol.* 68, 109–120.

Mangiamele, L. A. and Thompson, R. R. (2012). Testosterone rapidly increases ejaculate volume and sperm density in competitively breeding goldfish through an estrogenic membrane receptor mechanism. *Horm. Behav.* 62, 107–112.

Marlatt, V. L., Martyniuk, C. J., Zhang, D., Xiong, H., Watt, J., Xia, X., Moon, T. and Trudeau, V. L. (2008). Auto-regulation of estrogen receptor subtypes and gene expression profiling of 17β-estradiol action in the neuroendocrine axis of male goldfish. *Mol. Cell. Endocrinol.* 283, 38–48.

Martinovic, D., Hogarth, W. T., Jones, R. E. and Sorensen, P. W. (2007). Environmental estrogens suppress hormones, behavior, and reproductive fitness in male fathead minnows. *Environ. Toxicol. Chem.* 26, 271–278.

Matthews, H. B. and Dedrick, R. L. (1984). Pharmacokinetics of PCBs. *Annu. Rev. Pharmacol. Toxicol.* 24, 85–103.

McCormick, S. D., O'Dea, M. F., Moeckel, A. M., Lerner, D. T. and Björnsson, B. T. (2005). Endocrine disruption of parr-smolt transformation and seawater tolerance of Atlantic salmon by 4-nonylphenol and 17β-estradiol. *Gen. Comp. Endocrinol.* 142, 280–288.

McKim, J., Schmieder, P. and Vieth, G. (1985). Absorption dynamics of organic chemical transport across trout gills as related to octanol-water partition coefficient. *Toxicol. Appl. Pharmacol.* 77, 1–10.

Meier, S., Andersen, T. E., Norberg, B., Thorsen, A., Taranger, G. L., Kjesbu, O. S., Dale, R., Morton, H. C., Klungsøyr, J. and Svardal, A. (2007). Effects of alkylphenols on the reproductive system of Atlantic cod (*Gadus morhua*). *Aquat. Toxicol.* 81, 207–218.

Metcalfe, C. D., Metcalfe, T. L., Kiparissis, Y., Koenig, B. G., Khan, C., Hughes, R. J., Croley, T. R., March, R. E. and Potter, T. (2001). Estrogenic potency of chemicals detected in sewage treatment plant effluents as determined by in vivo assays with Japanese medaka (*Oryzias latipes*). *Environ. Toxicol. Chem.* 20, 297–308.

Meucci, V. and Arukwe, A. (2006a). Transcriptional modulation of brain and hepatic estrogen receptor and P450arom isotypes in juvenile Atlantic salmon (*Salmo salar*) after waterborne exposure to the xenoestrogen, 4-nonylphenol. *Aquat. Toxicol.* 77, 167–177.

Meucci, V. and Arukwe, A. (2006b). The xenoestrogen 4-nonylphenol modulates hepatic gene expression of pregnane X receptor, aryl hydrocarbon receptor, CYP3A and CYP1A1 in juvenile Atlantic salmon (*Salmo salar*). *Comp. Biochem. Physiol. C* 142, 142–150.

Milla, S., Depiereux, S. and Kestemont, P. (2011). The effects of estrogenic and androgenic endocrine disruptors on the immune system of fish. *Ecotoxicology* 20, 305–319.

Milston, R. H., Fitzpatrick, M. S., Vella, A. T., Clements, S., Gundersen, D., Fiest, G., Crippen, T. L., Leong, J. and Schreck, C. B. (2003). Short term exposure of Chinook salmon (*Oncorhynchus tshawytscha*) to o,p'-DDE or DMSO during early life-history stages causes long-term humoral immunosuppression. *Environ. Health Perspect.* 111, 1601–1607.

Mitksicek, R. J. (1995). Estrogenic flavonoids: structural requirements for biological activity. *Proc. Exp. Biol. Med.* 208, 44–50.

Moens, L. N., van der Ven, K., Van Remortel, P., Del-Favero, J. and De Coen, W. M. (2007). Gene expression analysis of estrogenic compounds in the liver of common carp (*Cyprinus carpio*) using a custom cDNA microarray. *J. Biochem. Mol. Toxicol.* 21, 299–311.

Mommsen, T. P. and Walsh, P. J. (1988). Vitellogenesis and oocyte assembly. In *Fish Physiology XI A* (eds. W. S. Hoar and D. J. Randall), pp. 347–406. San Diego, CA: Academic Press.

Mommsen, T. P. and Korsgaard, B. (2008). Vitellogenesis. In *Fish Reproduction* (eds. M. Rocha, A. Arukwe and B. Kapoor). Enfield, New Hampshire, USA: Science Publishers, Chapter 4.

Mortensen, A. S. and Arukwe, A. (2007). Interactions between estrogen- and Ah-receptor signalling pathways in primary cultures of salmon hepatocytes exposed to nonylphenol and 3,3',4,4'-tetrachlorobiphenyl (congener 77). *Comp. Hepatol.* 6, 2.

Nagler, J. J., Cavileer, T., Sullivan, J., Cyr, D. G. and Rexroad, C., III (2007). The complete nuclear estrogen receptor family in the rainbow trout: Discovery of the novel ERα2 and both ERβ isoforms. *Gene* 392, 164–173.

Nakamura, I., Kusakabe, M. and Young, G. (2009). Differential suppressive effects of low physiological doses of estradiol-17β in vivo on levels of mRNAs encoding steroidogenic acute regulatory protein and three steroidogenic enzymes in previtellogenic ovarian follicles of rainbow trout. *Gen. Comp. Endocrinol.* 163, 318–323.

Nash, J. P., Kime, D. E., Van, D. V., Wester, P. W., Brion, F., Maack, G., Stahlschmidt-Allner, P. and Tyler, C. R. (2004). Long-term exposure to environmental concentrations of the pharmaceutical ethynylestradiol causes reproductive failure in fish. *Environ. Health Perspect.* 112, 1725–1733.

Navas, J. M. and Segner, H. (2001). Estrogen-mediated suppression of cytochrome P4501A (CYP1A) expression in rainbow trout hepatocytes: role of estrogen receptor. *Chemico-Biol. Interact.* 138, 285–298.

Nelson, E. R. and Habibi, H. R. (2010). Functional significance of nuclear estrogen receptor subtypes in the liver of goldfish. *Endocrinology* 151, 1668–1676.

Niimi, A. J. and Oliver, B. G. (1983). Biological half-lives of polychlorinated biphenyl (PCB) congeners in whole fish muscle of rainbow trout (*Salmo gairdneri*). *Can. J. Fish. Aquat. Sci.* 40, 1388–1394.

Nyagode, B.A. (2007). Biotransformation of methoxychlor and selected xenobiotics by channel catfish (*Ictalurus punctatus*) and largemouth bass (*Micropterus salmoides*). Doctoral Dissertation, University of Florida, Gainesville, FL.

Oehlmann, J., Schulte-Oehlmann, U., Kloas, W., Jagnytsch, O., Lutz, I., Kusk, K. O., Wollenberger, L., Santos, E. M., Paull, G. C., Van Look, K. J. W. and Tyler, C. R. (2009). A critical analysis of the biological impacts of plasticizers on wildlife. *Phil. Trans. Soc. B* 364, 2047–2062.

Olsson, P., Kling, P., Petterson, C. and Silversand, C. (1995). Interaction of cadmium and oestradiol-17 beta on metallothionein and vitellogenin synthesis in rainbow trout (Oncorhynchus mykiss). *Biochem. J.* 307, 197–203.

Orn, S., Norman, A., Holbech, H., Gessbo, A., Petersen, G. I. and Norrgren, L. (2001). Short-term exposure of zebrafish to pulp mill effluent and 17α-ethinylestradiol: A comparison between juvenile and adult fish. In *Suitability of Zebrafish as Test Organism for Detection of Endocrine Disrupting Chemicals* (eds. G. I. Petersen, L. Norrgren, H. Holbech, A. Lundgren and S. Koivisto), vol. 597, pp. 26–34. Copenhagen, Denmark: TemaNord.

Orrego, R., Burgos, A., Moraga-Cid, G., Inzunza, B., Gonzalez, M., Valenzuela, A., Barra, R. and Gavilan, J. F. (2006). Effects of pulp and paper mill discharges on caged rainbow trout (*Oncorhynchus mykiss*): Biomarker responses along a pollution gradient in the Biobio River, Chile. *Environ. Toxicol. Chem.* 25, 2280–2287.

Orrego, R., Guchardi, J., Krause, R. and Holdway, D. (2010). Estrogenic and anti-estrogenic effects of wood extractives present in pulp and paper mill effluents on rainbow trout. *Aquat. Toxicol.* 99, 160–167.

Osachoff, H. L., van Aggelen, G. C., Mommsen, T. P. and Kennedy, C. J. (2013). Concentration–response relationships and temporal patterns in hepatic gene expression of Chinook salmon (*Oncorhynchus tshawytscha*) exposed to sewage. *Comp. Biochem. Physiol. Part D. Genomics Proteomics* 8, 32–44.

Oulton, R. L., Kohn, T. and Cwiertny, D. M. (2010). Pharmaceuticals and personal care products in effluent matrices: A survey of transformation and removal during wastewater treatment and implications for wastewater management. *J. Environ. Monit.* 12, 1956–1978.

Page, Y. L., Vosges, M., Servili, A., Brion, F. and Kah, O. (2011). Neuroendocrine effects of endocrine disrupters in teleost fish. *J. Toxicol. Environ. Health B* 14, 370–386.

Pang, Y. and Thomas, P. (2010). Role of G protein-coupled estrogen receptor 1, GPER, in inhibition of oocyte maturation by endogenous estrogens in zebrafish. *Develop. Biol.* 342, 194–206.

Parrott, J. L., Wood, C. S., Boutot, P. and Dunn, S. (2003). Changes in growth and secondary sex characteristics of fathead minnows exposed to bleached sulfite mill effluent. *Environ. Toxicol. Chem.* 22, 2908–2915.

Parrott, J. L., Wood, C. S., Boutot, P. and Dunn, S. (2004). Changes in growth, secondary sex characteristics, and reproduction of fathead minnows exposed for a life cycle to bleached sulfite mill effluent. *J. Toxicol. Environ. Health A* 67, 1755–1764.

Parrott, J. L. and Blunt, B. R. (2005). Life-cycle exposure of fathead minnows (*Pimephales promelas*) to an ethinylestradiol concentration below 1 ng/L reduces egg fertilization success and demasculinizes males. *Environ. Toxicol.* 20, 131–141.

Parrott, J. L., McMaster, M. E. and Mark Hewitt, L. (2006). A decade of research on the environmental impacts of pulp and paper mill effluents in Canada: development and application of fish bioassays. *J. Toxicol. Environ. Health, Pt. B* 9, 297–317.

Paterson, G., Drouillard, K. G., Leadley, T. A. and Haffner, G. D. (2007). Longterm polychlorinated biphenyl elimination by three size classes of yellow perch (*Perca flavescens*). *Can. J. Fish. Aquat. Sci.* 64, 1222–1233.

Patlak, M. (1996). A testing deadline for endocrine disrupters. *Environ. Sci. Technol. News* 30, 542.

Pavlov, D. S. and Kasumyan, A. O. (2000). Patterns and mechanisms of schooling behaviour in fish: a review. *J. Ichthyol.* 40, 163–231.

Persson, P., Johannsson, S. H., Takagi, Y. and Björnsson, B. T. (1997). Estradiol-17β and nutritional status affect calcium balance, scale and bone resorption, and bone formation in rainbow trout, Oncorhynchus mykiss. *J. Comp. Phys. B* 167, 468–473.

Purdom, C. E., Hardiman, P. A., Bye, V. V. J., Eno, N. C., Tyler, C. R. and Sumpter, J. P. (1994). Estrogenic effects of effluents from sewage treatment works. *Chem. Ecol.* 8, 275–285.

Purkey, H. E., Palaninathan, S. K., Kent, K. C., Smith, C., Safe, S. H., Sacchettini, J. C. and Kelly, J. W. (2004). Hydroxylated polychlorinated biphenyls selectively bind transthyretin in blood and inhibit amyloidogenesis: rationalizing rodent PCB toxicity. *Chem. Biol.* 11, 1719–1728.

Rempel, M. A. and Schlenk, D. (2008). Effects of environmental estrogens and antiandrogens on endocrine function, gene regulation, and health in fish. *Int. Rev. Cell Mol. Biol.* 267, 207–252.

Reyhanian, N., Bolkova, K., Hallgren, S., Bollner, T., Olsson, P. E., Olsén, H. and Hällström, I. P. (2011). 17α-ethinyl estradiol affects anxiety and shoaling behaviour in adult male zebra fish (*Danio rerio*). *Aquat. Toxicol.* 105, 41–48.

Robertson, L. S., Iwanowicz, L. R. and Marranca, J. M. (2009). Identification of centrarchid hepcidins and evidence that 17β-estradiol disrupts constitutive expression of hepcidin-1 and inducible expression of hepcidin-2 in largemouth bass (*Micropterus salmoides*). *Fish Shell. Immunol.* 26, 898–907.

Rodgers-Gray, T. P., Jobling, S., Kelly, C., Morris, S., Brighty, G., Waldock, M. J., Sumpter, J. P. and Tyler, C. R. (2001). Exposure of juvenile roach (*Rutilus rutilus*) to treated sewage effluent induces dose-dependent and persistent disruption in gonadal duct development. *Environ. Sci. Technol.* 35, 462–470.

Saili, K. S., Corvi, M. M., Weber, D. N., Patel, A. U., Das, S. R., Przybyla, J., Anderson, K. A. and Tanguay, R. L. (2012). Neurodevelopmental low-dose bisphenol A exposure leads to early life-stage hyperactivity and learning deficits in adult zebrafish. *Toxicology* 291, 83–92.

Schlenk, D., Stresser, D. M., Rimoldi, J., Arcand, L., McCants, J., Nimrod, A. C. and Benson, W. H. (1998). Biotransformation and estrogenic activity of methoxychlor and its metabolites in channel catfish (*Ictalurus punctatus*). *Mar. Environ. Res.* 46, 159–162.

Scholz, S. and Gutzeit, H. O. (2000). 17-a-ethinylestradiol affects reproduction, sexual differentiation and aromatase gene expression of the medaka (*Oryzias latipes*). *Aquat. Toxicol.* 50, 363–373.

Schultz, I. R., Skillman, A., Nicolas, J., Cyr, D. G. and Nagler, J. J. (2003). Short-term exposure to 17b-ethynylestradiol decreases the fertility of sexually maturing male rainbow trout (*Oncorhynchus mykiss*). *Environ. Toxicol. Chem.* 22, 1272–1280.

Schwaiger, J., Speiser, O. H., Bauer, C., Ferling, H., Mallow, U., Kalbfus, W. and Negele, R. D. (2000). Chronic toxicity of nonylphenol and ethinylestradiol: Hematological and histopathological effects in juvenile common carp (*Cyprinus carpio*). *Aquat. Toxicol.* 51, 69–78.

Seo, J. S., Lee, Y. M., Jung, S. O., Kim, I. C., Yoon, Y. D. and Lee, J. S. (2006). Nonylphenol modulates expression of androgen receptor and estrogen receptor genes differently in different gender types of the hermaphroditic fish Rivulus marmoratus. *Biochem. Biophys. Res. Commun.* 346, 213–223.

Shelley, L. K., Ross, P. S., Miller, K. M., Kaukinen, K. H. and Kennedy, C. J. (2012). Toxicity of atrazine and nonylphenol in juvenile rainbow trout (*Oncorhynchus mykiss*): Effects on general health, disease susceptibility and gene expression. *Aquat. Toxicol.* 124–125, 217–226.

Shelley, L. K., Osachoff, H. L., van Aggelen, G. C., Ross, P. S. and Kennedy, C. J. (2013). Alteration of immune function endpoints and differential expression of estrogen receptor isoforms in leukocytes from 17β-estradiol exposed rainbow trout (*Oncorhynchus mykiss*). *Gen. Comp. Endocrinol.* 180, 24–32.

Shved, N., Berishvili, G., Hausermann, E., D'Cotta, H., Baroiller, J. F. and Eppler, E. (2009). Challenge with 17α-ethinylestradiol (EE2) during early development persistently impairs growth, differentiation, and local expression of IGF-I and IGF-II in immune organs of tilapia. *Fish Shell. Immunol.* 26, 524–530.

Snyder, S. A., Keith, T. L., Pierens, S. L., Snyder, E. M. and Giesy, J. P. (2001a). Bioconcentration of nonylphenol in fathead minnows (*Pimephales promelas*). *Chemosphere* 44, 1697–1702.

Snyder, S. A., Villeneuve, D. L., Snyder, E. M. and Giesy, J. P. (2001b). Identification and quantification of estrogen receptor agonists in wastewater effluents. *Environ. Sci. Technol.* 35, 3620–3625.

Sohoni, P. and Sumpter, J. P. (1998). Several environmental estrogens are also anti-androgens. *J. Endocrinol.* 158, 327–339.

Soto, A. M., Calabro, J. M., Prechtl, N. V., Vau, A. V., Orlando, E. F., Daxenberger, A., Kolok, A. S., Guillette, L. J., Jr., le Bizec, B., Lange, I. G. and Sonnenschein, C. (2004). Androgenic and estrogenic activity in water bodies receiving cattle feedlot effluent in Eastern Nebraska, USA. *Environ. Health Perspect.* 112, 346–352.

Sousa, A., Schonenberger, R., Jonkers, N., Suter, M., Tanabe, S. and Barroso, C. (2010). Chemical and biological characterization of estrogenicity in effluents from WWTPs in Ria de Aveiro (NW Portugal). *Arch. Environ. Contam. Toxicol.* 58, 1–8.

Spengler, P., Korner, W. and Metzger, J. W. (2001). Substances with estrogenic activity in effluents of sewage treatment plants in southwestern Germany. 1. Chemical analysis. *Environ. Toxicol. Chem.* 20, 2133–2141.

Staples, C. A., Dorn, P. B., Klecks, G. M., O'Block, S. T. and Harris, L. R. (1998). A review of the environmental fate, effects, and exposures of bisphenol A. *Chemosphere* 36, 2149–2173.

Stegeman, J. J. (1993). Cytochrome P450 forms in fish. In *Cytochrome P450, Handbook in Experimental Pharmacology* (eds. J. B. Schenkman and H. Greim), Vol. 105, pp. 279–310. Berlin: Springer-Verlag.

Sturm, A. and Segner, H. (2005). *P-glycoproteins and Xenobtiotic Efflux Transport in Fish Biochemistry and Molecular Biology of Fishes*, Vol. 6. Elsevier, B.V.

Swann, R. L., McCall, P. J. and Laskowski, D. A. (1981). Estimation of soil sorption constants of organic chemicals by high-performance liquid chromatography. *ASTM Spec. Tech. Pub.* 737, 43–48.

Takayanagi, S., Tokunaga, T., Liu, X., Okada, H., Matushima, A. and Shimohigashi, Y. (2006). Endocrine disruptor bisphenol A strongly binds to human estrogen-related receptor γ (ERRγ) with high constitutive activity. *Toxicol. Lett.* 167, 95–105.

Tarrant, A. M., Greytak, S. R., Callard, G. V. and Hahn, M. E. (2006). Estrogen receptor-related receptors in the killifish Fundulus heteroclitus: diversity, expression and estrogen responsiveness. *J. Mol. Endocrinol.* 37, 105–120.

Ternes, T. A., Stumpf, M., Mueller, J., Haberer, K., Wilken, R. D. and Servos, M. (1999). Behavior and occurrence of estrogens in municipal sewage treatment plants—I. Investigations in Germany, Canada and Brazil. *Sci. Total Environ.* 225, 81–90.

Thibault, R., Perdu, E. and Cravedi, J. P. (2001). Testosterone metabolism as a possible biomarker of exposure of fish to nonylphenol. *Rev. Med. Vet.* 153, 589–590.

Thibault, R. and Porte, C. (2004). Effects of endocrine disrupters on sex steroid synthesis and metabolism pathways in fish. *J. Ster. Biochem. Mol. Biol.* 92, 485–494.

Thilagam, H., Gopalakrishnan, S., Bo, J. and Wang, K. E. (2009). Effect of 17β-estradiol on the immunocompetence of Japanese sea bass (Lateolabrax japonicas). *Environ. Toxicol. Chem.* 28, 1722–1731.

Thomas, P. (2012). Rapid steroid hormone actions initiated at the cell surface and the receptors that mediate them with an emphasis on recent progress in fish models. *Gen. Comp. Endocrinol.* 175, 367–383.

Thorpe, K. L., Cummings, R. I., Hutchinson, T. H., Scholze, M., Brighty, G., Sumpter, J. P. and Tyler, C. R. (2003). Relative potencies and combination effects of steroidal estrogens in fish. *Environ. Sci. Technol.* 37, 1142–1149.

Thorpe, K. L., Gross-Sorokin, M., Johnson, I., Brighty, G. and Tyler, C. R. (2006). An assessment of the model of concentration addition for predicting the estrogenic activity of chemical mixtures in wastewater treatment works effluents. *Environ. Health Perspect.* 114, 90–97.

Thorpe, K. L., Benstead, R., Hutchinson, T. H. and Tyler, C. R. (2007). Associations between altered vitellogenin concentrations and adverse health effects in fathead minnow (*Pimephales promelas*). *Aquat. Toxicol.* 85, 176–183.

Tollefsen, K. E., Meys, J. F. A., Frydenlund, J. and Stenersen, J. (2002). Environmental estrogens interact with and modulate the properties of plasma sex steroid-binding proteins in juvenile Atlantic salmon (Salmo salar). *Mar. Environ. Res.* 54, 697–701.

Tollefsen, K. E. (2007). Binding of alkylphenols and alkylated non-phenolics to the rainbow trout (*Oncorhynchus mykiss*) plasma sex steroid-binding protein. *Ecotoxicol. Environ. Safe.* 68, 40–48.

Tremblay, L. and Van Der Kraak, G. (1999). Comparison between the effects of the phytosterol b-sitosterol and pulp and paper mill effluents on sexually immature rainbow trout. *Environ. Toxicol. Chem.* 18, 329–336.

Ueda, K., Taguchi, Y. and Morishima, M. (1997). How does P-glycoprotein recognize its substrates? *Semin. Cancer Biol.* 8, 151–159.

U.S. Environmental Protection Agency (EPA). (1989). Environmental Fate and Effects Division, Pesticide Environmental Fate One Line Summary: DDT (p,p'). Washington, DC.

U.S. Environmental Protection Agency (EPA). (2005). Aquatic Life Ambient Water Quality Criteria—Nonylphenol EPA-822-R-05-005, 1–96.

Vaccaro, E., Meucci, V., Intorre, L., Soldani, G., Di Bello, D., Longo, V., Gervasi, P. G. and Pretti, C. (2005). Effects of 17β-estradiol, 4-nonylphenol and PCB 126 on the estrogenic activity and phase 1 and 2 biotransformation enzymes in male sea bass (*Dicentrarchus labrax*). *Aquat. Toxicol.* 75, 293–305.

Vajda, A. M., Barber, L. B., Gray, J. L., Lopez, E. M., Woodling, J. D. and Norris, D. O. (2008). Reproductive disruption in fish downstream from an estrogenic wastewater effluent. *Environ. Sci. Technol.* 42, 3407–3414.

Vajda, A. M., Barber, L. B., Gray, J. L., Lopez, E. M., Bolden, A. M., Schoenfuss, H. L. and Norris, D. O. (2011). Demasculinization of male fish by wastewater treatment plant effluent. *Aquat. Toxicol.* 103, 213–221.

van Aerle, R., Pounds, N., Hutchinson, T., Maddix, S. and Tyler, C. (2002). Window of sensitivity for the estrogenic effects of ethinylestradiol in early life-stages of fathead minnow, Pimephales promelas. *Ecotoxicology* 11, 423–434.

Vetillard, A. and Bailhache, T. (2005). Cadmium: an endocrine disrupter that affects gene expression in the liver and brain of juvenile Rainbow trout. *Biol. Reprod.* 72, 119–126.

Volkova, K., Reyhanian, N., Kot-Wasik, A., Olsén, H., Hällström, I. P. and Hallgren, S. (2012). Brain circuit imprints of developmental 17α-ethinylestradiol exposure in guppies (*Poecilia reticulata*): Persistent effects on anxiety but not on reproductive behaviour. *Gen. Comp. Endocrinol.* 178, 282–290.

Vosges, M., Kah, O., Hinfray, N., Chadili, E., Page, Y. L., Combarnous, J. M. P. and Brion, F. (2012). 17α-ethinylestradiol and nonylphenol affect the development of forebrain GnRH neurons through an estrogen receptors-dependent pathway. *Reprod. Toxicol.* 33, 198–204.

Wang, R. and Belosevic, M. (1997). Estradiol increases susceptibility of goldfish to Trypanosoma danilewskyi. *Develop. Comp. Immunol.* 18, 337–387.

Wang, X. H. and Wang, W. X. (2005). Uptake, absorption efficiency and elimination of DDT in marine phytoplankton, copepods and fish. *Environ. Pollut.* 136, 453–464.

Wang, D. S., Senthilkumaran, B., Sudhakumari, C. C., Sakai, F., Matsuda, M., Kobayashi, T., Yoshikuni, M. and Nagahama, Y. (2005). Molecular cloning, gene expression and characterization of the third estrogen receptor of the Nile tilapia, Oreochromis niloticus. *Fish Physiol. Biochem.* 31, 255–266.

Wang, H., Wang, J., Wu, T., Qin, F., Hu, X., Wang, L. and Wang, Z. (2011). Molecular characterization of estrogen receptor genes in Gobiocypris rarus and their expression upon endocrine disrupting chemicals exposure in juveniles. *Aquat. Toxicol.* 101, 276–287.

Ward, A. J. W., Duff, A. J. and Currie, S. (2006). The effects of the endocrine disrupter 4-nonylphenol on the behaviour of juvenile rainbow trout (*Oncorhynchus mykiss*). *Can. J., Fish. Aquat. Sci.* 63, 377–382.

Waxman, D. J., LeBlanc, G. A., Morrissey, J. J., Staunton, J. and Lapenson, D. P. (1988). Adult male-specific and neonatally programmed rat hepatic P-450 forms RLM2 and 2a are not dependent on pulsatile plasma growth hormone for expression. *J. Biol. Chem.* 263, 11396–11406.

Wenger, M., Sattler, U., Goldschmidt-Clermont, E. and Segner, H. (2011). 17Beta-estradiol affects the response of complement components of rainbow trout (Oncorhynchus mykiss) challenged by bacterial infection. *Fish Shell. Immunol.* 31, 90–97.

Wenger, M., Krasnov, A., Skugor, S., Goldschmidt-Clermont, E., Sattler, U., Afanasyev, S. and Segner, H. (2012). Estrogen modulates hepatic gene expression and survival of rainbow trout infected with pathogenic bacteria Yersinia ruckeri. *Mar. Biotechnol.* 14, 530–543.

Williams, R. J., Churchley, J. H., Kanda, R. and Johnson, A. C. (2012). Comparing predicted against measured steroid estrogen concentrations and the associated risk in two United Kingdom river catchments. *Environ. Toxicol. Chem.* 31, 892–898.

Williams, T. D., Diab, A. M., George, S. G., Sabine, V. and Chipman, J. K. (2007). Gene expression responses of European flounder (*Platichthys flesus*) to 17β-estradiol. *Toxicol. Lett.* 168, 236–248.

Wong, C. S., Lau, F., Clark, M., Mabury, S. A. and Muir, D. C. G. (2002). Rainbow trout (*Oncorhynchus mykiss*) can eliminate chiral organochlorine compounds enantioselectively. *Environ. Sci. Technol.* 36, 1257–1262.

Wong, C. S., Mabury, S. A., Whittle, D. M., Backus, S. M., Teixeira, C., Devault, D. S., Bronte, C. R. and Muir, D. C. G. (2004). Organochlorine compounds in Lake Superior: chiral polychlorinated biphenyls and biotransformation in the aquatic food web. *Environ. Sci. Technol.* 38, 84–92.

Xia, Z., Gale, W., Chang, X., Langenau, D., Patino, R., Maule, A. G. and Densmore, L. D. (2000). Phylogenetic sequence analysis, recombinant expression and tissue distribution of a channel catfish estrogen receptor β. *Gen. Comp. Endocrinol.* 118, 139–149.

Yadetie, F. and Male, R. (2002). Effects of 4-nonylphenol on gene expression of pituitary hormones in juvenile Atlantic salmon (*Salmo salar*). *Aquat. Toxicol.* 58, 113–129.

Yan, Z., Lu, G. and He, J. (2012). Reciprocal inhibiting interactive mechanism between the estrogen receptor and aryl hydrocarbon receptor signaling pathways in goldfish (*Carassium auratus*) exposed to 17β-estradiol and benzo[a]pyrene. *Comp. Biochem. Physiol. C* 156, 17–23.

Yeh, G. C., Lopaczynska, J., Poore, C. M. and Phang, J. M. (1992). A new functional role for P-glycoprotein: efflux pump for benzo(a)pyrene in human breast cancer MCF-cells. *Cancer Res.* 52, 6692–6695.

Yoshimura, J., Yonemoto, Y., Yamada, H., Koga, M., Oguri, K. and Saeki, S. (1987). Metabolism in vivo of 3,4,3',4'-tetrachlorobiphenyl and toxicological assessment of the metabolites in rats. *Xenobiotica* 17, 897–910.

Zha, J., Wang, Z., Wang, N. and Ingersoll, C. (2007). Histological alternation and vitellogenin induction in adult rare minnow (Gobiocypris rarus) after exposure to ethynylestradiol and nonylphenol. *Chemosphere* 66, 488–495.

Zimniak, P. and Waxman, D. J. (1993). Liver cytochrome P450 metabolism of endogenous steroid hormones, bile acids and fatty acids. In *Cytochrome P450; Handbook in Experimental Pharmacology* (eds. J. B. Schenkman and H. Greim), Vol. 105, pp. 125–144. Berlin, Germany: Springer-Verlag.

6

INSECTICIDE TOXICITY IN FISH

M.H. FULTON

P.B. KEY

M.E. DELORENZO

Organic Chemical Toxicology of Fishes: Volume 33
FISH PHYSIOLOGY

1. ABSTRACT/INTRODUCTION

Insecticides are unique among organic pollutants because they are designed to cause toxicity in target organisms. Since many of their target sites are largely conserved across many taxa, they have the potential to cause toxicity in nontarget organisms including fish. Natural insecticides were used as early as 4500 years ago, and synthetic insecticide development began in the late nineteenth and early twentieth centuries. The first major class of synthetic insecticides to be developed was the organochlorines (OCs). Most of these compounds were highly persistent, bioaccumulative, and highly toxic to fish. Most have now been banned in most parts of the world. The next major class to appear was the organophosphates (OPs). These compounds were first developed during World War II, and their production and use expanded throughout most of the twentieth century. They are much less persistent than the OCs, but many are highly toxic to fish and other nontarget organisms including humans. Their use began to decline in the late twentieth century because of these concerns. Carbamate insecticides were first marketed in the 1950s, and they continue to be extensively used on crops, lawns, and gardens. Like the OPs, carbamates are relatively short-lived and do not bioaccumulate. Generally, they are less toxic to fish than either the OCs or OPs. Synthetic pyrethroid insecticides were first developed in the 1940s, and their utilization has continued to increase as they are used as replacements for the OCs and OPs.

Of the currently used insecticides, the pyrethroids have the greatest acute toxicity to fish; however, pyrethroids generally degrade rapidly in the water column and have a high affinity for sediment binding, which reduces their bioavailability to fish. Neonicotinoid insecticides were first marketed in the 1980s, and like the synthetic pyrethroids their use is increasing. They degrade rapidly in water, and their toxicity to fish is generally low. Fipronil was first marketed in 1996. It is presently the only phenylpyrazole in use. It is highly toxic to fish, and some of its degradation products are more toxic than the parent. More than 1500 microbial products have been identified as possible insecticides. In general, microbial insecticides have very low toxicities to fish and degrade rapidly in the aquatic environment. Insect growth regulators (IGRs) are insecticides that inhibit the life cycle of insects. This class of insecticides was first marketed in the 1970s. They generally have low toxicity in fish. Most contemporary-use insecticides have been developed to be more selective and less persistent, and thus they have lower toxicity to fish than their precursors. Since many of these modern insecticides are highly toxic to aquatic invertebrates, however, there is a need for more research to examine possible indirect effects on fish populations. There is also a need to better characterize possible sublethal effects in fish as well as to determine how environmental variables such as temperature and salinity may modify toxicity. Understanding the influence of these factors on the bioavailability, uptake, and toxicity may be particularly important under predicted climate change scenarios.

2. BACKGROUND

The use of chemicals with insecticidal properties has led to vast improvements in agricultural yields and better quality of life through pest control. Insecticides are pollutants of unique concern because the majority of their mechanisms of toxicity in insect pests overlap with those of nontarget organisms. Thus the regulation of insecticides must balance the benefits of pest management with both ecological and human health risks. In this chapter we provide an overview of the available fish toxicity information for the most important insecticide classes. For each chemical class, we discuss the history of use; describe target sites and modes of action; characterize properties of uptake and metabolism; and present available information on acute, chronic, and sublethal toxicity. Finally, we discuss data gaps and priorities for future research.

2.1. Early History of Insecticide Use

From its earliest history about 10,000 years ago in the fertile crescent of Mesopotamia, agriculture suffered from pests and diseases. The earliest known use of insecticides was by the Sumerians about 4500 years ago who used sulfur to control insects and mites. About 3000 years ago the Chinese began using arsenic and mercury compounds to control body lice (Ecobichon, 1996). Pyrethrum, an extract derived from the dried flowers of *Chrysanthemum cineraiaefolium*, has been used as an insecticide for more than 2000 years (Ecobichon, 1996). Modern synthetic insecticide classes include organochlorines (OCs), organophosphates (OPs), carbamates, synthetic pyrethroids, neonicotinoids, phenylpyrazoles, microbial insecticides, and insect growth regulators (Table 6.1).

2.2. Nontarget Effects of Insecticides on Fish

Insecticides can reach surface waters as a result of spray drift, runoff, or intentional or accidental misapplication. Due to conservation across many systems (e.g., neurological targets), fish can suffer negative effects from insecticide exposure that are similar to effects in target insects (Table 6.1). If concentrations are sufficient, fish kills may result. At lower concentrations, more sensitive invertebrates may be impacted, thus reducing the food available for fish consumption (Pimental, 2005). Pesticides are reported to kill 6–14 million fish annually in the United States (Pimental, 1992). In addition to mortality, insecticide exposure may cause sublethal effects in fish. Insecticides have been shown to impair fish growth, behavior, reproduction, and immune function. The nature of these effects varies widely and is dependent on the species and life stage of fish exposed, as well as the type of insecticide involved.

3. INSECTICIDE CLASSES

3.1. Organochlorines

Organochlorines (OCs) include DDT and its analogues; cyclodienes; lindane (benzene hexachloride-BHC); toxaphene and similar chemicals; and

Table 6.1

Review of the insecticide classes discussed in this chapter including the year the first active ingredient was introduced, current number of active ingredients registered, and mode or modes of action

Class	Year Commercially Introduced in United States	Current Number of Active Ingredients Registered in United States[*]	Mode(s) of Action
Organochlorine	1945 (DDT)	3	Deactivates sodium gates of the nerve axon. Inhibits chloride flux resulting in partial repolarization of the neuron.
Organophosphate	1947 (TEPP)	27	Inhibition of cholinesterase enzymes of the central nervous system.
Carbamate	1953 (carbaryl)	15	Reversible inhibition of the AChE enzyme at parasympathetic junctions, skeletal muscle myoneural junctions, autonomic ganglia, and central nervous system.
Pyrethroid	1949 (allethrin)	20	Hyperstimulation of the nervous system due to the blocking of sodium ion movement through the nerve membrane.
Neonicotinoid	1994 (imidacloprid)	7	Irreversible blocking of acetylcholine receptors.
Phenylpyrazole	1996 (fipronil)	1	Disruption of the passage of chloride ions through the γ-aminobutyric acid (GABA)-regulated chloride channel
Microbial	1998 (Bti)	8	Synergistic interaction of insecticidal crystal proteins which dissolve, then release the endotoxin that binds to and lyses midgut epithelial cells.
Insect Growth Regulator	1975 (methoprene)	10	Interferes with sexual maturation. Prevents the formation of chitin.

[*]U.S. EPA (http://www.epa.gov/pesticides/index.htm; accessed February 14, 2013)

the caged structures (mirex and chlordecone) (Smith, 1991). Although similar in structure and bioeffects, OCs vary widely in their toxicity and bioaccumulation potential.

3.1.1. DDT

DDT was synthesized in 1874, but its insecticidal properties were not recognized until 1939 by Paul Muller, who was awarded a Nobel Prize for this discovery in 1948 (Ecobichon, 1996). It was first used during World War II to control vector-borne diseases such as malaria, typhus, and dengue fever. From the 1950s through the 1970s DDT was widely used in agriculture around the world, with more than 36,287 metric tons applied annually. In the United States, usage peaked in 1962 at 77 million kg (ATSDR, 2002). Concerns arose about the persistence and potential ecological effects of DDT and other persistent chlorinated insecticides. These concerns culminated in 1962 with the publication of Rachel Carson's *Silent Spring*. In 1972 the EPA announced the cancellation of most uses of DDT; however, an exemption remains in effect for emergency uses related to human health (ATSDR, 2002). The Stockholm Convention (an international treaty) in 2004 outlawed several persistent organic compounds and restricted the use of DDT to vector control (van den Berg, 2009).

3.1.2. CYCLODIENES

Examples of cyclodiene insecticides include chlordane, heptachlor, aldrin, dieldrin, endrin, and endosulfan. Most of the cyclodienes were developed in the 1940s and 1950s and were widely used in agriculture and for disease vector and termite control through the early 1970s (Smith, 1991). Because of their persistence and high propensity for bioaccumulation, most cyclodienes have been banned in most developed countries. Some of the cyclodienes such as chlordane and endosulfan continue to be widely used in China, India, and parts of Africa. Endosulfan is currently being phased out in the United States, with the EPA terminating all uses by 2016 (U.S. EPA, 2012a).

3.1.3. LINDANE

The insecticidal properties of lindane were first discovered in the 1940s. It has been widely used in agriculture, for disease vector control, and as a pharmaceutical treatment for lice and scabies. In 1997 alone, an estimated 425,016 kg of lindane were used in U.S. agriculture (U.S. EPA, 2002). In the United States, all agricultural uses of lindane were canceled by 2007 (U.S. EPA, 2006a). Lindane medications for the control of lice and scabies are still available in the United States, but they are now designated as "second-line" treatments and can only be used when other treatments have failed (U.S. EPA, 2006a).

3.1.4. TOXAPHENE

Toxaphene was first introduced as an insecticide in the 1940s. It was used extensively on cotton and soybeans as well as to control insect pests on cattle, sheep, goats, and swine (Smith, 1991). In the early to mid-1970s, it was the most widely used insecticide in the United States (NTP, 2011). Because of its nontarget toxicity, persistence, and bioaccumulative potential, it has now been banned in many countries. It was banned for all uses in the United States in 1990 (NTP, 2011).

3.1.5. CAGED STRUCTURES (MIREX)

Mirex was first introduced as an insecticide in 1959. Its main use was to control the imported fire ant in the Southeastern United States. Between 1962 and 1975, approximately 250,000 kg were applied to fields, primarily as a bait consisting of mirex mixed with soybean oil and corncob grits (U.S. EPA, 1979). Because of its persistence and potential for bioaccumulation, all uses were canceled in the United States in 1978 (U.S. EPA, 1979).

3.2. Structure

The OCs are grouped into three distinct chemical classes based on structure: dichlorodiphenylethanes, chlorinated cyclodienes, and chlorinated benzenes/cyclohexanes (Ecobichon, 2001). The dichlorodiphenylethanes include DDT and at least 10 analogues, including DDD, methoxychlor, dicofol, chlorfenethol, ethylan, and chlorobenzilate. This group has very diverse properties ranging from the well-known environmental persistence and bioconcentration of DDT to the rapid metabolism and negligible persistence of methoxychlor (Blus, 1995; Ecobichon, 2001). This class is denoted by two *p*-substituted phenyl rings connected by a monosubstituted methylene bridge (Figure 6.1). The *para* substituents must be nonpolar and of relatively low molecular volume to allow penetration of the insect cuticle and the nerve sheath. Any additional or alternative groups will result in lower insecticidal activity. The DDT analogues are created in part by the substituents attached to the methylene bridge. These are normally slightly polar nitro groups that contribute to the toxicity of the analogue (Coats, 1990).

The cyclodienes and related compounds include aldrin, dieldrin, isodrin, endrin, chlordane, heptachlor, and endosulfan. All have the presence of a hexachlorocyclopentadiene ring, which contributes to their toxicity. This toxicity varies based on different moieties attached to the ring and different stereochemical positions. In terms of acute effects, the insecticides in this

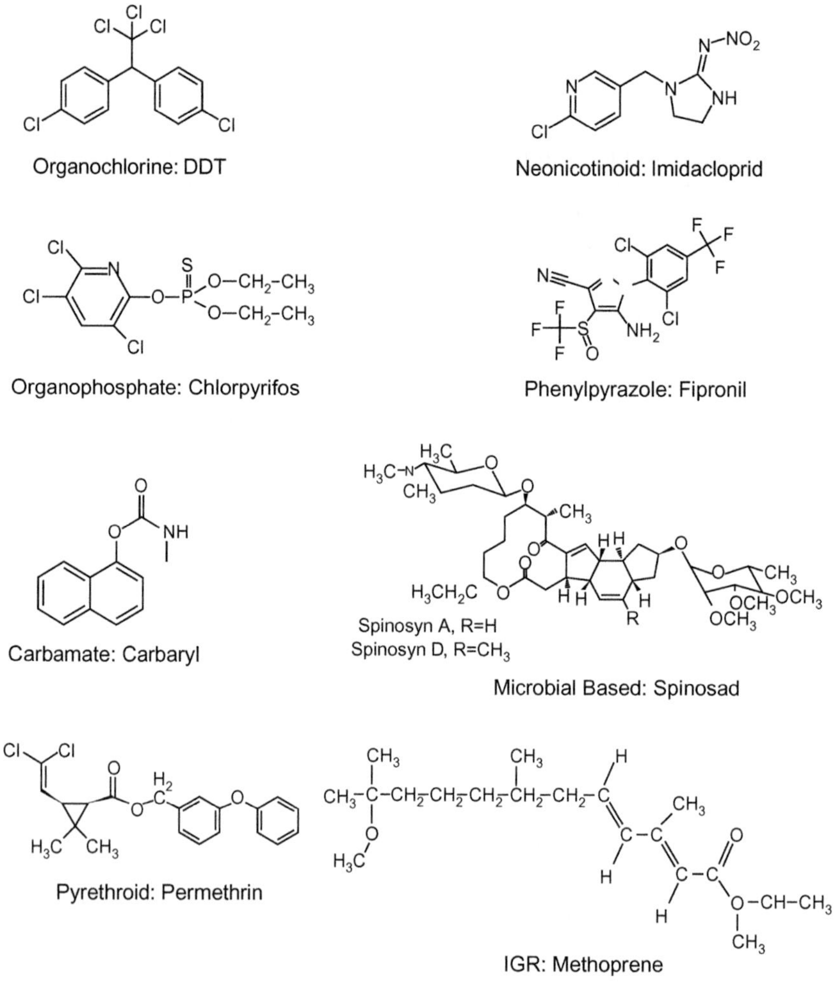

Organochlorine: DDT

Neonicotinoid: Imidacloprid

Organophosphate: Chlorpyrifos

Phenylpyrazole: Fipronil

Carbamate: Carbaryl

Spinosyn A, R=H
Spinosyn D, R=CH$_3$

Microbial Based: Spinosad

Pyrethroid: Permethrin

IGR: Methoprene

Figure 6.1. Chemical structures of representative compounds in each insecticide class.

group are considered the most toxic of the OCs. Endrin and isodrin are stereoisomers of aldrin. A double bond in the ring increases the toxicity while conversion to an epoxide product, dieldrin, slightly decreases toxicity. Endosulfan has two isomers (α and β) and a metabolite (endosulfan sulfate) with insecticidal properties. Chlordane is a mixture of 14 components and, similar to endosulfan, contains α and β isomers that determine the toxicity of the insecticide. Heptachlor's toxicity is determined by its conversion to an epoxide after an oxidation reaction (Kaushik and Kaushik, 2007).

The chlorinated benzenes/cyclohexanes consist of hexachlorocyclohexane (HCH) and its eight stereoisomers which are as diverse in properties as the dichlorodiphenylethanes. The most well-known HCH is the γ isomer lindane (Blus, 1995; Ecobichon, 2001). Lindane is the only isomer of this group with pronounced insecticidal properties (Kaushik and Kaushik, 2007).

3.3. Mechanism of Action

Organochlorines have two basic modes of action: effects on ion permeability as found with DDT and its analogues; and effects on nerve receptors as found with cyclodienes and lindane. All OC compounds are absorbed by an organism through the respiratory system or a dermal route (Blus, 1995). DDT insecticides act on the peripheral nervous system by preventing the deactivation of the sodium gates of the nerve axon. This results in sodium ions continuing to leak through the nerve membrane leading to repetitive discharges in the neuron (Coats, 1990).

Martyniuk et al. (2011) injected methoxychlor, a DDT analogue, into largemouth bass (*Micropterus salmoides*) to characterize its mode of action. They showed that this OC elicits transcriptional effects through the estrogen receptor as well as androgen receptor-mediated pathways in the liver. The authors concluded that methoxychlor has estrogenic, anti-estrogenic, and anti-androgenic activities in fish.

The cyclodiene mechanism of action is localized more within the central nervous system. This group binds to the picrotoxinin site in the γ-aminobutyric acid (GABA) chloride ionophore complex. The binding inhibits chloride flux into the nerve, resulting in partial repolarization of the neuron leading to hyperexcitation (Coats, 1990; Kaushik and Kaushik, 2007). The lindane mechanism is similar to that of the cyclodienes. It acts by blocking the chloride flux through the GABA receptors, resulting in hyperexcitation. Lindane can also alter calcium homeostasis by elevating free calcium ions with the release of neurotransmitters (Ecobichon, 2001).

3.4. Uptake and Metabolism

All OCs have characteristics of high solubility in lipids, low solubility in water, and persistence in the environment for months to years. Heptachlor and aldrin are rapidly metabolized by organisms, but their metabolites, heptachlor epoxide and dieldrin, are just as toxic and less susceptible to degradation. Endrin has a relatively short half-life in organisms, but its metabolite, 12-ketoendrin, is highly toxic (Coats, 1990; Blus, 1995).

Resistance by insects to DDT was documented soon after it was marketed (Blus, 1995). Resistance of fish to normally lethal doses of OCs was commonly reported in the 1960s. Johnson (1968) summarized this research, which dealt mainly with mosquitofish (*Gambusia affinis*) resistance to DDT and cyclodiene compounds. Mosquitofish displayed a greater tolerance for cyclodienes than DDT (Johnson, 1968). In addition to resistance being explained by heritable genetic variation, other factors to be considered are the rate of uptake, rate of detoxification, efficiency of binding to membranes, and lipid content of the exposed fish (Murty, 1986). In-depth research of resistance of fish to OCs can be found in Dustman and Stickel (1969), Murty (1986), and Andreasen (1985) among others.

3.5. Acute Toxicity

Among the OCs, the cyclodienes such as endosulfan, endrin, and dieldrin are usually the most acutely toxic to fish (Table 6.2). Johnson (1968) summarized OC acute toxicity data that was published during the 1950s. The most toxic OC was endrin, which was consistently two orders of magnitude more toxic than DDT and one order of magnitude more toxic than dieldrin to the range of fish species tested. The three saltwater fish species listed (Atlantic salmon [*Salmo salar*], Coho salmon [*Oncorhynchus kisutch*], and northern puffer [*Sphaeroides maculatus*]) displayed similar sensitivities to the freshwater species (Johnson, 1968).

Beginning in the mid-1960s, the EPA's Environmental Research Laboratory in Gulf Breeze, Florida, investigated the effects of 15 OCs in multiple saltwater species. Among the species tested, none were more sensitive than the others; but two OCs, endosulfan and endrin, were consistently more toxic (Mayer, 1987).

Schimmel et al. (1977) examined the effects of the cyclodiene, endosulfan, on juveniles from three saltwater fish species. The most sensitive was spot (*Leiostomus xanthurus*) with a 96 h LC50 of 0.14 µg/L followed by pinfish (*Lagodon rhomboides*) with a 96 h LC50 of 0.44 µg/L and striped mullet (*Mugil cephalus*) with a 96 h LC50 of 0.51 µg/L. Another saltwater fish species tested for endosulfan toxicity was the mummichog (*Fundulus heteroclitus*). A 96 h LC50 of 0.25 µg/L was reported, which is comparable to the previously mentioned values for other saltwater species (Scott et al., 1987). In an estuarine mesocosm, the 96 h LC50 for mummichog exposed to endosulfan was higher at 0.64 µg/L compared to acute laboratory toxicity tests (Pennington et al., 2004). When comparing results with those of freshwater fish species exposed to endosulfan, Schimmel et al. (1977) reported that sensitivity was lower in the freshwater fish with 48 h LC50s ranging from 1.0 to 10.0 µg/L for goldfish (*Carassius auratus*) and guppy

Table 6.2
Acute toxicity values (µg/L) for fish by compound. Values derived from sources cited within text and U.S. EPA EcotoxicologyDatabase (http://cfpub.epa.gov/ecotox/advanced_query.htm; accessed February 12, 2013)

Insecticide Class	Compound	Lowest LC50	Highest LC50	Number of Species
carbamate	aldicarb	41.00	1370.00	6
carbamate	bendiocarb	470.00	1650.00	3
carbamate	carbaryl	250.00	20,000.00	22
carbamate	carbofuran	33.00	872.00	12
carbamate	isoprocarb	3500.00	12,000.00	4
carbamate	methiocarb	190.00	750.00	3
carbamate	methomyl	300.00	4050.00	9
carbamate	oxyamyl	2600.00	27,500.00	5
carbamate	propoxur	1000.00	50,000.00	7
carbamate	thiodicarb	470.00	4450.00	5
carbamate	trimethacarb	1000.00	4700.00	3
insect growth regulator	buprofezin	260.00	260.00	2
insect growth regulator	diflubenzuron	13.00	3,89,000.00	11
insect growth regulator	fenoxycarb	1070.00	2700.00	4
insect growth regulator	halofenozide	8400.00	8800.00	3
insect growth regulator	methoprene	1010.00	1,25,000.00	4
insect growth regulator	methoxyfenozide	2800.00	4300.00	3
insect growth regulator	pyriproxyfen	270.00	325.00	3
insect growth regulator	tebufenozide	720.00	5380.00	3
microbial insecticide	Bti	656.00	9,80,000.00	4
microbial insecticide	spinosad	4990.00	30,000.00	4
neonicotinoid	clothianidin	93,600.00	1,17,000.00	3
neonicotinoid	dinotefuran	99,300.00	1,09,000.00	3
neonicotinoid	imidacloprid	1,05,000.00	2,41,000.00	4
neonicotinoid	thiamethoxam	1,00,000.00	1,14,000.00	3
organochlorine	aldrin	2.60	6.20	2
organochlorine	chlordane	12.55	42.00	3
organochlorine	DDT	0.30	20.30	24
organochlorine	dieldrin	1.20	10.00	12
organochlorine	endosulfan	0.01	10.00	8
organochlorine	endrin	0.09	5.60	18
organochlorine	heptachlor	0.85	63.00	12
organochlorine	kepone	6.60	340.00	9
organochlorine	lindane	1.70	190.00	18
organochlorine	methoxychlor	1.70	76.00	15
organochlorine	mirex	23.00	100.00	7
organochlorine	toxaphene	0.53	18.00	17
organophosphate	acephate	1000.00	32,00,000.00	14
organophosphate	azinphosmethyl	1.20	4270.00	20
organophosphate	chlorethoxyfos	1.80	89.00	3
organophosphate	chlorpyrifos	0.42	1000.00	20
organophosphate	coumaphos	280.00	1100.00	8
organophosphate	demeton	42.00	14,300.00	12

(*Continued*)

Table 6.2 (continued)

Insecticide Class	Compound	Lowest LC50	Highest LC50	Number of Species
organophosphate	diazinon	90.00	7800.00	9
organophosphate	dichlorvos	100.00	11,600.00	12
organophosphate	dicrotophos	6300.00	83,800.00	5
organophosphate	dimethoate	6000.00	1,80,000.00	5
organophosphate	disulfoton	8.20	7200.00	8
organophosphate	ethion	49.00	7600.00	10
organophosphate	fenamiphos	4.50	68.00	3
organophosphate	fenitrothion	330.00	12,200.00	15
organophosphate	fenthion	453.00	2780.00	19
organophosphate	fonofos	6.80	240.00	4
organophosphate	isofenphos	1400.00	2100.00	4
organophosphate	malathion	4.00	11,700.00	22
organophosphate	methamidophos	1280.00	68,000.00	4
organophosphate	methidathion	2.20	32.00	5
organophosphate	methyl parathion	2750.00	12,000.00	3
organophosphate	mevinphos	11.90	810.00	5
organophosphate	moncrotophos	4930.00	50,000.00	4
organophosphate	naled	87.00	3300.00	8
organophosphate	parathion	17.80	2650.00	15
organophosphate	phorate	0.36	160.40	10
organophosphate	phosmet	22.00	7500.00	10
organophosphate	phosphamidon	1000.00	1,00,000.00	7
organophosphate	phostebupirim	7.70	2200.00	2
organophosphate	profenofos	7.70	800.00	8
organophosphate	propetamphos	190.00	940.00	2
organophosphate	temephos	158.00	34,100.00	16
organophosphate	terbufos	0.77	150.00	6
organophosphate	trichlorfon	230.00	7900.00	12
phenylpyrazole	fipronil	42.00	430.00	6
pyrethroid	allethrin	4.00	80.00	6
pyrethroid	bifenthrin	0.15	17.50	3
pyrethroid	cyfluthrin	0.30	4.05	3
pyrethroid	cypermethrin	0.20	2.20	8
pyrethroid	deltamethrin	0.17	1.97	6
pyrethroid	esfenvalerate	0.07	1.00	6
pyrethroid	etofenprox	2.00	13.00	5
pyrethroid	fenpropathrin	1.95	5.50	5
pyrethroid	fenvalerate	0.18	5.40	11
pyrethroid	flucythrinate	0.22	1.60	6
pyrethroid	fluvalinate	0.90	10.80	4
pyrethroid	λ-cyhalothrin	0.15	0.70	6
pyrethroid	permethrin	0.62	75.00	18
pyrethroid	phenothrin	1.40	94.20	3
pyrethroid	resmethrin	0.28	11.00	14
pyrethroid	tefluthrin	0.06	0.13	3
pyrethroid	tetramethrin	3.70	21.00	2
pyrethroid	tralomethrin	1.60	2.80	3

(*Lebistes reticulatus*). The freshwater Florida flagfish (*Jordanella floridae*) was also less sensitive than saltwater species. Larval flagfish exposure to endosulfan resulted in a 96 h LC50 of 4.35 μg/L (Beyger et al., 2012).

Both Schimmel et al. (1977) and Carriger and Rand (2008) commented on the toxicity of the endosulfan metabolite, endosulfan sulfate, and concluded that for both freshwater and saltwater fish the metabolite was equally as toxic as the parent insecticide. Carriger et al. (2011) exposed four freshwater fish species to endosulfan sulfate for 96 h. Toxicities were similar, ranging from 2.06 μg/L for least killifish (*Heterandria formosa*) to 3.5 μg/L for sailfin molly (*Poecilia latipinna*). Compared with endosulfan, Carriger et al. (2011) concluded that endosulfan sulfate has similar toxicity to the parent compound.

Acute toxicity to OCs can also be dependent on the age or life stage. In general, eggs and larvae with yolk sacs are less sensitive than the juveniles. As age increases, toxicity once again decreases (Murty, 1986). Carp (*Cyprinus carpio*) sensitivity to endrin ranged from a 24 h LC50 of 19.9 mg/kg for eggs down to 0.061 mg/kg for 6 d old larvae (Iyatomi et al., 1958). Chlordane at 17 and 36 μg/L did not affect the eggs and embryos of sheepshead minnows (*Cyprinodon variegatus*), but did kill significant numbers of fry (Parrish et al., 1976). Anadu et al. (1999) reported a significantly lower 96 h LC50 of 11.3 μg/L in 2 d old mummichog larvae after exposure to DDT as compared to the 96 h LC50 of 20.3 μg/L in adults.

3.6. Chronic Toxicity

Beyger et al. (2012) exposed Florida flagfish to pulse exposures of endosulfan for one complete life cycle (~170 d), resulting in significant mortality at the highest concentration of 10.8 μg/L. There were no significant differences seen in sublethal parameters measured (growth, reproductive capacity, and hatchability). Toxaphene was fed to Arctic char (*Salvelinus alpinus*) for 104 d. Researchers saw decreased growth rate, muscle lipid, and protein content in the char, with no increase in parasite infections (Blanar et al., 2005).

Immune function of rainbow trout (*Oncorhynchus mykiss*) was studied after exposure to two levels of lindane and a high and low dose of vitamin C. One month after exposure to 10 mg/kg lindane, proliferation of B lymphocytes in fish decreased, but no effect was seen at 50 mg/kg lindane. However, vitamin C stimulated B lymphocytes proliferation at the higher lindane exposure (Dunier et al., 1995). Carp exposed to lindane at 0.038 μg/L for 25 d showed decreased red blood cell counts and increased white blood cell counts. Plasma glucose levels increased while plasma protein levels decreased (Saravanan et al., 2011). Jarvinen et al. (1988) observed reduced growth in larval fathead minnows (*Pimephales promelas*) exposed to endrin for 30 d.

3.7. Sublethal Effects/Biomarkers

As previously stated, organochlorines are present in fish food chains due to their high lipid solubility. Cope (1966) reviewed research from the early 1960s dealing with OC effects on fish in terms of behavior, growth, reproduction, pathology, and resistance. The majority of these studies focused on DDT. Two reports observed a lateral curvature of the spine (scoliosis) in sheepshead minnows exposed to kepone (Couch et al., 1977; Hansen et al., 1977). The induction of scoliosis with kepone was dose and time dependent, with significant effects occurring after one d at 24 μg/L and after 17 d at 4 μg/L. It was not until the early 1970s that Janicki and Kinter (1971) determined that DDT was a sodium channel blocker in a set of experiments with eels (*Anguilla rostrata*). Later, Murty (1986) reviewed the literature concerning OC sublethal effects on fish. Organochlorine-induced enzymatic changes were reported in sailfin mollies from dieldrin, in Asian stinging catfish (*Heteropneustes fossilis*) from chlordane, and in channel catfish (*Ictalurus punctatus*) from mirex (Murty, 1986). Other research has shown changes in lipid, protein, and glycogen content in fish; effects on fish endocrine systems; hematological changes; and detrimental effects on respiration, growth, and reproduction (Murty, 1986).

Scott et al. (1987) exposed mummichogs to a sublethal concentration of endosulfan (0.60 μg/L) at high and low salinities and measured respiration and ammonia levels. After 96 h, exposed fish at both salinities had significantly higher respiration rates. After 168 h of depuration, respiration rates were still significantly higher. Ammonia excretion rates were significantly reduced after 96 h in both scenarios. These levels returned to normal after depuration. Brown et al. (2004) reviewed the effects of contaminants on the fish thyroid. OCs, including mirex, DDT, endrin, endosulfan, and lindane, were found to alter thyroid function. After reviewing 19 studies, the authors concluded that both short and long term exposure to high doses of OCs altered fish thyroid activity and impaired thyroid hormone synthesis (Brown et al., 2004). Saravanan et al. (2011) exposed carp to 38 μg/L lindane (1/10th of its 24 h LC50 value) for 25 d. Red blood cell counts were significantly depressed, while leucocyte count was significantly elevated throughout the 25 d exposure. Hemoglobin and hematocrit were also significantly depressed, but after the 10th d, significant recovery occurred. Glycogen levels were mixed after 25 d exposure (Saravanan et al., 2011). Juvenile walking catfish (*Clarias batrachus*) were exposed to 2.5 μg/L of endosulfan for 50 d. This exposure resulted in a decreased testicular differentiation in the juvenile fish and decreased the expression of several genes, including testis-related transcription factors (Rajakumar et al., 2012).

Research reviewed by Martyniuk et al. (2011) showed that methoxychlor negatively impacted normal reproductive processes in fathead minnows, along with an induction of plasma vitellogenin in males at a concentration of 5.0 µg/L methoxychlor. Following exposure to 2.5 µg/L methoxychlor, male sheepshead minnows showed an induction of vitellogenin (Martyniuk et al., 2011).

3.8. Field Studies

Field studies with OCs began in the 1940s to determine the best application rates that would control insect pests without detrimental wildlife effects (Cottam and Higgins, 1946). By the 1960s, interest in OC occurrence in water bodies, sediment, and fish tissue was growing, with many published reports reviewed by Johnson (1968).

By the 1990s most OC insecticides were no longer in use, but environmental concentrations were still posing threats as sediments remained a reservoir for these chemicals. Olivares et al. (2010) sampled carp and sediment from the Ebro River in Spain. DDT and HCH metabolites were detected in sediment up to 1300 ng/g along with high levels of dioxins and PCBs. Carp were measured for condition factor, hepatosomatic index, micronuclei index, EROD, and Cyp1A gene expression. All measurements were significant for fish sampled from the most polluted site as compared to the least polluted site (Olivares et al., 2010).

Other sampling efforts have documented the still common occurrence of DDT metabolites and other OCs (chlordane compounds, dieldrin, and toxaphene) in U.S. fish tissues (Munn and Gruber, 1997; Loganathan et al., 2001; Schmitt et al., 2005; Karouna-Renier et al., 2011). Toxicity thresholds, however, were rarely exceeded (Loganathan et al., 2001; Schmitt et al., 2005; Karouna-Renier et al., 2011).

OCs also persist in fish tissues worldwide. For example, Berg et al. (1997) found low to moderate OC levels off the coast of Greenland, and fish in coastal Argentina were contaminated with endosulfan sulfate, chlordane, HCH, and DDT metabolites (Lanfranchi et al., 2006).

4. ORGANOPHOSPHATES

The first OP to be commercially produced was tetraethylpyrophosphate (TEPP). It was developed during World War II as a replacement for nicotine, a naturally occurring insecticide used since the late seventeenth century. Other extremely toxic OPs were developed for military purposes, including isopropylmethylphosphonofluoridate (sarin) and ethyl-N-dimethylphosphoroamidocyanidate (tabun). These compounds and others later became known as nerve gases (Murphy, 1980).

Because OPs were much less persistent than OCs, their manufacture and use increased as OC insecticide use was phased out. Widely used OP insecticides have included parathion, malathion, chlorpyrifos, azinphosmethyl, acephate, and diazinon. OPs have been used in agriculture, in disease vector control (e.g., mosquitoes), and in the control of residential and household insect pests. Near the end of the twentieth century, increasing concerns were raised about the potential for OP insecticides to pose a significant human health risk, especially in young children and developing fetuses. The EPA has now banned the use of many OPs in residential settings; however, OPs continue to be extensively used in agriculture and for mosquito control (U.S. EPA, 2011a). Overall, the use of OPs declined by 60% between 1990 and 2007; from 85 million pounds in 1990 to 33 million pounds in 2007 (U.S. EPA, 2011a).

4.1. Structure

OPs are esters, amides or thiol derivatives of either phosphoric acid or thiophosphoric acid (Storm, 2012). The three basic classes of OPs are phosphorothionate esters, phosphorodithioate esters, and phosphate esters. The phosphorothionate esters have a $P = S$ (thiono) group with a general formula of $(RO)_3PS$, where R is a hydrocarbon or substituted hydrocarbon moiety. Typical phosphorothionate esters are diazinon and parathion. The majority of OPs now in use are the phosphorodithioate esters also with the thiono group but with a general formula of $(RO)_2PS \cdot SR$ (Figure 6.1). These are represented by azinphosmethyl, chlorpyrifos, and malathion.

The phosphorothionate and phosphorodithioate esters are also grouped together and are known as thion-type compounds. The phosphate esters do not contain sulfur but have a $P = O$ functional group. These are the most toxic OP insecticides as represented by paraoxon (active metabolite of parathion), mevinphos, and dichlorvos (Manahan, 1992). This group is also known as the oxon-type compounds.

4.2. Mechanism of Action

The toxicity of OPs results from the inhibition of certain cholinesterase enzymes of the nervous system. In vertebrates, cholinesterases act by removing the neurotransmitter, acetylcholine (ACh), from the synaptic cleft through hydrolysis (Habig and DiGiulio, 1991). ACh acts as an excitatory transmitter for voluntary muscle in the somatic nervous system. ACh also serves as both a preganglionic and postganglionic transmitter in the parasympathetic system and as a preganglionic transmitter in the sympathetic system. In critical regions of the central nervous system, ACh serves as an excitatory transmitter. When cholinesterases are inactivated by the binding of OPs, an accumulation of ACh occurs at the nerve synapse interfering with the normal nervous system function. This produces rapid twitching of voluntary muscles followed by paralysis (Habig and DiGiulio, 1991; Ware, 1989). Once bound, organophosphorus compounds are considered irreversible inhibitors as recovery usually depends on new enzyme synthesis (Habig and Di Giulio, 1991). The thion-type compounds do not inhibit acetylcholinesterase (AChE) directly but must have the sulfur atom replaced with oxygen to become oxon-type compounds; such as is the case with malathion and its oxygen analogue, malaoxon (Kobayashi et al., 2010). This metabolic conversion allows the OP compound to phosphorylate the active site serine of AChE (Chambers et al., 2010).

Fish have two predominant cholinesterase enzymes: AChE generally found in brain and muscle; and butyrylcholinesterase (BChE) localized in liver and plasma with some in muscle (Fulton and Key, 2001). Toxicity results from AChE inhibition, and it may also result from alkylation of critical macromolecules, with relative contributions varying according to the site-specific reactivities of the OPs (Schuurmann, 1998). It has also been demonstrated in zebrafish (*Danio rerio*) embryos that chlorpyrifos-oxon significantly inhibits axonal growth of sensory neurons along with primary and secondary motoneurons (Yang et al., 2011).

4.3. Uptake and Metabolism

Unlike organochlorine insecticides, OPs do not generally bioaccumulate but undergo hydrolysis and biodegradation in the environment. Hydrolytic degradation of OPs proceeds by two pathways—either PO bond cleavage or CO bond cleavage (Schuurmann, 1998). In mammals, some OPs, such as malathion, are rapidly degraded and excreted via urine. Metabolic enzymes (carboxyesterases, carboxyamidases, and microsomal oxidases) are responsible for this action, with only small proportions being converted into the toxic oxon (Derache, 1977). In insects, crustaceans, and fish, this ability is reduced since they possess a very active oxidative enzyme system and relatively less of the hydrolytic, metabolizing enzymes (Kurtz et al., 1989). Once the oxon is produced, however, a large portion reacts with nontarget molecules at locations such as the liver and in the blood. Only a small portion actually reaches the target AChE in the nervous system (Chambers et al., 2010).

Generally, fish will rapidly absorb, metabolize, and excrete OP compounds. Chlorpyrifos is a prime example of an OP that is hydrophobic but relatively nonpersistent and susceptible to metabolic degradation (Barron et al., 1991). Most OP insecticides will degrade rapidly in the environment and fall below detectable limits in hours to days, but OP-induced AChE inhibition in fish can last from days to weeks (Fulton and Key, 2001).

4.4. Acute Toxicity

Much acute toxicity data for OP effects in fish have been gathered over the years and are well documented for malathion and parathion (Mulla and Mian, 1981); azinphosmethyl (Labat-Anderson, 1992); and chlorpyrifos (Marshall and Roberts, 1978; Giesy et al., 1999). The acute toxicity of a larger number of OPs to freshwater fish has been also well documented in the past by Johnson and Finley (1980) and toxicity to estuarine fish by Mayer (1987) (Table 6.2).

Fulton and Scott (1991) determined 96 h LC50 values of 3078 mg/L acephate and 32.16 µg/L azinphosmethyl for the estuarine mummichog. Johnson and Finley (1980) reported acephate 96 h LC50s of over 1000 mg/L for seven freshwater fish. Azinphosmethyl was also used in exposures with two common estuarine fish, adult mummichog and juvenile red drum (*Sciaenops ocellatus*). The juvenile red drum were significantly more sensitive to azinphosmethyl than mummichog after 96 h exposure (6.2 µg/L versus 64.5 µg/L). The researchers concluded that age-specific effects were, in part, responsible for the differences in toxicity since size and age can influence fish susceptibility to insecticides (Van Dolah et al., 1997).

More recently, the concern with OP effects on fish has been with the salmon industry in Europe and the Americas and with farmed fish generally worldwide. Control of sea lice (ectoparasitic copepods) in aquaculture is a major concern, with the OPs malathion, trichlorfon, diclorvos (a trichlorfon metabolite), and azamethiphos all having been used as treatments (Burridge et al., 2010). Herring (*Clupea harengus*) were found to have a 96 h LC50 of 122 µg/L after exposure to dichlorvos (McHenery et al., 1991). The dichlorvos parent compound, trichlorfon, may be less toxic to nontarget fish with 96 h LC50s to Tra catfish (*Pangasianodon hypophthalmus*) of 1.0 mg/L (Sinha et al., 2010) and to zebrafish of 28.8 mg/L (Coelho et al., 2011). Azamethiphos was determined to be lethal to nontarget species only if concentrations were greater than 100 µg/L (Burridge et al., 2010). Three-spined sticklebacks (*Gasterosteus aculeatus*) were exposed to azamethiphos for 96 h with a resulting LC50 of 190 µg/L (Ernst et al., 2001). Due to the short exposure period of these OPs to the farmed fish, these compounds are not considered toxic to them, but effects on nontarget species and a buildup of resistant sea lice have greatly reduced the use of OPs.

In the U.S. Pacific Northwest, the concern is runoff of OPs from agricultural fields into wild salmon breeding streams and farming operations, causing not only mortality from acute exposures but depressed levels of the AChE enzyme. Juvenile Chinook salmon (*Oncorhynchus tshawytscha*) mortality reached 20% after 96 h exposure to 10 µg/L chlorpyrifos and 100% mortality after 96 h exposure to 100 µg/L (Wheelock et al., 2005). In juvenile Coho salmon exposed to chlorpyrifos for 7 d, olfactory response was reduced by 20% at a concentration of 0.72 µg/L (Sandahl et al., 2004). Sandahl and Jenkins (2002) reported that 96 h LC50 values for chlorpyrifos (commonly used in fruit orchards in Oregon) in Pacific steelhead salmon ranged from 6.0 to 9.4 µg/L. These levels were well above concentrations measured in area surface waters. Moore and Teed (2012) evaluated the risks of OPs and OP/carbamate mixtures to Pacific salmon in nonpoint source runoff waters. They concluded that surface-water monitoring data showed OP levels were below those that could cause mortality or significant AChE inhibition in Coho salmon.

4.5. Chronic Toxicity

Due to the chemical nature of OPs, they have negligible chronic toxicity unless exposure is continuous. A few early studies investigated effects on fish from long-term exposures to azinphosmethyl, chlorpyrifos, and malathion (Adelman et al., 1976; Cripe et al., 1984; Mulla et al., 1981).

Biomarkers were the subject of two recent chronic studies. Carp were exposed to three concentrations of diazinon for up to 30 d (Oruc, 2011). All doses showed significantly decreased AChE activity in the liver throughout the exposures. Antioxidant enzyme activities in the liver (superoxide dismutase, glutathione peroxidase, and catalase) were all induced by diazinon. There were no significant changes in lipid peroxidation indicating that liver tissue induced antioxidant levels to counteract oxidative stress (Oruc, 2011). Sinha et al. (2010) exposed Tra catfish to trichlorfon at levels up to 500 µg/L for 56 d to detect expression of various biomarker genes in liver and gills. A time- and dose-dependent increase was observed in the expression of heat shock protein70, cytochrome P4501B, and cytochrome oxidase subunit 1 (Sinha et al., 2010). Snakehead fish (*Channa striata*) were exposed twice to 4 d pulses of diazinon ranging from 16 to 350 µg/L over a two-month period. Brain AChE was inhibited up to 91%, but mortality remained below 7%. AChE remained significantly inhibited in the fish more than 30 d after the dosing (Cong et al., 2009).

4.6. Sublethal Effects/Biomarkers

The measurement of AChE enzyme levels is one of the main biomarkers used in OP research. Hundreds of research papers and many books have been published on this subject dealing with fish alone (summarized by Murty, 1986; Mineau, 1991; Fulton and Key, 2001, among others). An example of AChE research can be illustrated with malathion. Coppage et al. (1975) exposed pinfish to malathion for 72 h at up to 575 µg/L. Levels of malathion that caused 40% to 60% mortality resulted in 72–79% brain AChE inhibition. A sublethal mean concentration of 31 µg/L caused 34% AChE inhibition (Coppage et al., 1975). Richmonds and Dutta (1992) exposed bluegill sunfish (*Lepomis macrochirus*) to six concentrations of malathion up to 48 µg/L for 24 h. As with the research previously described, brain AChE significantly decreased as the malathion concentration increased. Laetz et al. (2009) exposed Coho salmon to diazinon, malathion, and chlorpyrifos individually and in mixture for 96 h. All binary mixtures showed additive or synergistic AChE inhibition and lethality. The researchers stated that OP mixtures, which have been reported in salmon habitats, may be more of a challenge to species recovery than single chemical exposures.

Sublethal effects in fish other than AChE inhibition has been the subject of significant research as well. Weis and Weis (1976) exposed eggs of the sheepshead minnow during development to 1, 3, and 10 mg/L malathion. The two highest concentrations caused skeletal malformations in fry, with greater abnormalities present at the highest concentration. This was attributed to the OP interfering with normal neuronal function (Weis and Weis, 1976). It is now known that OPs can inhibit axonal growth of sensory neurons along with primary and secondary motoneurons (Yang et al., 2011). Adult mummichog were exposed to chlorpyrifos in pulsed doses for 96 h and 28 d at concentrations up to 10 µg/L (Karen et al., 1998). In both exposures, caudal vertebral strength was measured throughout the tests. After 96 h exposure, vertebral strength was significantly weaker in chlorpyrifos-exposed fish. The 28 d test found significantly weaker vertebrae after two weeks' exposure but not after four weeks' exposure. It was not known if the vertebral weakening was a mechanical or biochemical process (Karen et al., 1998).

Swimming ability in fish exposed to OPs has been examined in several studies. Sheepshead minnows were exposed to azinphosmethyl for 219 d and to *O*-ethyl *O*-(4-nitrophenyl) *P*-phenylphosphonothioate (EPN) for 265 d

(Cripe et al., 1984). Azinphosmethyl significantly affected reproduction at the highest concentration (0.5 µg/L), but swimming performances were not significantly affected. EPN significantly affected swimming performances at 2.2 µg/L, while survival and growth were only affected at the highest concentration (7.9 µg/L). Van Dolah et al. (1997) also exposed juvenile red drum and mummichog to sublethal levels of azinphosmethyl. Swimming ability in red drum significantly decreased after a 6 h exposure to 12 µg/L azinphosmethyl, with no effects on mummichog up to 24 µg/L azinphosmethyl. Beauvais et al. (2000) exposed larval rainbow trout for 96 h to diazinon at up to 1000 µg/L and malathion at up to 40 µg/L. For both OPs, swimming speed and distance decreased while tortuosity of path increased. It was concluded that AChE inhibition correlated well with these changes in swimming behavior.

Respiration was studied in mummichog and spot exposed to azinphosmethyl for 24 h at 10 µg/L by Cochran and Burnett (1996). While AChE activity was significantly decreased for both species, there was no effect on oxygen uptake. Goldfish were exposed to chlorpyrifos for up to 15 d at five sublethal concentrations (Wang et al., 2010). Brain and muscle AChE activity was significantly inhibited in a dose- and time-dependent manner. Gill glutathione S-transferase (GST) activity was also significantly inhibited by chlorpyrifos. Since elevated levels of GST have a protective effect against contaminants, any reductions in GST levels may affect metabolic detoxification of OPs (Wang et al., 2010).

Oxidative stress can be caused by the activation of proteolytic enzymes, nitric oxide synthase, and free radicals due to the break in neuronal signaling after OP exposure leading to an influx of calcium. Glutathione consumption resulting from the metabolism of OPs in the liver may also contribute to oxidative stress (Lushchak, 2011). Nile tilapia (*Oreochromis niloticus*) were exposed to chlorpyrifos at three sublethal concentrations for up to 30 d (Oruc, 2012). The results indicated that chlorpyrifos induced oxidative stress in the liver, but this was not related to any inhibition of AChE activity (Oruc, 2012). Other examples of oxidative stress caused by OP exposure were mentioned by Lushchak (2011), with dichlorvos inducing stress in carp and brown bullhead catfish (*Ictalurus nebulosus*), malathion inducing stress in sea bream (*Sparus aurata*), and trichlorfon inducing stress in Nile tilapia.

Jacobson et al. (2010) investigated the development of zebrafish embryos after a three-day exposure to a sublethal concentration of chlorpyrifos-oxon of 300 nM (100.35 µg/L). After one day, AChE was inhibited by 57% and by 89% after two days. Interestingly, this did not kill the embryos, leading the researchers to state that 100% AChE activity is not necessary for embryonic survival. The development of muscle fibers and nicotinic acetylcholine receptors was also not affected. However, Rohon-Beard neuron (a subset of early sensory neurons) formation was significantly depressed. This research points to the future of OP studies with fish in discovering how these insecticides can alter neuronal development during embryogenesis (Jacobson et al., 2010).

4.7. Field Studies

Monitoring of AChE levels in fish has been widely used in field studies as an indication of OP exposure. By the 1950s, researchers called for monitoring brain AChE activity in fish collected from field-exposed sites (Weiss, 1958). This has proved quite useful since significant inhibition indicates that a sufficient dose of the OP has reached the target site of the fish to produce an effect. Also, AChE inhibition can persist in fish for days after the OP has degraded and is no longer detectable (Fulton and Key, 2001).

Lockhart et al. (1985) sampled juvenile walleye (*Stizostedion vitreum vitreum*) from a pond after two aerial applications of malathion (210 g/ha) over a four-month period. After the first application, brain AChE was reduced to 25% of pre-spray values and then recovered to 80% of pre-spray values after two weeks. This same pattern was evident after the second application as well. Walleye growth was reduced after the first malathion application but not after the second. Sturm et al. (2000) collected three-spined stickleback from eight pesticide-contaminated streams in Germany. After analyzing for both muscle BChE and muscle AChE, the researchers determined that muscle BChE was more sensitive to OPs than AChE. However, they noted that BChE does not occur in the muscle of all fish, namely, some salmonids and cyprinids.

Since fish will rapidly absorb, metabolize, and excrete OP compounds, OP residues are generally not found in field collected fish. Stahl et al. (2009) searched a database of fish tissue contaminants collected from 500 lakes and reservoirs in the United States. There were no detections of nine OPs (chlorpyrifos, paraoxon, parathion (ethyl), diazinon, disulfoton, disulfoton sulfone, terbufos, terbufos sulfone, and ethion) or their metabolites. Macek et al. (1972) sampled bluegill sunfish and largemouth bass from a pond after two applications of chlorpyrifos at a rate, used at the time, for mosquito control (0.01 to 0.05 lb/ac). Maximum chlorpyrifos residues in bluegill sunfish were 3.82 mg/L and 2.55 mg/L in bass three days after application. Residues were not detected four weeks after application, but AChE activity was still depressed.

Microcosms and mesocosms have been used to simulate field conditions of OP exposure. A mesocosm constructed to simulate a tidal marsh was dosed with 2 and 8 µg/L azinphosmethyl for 96 h after stocking with mummichog and Atlantic silverside minnows. After 24 h, all fish were dead in the high treatment. No mummichog mortality occurred in the low treatment after 96 h, but a 96 h LC50 of 1.19 µg/L was determined for the Atlantic silverside (Lauth et al., 1996). Outdoor pond microcosms containing juvenile bluegill sunfish were treated with chlorpyrifos ranging from 0.03 to 2.58 µg/L (Giddings et al., 1997). Chlorpyrifos half-lives ranged from one to six days after applications ended, and fish survival and total biomass was not affected. AChE values were not obtained. Sibley et al. (2000) used outdoor microcosms to simulate a freshwater community, including fathead minnows. Three different mixtures of azinphosmethyl, chlorpyrifos, and the carbamate diazinon were used in a 7-d study, with concentrations ranging from 50 to 1750 µg/l. Mortality was significantly correlated with AChE activity with 10 to 50% mortality corresponding to a 50 to 90% reduction in AChE activity (Sibley et al., 2000). This relationship between AChE and mortality was also reported by Fulton and Key (2001) whose review determined that a 70% to 80% reduction in AChE led to mortality in fish.

5. CARBAMATES

The first carbamate to be marketed as an insecticide was carbaryl, which was first synthesized in 1953. A variety of other carbamates were developed and marketed throughout the 1960s and 1970s. Examples include aldicarb, carbofuran, and carbosulfan (Baron, 1991). Carbamates have been used to control a variety of insects on agricultural crops, lawns, and gardens. In 2007, carbaryl was the most extensively used insecticide in the home and garden market sector (Grube et al., 2011).

5.1. Structure

Carbamate insecticides have many similarities with organophosphate insecticides, even though their structures are quite dissimilar. Carbamates, like OPs, are used as replacements for organochlorine insecticides, are relatively short-lived in the environment, are rapidly metabolized or excreted, and do not bioconcentrate in food webs (Hill, 1995). Currently, at least 15 carbamates are registered for use by the U.S. EPA (U.S. EPA, 2012b). Carbamates can be divided into several groups depending on their structure (Fukuto, 1990; Baron, 1991): (1) benzofuranyl methylcarbamates (i.e., carbofuran); (2) dimethylcarbamates (i.e., pirimicarb); (3) oxime methylcarbamates (i.e., aldicarb); (4) phenyl methylcarbamates (i.e., propoxur); (5) procarbamates (i.e., carbosulfan); and (6) naphthyl methyl-carbamate (i.e., carbaryl, Figure 6.1).

5.2. Mechanism of Action

Like the OPs, carbamates cause toxicity by inhibiting cholinesterases. With carbamates, a reversible carbamylation of the AChE enzyme occurs at parasympathetic junctions, skeletal muscle myoneural junctions, and autonomic ganglia, in addition to the central nervous system in the brain. In contrast to OP inhibition of AChE, the carbamylation is more readily reversible (i.e., decarbamylation is rapid) than the OP phosphorylation, which can limit the effect that carbamates have on fish and account for a greater span between symptoms and a lethal dose (Reigart and Roberts, 1999).

5.3. Uptake and Metabolism

Carbamates undergo hydrolysis in two stages: (1) the aryl or alkyl group is removed with the formation of a carbamylated enzyme; and (2) the inhibited enzyme is then decarbamylated with a free active enzyme generated (Ecobichon, 2001). Carbamates are rapidly biotransformed *in vivo* and excreted by the kidneys and the liver, which can lessen the duration and intensity of the toxicity. This also leads to their low risk for bioaccumulation in fish from sublethal exposures. However, fish feeding on carbofuran-contaminated food can be affected due to any unabsorbed insecticide on the cuticle of arthropods or in the gut of worms (Hill, 1995). Bluegill sunfish were exposed to aldicarb at levels of 10 µg/L and 100 µg/L for 8 weeks. After one week, tissues levels of aldicarb had equilibrated at up to 4.5 times the concentrations; however, after transferring to clean water, levels were undetectable after three weeks (Baron and Merriam, 1988).

The methylcarbamate groups are directly toxic in that they inhibit AChE without metabolic activation as is required for most OPs. The oxime carbamates need to be metabolically activated and are structurally similar to acetylcholine, which makes them good inhibitors of AChE. The toxicity of oxime carbamates to fish depends on the particular species of fish. It has been found that channel catfish do not express the flavin monooxygenase system used in the biotransformation of aldicarb to its potent oxide aldicarb sulfoxide. Therefore, channel catfish, and some other freshwater fish as well, are resistant to aldicarb effects. Rainbow trout, in contrast, are very sensitive to aldicarb and other carbamates since they do express the flavin monooxygenase system. Differences in the expression of this system will therefore lead to differences in carbamate metabolism in fish (Fukuto, 1990; Perkins and Schlenk, 2000).

5.4. Acute Toxicity

Compared to OPs, carbamates are considered moderately toxic to fish (Table 6.2). Acute studies of carbamate effects on fish have been well reviewed (Macek and McAllister, 1970; Mayer, 1987). Morgan and Brunson (2002) reviewed agricultural pesticide effects on bluegill sunfish, channel catfish, and rainbow trout with 96 h LC50 results for 10 carbamates. In bluegill sunfish, the aldicarb LC50 was 50 μg/L, while it was 560 μg/L for rainbow trout. Carbaryl was much less toxic with an LC50 of 676 μg/L for bluegill sunfish; 15,800 μg/L for channel catfish and 1950 μg/L for rainbow trout (Morgan and Brunson, 2002). A carbaryl 96 h LC50 for goldfish was 13,900 μg/L (Ferrari et al., 2004). Dwyer et al. (2005) exposed 18 endangered or threatened fish species to carbaryl for 96 h. Atlantic sturgeon (*Acipenser oxyrhynchus*) were the most sensitive with an LC50 of less than 800 μg/L, and desert pupfish (*Cyprinodon macularius*) were the least sensitive with an LC50 of 7710 μg/L. Isoprocarb, widely used in rice production, was more toxic to goldfish with a 96 h LC50 of 4610 μg/L (Wang et al., 2012). Carbofuran was even more toxic to Nile tilapia larvae with a 96 h LC50 of 214.7 μg/L to 220.7 μg/L, depending on formulation (Pessoa et al., 2011). Gül et al. (2012) reported 96 h LC50 values for propoxur of 7340 μg/L to carp fry and 6500 μg/L to Asian stinging catfish fry.

5.5. Chronic Toxicity

As with organophosphates, the chemical nature of carbamates means they have negligible chronic toxicity unless exposure is continuous. Many of these chronic research studies found organ and tissue damage in fish at levels considered nonlethal in acute studies. Snakehead fish were exposed to carbofuran at 4500 μg/L for six months (Ram and Singh, 1988). Juveniles were more susceptible to thyroid damage (fibrosis, cystic cellular masses, hemorrhage) than adult fish. This suggests that carbofuran induced adverse changes in fish thyroid in an age- and size-dependent manner. Snakehead liver was also examined after six months' exposure to 4500 μg/L carbofuran (Ram and Singh, 1988). Livers exhibited cytoplasmolysis, a significant decrease in hepatosomatic index, and extensive degeneration of proliferated hepatocytes. An induction of liver tumors was also observed.

The walking catfish was exposed for up to 168 h to 15,300 μg/L carbaryl (Jyothi and Narayan, 1999). Serum levels of glucose, alkaline phosphatase, and bilirubin significantly increased over the exposure period compared to controls. This was seen as evidence of liver damage due to the carbamate exposure. The Asian stinging catfish was exposed to carbofuran at levels ranging from 500 to 2000 μg/L for 30 d (Chatterjee et al., 2001). 17β-Estradiol was significantly reduced in serum and ovaries in a dose-dependent manner in both pre-spawning and spawning fish. Vitellogenin levels were also affected but only in pre-spawning fish, leading the authors to conclude that carbofuran can act as an anti-estrogenic, endocrine-disrupting chemical in these fish.

Todd and Van Leeuwen (2002) continuously exposed zebrafish embryos for 144 h until hatch to four carbaryl levels up to 17.0 μg/L. The average mortality rate was low, but embryos were consistently smaller than controls starting at 24 h after spawning until hatching. Embryos in the highest concentration took up to twice as long to hatch as the controls, which might increase vulnerability to predation. Brain AChE, malondialdehyde (MDA), and glutathione (GSH) levels were studied in tench (*Tinca tinca*) exposed to carbofuran for 60 d, then allowed to depurate in clean water for 30 d (Hernández-Moreno et al., 2010). At the highest dose of 100 μg/L carbofuran, AChE was significantly inhibited after 30 d exposure but returned to normal levels after this period. Significant induction of MDA levels were only observed during the depuration period at 70 and 80 d after an initial exposure to the high dose. GSH levels were variable with a significant increase again at 80 d, leading the authors to conclude that MDA and GSH levels would be more associated with chronic exposures (Hernández-Moreno et al., 2010). Hernández-Moreno et al. (2008) had previously exposed tench to carbofuran at the same concentrations and time period. CYP1A and CYP3A activity along with the phase II enzyme uridine diphospho-glucuronosyltransferase (UDPGT) were measured with a clear time-dependent inhibition of CYP1A and UDPGT seen in the highest concentration. CYP3A activity was not as noticeably altered during the exposure.

Similarly, Ensibi et al., (2012) exposed carp to carbofuran for 30 d at up to 100 μg/L. Oxidative stress was found in the liver at 4 and 30 d time points. In contrast to tench, MDA levels were decreased after 30 d exposure to 100 μg/L. AChE levels were reduced by 74% in brain and 67% in muscle tissue.

5.6. Sublethal Effects/Biomarkers

The toxicity of carbamate insecticides is produced through the inhibition of cholinesterase enzymes of the nervous system and as with OPs, the detection of AChE enzyme levels is one of the main biomarkers used in carbamate research. Ferrari et al. (2004) exposed goldfish to carbaryl and two OPs. At the 96 h LC50 for carbaryl, 13,900 μg/L, AChE levels were reduced by 86%. The IC50 (amount inhibiting AChE by 50%) was 2620 μg/L, over five times lower than the LC50. After depurating for 96 h, AChE levels recovered to 76% of the control values. The authors concluded that goldfish do not require as much AChE to sustain life as other fish and toxicity from carbaryl may be due to effects on other target sites.

Four carbamates (methomyl, thiodicarb, carbofuran, and carbosulfan) were studied for AChE inhibition in male and female goldfish (Yi et al., 2006). Carbofuran was the strongest inhibitor of brain AChE in both male and female fish. Rates of carbamylation for all four insecticides were higher in females than males. The rates of decarbamylation were similar for males and females except for carbosulfan in which females recovered more quickly than males. The results indicated that sex differences may exist in carbamate inhibition of AChE (Yi et al., 2006). Carbofuran was also a strong inhibitor *in vitro* to the Amazonian fish, tambaqui (*Colossoma macropomum*). The carbofuran IC50 was 0.92 μmol/L compared to the carbaryl IC50 of 33.8 μmol/L (Assis et al., 2010).

Studies have also documented the effects of carbamates on biotransformation enzymes and organ and tissue condition. Ferrari et al. (2007) exposed juvenile rainbow trout to carbaryl (1000 and 3000 μg/L for up to 96 h) and determined levels of five different xenobiotic-biotransforming enzymes. Carbaryl caused an initial induction followed by inhibition of CYP1A, GST, and CAT activity, and carboxyesterases were significantly inhibited in the liver (Ferrari et al., 2007). Cartap (S,S'-[2-(dimethylamino)-1,3-propanediyl] dicarbamothioate) was used in exposures with green snakehead fish (*Channa punctatus*) at a concentration of 180 μg/L for 96 h. Hypertrophy was induced in neurosecretory cells, and varying degrees of cytoplasmic vacuolization and necrosis were observed in the brain (Mishra et al., 2008). Gül et al. (2012) exposed carp to sublethal levels (5000 μg/L) of propoxur for 96 h. Histopathological examinations of spleen, intestine, muscle, or skin showed no significant changes. However, many other effects were observed, including edema in the kidney, hydropic degeneration of the liver, damage to gill tissues, as well as several hematological changes.

The behavior of fish after exposure to carbamates has been the focus of several studies. Bretaud et al. (2002) exposed goldfish to carbofuran at three levels up to 500 μg/L for 24 h and 48 h. Levels of brain catecholamines (a group of amines that have important physiological effects as neurotransmitters and hormones) were then measured. Norepinephrine and dopamine levels increased after 48 h of exposure. This may influence locomotor activity and aggression. At even the lowest concentration tested (5 μg/L), swimming patterns and social interactions were affected (Bretaud et al., 2002). Agonistic acts (threats, nips, chases) were observed in goldfish exposed to carbofuran in an earlier study by Saglio et al. (1996). Behavior can also be affected by neurogenesis and apoptosis in the brain of fish. Zhou et al. (2009) exposed zebrafish during embryogenesis to the carbamate cartap at concentrations up to 100 μg/L. This resulted in fewer differentiated neurons in the brain in the regions of the telencephalon, diencephalon, hindbrain, and spinal cord. Carbofuran also affected breeding in the rohu (*Labeo rohita*). The total number of eggs, total weight of eggs, and hatching percentage were all significantly reduced after 96 h exposure to 150 and 300 μg/L carbofuran (Adhikari et al., 2008). Nine-day-old Nile tilapia were exposed to carbofuran, giving 68% inhibition at an exposure level of 297 μg/L. Behavioral parameters were also observed, with significant reductions in swimming speed, visual acuity, prey capture, and weight gain (Pessoa et al., 2011). Behavior and biomarker assessments were also investigated in sea bass (*Dicentrarchus labrax*) exposed to carbofuran. Swimming velocity was significantly decreased in sea bass after 96 h exposure to 250 μg/L. Brain and muscle AChE and EROD activities were all reduced as well (Hernández-Moreno et al., 2011).

As with OPs, the effects of carbamates on salmon species are a major concern in the U.S. Northwest. This is due not only to nonpoint-source runoff, but also to the practice in Washington State of directly treating estuarine flats with carbaryl to control certain crustaceans (ended in 2012). Labenia et al. (2007) exposed cutthroat trout to carbaryl concentrations ranging from 1 to 2000 μg/L for 2 to 6 h. AChE activity was significantly reduced after 6 h but recovered after 42 h depuration. At concentrations of 750 μg/L and above, trout swimming performance was affected while predator avoidance was affected at concentrations of 500 μg/L and above. The researchers concluded that estuarine applications of carbaryl would likely impair trout behavior and increase mortality.

5.7. Field Studies

In the early 1990s, Bailey et al. (1994) found that striped bass larval recruitment variations could be explained 23% to 63% of the time by the presence of carbamates (carbaryl and carbofuran) and OPs in runoff water in a California drainage basin. Kirby et al. (2000) investigated continuous seasonal inputs of insecticides in UK estuaries and its effects on flounder (*Platichthys flesus*). Six estuaries were found to contain nine OPs and six carbamates, including carbaryl and carbofuran in amounts ranging from <2 to 21 ng/L. Flounder muscle tissue from these estuaries were assayed for AChE levels, with 12 out of 16 sites showing significantly depressed AChE activity compared with a reference site. The authors concluded that the presence of OP and carbamate insecticides contributed to the lower AChE levels (Kirby et al., 2000). Villa et al. (2003) saw similar results when sampling the grass goby (*Zosterisessor ophiocephalus*) from an Italian lagoon that received pesticides draining from an agricultural region. A significant reduction in AChE activity was observed in spring after pesticide applications compared to the AChE activity in those fish sampled in winter. Furathiocarb was the only carbamate of seven targeted insecticides measured. Other research by Whitehead et al. (2004), Muñoz et al. (2010), and Sumith et al. (2012) have all seen effects on fish from agricultural runoff containing a pesticide suite, but carbamates (carbaryl and carbofuran ranging from <1 to 31.6 ng/L) constituted just one or two of the pesticides measured in the suite.

6. PYRETHROIDS

The synthetic pyrethroids were developed based on an understanding of the chemical structure of naturally occurring plant derivatives with insecticidal activity, pyrethrum, and the pyrethrins (Davies, 1985). The precise location of the earliest discovery of the insecticidal activity of these plant derivatives is unknown, but it is likely that it occurred somewhere in Asia, and some have suggested that it was known in China as early as the first century AD. These compounds were first introduced into Europe in the nineteenth century (Ecobichon, 1996).

6.1. Structure

Six chemicals with insecticidal properties can be derived from the *Chrysanthemum* flower. These natural products are called pyrethrins. Their chemical structure contains a cyclopropane-carboxylic acid group and an alcohol group joined by an ester linkage (Gan, 2008). While having good "knock-down" power, pyrethrins are very unstable in the environment and break down quickly in sunlight. To improve the stability of pyrethrins, synthetic analogues called pyrethroids were created. The first synthetic pyrethroid to be marketed commercially was allethrin in the late 1940s. The earliest pyrethroids (allethrin, resmethrin, tetramethrin, phenothrin) were based on substitutions on the alcohol portion of the molecule. These compounds still had a high rate of photolysis. Newer pyrethroids, such as permethrin, were formed by making modifications to the acid portion of the molecule (Figure 6.1). These compounds have improved photostability (Gan, 2008). Pyrethroids are chiral compounds, some having many isomers. Type II pyrethroids (cypermethrin, cyfluthrin, deltamethrin, lambda-cyhalothrin), which have an alpha cyano group display, increased biological activity relative to their Type I analogues (permethrin, resmethrin, and phenothrin) (Gan, 2008).

Pyrethroids generally have low water solubility (<0.2 mg/L), high affinity for sediments and organic carbon (log Kow [octanol-water partition coefficient] values typically 4–6), and are moderately persistent (Oros and Werner, 2005). Pyrethroids generally have hydrolysis half-lives of days to weeks in aquatic environments (Oros and Werner, 2005). The aerobic half-lives of most pyrethroids in soil are approximately 30–100 d (Oros and Werner, 2005). These compounds undergo photochemical transformation and sediment sorption. Degradation occurs primarily via microbial activity and photolysis (Cox, 1998). The majority of pyrethroids are pyrethroid esters. Etofenprox is a pyrethroid ether compound commonly used to control insect pests in residential settings, on rice crops, and for mosquito control. Etofenprox has very low water solubility (approximately 0.02 mg/L), with high affinity for soil and sediments (log Kow value 7.05). An aqueous half-life based on photolysis in natural pond water was determined to be 7.9 d (U.S. EPA, 2007). The structural modification of pyrethroid ether compounds results in lower toxicity to humans and other mammals compared to pyrethroid esters (U.S. EPA, 2007).

6.2. Mechanism of Action

In target invertebrates, the principal mechanism of action of pyrethroids is disruption of sodium channel function in the nervous system (Miller and Salgado, 1985), specifically causing hyperstimulation of the nervous system due to the blocking of sodium ion movement through the nerve membrane (Cox, 1998). *In vivo* and *in vitro* studies have demonstrated a mechanistic difference in binding sites on the sodium channel for Type I and Type II pyrethroids (Breckenridge et al., 2009). Generally, pyrethroid toxicity to fish increases with log Kow; however, exceptions to this relationship can be attributed to the α-cyano difference in toxic mechanism (Haya, 1989). Pyrethroids have also been shown to act outside the sodium channel, on isoforms of voltage-sensitive calcium channels, thereby contributing to the release of neurotransmitters (Breckenridge et al., 2009).

The synergist piperonyl butoxide (PBO) is commonly added to pyrethroid and pyrethrin formulations to enhance the toxic effects of the active ingredient. PBO functions by inhibiting a group of enzymes (mixed-function oxidases), which are involved in pyrethroid detoxification. PBO can enhance the toxicity of pyrethroids by 10 to 150 times (Wheelock et al., 2004). Toxicity values (96 h LC50) reported for PBO include 1.8 mg/L for rainbow trout and sheepshead minnow, 4.0 mg/L for bluegill sunfish and carp, and 6.0 mg/L for fathead minnow and catfish (U.S. EPA, 2000).

6.3. Uptake and Metabolism

In most vertebrates, pyrethroid metabolism occurs through hydrolysis of the ester bond, followed by conjugation and excretion in the bile (Haya, 1989). Fish may be deficient in the enzyme system that hydrolyzes pyrethroids (Haya, 1989). The elimination half-lives of pyrethroids reported in trout were approximately 48 h, compared to 6–12 h in mammals and birds. In addition only about 20% to 30% of the original dose was metabolized compared to 90% for terrestrial vertebrates in which up to 70% of the metabolites were ester hydrolysis products (Bradbury and Coats, 1989). The main reaction involved in the metabolism of deltamethrin, cypermethrin, or cyhalothrin in mice and rats is ester cleavage mainly due to the action of carboxyesterase. Metabolism in fish is largely oxidative (Di Giulio and Hinton, 2008). Despite low metabolic activity, pyrethroid bioconcentration factors (BCF) are generally low in fish (e.g., 300–3000 for fenvalerate) (Bradbury and Coats, 1989; Goodman et al., 1992).

6.4. Acute Toxicity

Pyrethroids have very high toxicity to fish relative to all current-use insecticides (Table 6.2). Allethrin had 96 h LC50 toxicity values of 17.5 μg/L, 22.2 μg/L >30.1 μg/L, and 80 μg/L for rainbow trout, Coho salmon, channel catfish, and fathead minnow (bio-allethrin), respectively (Smith and Stratton, 1986). Bifenthrin had 96 h LC50 toxicity values of 17.8 μg/L (Werner and Moran, 2008) and 19.81 μg/L (Harper et al., 2008) for sheepshead minnow. Bifenthrin toxicity (96 h LC50) to rainbow trout was 0.15 μg/L (Werner and Moran, 2008). Cyfluthrin had toxicity values (96 h LC50) of 2.49–4.05 μg/L for sheepshead minnow, 0.3 μg/L for rainbow trout, and 0.87 μg/L for bluegill sunfish (Werner and Moran, 2008).

Cypermethrin had a 96 h LC50 toxicity value of 0.2 μg/L for the South American silversides (*Odontesthes bonariensis*) (Carriquiriborde et al., 2012). Cypermethrin 96 h LC50 toxicity values of 0.5 μg/L and 0.4–1.1 μg/L were determined for rainbow trout and carp, respectively (Smith and Stratton, 1986). Cis-cypermethrin was less toxic to rainbow trout, with a 96 h LC50 toxicity value of 6.0 μg/L (Smith and Stratton, 1986). Cypermethrin had 96 h LC50 toxicity values of 0.73 μg/L for sheepshead minnow and 1.78 μg/L for bluegill sunfish (Werner and Moran, 2008), and 2.2 μg/L for Atlantic salmon (Smith and Stratton, 1986). Toxicity values for cypermethrin were determined with two species of Australian freshwater riverine fish, common jollytail (*Galaxias maculatus*), and tupong (*Pseudaphritis urvillii*) (Davies et al., 1994). Four-day and ten-day LC50 values determined were 2.19 μg/L and 1.47 μg/L for tupong and 2.34 μg/L and 1.98 μg/L for common jollytail.

These were compared to rainbow trout values derived from the same exposure, which yielded a 96 h LC50 and 10 d LC50 of 1.47 µg/L for both time points (Davies et al., 1994).

Deltamethrin 96 h LC50 toxicity values of 0.5 µg/L, 0.86 µg/L, and 1.0 µg/L were reported for rainbow trout, carp, and mosquito fish, respectively (Smith and Stratton, 1986). Toxicity values of 0.59–1.97 µg/L (96 h LC50) were reported for deltamethrin and Atlantic salmon (Smith and Stratton, 1986). Sheepshead minnow and bluegill sunfish both had 96 h LC50 values of 0.36 µg/L (Werner and Moran, 2008). Carp embryos were less sensitive to deltamethrin (48-h LC50 of 0.21 µg/L) than larval forms (48 h LC50 of 0.074 µg/L) (Köprücü and Aydın, 2004).

Esfenvalerate had 96 h LC50 toxicity values of 0.07 µg/L for rainbow trout, 0.1–1.0 µg/L for Chinook salmon, 0.22 µg/L for sheepshead minnow, and 0.26 µg/L for bluegill sunfish (Werner and Moran, 2008). In a study with Japanese medaka (*Oryzias latipes*), no mortality was noted after a 96-h exposure to 0.19 µg/L esfenvalerate, whereas 100% mortality was observed at the next concentration tested (9.4 µg/L; Werner et al., 2002). A study with larval Australian crimson-spotted rainbow fish (*Melanotaenia fluviatilis*) found very short-term exposures to esfenvalerate (pulse-dose for 1 h) caused significant mortality at concentrations as low as 0.060 µg/L, and two-day-old fish were more sensitive to esfenvalerate than 14-day-old fish (Barry et al., 1995).

Etofenprox (a pyrethroid ether) was found to have 24 h LC50 values of 8.4 mg/L and 5.0 mg/L for the tropical fish species Nile tilapia and redbelly tilapia (*Tilapia zillii*), respectively (Yameogo et al., 2001). The 96 h LC50 values determined for rainbow trout and bluegill sunfish were 2.7 µg/L and 13.0 µg/L, respectively (U.S. EPA, 2007).

Fenvalerate 96 h LC50 values of 1.2 µg/L, 2.1 µg/L, 5.0 µg/L, and 5.4 µg/L have been reported for Atlantic salmon, rainbow trout, sheepshead minnow, and fathead minnow, respectively (summarized in Smith and Stratton, 1986). Topsmelt (*Atherinops affinis*) embryos were less sensitive to fenvalerate than juveniles, with no mortality of embryos in a 30 d exposure to 3.2 µg/L but complete mortality of topsmelt fry at ≥0.82 µg/L fenvalerate (Goodman et al., 1992).

A 96 h LC50 toxicity value of 1.95 µg/L was reported for fenpropathrin in bluegill sunfish (Smith and Stratton, 1986). The 96 h LC50 toxicity values for flucythrinate were 0.22 µg/L (fathead minnow; Spehar et al., 1983), 0.32 µg/L (rainbow trout; Worthing and Walker, 1983), and 1.6 µg/L (sheepshead minnow; Schimmel et al., 1983).

Werner and Moran (2008) reported lambda-cyhalothrin 96 h LC50 values of 0.7 µg/L for sheepshead minnow, 0.24–0.54 µg/L for rainbow trout, 0.5 µg/L for common carp, and 0.21–0.42 µg/L for bluegill sunfish. A study of Chinook salmon showed fish embryos were less sensitive to lambda-cyhalothrin than larvae. There was no effect on mortality, hatching success, or larval survival when embryos were exposed to concentrations ≤5.0 µg/L. The estimated 96 h LC50 for Chinook salmon fry was 0.15 µg/L (Phillips, 2006).

Permethrin acute toxicity values (96 h LC50) for various freshwater and saltwater fish species were summarized by Smith and Stratton (1986) as follows: 1.1 µg/L for channel catfish, 6.4, 7.0, and 9.0 µg/L for rainbow trout, 8.5 µg/L for largemouth bass, 12.0 µg/L for Atlantic salmon, 15.0 µg/L for mosquitofish, and 15.6 µg/L for fathead minnow. Other fish 96 h LC50 values reported include <1.2 µg/L for Atlantic sturgeon and juvenile shortnose sturgeon (*Acipenser brevirostrum*) (Dwyer et al., 2005), 2.2 µg/L for juvenile Atlantic silverside (*Menidia menidia*) and 7.8 µg/L for adult sheepshead minnow (Schimmel, 1983), 5.5 µg/L for juvenile striped mullet (Mayer, 1987), 6.4 µg/L for larval inland silverside (*Menidia beryllina*) and 17.0 µg/L for Coho salmon (U.S. EPA, 2005), 8.3 µg/L for juvenile red drum and 23 µg/L for adult mummichog (Parent et al., 2011). Permethrin was also found to have 24 h and 48 h LC50 values of 40 µg/L and 27 µg/L for Nile tilapia, respectively, and 24 h and 48 h LC50 values of 75 µg/L and 49 µg/L for redbelly tilapia, respectively (Yameogo et al., 2001). A study on juvenile (90 d old) hybrid striped bass (*Morone saxitalis*) determined 24 h and 96 h LC50s of 32.9 µg/L and 16.4 µg/L for a 1:1 active ingredient mixture of PBO and permethrin; however, a toxicity value for permethrin alone was not determined (Rebach, 1999).

Phenothrin (sumithrin) had 96 h LC50 values of 16.7 µg/L for rainbow trout and 18.0 µg/L for bluegill sunfish (Smith and Stratton, 1986). The inland silverside had a 96 h LC50 value of 94.23 µg/L phenothrin (U.S. EPA, 2008).

Resmethrin toxicity is described by 96 h LC50 values of 0.45 µg/L for rainbow trout, 2.36 µg/L for yellow perch, and 2.62 µg/L for bluegill sunfish (Smith and Stratton, 1986). Additional toxicity values (96 h LC50) reported for resmethrin include 1.68 µg/L (lake trout), 1.05 µg/L (Northern pike), 1.7 µg/L (Coho salmon), and 2.96 µg/L (fathead minnow) (U.S. EPA, 2000). Earlier life stages (fingerlings) were more sensitive to resmethrin, with 96 h LC50 values of 0.51 µg/L, 0.66 µg/L, 0.74 µg/L, 0.75 µg/L, and 0.76 µg/L for yellow perch, largemouth bass, brown trout, Coho salmon, and lake trout, respectively (U.S. EPA, 2000).

Tetramethrin had a 96 h LC50 value of 21 µg/L for bluegill sunfish (Smith and Stratton, 1986). Reported tralomethrin 96 h LC50 values were 1.6 µg/L for rainbow trout, 2.48 µg/L sheepshead minnow, and 2.8 µg/L for bluegill sunfish (U.S. EPA, 2000).

The presence of sediment is an important variable in determining pyrethroid toxicity, usually resulting in lower acute toxicity (Clark et al.,

1989; Coats et al., 1989; DeLorenzo and De Leon, 2010; DeLorenzo et al., 2006; Key et al., 2005, 2011; Stueckle et al., 2008). Salinity and water hardness can also affect pyrethroid toxicity. A study with bluegill sunfish fry found an increase in salinity from 4.25 ppt to >8.5 ppt increased fenvalerate toxicity by 50%, and increasing water hardness from 6 mg $CaCO_3$ per L to >36 mg/L increased toxicity by 50% (Dyer et al., 1989). It is possible that stress to the osmoregulatory system may be a secondary mode of toxicity of pyrethroids in fish (Di Giulio and Hinton, 2008). Contrary to many chemicals, pyrethroid toxicity typically increases at lower temperatures (Di Giulio and Hinton, 2008). The pyrethroid toxicity change is a result of increased accumulation of parent compound and increased nerve sensitivity at lower temperatures (Harwood et al., 2009).

6.5. Chronic Toxicity

A 28-d early life stage chronic study with permethrin and sheepshead minnows yielded a no observable effect concentration (NOEC) of 10 µg/L, and hatchling survival was reduced at 22 µg/L (Hansen et al., 1983). A BCF of 480 was determined for permethrin and sheepshead minnows (Hansen et al., 1983). In the same study, fenvalerate had a 28-d NOEC of 0.56 µg/L and a BCF of 570 (Hansen et al., 1983). Fenvalerate reduced hatchling survival at 3.9 µg/L, and fish length and weight were reduced at 2.2 µg/L (Hansen et al., 1983). Tanner and Knuth (1996) found that young-of-the-year bluegill sunfish growth was reduced by 57, 62, and 86% after two applications of 0.08 µg/L, 0.2 µg/L, and 1 µg/L esfenvalerate, respectively; and delayed spawning was observed after exposure to 1 µg/L. In another study, a 30-d exposure to cypermethrin with the fathead minnow yielded a NOEC of 0.14 µg/L for growth (U.S. EPA, 2006b). A 60-d post-hatch survival NOEC value of 1.1 µg/L phenothrin was determined for rainbow trout (U.S. EPA, 2008). In an early life stage test with zebrafish, the 40 d NOEC and LOEC (lowest observable effect concentration) determined for the pyrethroid ether etofenprox were 25 µg/L and 50 µg/L, respectively (U.S. EPA, 2007).

6.6. Sublethal Effects

Sublethal effects such as erratic swimming, loss of equilibrium, jaw spasms, gulping respiration, lethargy, and darkened pigmentation were observed in various fish studies after pyrethroid exposure, and these effects typically occurred at concentrations that were less than concentrations where mortality occurred. For example, behavioral 96 h NOEC values determined for rainbow trout and bluegill sunfish with cypermethrin were 0.00068 µg/L and <0.0022 µg/L, respectively, compared to their 96 h LC50 values of 0.8 µg/L and 2.2 µg/L (U.S. EPA, 2006b). Similar effects were noted with sheepshead minnow, where acute NOEC values for sublethal behavioral effects ranged from 0.84 µg/L to 1.4 µg/L and were approximately two- to threefold lower than the corresponding LC50 values of 2.7 µg/L and 2.4 µg/L, respectively (U.S. EPA, 2006b). Lethargy and abnormal swimming behavior were observed with sheepshead minnows exposed to both permethrin and fenvalerate (Hansen et al., 1983). Abnormal swimming behavior, reduced growth, and increased predation risk were noted in larval fathead minnows in response to a 40 h sublethal (0.455–1.142 µg/L) esfenvalerate exposure (Floyd et al., 2008).

While many acute toxicity values for pyrethroids and fish are less than 1 µg/L, the sublethal effects values available are often one to two orders of magnitude lower (Bradbury and Coats, 1989). For example, low levels of cypermethrin (<0.004 µg/L for 5 d) were found to impair reproductive function (specifically, the pheromonal-mediated endocrine system) in Atlantic salmon, as demonstrated by inhibited male olfactory response to the female mating pheromone. In addition, exposure to 0.1 µg/L cypermethrin was found to decrease the number of fertilized eggs (Moore and Waring, 2001).

Delayed spawning and reduced larval survival were observed with bluegill sunfish after two applications of 1 µg/L esfenvalerate in littoral enclosures (Tanner and Knuth, 1996). Some evidence of decreased fecundity in adult Japanese medaka and a decrease in percentage of viable larvae were shown in response to a 7-d esfenvalerate exposure through food (148 mg/kg), although this was a study with low sample size (Werner et al., 2002). Another study found carp embryo hatching success was decreased by deltamethrin exposure (<50% at ≥0.5 µg/L) (Köprücü and Aydın, 2004).

The induction of heat shock proteins (hsp) indicates the occurrence of protein damage in cells and tissues. Werner et al. (2002) found that hsp were good indicators of esfenvalerate exposure in Japanese medaka. Dietary exposure to 21 mg/kg and 148 mg/kg esfenvalerate significantly elevated heat shock proteins in a 7-d exposure. Heat shock proteins were also found to be a sensitive indicator of esfenvalerate exposure in Chinook salmon, with hsp levels in the muscle increasing significantly at concentrations ≥0.01 µg/L after 96 h (Eder et al., 2009).

A study of the effects of cypermethrin on Korean rock fish (*Sebastes schlegeli*) found a significant decrease in red blood cell count, hemoglobin, and hematocrit after an 8-week exposure to 0.04 µg/L (Jee et al., 2005). The same study also documented decreases in serum total protein, albumin, cholesterol, and lysozyme activity and found significantly higher serum concentrations of glucose, bilirubin, and malondialdehyde. The activity of several enzymes (glutamic oxalate transaminase, glutamic pyruvate transaminase, alkaline phosphatase) and serum osmolality were also altered. These changes were attributed to cellular stress and tissue damage, and subsequent physiological response to increased energy demand (Jee et al., 2005). Biochemical responses measured in the common carp exposed for 96 h to formulated bifenthrin (Talstar EC10) were increased levels of plasma glucose, ammonia, aspartate aminotransferase, and creatine kinase (Velisek et al., 2009). Cypermethrin was found to increase brain acetylcholinesterase activity in rainbow trout exposed to 0.17 µg/L, 0.33 µg/L, and 0.49 µg/L for 10 days (Davies et al., 1994). In the same study, rainbow trout hepatic glutathione-S-transferase activity was found to increase at 0.87 µg/L cypermethrin after 10 days (Davies et al., 1994).

Oxygen consumption in carp was affected by exposure to fenvalerate (Reddy et al., 1992). Increased oxygen consumption was also measured in trout exposed to permethrin (Haya, 1989). Lipid peroxidation activity of the liver in juvenile red drum and the mummichog was significantly altered by permethrin exposures (Parent et al., 2011). There was an increase in lipid peroxidation activity of the liver after 24 h, followed by a decreasing trend after 96 h. The minimum effective dose was 1.23 µg/L for juvenile red drum and 11.1 µg/L for mummichog (Parent et al., 2011). Sayeed et al. (2003) reported sublethal effects of deltamethrin on the green snakehead fish, including increased lipid peroxidation activity, increased glutathione, and decreased catalase activity.

A significant reduction in liver glycogen levels of fathead minnow was observed after 96 h exposure to 0.20 µg/L esfenvalerate (Denton et al., 2003). Carp exposed to formulated bifenthrin had severe histological damage to the gills and degeneration of hepatocytes, and hepatocyte vacuole shape was typical of fatty degeneration of liver (Velisek et al., 2009).

Parent et al. (2011) demonstrated decreased spleen cell proliferation in response to permethrin (≥ 1.2 µg/L) using juvenile red drum. Higher levels of permethrin (33.3 µg/L) also inhibited mummichog spleen cell proliferation (Parent et al., 2011). Zelikoff et al. (2000) found that Japanese medaka exposed in the water for 48 h to a nonlethal concentration (i.e., 0.05 µg/L) of permethrin and then injected with a dose of the Gram-negative bacterial pathogen, *Yersinia ruckerii*, exhibited a 30% increase in mortality after 96 h than did unexposed, bacterially injected control fish. Eder et al. (2008) found that a 96-h exposure to 0.08 µg/L esfenvalerate altered the transcription of immune-system messenger molecules (cytokines) in juvenile Chinook salmon.

6.7. Mesocosm/Field Studies

A study conducted in 0.1 ha mesocosms found bluegill sunfish survival, biomass, adult male survival, and reproductive success were significantly decreased with esfenvalerate exposure (six 0.67 μg/L doses at two-week intervals) (Fairchild et al., 1992). A freshwater enclosure (5 × 10 m) study with esfenvalerate found 100% mortality of juvenile bluegill sunfish at 1 μg/L and 5 μg/L, and 45% mortality of larval fathead minnows and larval northern redbelly dace (*Phoxinus eos*) at 1 μg/L after 30 d (Lozano et al., 1992). A 28-d saltwater mesocosm study with bifenthrin found no significant effect on juvenile sheepshead minnow survival at concentrations up to 0.2 μg/L (dosed weekly) (Harper, 2007).

7. NEONICOTINOIDS

The neonicotinoid insecticides were developed as a result of research to understand the mechanism involved in the insecticidal properties of the naturally occurring compound nicotine, which had been used for several centuries to control insect pests. Research on the synthetic neonicotinoids began in the 1970s, and in 1985 the first neonicotinoid to be developed commercially, imidacloprid, was discovered (Silcox and Vittum, 2008). It was marketed in the 1990s and is now the top-selling insecticide around the world (Silcox and Vittum, 2008). Recently, concerns have been raised about the possibility that neonicotinoid exposure may be a factor in "colony collapse disorder" in bees (Henry et al., 2012).

7.1. Structure

Neonicotinoid compounds possess either a nitromethylene, nitroimine, or cyanoimine group. Insecticides in this group include imidacloprid, nitenpyram, acetamiprid, dinotefuran, clothianidin, and thiamethoxam (Figure 6.1). The neonicotinoid structure resembles nicotine and epibatidine, both of which are potent agonists of postsynaptic nicotinic acetylcholine receptors (nAChRs) (Matusda et al., 2001).

7.2. Mechanism of Action

Neonicotinoids irreversibly block acetylcholine receptors. These compounds have little or no affinity for mammalian nicotinic acetylcholine receptors (nAChRs), but are highly selective for target insect nAChRs (Matusda et al., 2001).

7.3. Uptake and Metabolism

Imidacloprid has a photodegradation half-life in water of less than 3 h (Moza et al., 1998). Without light, hydrolysis can range from 33 d to 44 d and imidacloprid has an estimated half-life on soil of 39 d (Moza et al., 1998). Imidacloprid has a high water solubility (0.51 g/L at 20°C) and was detected in groundwater near potato farms at a maximum concentration of 6.4 μg/L (CCME, 2007a). Potential for imidacloprid to leach into groundwater was also demonstrated by Armbrust and Peeler (2002). In surface runoff collected from potato farms in Prince Edward Island following rainfall events in 2001 and 2002, concentrations ranged from below the detection limit of 0.5 μg/L to 11.9 μg/L (CCME, 2007a). Imidacloprid has a low log Kow value (0.57), indicating a low potential for bioaccumulation (Moza et al., 1998).

7.4. Acute Toxicity

Neonicotinoids have low acute toxicity to fish (Table 6.2). Barbee and Stout (2009) summarize available toxicity data for several neonicotinoid compounds. The 96 h LC50 values reported for bluegill sunfish were 117 mg/L clothianidin; >99.3 mg/L dinotefuran; >114 mg/L thiamethoxam. The 96 h LC50 values reported for rainbow trout were 105 mg/L clothianidin; >99.5 mg/L dinotefuran; >100 mg/L thiamethoxam (Barbee and Stout, 2009).

A 96 h LC50 of 241 mg/L imidacloprid was determined for adult zebrafish, and a 48 h EC50 value of > 320 mg/L imidacloprid was determined for zebrafish embryo development (Tišler et al., 2009). The same study found increased toxicity due to the mixture of solvents and imidacloprid when testing the formulated product Confidor SL 200 (Tišler et al., 2009). A summary report by the USDA Forest Service cites imidacloprid 96 h LC50 values of >105 mg/L for bluegill sunfish, 211 mg/L for rainbow trout, and 161 mg/L for sheepshead minnow (Anatra-Cordone and Durkin, 2005). A 7 d LC50 value of 77 mg/L and a 7 d LOEC value of 34 mg/L for growth inhibition were reported for larval inland silverside (CCME, 2007a). A 96 h LC50 value of 161 mg/L was reported for adult sheepshead minnow (CCME, 2007a). Against target insect species, acetamiprid, imidacloprid, and thiamethoxam were found to increase in toxicity with an increase in temperature (from 27 to 37°C) (Boina et al., 2009).

7.5. Chronic/Sublethal Effects/Biomarkers

A chronic 60-d imidacloprid exposure beginning with newly fertilized rainbow trout eggs yielded an LOEC of 2.3 mg/L for growth, while no effects on hatching or survival were observed at the highest test concentration of 19 mg/L (Anatra-Cordone and Durkin, 2005; CCME, 2007a). Behavioral effects were observed in a 96-h imidacloprid exposure to juvenile rainbow trout, with a reported LOEC of 64 mg/L (CCME, 2007a). Development of zebrafish embryos was not impaired at imidacloprid concentrations ≤320 mg/L; however, equivalent concentrations of imidacloprid in the formulated product Confidor SL 200 caused a significant decrease in several developmental end-points, the most sensitive of which were blood circulation and heartbeat (Tišler et al., 2009).

7.6. Field Studies/Effects

A three-month study of imidacloprid-treated rice cultivation fields found significant sublethal effects on juvenile Japanese medaka (Sanchez-Bayo and Goka, 2005). Fish in the imidacloprid-treated paddies had significantly higher rates of ectoparasite (*Trichodina domerguei*) infection than the controls (Sanchez-Bayo and Goka, 2005). It was hypothesized that insecticide-induced physiological stress made the fish more susceptible to infection, and severely parasitized fish in the imidacloprid treatments did have significantly lower weight/length ratios than control fish (Sanchez-Bayo and Goka, 2005).

8. PHENYLPYRAZOLES

The only currently marketed phenylpyrazole is fipronil. It was first developed in 1987 and registered and marketed in 1996. Fipronil is used to control a variety of pests, including ants, fleas, ticks, termites, and mole crickets. It is used in seed treatments (a particular exposure to fish in Brazil where carp are co-farmed in rice fields; Clasen et al., 2012), granular turf products, liquid termiticides, agricultural products, and topical pet products (Silcox and Vittum, 2008).

8.1. Structure

The phenylpyrazole insecticide, fipronil, is an N-phenylpyrazole with a trifluoromethyl-sulfinyl substituent (Figure 6.1). Fipronil is chiral and is formulated as a racemate (50% each of the + and − enantiomers).

8.2. Mechanism of Action

The primary mode of action of fipronil is disruption of the passage of chloride ions through the γ-aminobutyric acid (GABA)-regulated chloride channel, resulting in loss of neuronal signaling, hyperexcitation, and subsequent mortality (Gant et al., 1998). Fipronil is highly selective in its toxicity, having greater affinity for arthropod GABA receptors than mammalian receptors (Hainzl et al., 1998; Tingle et al., 2003).

8.3. Uptake and Metabolism

Although enantiomers have identical physical and chemical properties and abiotic degradation rates (Garrison, 2006), their individual toxicity, biological activity, and microbial degradation rates have been shown to differ (Jones et al., 2007; Konwick et al., 2006a, 2006b). For example, biodegradation experiments in anoxic sediments indicated that the fipronil (+) enantiomer degrades faster under sulfidogenic conditions, while the (−) enantiomer degrades faster under methanogenic conditions (Jones et al., 2007). Fipronil undergoes photolysis to form the photoproduct desulfinyl, which is more stable to photolysis but with similar toxicity (Hainzl et al., 1998), leading to potential for environmental persistence and nontarget toxicity (Ngim et al., 2000). Moreover, the primary fipronil metabolite sulfone is also more persistent in the environment and more toxic to fish than the parent compound (Hainzl et al., 1998).

Juvenile rainbow trout rapidly accumulated fipronil, but it was also rapidly depurated (half-life 0.6 d) (Konwick et al., 2006b). The parent compound was converted to fipronil sulfone, which persisted longer (2 d) in the fish (Konwick et al., 2006b). Demcheck and Skrobialowski (2003) reported parent fipronil concentrations as high as 5.29 μg/L in surface waters surrounded by rice agriculture. They also found the degradation product fipronil sulfone at concentrations as high as 10.5 μg/kg.

8.4. Acute Toxicity

As summarized by Tingle et al. (2003), fipronil has 96 h LC50 values of 42 μg/L for Nile tilapia, 83 μg/L for bluegill sunfish, 246 μg/L for rainbow trout, 340 μg/L for Japanese carp, and 430 μg/L for carp. A 96 h LC50 of 130 μg/L was reported for sheepshead minnows (U.S. EPA, 1999) (Table 6.2). The metabolite fipronil sulfone was found to have greater toxicity than the parent compound to rainbow trout (6.3 times more toxic) and bluegill (3.3 times more toxic; Tingle et al., 2003).

The presence of dissolved organic matter has been shown to increase the acute toxicity of fipronil to copepods (Bejarano et al., 2005). While species-specific differences in sensitivity to fipronil have been documented, there is no apparent difference in sensitivity between freshwater and saltwater species (Overmyer et al., 2007; U.S. EPA, 1999).

8.5. Chronic/Sublethal Effects/Biomarkers

In a chronic study, carp were exposed to an initial fipronil concentration of 0.65 mg/L for 7, 30, and 90 d. Sublethal biomarker results showed an increase in liver superoxide dismutase activity, lipid peroxidation activity (liver, muscle and brain), and protein carbonyl; an inhibition of liver catalase activity; and no change in glutathione-S-transferase activity (Clasen et al., 2012).

Zebrafish embryos exposed to fipronil concentrations $\geq$333 µg/L had nervous system developmental effects, including notochord degeneration, shorter rostral-caudal body size, and impaired mobility in response to stimuli (Stehr et al., 2006). In early life stage testing with rainbow trout, fipronil decreased larval growth (NOEC of 6.6 µg/L and LOEC of 15 µg/L) (Tingle et al., 2003). Fipronil was found to impair fathead minnow swimming ability at 142 µg/L (Beggel et al., 2010).

9. MICROBIAL-BASED INSECTICIDES

9.1. Bt

Over 1500 microbial products have been identified as possible insecticides (Miller et al., 1983). Bacterial insecticides represent the most widely used group of microbial products. *Bacillus thuringiensis* (Bt) is the most widely used bacterial insecticide. It was first isolated in 1901 from dying silkworm larvae. Twenty-two different varieties have been identified since that time, and some have become widely used. Bti (*Bacillus thuringiensis* subspecies *israelensis*) and *Bacillus sphaericus* are used as larvicides in mosquito control. They are toxic to a number of insect pests, particularly mosquitoes and biting flies (Ray, 1991).

9.1.1. STRUCTURE AND MECHANISM OF ACTION

Bacillus thuringiensis is a Gram-positive, rod-shaped, spore-forming bacterium. The mode of action of Bti involves the synergistic interaction of insecticidal crystal proteins (Cry4A, Cry4B, Cry11Aa, and Cyt1Aa) (Glare and O'Callaghan, 1998). Bti products must be ingested by the larval stage of the mosquito to cause mortality. Following ingestion, the parasporal crystals are dissolved in the alkaline larval insect midgut, releasing the endotoxin that binds to and lyses midgut epithelial cells, causing digestive paralysis. Larval mortality usually occurs 1–6 h after ingestion (Glare and O'Callaghan, 1998).

9.1.2. UPTAKE AND METABOLISM

Bti does not persist in the environment after application; reports of activity after application show a decline in efficacy within days and little residual activity after several weeks (Wirth et al., 1997). The persistence of Bti after application is dependent on the type of formulated product used, with some formulations (pellets/briquettes) designed specifically to enhance residual activity. VectoBac® is a common Bti product.

For microbial-based insecticides, transfer of genetic material must also be considered in the environmental risk assessment. Some of the toxic proteins of Bti are encoded in genes on plasmids which can be exchanged between strains and species. While genetic transfer between Bti and other soil bacteria has been demonstrated in the laboratory (in culture, insects, and sterile soils), it has not been shown to occur in the field (Glare and O'Callaghan, 1998).

9.1.3. TOXICITY

Fish toxicity is generally low for *Bacillus* insecticides (Table 6.2), with little effects noted either in the laboratory or after field application (U.S. EPA, 1998). The sheepshead minnow had a reported 96 h LC50 value of 7.9 mg/L (U.S. EPA, 2000). A 96 h LC50 of 980 mg/L Bti was reported for the mummichog (Lee and Scott, 1989), whereas a 96 h LC50 of 423.17 mg/L was reported for the mummichog exposed to *B. sphaericus* (Key and Scott, 1992). No mortality was observed in zebrafish or Nile tilapia exposed to *B. thuringiensis* or *B. sphaericus* at 10^6 spores/mL, and no genotoxicity (as measured by cellular apoptosis in Nile tilapia) was detected (Grisolia et al., 2009). Bti was applied to bait ponds containing the golden shiner (*Notemigonus crysoleucas*), with no adverse effects on fish noted at concentrations of 2 kg/ha (USFWS, 2003).

Bti toxicity is positively correlated with temperature, with toxicity increasing in target insects as water temperature increased (Atwood et al., 1992). The increased presence of organic matter has been shown to decrease the toxicity of Bti, most likely due to decreased bioavailability of Bti in the water column (Glare and O'Callaghan, 1998). Bti effectiveness was not affected by pH within the range of approximately 5–8.6 (Glare and O'Callaghan, 1998).

9.2. Spinosad

9.2.1. STRUCTURE

Spinosad is a macrolide derived from the aerobic fermentation of the actinomycete *Saccharopolyspora spinosa*, a bacterial organism isolated from soil. Spinosad is composed of spinosyns A and D, found in a ratio of approximately 85:15 in commercial formulations (Figure 6.1).

9.2.2. MECHANISM OF ACTION

The specific mode of action of spinosad is to alter the function of nicotinic and GABA-gated ion channels, causing rapid excitation of the insect nervous system, leading to involuntary muscle contractions, tremors, paralysis, and death (Salgado, 1998).

9.2.3. UPTAKE AND METABOLISM

Spinosad breaks down in the aquatic environment within 1–2 d in the presence of sunlight but can persist approximately 200 d if photolysis does not occur (Cleveland et al., 2002). A bioaccumulation factor for spinosad A in rainbow trout of 114 was reported (U.S. EPA, 2011b). Several spinosad degradation compounds have been shown to have equal or greater toxicity than the parent compound (U.S. EPA, 2011b).

9.2.4. TOXICITY

Cleveland et al. (2002) summarize acute toxicity (96 h LC50) values for spinosad and various fish species: 4.99 mg/L (carp), 7.87 mg/L (sheepshead minnow), 5.9 mg/L (bluegill sunfish), and 30 mg/L (rainbow trout) (Table 6.2). Chronic values (NOEC) for early life stage testing with sheepshead minnow and bluegill sunfish were reported as 1.15 mg/L and 1.12 mg/L, respectively (Cleveland et al., 2002). Early life stage testing with rainbow trout found significant reductions in egg hatchability, growth, and survival at spinosad concentrations of 0.962 mg/L, 3.67 mg/L, and 1.89 mg/L, respectively (U.S. EPA, 2011b).

10. INSECT GROWTH REGULATORS

Insect growth and development is characterized by periodic molting during which the old exoskeleton is replaced with a new one. Insect growth regulators (IGRs) are chemicals that inhibit the molting life cycle of insects and are used to control various insect pests including mosquitoes, cockroaches, and fleas.

10.1. Structure and Mechanism of Action

Some IGRs are designed to mimic the juvenile hormone in insects. An example of this class of IGR is methoprene (Figure 6.1), which was first registered and marketed in 1975 (Henrick, 2007). Juvenile growth hormone mimics, such as methoprene, fenoxycarb, and pyriproxyfen, interfere with insect maturation and prevent the insect from reaching the adult stage. Altosid® is a common methoprene product used for larval mosquito control. Molting accelerating compounds (MACs) mimic the natural function of the molting hormone ecdysone and cause early and repeated molts, thus preventing development and leading to insect death (Nauen and Bretschneider, 2002). Tebufenozide, methoxyfenozide, halofenozide, and chromafenozide are examples of this type of IGR.

Other IGRs inhibit chitin synthesis (necessary for exoskeleton formation). Chitin synthesis inhibitors can also kill eggs by disrupting normal embryonic development. Diflubenzuron works in this way and was first marketed in 1976. Buprofezin is another example of a chitin synthesis inhibiting compound (Nauen and Bretschneider, 2002).

10.2. Toxicity

The 96 h LC50 toxicity value determined for the mummichog was 124.95 mg/L methoprene EC® (Lee and Scott, 1989). Acute toxicity values (96 h LC50) reported for rainbow trout, fathead minnow, and bluegill sunfish were >50 mg/L, >10 mg/L, and 1.52 mg/L methoprene, respectively (U.S. EPA, 2000) (Table 6.2). Additional methoprene 96 h LC50 values of 86 mg/L and >100 mg/L were reported for Coho salmon and channel catfish, respectively (CCME, 2007b). Early life stages were found to be more sensitive to methoprene, with a 7 d EC50 for embryo survival of 0.65 mg/L for rainbow trout (CCME, 2007b).

No behavioral effects on swimming ability were found in either the goldfish or mosquitofish after a two-week exposure to 2 mg/L methoprene (CCME, 2007b). Two field studies with mosquitofish found no effects of methoprene after repeated aerial applications (every 2 d for 12 d and 6 times in an 18-month period) (USFWS, 2003).

10.3. Uptake and Metabolism

Methoprene degradation occurs by both photolysis and bacterial metabolism (80% loss within 13 d) in both fresh and saline waters (U.S. EPA, 1991). The reported bioconcentration factor for fish (bluegill sunfish) is 457 (U.S. EPA, 1991).

Salinity was not found to affect the degradation rate of methoprene, but methoprene does break down more quickly at higher temperatures. The presence of organic matter was not found to alter methoprene toxicity (CCME, 2007b).

11. MIXTURES

An assessment of organophosphate mixtures found synergistic toxicity to Coho salmon; with combinations of chlorpyrifos and malathion, and diazinon and malathion exhibiting greater than predicted acetylcholinesterase inhibition (Laetz et al., 2009). A previous study with Chinook salmon, however, found only additive effects on acetylcholinesterase with binary mixtures of organophosphates (chlorpyrifos, malathion, and diazinon) and carbamates (carbaryl and carbofuran) (Scholz et al., 2006). Based on surface-water monitoring data for organophosphates and carbamates in the Pacific Northwest, Moore and Teed (2012) concluded that mixtures of these compounds rarely occur at levels that would cause toxicity in salmon.

The carbamate isoprocarb and the organophosphate chlorpyrifos displayed additive toxicity to goldfish for acetylcholinesterase and glutathione-S-transferase activities (Wang et al., 2010). Pyrethroid and organophosphate mixtures (binary combinations of dichlorvos and permethrin or tetramethrin, phoxim, and permethrin or tetramethrin or bifenthrin) displayed additive toxicity to zebrafish, although synergistic toxicity was noted for the mixture of phoxim and etofenprox, and an antagonistic effect was observed with dichlorvos/etofenprox and diclorvos/bifenthrin mixtures (Zhang et al., 2010).

Chlorpyrifos combined with the heavy metal nickel was found to cause synergistic increases in metallothionein accumulation in the sea bass (Banni et al., 2011). Contrary to findings in invertebrate species, the herbicide atrazine in mixture with chlorpyrifos did not cause synergistic toxicity in bluegill sunfish at environmentally relevant concentrations (Wacksman et al., 2006).

12. DATA GAPS

Many contemporary-use insecticides, particularly those used to replace OCs and OPs, still have gaps in their environmental fate and effects data. For example, limited toxicity studies are available for imidacloprid and fipronil. Fewer data are available for marine and estuarine fish species compared to freshwater forms. A limited number of field studies have been conducted examining the direct and indirect effects of insecticides on fish populations. Mesocosm studies could supplement laboratory and field testing for many insecticides by providing environmental fate data, sublethal toxicity data, and indirect effect data as may be related to loss of invertebrate prey. More information is needed to describe how abiotic variables such as organic carbon, temperature, pH, and salinity may alter insecticide toxicity to fish.

13. SUMMARY

The insecticides that pose the greatest risk to fish have changed over time. The OCs, which are now largely banned, remain a concern due to their persistence in sediments and their potential for bioaccumulation. Of the more recently developed insecticide classes (pyrethroids, neonicotinoids, and phenylpyrazoles), only the pyrethroids are considered highly toxic to fish (Figure 6.2). The use of OPs is declining; however, they continue to be

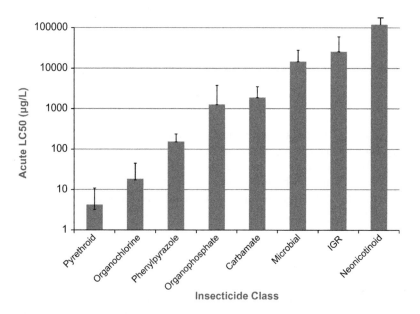

Figure 6.2. Acute toxicity values for fish by insecticide class. Values are average of rainbow trout, bluegill sunfish, and sheepshead minnow LC50 values on a logarithmic scale. Error bars denote standard deviation. See color plate at the back of the book.

widely used in agriculture. The use of pyrethroid insecticides continues to increase as they are being used to replace OPs and carbamates, particularly in nonagricultural applications. Insufficient monitoring data are available to fully describe pyrethroid exposure to fish. Furthermore, pyrethroids and other modern insecticides such as the phenylpyrazole fipronil are highly toxic to invertebrate species, and this may decrease food availability for fish. Other data gaps include the lack of information on the effects of modern insecticides in mixtures as well as how their toxicity may be affected by environmental variables such as temperature, salinity, and dissolved oxygen. Additional work is also needed to better characterize the potential sublethal effects of modern insecticides. Continued development of less toxic insecticides and improved regulation of existing compounds, along with runoff-mitigating techniques such as buffers, will decrease the risk of insecticides to fish.

REFERENCES

Adelman, I. R., Smith, L. L., Jr. and Siesennop, G. D. (1976). Chronic toxicity of guthion to the fathead minnow (*Pimephales promelas* Rafinesque). *Bull. Environ. Contam. Toxicol.* 15, 726–733.
Adhikari, S., Sarkar, B., Chattopadhyay, A., Chattopadhyay, D. N., Sarkar, S. K. and Ayyappan, S. (2008). Carbofuran induced changes in breeding of a freshwater fish, *Labeo rohita* (Hamilton). *Toxicol. Environ. Chem.* 90, 457–465.
Anadu, D. I., Scott, G. I. and Fulton, M. H. (1999). Toxicity of DDT to the different life stages of the mummichog *Fundulus heteroclitus* (Wabum). *Bull. Environ. Contam. Toxicol.* 63, 181–187.
Anatra-Cordone, M. and Durkin, P. (2005). Imidacloprid—Human Health and Ecological Risk Assessment—Final Report. In *SERA TR 05-43-24-03a* (ed. USDA), p. 283. Arlington, VA: United States Department of Agriculture, Forest Service.
Andreasen, J. K. (1985). Insecticide resistance in mosquitofish of the Lower Rio Grande valley of Texas—An ecological hazard? *Arch. Environ. Contam. Toxicol.* 14, 573–577.
Armbrust, K. L. and Peeler, H. B. (2002). Effects of formulation on the run-off of imidacloprid from turf. *Pest Manag. Sci.* 58, 702–706.
ATSDR (2002). *Toxicological Profile for 4,4'-DDT, 4,4'-DDE, 4, 4'-DDD (Update)*. Atlanta, GA: U.S. Department of Health and Human Services, Public Health Service, Agency for Toxic Substances and Diseases Registry, p. 497.
Assis, C. R. D., Castro, P. F., Amaral, I. P. G., Carvalho, E. V. M. M., Carvalho, L. B. and Bezerra, R. S. (2010). Characterization of acetylcholinesterase from the brain of the Amazonian tambaqui (*Colossoma macropomum*) and *in vitro* effect of organophosphorus and carbamate pesticides. *Environ. Toxicol. Chem.* 29, 2243–2248.
Atwood, D. W., Robinson, J. V., Meisch, M. V., Olson, J. K. and Johnson, D. R. (1992). Efficacy of Bacillus thuringiensis var. israelensis against larvae of the southern buffalo gnat, Cnephia pecuarum (Diptera: Simuliidae), and the influence of water temperature. *J. Am. Mosq. Control Assoc.* 8, 126–130.

Bailey, H. C., Alexander, C., Digiorgio, C., Miller, M., Doroshov, S. I. and Hinton, D. E. (1994). The effect of agricultural discharge on striped bass (*Morone saxatilis*) in California's Sacramento- San Joaquin drainage. *Ecotoxicology* 3, 123–142.

Banni, M., Jebali, J., Guerbej, H., Dondero, F., Boussetta, H. and Viarengo, A. (2011). Mixture toxicity assessment of nickel and chlorpyrifos in the sea bass *Dicentrarchus labrax*. *Arch. Environ. Contam. Toxicol.* 60, 124–131.

Barbee, G. C. and Stout, M. J. (2009). Comparative acute toxicity of neonicotinoid and pyrethroid insecticides to non-target crayfish (Procambarus clarkii) associated with rice–crayfish crop rotations. *Pest Manag. Sci.* 65, 1250–1256.

Baron, R. L. and Merriam, T. L. (1988). Toxicology of aldicarb. *Rev. Environ. Cont. Tox.* 105, 2–70.

Baron, R. L. (1991). Carbamate insecticides. In *Handbook of Pesticide Toxicology Volume 3 Classes of Pesticides* (eds. W. J. Hayes, Jr. and E. R. Laws, Jr), Chapter 17, pp. 1125–1189. San Diego: Academic Press.

Barron, M. G., Plakas, S. M. and Wilga, P. C. (1991). Chlorpyrifos pharmacokinetics and metabolism following intravascular and dietary administration in channel catfish. *Toxicol. Appl. Pharm.* 108, 474–482.

Barry, M. J., Logan, D. C., Ahokas, J. T. and Holdway, D. A. (1995). Effects of esfenvalerate pulse-exposure on the survival and growth of larval Australian crimson-spotted rainbow fish (*Melanotaenia fluviatilis*). *Environ. Toxicol. Water Qual.* 10, 267–274.

Beauvais, S. L., Jones, S. B., Brewer, S. K. and Little, E. E. (2000). Physiological measures of neurotoxicity of diazinon and malathion to larval rainbow trout (*Oncorhynchus mykiss*) and their correlation with behavioral measures. *Environ. Toxicol. Chem.* 19, 1875–1880.

Beggel, S., Werner, I., Connon, R. E. and Geist, J. P. (2010). Sublethal toxicity of commercial insecticide formulations and their active ingredients to larval fathead minnow (*Pimephales promelas*). *Sci. Total Environ.* 408, 3169–3175.

Bejarano, A. C., Chandler, G. T. and Decho, A. W. (2005). Influence of natural dissolved organic matter (DOM) on acute and chronic toxicity of the pesticides chlorothalonil, chlorpyrifos, and fipronil on the meiobenthic estuarine copepod *Amphiascus tenuiremis*. *J. Exp. Mar. Biol. Ecol.* 321, 43–57.

Berg, V., Ugland, K. I., Hareide, N. R., Aspholm, P. E., Polder, A. and Skaare, J. U. (1997). Organochlorine contamination in deep-sea fish from the Davis Strait. *Mar. Environ. Res.* 44, 135–148.

Beyger, L., Orrego, R., Guchardi, J. and Holdway, D. (2012). The acute and chronic effects of endosulfan pulse-exposure on *Jordanella floridae* (Florida flagfish) over one complete life-cycle. *Ecotox. Environ. Saf.* 76, 71–78.

Blanar, C. A., Curtis, M. A. and Chan, H. M. (2005). Growth, nutritional composition, and hematology of arctic charr (*Salvelinus alpinus*) exposed to toxaphene and tapeworm (*Diphyllobothrium dendriticum*) larvae. *Arch. Environ. Contam. Toxicol.* 48, 397–404.

Blus, L. J. (1995). Organochlorine pesticides. In *Handbook of Ecotoxicology* (eds. D. J. Hoffman, B. A. Rattner, G. A. Burton and J. Cairns), p. 733. Boca Raton. FL: Lewis Publishers.

Boina, D. R., Onagbola, E. O., Salyani, M. and Stelinski, L. L. (2009). Influence of posttreatment temperature on the toxicity of insecticides against *Diaphorina citri* (Hemiptera: Psyllidae). *J. Econ. Entomol.* 102, 685–691.

Bradbury, S. P. and Coats, J. R. (1989). Toxicokinetics and toxicodynamics of pyrethroid insecticides in fish. *Environ. Toxicol. Chem.* 8, 373–380.

Breckenridge, C. B., Holden, L., Sturgess, N., Weiner, M., Sheets, L., Sargent, D., Soderlund, D. M., Choi, J.-S., Symington, S., Clark, J. M., et al. (2009). Evidence for a separate mechanism of toxicity for the Type I and the Type II pyrethroid insecticides. *NeuroToxicology* 30, S17–S31.

Bretaud, S., Saglio, P., Saligaut, C. and Auperin, B. (2002). Biochemical and behavioral effects of carbofuran in goldfish (*Carassius auratus*). *Environ. Toxicol. Chem.* 21, 175–181.

Brown, S. B., Adams, B. A., Cyr, D. G. and Eales, J. G. (2004). Contaminant effects on the teleost fish thyroid. *Environ. Toxicol. Chem.* 23, 1680–1701.

Burridge, L., Weis, J. S., Cabello, F., Pizarro, J. and Bostick, K. (2010). Chemical use in salmon aquaculture: A review of current practices and possible environmental effects. *Aquaculture* 306, 7–23.

Carriger, J. F. and Rand, G. M. (2008). Aquatic risk assessment of pesticides in surface waters in and adjacent to the Everglades and Biscayne National Parks: I. Hazard assessment and problem formulation. *Ecotoxicology* 17, 660–679.

Carriger, J. F., Hoang, T. C., Rand, G. M., Gardinali, P. R. and Castro, J. (2011). Acute toxicity and effects analysis of endosulfan sulfate to freshwater fish species. *Arch. Environ. Contam. Toxicol.* 60, 281–289.

Carriquiriborde, P., Marino, D. J., Giachero, G., Castro, E. A. and Ronco, A. E. (2012). Global metabolic response in the bile of pejerrey (*Odontesthes bonariensis*, Pisces) sublethally exposed to the pyrethroid cypermethrin. *Ecotox. Environ. Saf.* 76, 46–54.

CCME (2007a). *Canadian water quality guidelines for the protection of aquatic life: Imidacloprid. Canadian Environmental Quality Guidelines.* Winnipeg: Canadian Council of Ministers of the Environment, p. 8.

CCME (2007b). *Canadian water quality guidelines for the protection of aquatic life: Methoprene. Canadian Environmental Quality Guidelines.* Winnipeg: Canadian Council of Ministers of the Environment, p. 11.

Chambers, J. E., Meek, E. C. and Ross, M. (2010). The metabolic activation and detoxication of anticholinesterase insecticides. In *Anticholinesterase Pesticides: Metabolism, Neurotoxicity, and Epidemiology* (eds. T. Satoh and R. Gupta), p. 600. Hoboken, NJ: John Wiley & Sons.

Chatterjee, S., Kumar Dasmahapatra, A. and Ghosh, R. (2001). Disruption of pituitary-ovarian axis by carbofuran in catfish, *Heteropneustes fossilis* (Bloch). *Comp. Biochem. Physiol. Part C: Toxicol. Pharm.* 129, 265–273.

Clark, J. R., Goodman, L. R., Borthwick, P. W., Patrick, J. M., Cripe, G. M., Moody, P. M., Moore, J. C. and Lores, E. M. (1989). Toxicity of pyrethroids to marine invertebrates and fish: a literature review and test results with sediment-sorbed chemicals. *Environ. Toxicol. Chem.* 8, 393–401.

Clasen, B., Loro, V. L., Cattaneo, R., Moraes, B., Lópes, T., de Avila, L. A., Zanella, R., Reimche, G. B. and Baldisserotto, B. (2012). Effects of the commercial formulation containing fipronil on the non-target organism *Cyprinus carpio*: Implications for rice–fish cultivation. *Ecotox. Environ. Saf.* 77, 45–51.

Cleveland, C. B., Mayes, M. A. and Cryer, S. A. (2002). An ecological risk assessment for spinosad use on cotton. *Pest Manag. Sci.* 58, 70–84.

Coats, J. R., Symonik, D. M., Bradbury, S. P., Dyer, S. D., Timson, L. K. and Atchison, G. J. (1989). Toxicology of synthetic pyrethroids in aquatic organisms: An overview. *Environ. Toxicol. Chem.* 8, 671–679.

Coats, J. R. (1990). Mechanisms of toxic action and structure-activity relationships for organochlorine and synthetic pyrethroid insecticides. *Environ. Health Perspect.* 87, 255–262.

Cochran, R. E. and Burnett, L. E. (1996). Respiratory responses of the salt marsh animals, *Fundulus heteroclitus, Leiostomus xanthurus*, and *Palaemonetes pugio* to environmental hypoxia and hypercapnia and to the organophosphate pesticide, azinphosmethyl. *J. Exp. Mar. Biol. Ecol.* 195, 125–144.

Coelho, S., Oliveira, R., Pereira, S., Musso, C., Domingues, I., Bhujel, R. C., Soares, A. M. V. M. and Nogueira, A. J. A. (2011). Assessing lethal and sub-lethal effects of trichlorfon on different trophic levels. *Aquat. Toxicol.* 103, 191–198.

Cong, N. V., Phuong, N. T. and Bayley, M. (2009). Effects of repeated exposure of diazinon on cholinesterase activity and growth in snakehead fish (*Channa striata*). *Ecotox. Environ. Saf.* 72, 699–703.

Cope, O. B. (1966). Contamination of the freshwater ecosystem by pesticides. *J. Appl. Ecol.* 3, 33–44.

Coppage, D., Matthews, E., Cook, G. and Knight, J. (1975). Brain acetylcholinesterase inhibition in fish as a diagnosis of environmental poisoning by malathion, 0,0-dimethyl S-(1,2-dicarbethoxyethyl) phosphorodithioate. *Pest. Biochem. Physiol.* 5, 536–542.

Cottam, C. and Higgins, E. (1946). DDT and its effect on fish and wildlife. *J. Econ. Entomol.* 39, 44–52.

Couch, J. A., Winstead, J. T. and Goodman, L. R. (1977). Kepone-induced scoliosis and its histological consequences in fish. *Science* 197, 585–587.

Cox, C. (1998). Permethrin insecticide fact sheet. *J. Pest. Reform* 18, 1–20.

Cripe, G., Goodman, L. and Hansen, D. (1984). Effect of chronic exposure to EPN and to Guthion on the critical swimming speed and brain acetylcholinesterase activity of *Cyprinodon variegatus. Aquatic Tox.* 5, 255–266.

Davies, J. H. (1985). The pyrethroids: an historical introduction. In *The Pyrethroid Insecticides* (ed. J. P. Leahey), Chapter 1, p. 41. Boca Raton, FL: Taylor and Francis.

Davies, P. E., Cook, L. S. J. and Goenarso, D. (1994). Sublethal responses to pesticides of several species of australian freshwater fish and crustaceans and rainbow trout. *Environ. Toxicol. Chem.* 13, 1341–1354.

DeLorenzo, M. E. and De Leon, R. G. (2010). Toxicity of the insecticide etofenprox to three life stages of the grass shrimp. *Palaemonetes pugio. Arch. Environ. Contam. Toxicol.* 58, 985–990.

DeLorenzo, M. E., Serrano, L., Chung, K. W., Hoguet, J. and Key, P. B. (2006). Effects of the insecticide permethrin on three life stages of the grass shrimp, *Palaemonetes pugio. Ecotox. Environ. Saf.* 64, 122–127.

Demcheck, D.K. and Skrobialowski, S.C. (2003). Fipronil and degradation products in the rice-producing areas of Mermentau River Basin, Louisiana, February-September 2000. USGS Fact Sheet FS-010-03. USDOI, USGS. Pp. 3–7.

Denton, D. L., Wheelock, C. E., Murray, S. A., Deanovic, L. A., Hammock, B. D. and Hinton, D. E. (2003). Joint acute toxicity of esfenvalerate and diazinon to larval fathead minnows (*Pimephales promelas*). *Environ. Toxicol. Chem.* 22, 336–341.

Derache, R. (1977). *Organophosphorus Pesticides: Criteria (Dose/Effect Relationships) for Organophosphorus Pesticides.* Oxford: Pergamon Press.

Di Giulio, R. T. and Hinton, D. E. (2008). *The Toxicology of Fishes.* Boca Raton, FL: CRC Press.

Dunier, M., Vergnet, C., Siwicki, A. K. and Verlhac, V. (1995). Effect of lindane exposure on rainbow trout (*Oncorhynchus mykiss*) immunity: IV. Prevention of nonspecific and specific immunosuppression by dietary vitamin c (ascorbate-2-polyphosphate). *Ecotox. Environ. Saf.* 30, 259–268.

Dustman, E. H. and Stickel, L. F. (1969). The occurrence and significance of pesticide residues in wild animals. *Ann. N. Y. Acad. Sci.* 160, 162–172.

Dwyer, F. J., Mayer, F. L., Sappington, L. C., Buckler, D. R., Bridges, C. M., Greer, I. E., Hardesty, D. K., Henke, C. E., Ingersoll, C. G., Kunz, J. L., et al. (2005). Assessing contaminants sensitivity of endangered and threatened aquatic species: Acute toxicity of five chemicals. *Arch. Environ. Contam. Toxicol.* 48, 143–154.

Dyer, R. A., Coats, J. R., Bradbury, S. P., Atchison, G. J. and Clark, J. M. (1989). Effects of water hardness and salinity on the acute toxicity and uptake of fenvalerate by bluegill (*Lepomis macrochirus*). *Bull. Environ. Contam. Toxicol.* 42, 359–366.

Ecobichon, D. J. (1996). Toxic effects of pesticides. In *Casarett and Doull's Toxicology: The Basic Science of Poisons* (ed. C. D. Klaassen), 5th ed, p. 1111. New York: McGraw-Hill. (pp. 643–689)

Ecobichon, D. J. (2001). Toxic effects of pesticides. In *Casarett and Doull's Toxicology: The Basic Science of Poisons* (ed. C. D. Klaassen), 6th ed, p. 1236. New York: McGraw-Hill.

Eder, K. J., Clifford, M. A., Hedrick, R. P., Köhler, H.-R. and Werner, I. (2008). Expression of immune-regulatory genes in juvenile Chinook salmon following exposure to pesticides and infectious hematopoietic necrosis virus (IHNV). *Fish and Shellfish Immunol.* 25, 508–516.

Eder, K. J., Leutenegger, C. M., Köhler, H.-R. and Werner, I. (2009). Effects of neurotoxic insecticides on heat-shock proteins and cytokine transcription in Chinook salmon (*Oncorhynchus tshawytscha*). *Ecotox. Environ. Saf.* 72, 182–190.

Ensibi, C., Hernández-Moreno, D., Míguez Santiyán, M. P., Daly Yahya, M. N., Rodríguez, F. S. and Pérez-López, M. (2012). Effects of carbofuran and deltamethrin on acetylcholinesterase activity in brain and muscle of the common carp. *Environ. Toxicol.* Online, doi:10.1002/tox.21765.

Ernst, W., Jackman, P., Doe, K., Page, F., Julien, G., MacKay, K. and Sutherland, T. (2001). Dispersion and toxicity to non-target aquatic organisms of pesticides used to treat sea lice on salmon in net pen enclosures. *Mar. Poll. Bull.* 42, 432–443.

Fairchild, J. F., La Point, T. W., Zajicek, J. L., Nelson, M. K., Dwyer, F. J. and Lovely, P. A. (1992). Population-, community- and ecosystem-level responses of aquatic mesocosms to pulsed doses of a pyrethroid insecticide. *Environ. Toxicol. Chem.* 11, 115–129.

Ferrari, A., Venturino, A. and Pechen de D'Angelo, A. M. (2004). Time course of brain cholinesterase inhibition and recovery following acute and subacute azinphosmethyl, parathion and carbaryl exposure in the goldfish (*Carassius auratus*). *Ecotox. Environ. Saf.* 57, 420–425.

Ferrari, A., Venturino, A. and Pechén de D'Angelo, A. M. (2007). Effects of carbaryl and azinphos methyl on juvenile rainbow trout (*Oncorhynchus mykiss*) detoxifying enzymes. *Pest. Biochem. Physiol.* 88, 134–142.

Floyd, E. Y., Geist, J. P. and Werner, I. (2008). Acute, sublethal exposure to a pyrethroid insecticide alters behavior, growth, and predation risk in larvae of the fathead minnow (Pimephales promelas). *Environ. Toxicol. Chem.* 27, 1780–1787.

Fukuto, T. R. (1990). Mechanism of action of organophosphorus and carbamate insecticides. *Environ. Health Perspect.* 87, 245–254.

Fulton, M. and Scott, G. (1991). The effects of certain intrinsic and extrinsic variables on the acute toxicity of selected organophosphorus insecticides to the mummichog, *Fundulus heteroclitus*. *J. Environ. Sci. Health* B26, 459–478.

Fulton, M. H. and Key, P. B. (2001). Acetylcholinesterase inhibition in estuarine fish and invertebrates as an indicator of organophosphorus insecticides exposure and effects. *Environ. Toxicol. Chem.* 20, 37–45.

Gan, J. (2008). *Synthetic Pyrethroids: Occurrence and Behavior in Aquatic Environments.* Washington, DC: American Chemical Society.

Gant, D. B., Chalmers, A. E., Wolff, M. A., Hoffman, H. B. and Bushey, D. F. (1998). Fipronil: action at the GABA receptor. *Rev. Toxicol.* 2, 147–156.

Garrison, A. W. (2006). Probing the enantioselectivity of chiral pesticides. *Environ. Sci. Technol.* 40, 16–23.

Giddings, J. M., Biever, R. C. and Racke, K. D. (1997). Fate of chlorpyrifos in outdoor pond microcosms and effects on growth and survival of bluegill sunfish. *Environ. Toxicol. Chem.* 16, 2353–2362.

Giesy, J., Solomon, K., Coats, J. R., Dixon, K., Giddings, J. and Kenaga, E. (1999). Chlorpyrifos: ecological risk assessment in North American aquatic environments. *Rev. Environ. Cont. Tox.* 160, 1–129.

Glare, T.R. and O'Callaghan, M. (1998). Environmental and health impacts of *Bacillus thuringiensis israelensis* (ed. G. D. Ministry of Health Biocontrol and Biodiversity, AgResearch), p. 58. New Zealand.

Goodman, L. R., Hemmer, M. J., Middaugh, D. P. and Moore, J. C. (1992). Effects of fenvalerate on the early life stages of topsmelt (Atherinops Affinis). *Environ. Toxicol. Chem.* 11, 409–414.

Grisolia, C., Oliveira-Filho, E., Ramos, F., Lopes, M., Muniz, D. and Monnerat, R. (2009). Acute toxicity and cytotoxicity of *Bacillus thuringiensis* and *Bacillus sphaericus* strains on fish and mouse bone marrow. *Ecotoxicology* 18, 22–26.

Grube, A., Donaldson, D., Kiely, T. and Wu, L. (2011). Pesticide industry sales and usage: 2006 and 2007 market estimates. EPA733-R-11-001, 33 pp.

Gül, A., Benli, A.Ç. K., Ayhan, A., Memmi, B. K., Selvi, M., Sepici-Dinçel, A., Çakiroğullari, G.Ç. and Erkoç, F. (2012). Sublethal propoxur toxicity to juvenile common carp (*Cyprinus carpio* L., 1758): biochemical, hematological, histopathological, and genotoxicity effects. *Environ. Toxicol. Chem.* 31, 2085–2092.

Habig, C. and DiGiulio, R. (1991). Biochemical characteristics of cholinesterases in aquatic organisms. In *Cholinesterase-inhibiting Insecticides: Their Impact on Wildlife and the Environment* (ed. P. Mineau), pp. 19–33. New York: Elsevier Science Publishers.

Hainzl, D., Cole, L. M. and Casida, J. E. (1998). Mechanisms for selective toxicity of fipronil insecticide and its sulfone metabolite and desulfinyl photoproduct. *Chem. Res. Toxicol.* 11, 1529–1535.

Hansen, D., Goodman, L. and Wilson, A. (1977). Kepone®: Chronic effects on embryo, fry, juvenile, and adult sheepshead minnows (*Cyprinodon variegatus*). *Chesap. Sci.* 18, 227–232.

Hansen, D. J., Goodman, L. R., Moore, J. C. and Higdon, P. K. (1983). Effects of the synthetic pyrethroids AC 222,705, permethrin and fenvalerate on sheepshead minnows in early life stage toxicity tests. *Environ. Toxicol. Chem.* 2, 251–258.

Harper, H. (2007). The effects of the insecticide bifenthrin on grass shrimp, *Palaemonetes pugio* and sheepshead minnow, *Cyprinodon variegatus*. vol. Master of Science, p. 208, *Marine Biology*. Charleston, SC: College of Charleston

Harper, H. E., Pennington, P. L., Hoguet, J. and Fulton, M. H. (2008). Lethal and sublethal effects of the pyrethroid, bifenthrin, on grass shrimp (*Palaemonetes pugio*) and sheepshead minnow (*Cyprinodon variegatus*). *J. Environ. Sci. Health, Part B* 43, 476–483.

Harwood, A. D., You, J. and Lydy, M. J. (2009). Temperature as a toxicity identification evaluation tool for pyrethroid insecticides: toxicokinetic confirmation. *Environ. Toxicol. Chem.* 28, 1051–1058.

Haya, K. (1989). Toxicity of pyrethroid insecticides to fish. *Environ. Toxicol. Chem.* 8, 381–391.

Hernández-Moreno, D., Pérez-López, M., Soler, F., Gravato, C. and Guilhermino, L. (2011). Effects of carbofuran on the sea bass (*Dicentrarchus labrax* L.): Study of biomarkers and behaviour alterations. *Ecotox. Environ. Saf.* 74, 1905–1912.

Hernández-Moreno, D., Soler, F., Míguez, M. P. and Pérez-López, M. (2010). Brain acetylcholinesterase, malondialdehyde and reduced glutathione as biomarkers of continuous exposure of tench, *Tinca tinca*, to carbofuran or deltamethrin. *Sci. Total Environ.* 408, 4976–4983.

Hernández-Moreno, D., Soler-Rodríguez, F., Míguez-Santiyán, M. P. and Pérez-López, M. (2008). Hepatic monooxygenase (CYP1A and CYP3A) and UDPGT enzymatic activities as biomarkers for long-term carbofuran exposure in tench (*Tinca tinca* L). *J. Environ. Sci. Health, Part B* 43, 395–404.

Henrick, C. A. (2007). Methoprene: Biorational control of mosquitoes. *American Mosquito Control Association Bulletin, 7* 23, 225–239.

Henry, M., Béguin, M., Requier, F., Rollin, O., Odoux, J.-F., Aupinel, P., Aptel, J., Tchamitchian, S. and Decourtye, A. (2012). A common pesticide decreases foraging success and survival in honey bees. *Science* 336, 348–350.

Hill, E. F. (1995). Organophosphorus and carbamate pesticides. In *Handbook of Ecotoxicology* (eds. D. Hoffman, B. A. Rattner, G. Burton and J. Cairns, Jr), p. 733. Boca Raton, FL: Lewis Publishers.

Jacobson, S. M., Birkholz, D. A., McNamara, M. L., Bharate, S. B. and George, K. M. (2010). Subacute developmental exposure of zebrafish to the organophosphate pesticide metabolite, chlorpyrifos-oxon, results in defects in Rohon-Beard sensory neuron development. *Aquat. Toxicol.* 100, 101–111.

Janicki, R. H. and Kinter, W. B. (1971). DDT: disrupted osmoregulatory events in the intestine of the eel *Anguilla rostrata* adapted to seawater. *Science* 173, 1146–1148.

Jarvinen, A. W., Tanner, D. K. and Kline, E. R. (1988). Toxicity of chlorpyrifos, endrin, or fenvalerate to fathead minnows following episodic or continuous exposure. *Ecotox. Environ. Saf.* 15, 78–95.

Jee, J.-H., Masroor, F. and Kang, J.-C. (2005). Responses of cypermethrin-induced stress in haematological parameters of Korean rockfish, *Sebastes schlegeli* (Hilgendorf). *Aquaculture Res.* 36, 898–905.

Johnson, D. W. (1968). Pesticides and fishes—a review of selected literature. *Trans. Am. Fish. Soc.* 97, 398–424.

Johnson, W. W. and Finley, M. T. (1980). Handbook of acute toxicity of chemicals to fish and aquatic invertebrates: Summaries of toxicity tests conducted at Columbia National Fisheries Research Laboratory, 1965–1978. Washington, DC: U.S. Department of Interior.

Jones, W., Mazur, C. S., Kenneke, J. F. and Garrison, A. (2007). Enantioselective microbial transformation of the phenylpyrazole insecticide fipronil in anoxic sediments. *Environ. Sci. Technol.* 41, 8301–8307.

Jyothi, B. and Narayan, G. (1999). Certain pesticide-induced carbohydrate metabolic disorders in the serum of freshwater fish *Clarias batrachus* (Linn.). *Food Chem. Toxicol.* 37, 417–421.

Karen, D. J., Draughn, R., Fulton, M. and Ross, P. (1998). Bone strength and acetylcholinesterase inhibition as endpoints in chlorpyrifos toxicity to *Fundulus heteroclitus*. *Pest. Biochem. Physiol.* 60, 167–175.

Karouna-Renier, N. K., Snyder, R. A., Lange, T., Gibson, S., Allison, J. G., Wagner, M. E. and Ranga Rao, K. (2011). Largemouth bass (*Micropterus salmoides*) and striped mullet (*Mugil cephalus*) as vectors of contaminants to human consumers in northwest Florida. *Mar. Environ. Res.* 72, 96–104.

Kaushik, P. and Kaushik, G. (2007). An assessment of structure and toxicity correlation in organochlorine pesticides. *J. Hazard. Mat.* 143, 102–111.

Key, P. B., Chung, K., Hoguet, J., Sapozhnikova, Y. and DeLorenzo, M. E. (2011). Toxicity of the mosquito control insecticide phenothrin to three life stages of the grass shrimp (*Palaemonetes pugio*). *J. Environ. Sci. Health, Part B* 46, 426–431.

Key, P. B., DeLorenzo, M. E., Gross, K., Chung, K. and Clum, A. (2005). Toxicity of the mosquito control insecticide Scourge to adult and larval grass shrimp (*Palaemonetes pugio*). *J. Environ. Sci. Health, Part B* 40, 585–594.

Key, P. B. and Scott, G. I. (1992). Acute toxicity of the mosquito larvicide, *Bacillus sphaericus*, to the grass shrimp, *Palaemonetes pugio*, and mummichog, *Fundulus heteroclitus*. *Bull. Environ. Contam. Toxicol.* 49, 425–430.

Kirby, M., Morris, S., Hurst, M., Kirby, S., Neall, P., Tylor, T. and Fagg, A. (2000). The use of cholinesterase activity in flounder (*Platichhtys flesus*) muscle tissue as a biomarker of neurotoxic contamination in UK estuaries. *Mar. Poll. Bull.* 40, 780–791.

Kobayashi, H., Suzuki, T., Akahori, F. and Satoh, T. (2010). Acetylcholinesterase and acetylcholine receptors: brain regional heterogeneity. In *Anticholinesterase Pesticides: Metabolism, Neurotoxicity, and Epidemiology* (eds. T. Satoh and R. Gupta), p. 600. Hoboken, NJ: John Wiley & Sons.

Konwick, B. J., Fisk, A. T., Garrison, A. W., Avants, J. and Black, M. C. (2006a). Acute enantioselective toxicity of fipronil and its desulfinyl photoproduct to *Ceriodaphnia dubia*. *Environ. Toxicol. Chem.* 24, 2350–2355.

Konwick, B. J., Garrison, A. W., Black, M. C., Avants, J. K. and Fisk, A. T. (2006b). Bioaccumulation, Biotransformation, and Metabolite Formation of Fipronil and Chiral Legacy Pesticides in Rainbow Trout. *Environ. Sci. Technol.* 40, 2930–2936.

Köprücü, K. and Aydın, R. (2004). The toxic effects of pyrethroid deltamethrin on the common carp (*Cyprinus carpio* L.) embryos and larvae. *Pest. Biochem. Physiol.* 80, 47–53.

Kurtz, P. J., Deskin, R. and Harrington, R. (1989). Pesticides. In *Principles and Methods of Toxicology* (ed. A. W. Hayes), pp. 137–152. New York: Raven Press.

Labat-Anderson, I. (1992). *Final forest service seed orchard pesticide background statement: azinphosmethyl*. Arlington, VA: Labat-Anderson Incorporated.

Labenia, J., Baldwin, D. H., French, B. L., Davis, J. W. and Scholz, N. L. (2007). Behavioral impairment and increased predation mortality in cutthroat trout exposed to carbaryl. *Mar. Ecol. Progress Series* 329, 1–11.

Laetz, C., Baldwin, D., Collier, T., Hebert, V., Stark, J. and Scholz, N. (2009). The synergistic toxicity of pesticide mixtures: Implications for risk assessment and the conservation of endangered pacific salmon. *Environ. Health Perspect.* 117, 348–353.

Lanfranchi, A. L., Menone, M. L., Miglioranza, K. S. B., Janiot, L. J., Aizpún, J. E. and Moreno, V. J. (2006). Striped weakfish (*Cynoscion guatucupa*): A biomonitor of organochlorine pesticides in estuarine and near-coastal zones. *Mar. Poll. Bull.* 52, 74–80.

Lauth, J. R., Scott, G. I., Cherry, D. S. and Buikema, A. L., Jr. (1996). A modular estuarine mesocosm. *Environ. Toxicol. Chem.* 15, 630–637.

Lee, B. M. and Scott, G. I. (1989). Acute toxicity of temephos, fenoxycarb, diflubenzuron, and methoprene and *Bacillus thuringiensis* var. *israelensis* to the mummichog (*Fundulus heteroclitus*). *Bull. Environ. Contam. Toxicol.* 43, 827–832.

Lockhart, W. L., Metner, D. A., Ward, F. J. and Swanson, G. M. (1985). Population and cholinesterase responses in fish exposed to malathion sprays. *Pest. Biochem. Physiol.* 24, 12–18.

Loganathan, B., Sajwan, K., Richardson, J., Chetty, C. and Owen, D. (2001). Persistent organochlorine concentrations in sediment and fish from Atlantic coastal and brackish waters off Savannah, Georgia, USA. *Mar. Poll. Bull.* 42, 246–250.

Lozano, S. J., O'Halloran, S. L., Sargent, K. W. and Brazner, J. C. (1992). Effects of esfenvalerate on aquatic organisms in littoral enclosures. *Environ. Toxicol. Chem.* 11, 35–47.

Lushchak, V. I. (2011). Environmentally induced oxidative stress in aquatic animals. *Aquat. Toxicol.* 101, 13–30.

Macek, K. and McAllister, W. (1970). Insecticide susceptibility of some common fish family representatives. *Trans. Am. Fish. Soc.* 1, 20–27.

Macek, K. J., Walsh, D. F., Hogan, J. W. and Holz, D. D. (1972). Toxicity of the insecticide dursban to fish and aquatic invertebrates in ponds. *Trans. Am. Fish. Soc.* 101, 420–427.

Manahan, S. E. (1992). *Toxicological Chemistry.* Boca Raton, FL: Lewis Publishers.

Marshall, W. and Roberts, J. (1978). *Ecotoxicology of Chlorpyrifos.* Ottawa, Ontario, Canada: Natl. Res. Counc. Canada, Assoc. Comm. Sci. Crit. Environ. Qual. Publication 16059.

Martyniuk, C. J., Spade, D. J., Blum, J. L., Kroll, K. J. and Denslow, N. D. (2011). Methoxychlor affects multiple hormone signaling pathways in the largemouth bass (*Micropterus salmoides*) liver. *Aquat. Toxicol.* 101, 483–492.

Matusda, K., Buckingham, S. D., Kleier, D., Rauh, J. J., Grauso, M. and Sattelle, D. B. (2001). Neonicotinoids: insecticides acting on insect nicotinic acetylcholine receptors. *Trends Pharmacol. Sci.* 22, 573–580.

Mayer, F. L., Jr (1987). *Acute Toxicity Handbook of Chemicals to Estuarine Organisms.* Gulf Breeze, FL: U.S. Environmental Protection Agency, p. 274.

McHenery, J. G., Saward, D. and Seaton, D. D. (1991). Lethal and sub-lethal effects of the salmon delousing agent dichlorvos on the larvae of the lobster (*Homarus gammarus* L.) and herring (*Clupea harengus* L.). *Aquaculture* 98, 331–347.

Miller, L. K., Lingg, A. J. and Bulla, L. A. (1983). Bacterial, viral, and fungal insecticides. *Science* 219, 715–721.

Miller, T. A. and Salgado, V. L. (1985). The mode of action of pyrethroids on insects. In *The Pyrethroid Insecticides* (ed. J. P. Leahy), pp. 43–97. London: Taylor & Francis.

Mineau, P. (1991). *Cholinesterase-inhibiting Insecticides: Their Impact on Wildlife and the Environment.* New York: Elsevier Science Publishers.

Mishra, D. K., Bohidar, K. and Pandey, A. K. (2008). Effect of sublethal exposure of Cartap on hypothalamo-neurosecretory system of the freshwater spotted murrel, *Channa punctatus* (Bloch). *J. Environ. Biol.* 29, 917–922.

Moore, A. and Waring, C. P. (2001). The effects of a synthetic pyrethroid pesticide on some aspects of reproduction in Atlantic salmon (*Salmo salar* L.). *Aquat. Toxicol.* 52, 1–12.

Moore, D. R. J. and Teed, R. S. (2012). Risks of carbamate and organophosphate pesticide mixtures to salmon in the Pacific Northwest. *Integr. Environ. Assess. Manag.* Online Doi:10.1002/ieam.1329.

Morgan, E. R. and Brunson, M. W. (2002). *Toxicities of Agricultural Pesticides to Selected Aquatic Organisms.* Stoneville, MISS: Southern Regional Aquaculture Center, p. 27.

Moza, P. N., Hustert, K., Feicht, E. and Kettrup, A. (1998). Photolysis of imidacloprid in aqueous solution. *Chemosphere* 36, 497–502.

Mulla, M. and Mian, L. (1981). Biological and environmental impacts of the insecticides malathion and parathion on nontarget biota in aquatic ecosystems. *Residue Rev.* 78, 101–135.

Mulla, M., Mian, L. and Kawecki, J. (1981). Distribution, transport, and fate of the insecticides malathion and parathion in the environment. *Residue Rev.* 81, 1–171.

Munn, M. D. and Gruber, S. J. (1997). The relationship between land use and organochlorine compounds in streambed sediment and fish in the Central Columbia Plateau, Washington and Idaho, USA. *Environ. Toxicol. Chem.* 16, 1877–1887.

Muñoz, I., Martínez Bueno, M. J., Agüera, A. and Fernández-Alba, A. R. (2010). Environmental and human health risk assessment of organic micro-pollutants occurring in a Spanish marine fish farm. *Environ. Poll.* 158, 1809–1816.

Murphy, S. D. (1980). Pesticides. In *Casarett and Doull's Toxicology: The Basic Science of Poisons* (eds. J. Doull, C. Klaassen and M. Amdur), 2nd Edition, pp. 357–408. New York: Macmillan Publishing Co.

Murty, A. S. (1986). *Toxicity of Pesticides to Fish.* Boca Raton, FL: CRC Press.

Nauen, R. and Bretschneider, T. (2002). New modes of action of insecticides. *Pestic. Outlook* 13, 241–245.

Ngim, K. K., Mabury, S. A. and Crosby, D. G. (2000). Elucidation of fipronil photodegradation pathways. *J. Agric. Food Chem.* 48, 4661–4665.

NTP (2011). *Report on Carcinogens* (Twelfth Edition). Research Triangle Park, NC: U.S. Department of Health and Human Services, Public Health Service, National Toxicology Program, 499 pp.

Olivares, A., Quirós, L., Pelayo, S., Navarro, A., Bosch, C., Grimalt, J. O., Fabregat, M. D. C., Faria, M., Benejam, L., Benito, J., et al. (2010). Integrated biological and chemical analysis of organochlorine compound pollution and of its biological effects in a riverine system downstream the discharge point. *Sci. Total Environ.* 408, 5592–5599.

Oros, D. R. and Werner, I. (2005). Pyrethroid Insecticides: An Analysis of Use Patterns, Distributions, Potential Toxicity and Fate in the Sacramento-San Joaquin Delta and Central Valley. White Paper for the Interagency Ecological Program. In *SFEI Contribution 415*. Oakland, CA: San Francisco Estuary Institute.

Oruc, E. (2011). Effects of diazinon on antioxidant defense system and lipid peroxidation in the liver of *Cyprinus carpio* (L.). *Environ. Toxicol.* 26, 571–578.

Oruc, E. (2012). Oxidative stress responses and recovery patterns in the liver of *Oreochromis niloticus* exposed to chlorpyrifos-ethyl. *Bull. Environ. Contam. Toxicol.* 88, 678–684.

Overmyer, J. P., Rouse, D. R., Avants, J. K., Garrison, A. W., DeLorenzo, M. E., Chung, K. W., Key, P. B., Wilson, W. A. and Black, M. C. (2007). Toxicity of fipronil and its enantiomers to marine and freshwater non-targets. *J. Environ. Sci. Health, Part B* 42, 471–480.

Parent, L. M., DeLorenzo, M. E. and Fulton, M. H. (2011). Effects of the synthetic pyrethroid insecticide, permethrin, on two estuarine fish species. *J. Environ. Sci. Health, Part B* 46, 615–622.

Parrish, P. R., Schimmel, S. C., Hansen, D. J., Patrick, J. M. and Forester, J. (1976). Chlordane: Effects on several estuarine organisms. *J. Toxicol. Environ. Health* 1, 485–494.

Pennington, P. L., DeLorenzo, M. E., Lawton, J. C., Strozier, E. D., Fulton, M. H. and Scott, G. I. (2004). Modular estuarine mesocosm validation: I. Ecotoxicological assessment of direct effects with the model compound endosulfan. *J. Exp. Mar. Biol. Ecol.* 298, 369–387.

Perkins, E. J. and Schlenk, D. (2000). *In vivo* acetylcholinesterase inhibition, metabolism, and toxicokinetics of aldicarb in channel catfish: role of biotransformation in acute toxicity. *Toxicol. Sci.* 53, 308–315.

Pessoa, P. C., Luchmann, K. H., Ribeiro, A. B., Veras, M. M., Correa, J. R. M. B., Nogueira, A. J., Bainy, A. C. D. and Carvalho, P. S. M. (2011). Cholinesterase inhibition and behavioral toxicity of carbofuran on *Oreochromis niloticus* early life stages. *Aquat. Toxicol.* 105, 312–320.

Phillips, J. P. (2006). Acute and sublethal effects of lambda-cyhalothrin on early life stages of chinook salmon (*Oncorhynchus tshawytscha*): University of California, Davis.

Pimental, D. (2005). Environmental and economic costs of the application of pesticides primarily in the United States. *Environ. Develop. Sustain.* 7, 229–252.

Pimental, D. (1992). Environmental and economic costs of pesticide use. *Bioscience* 42, 750–760.

Rajakumar, A., Singh, R., Chakrabarty, S., Murugananthkumar, R., Laldinsangi, C., Prathibha, Y., Sudhakumari, C. C., Dutta-Gupta, A. and Senthilkumaran, B. (2012). Endosulfan and flutamide impair testicular development in the juvenile Asian catfish. *Clarias batrachus*. *Aquat. Toxicol.* 110–111, 123–132.

Ram, R. N. and Singh, S. K. (1988). Carbofuran-induced histopathological and biochemical changes in liver of the teleost fish, *Channa punctatus* (bloch). *Ecotox. Environ. Saf.* 16, 194–201.

Ray, D. E. (1991). Pesticides derived from plants and other organisms. In *Handbook of Pesticide Toxicology Volume 2 Classes of Pesticides* (eds. W. J. Hayes, Jr. and E. R. Laws, Jr), Chapter 13, pp. 585–636. San Diego: Academic Press.

Rebach, S. (1999). Acute toxicity of permethrin/piperonyl butoxide on hybrid striped bass. *Bull. Environ. Contam. Toxicol.* 62, 448–454.

Reddy, P. M., Philip, G. H. and Bashamohideen, M. (1992). Regulation of AChE system of freshwater fish, *Cyprinus carpio*, under fenvalerate toxicity. *Bull. Environ. Contam. Toxicol.* 48, 18–22.

Reigart, J. R. and Roberts, J. R. (1999). *Recognition and management of pesticides poisonings.* Washington, DC: U.S. EPA Office of Pesticide Programs.

Richmonds, C. R. and Dutta, H. M. (1992). Effect of malathion on the brain acetylcholinesterase activity of bluegill sunfish *Lepomis macrochirus. Bull. Environ. Contam. Toxicol.* 49, 431–435.

Saglio, P., Trijasse, S. and Azam, D. (1996). Behavioral effects of waterborne carbofuran in goldfish. *Arch. Environ. Contam. Toxicol.* 31, 232–238.

Salgado, V. L. (1998). Studies on the mode of action of spinosad: Insect symptoms and physiological correlates. *Pest. Biochem. Physiol.* 60, 91–102.

Sanchez-Bayo, F. and Goka, K. (2005). Unexpected effects of zinc pyrithione and imidacloprid on Japanese medaka fish (*Oryzias latipes*). *Aquat. Toxicol.* 74, 285–293.

Sandahl, J. F. and Jenkins, J. J. (2002). Pacific steelhead (*Oncorhynchus mykiss*) exposed to chlorpyrifos: benchmark concentration estimates for acetylcholinesterase inhibition. *Environ. Toxicol. Chem.* 21, 2452–2458.

Sandahl, J. F., Baldwin, D. H., Jenkins, J. J. and Scholz, N. L. (2004). Odor-evoked field potentials as indicators of sublethal neurotoxicity in juvenile coho salmon (*Oncorhynchus kisutch*) exposed to copper, chlorpyrifos, or esfenvalerate. *Can. J. Fish. Aquatic Sci.* 61, 404–413.

Saravanan, M., Prabhu Kumar, K. and Ramesh, M. (2011). Haematological and biochemical responses of freshwater teleost fish *Cyprinus carpio* (Actinopterygii: Cypriniformes) during acute and chronic sublethal exposure to lindane. *Pest. Biochem. Physiol.* 100, 206–211.

Sayeed, I., Parvez, S., Pandey, S., Bin-Hafeez, B., Haque, R. and Raisuddin, S. (2003). Oxidative stress biomarkers of exposure to deltamethrin in freshwater fish, *Channa punctatus* Bloch. *Ecotox. Environ. Saf.* 56, 295–301.

Schimmel, S. C., Patrick, J. M. and Wilson, A. (1977). Acute toxicity to and bioconcentration of endosulfan by estuarine animals. In *Aquatic Toxicology and Hazard Evaluation, ASTM STP 634* (eds. F. Mayer and J. Hamelink), pp. 241–252. Philadelphia, PA: American Society for Testing and Materials.

Schimmel, S. C., Garnas, R. L., Patrick, J. M., Jr. and Moore, J. C. (1983). Acute toxicity, bioconcentration, and persistence of AC 222,705, benthiocarb, chlorpyrifos, fenvalerate, methyl parathion, and permethrin in the estuarine environment. *J. Agric. Food Chem.* 31, 104–113.

Schmitt, C. J., Ellen Hinck, J., Blazer, V. S., Denslow, N. D., Dethloff, G. M., Bartish, T. M., Coyle, J. J. and Tillitt, D. E. (2005). Environmental contaminants and biomarker responses in fish from the Rio Grande and its U.S. tributaries: Spatial and temporal trends. *Sci. Total Environ.* 350, 161–193.

Scholz, N. L., Truelove, N. K., Labenia, J. S., Baldwin, D. H. and Collier, T. K. (2006). Dose-additive inhibition of chinook salmon acetylcholinesterase activity by mixtures of organophosphate and carbamate insecticides. *Environ. Toxicol. Chem.* 25, 1200–1207.

Schuurmann, G. (1998). Ecotoxic modes of action of chemical substances. In *Ecotoxicology: Ecological Fundamentals, Chemical Exposure, and Biological Effects* (eds. G. Schuurmann and B. Markert), p. 811. New York: John Wiley & Sons.

Scott, G., Baughman, D., Trim, A. and Dee, J. (1987). Lethal and sublethal effects of insecticides commonly found in nonpoint source agricultural runoff to estuarine fish and shellfish. In *Pollution Physiology of Estuarine Organisms* (eds. W. Vernberg, A. Calabrese, F. Thurberg and F. Vernberg), pp. 251–274. Columbia: University of South Carolina Press.

Sibley, P. K., Chappel, M. J., George, T. K., Solomon, K. R. and Liber, K. (2000). Integrating effects of stressors across levels of biological organization: examples using organophosphorus insecticide mixtures in field-level exposures. *J. Aquatic Ecosys. Stress Recov.* 7, 117–130.

Silcox, C.A. and Vittum, P.J. (2008). Turf insecticide classes and modes of action. *Golf Course Management.* September. pp. 82–90.

Sinha, A. K., Vanparys, C., De Boeck, G., Kestemont, P., Wang, N., Phuong, N. T., Scippo, M.-L., De Coen, W. and Robbens, J. (2010). Expression characteristics of potential biomarker genes in Tra catfish, *Pangasianodon hypophthalmus*, exposed to trichlorfon. *Comp. Biochem. Physio. Part D: Gen. Proteom.* 5, 207–216.

Smith, T. M. and Stratton, G. W. (1986). Effects of synthetic pyrethroid insecticides on nontarget organisms. *Residue Rev.* 97, 93–119.

Smith, A. G. (1991). Chlorinated hydrocarbon insecticides. In *Handbook of Pesticide Toxicology Volume 2 Classes of Pesticides* (eds. W. J. Hayes, Jr. and E. R. Laws, Jr), pp. 731–915. San Diego: Academic Press.

Stahl, L., Snyder, B., Olsen, A. and Pitt, J. (2009). Contaminants in fish tissue from US lakes and reservoirs: a national probabilistic study. *Environ. Monit. Assess.* 150, 3–19.

Stehr, C. M., Linbo, T. L., Incardona, J. P. and Scholz, N. L. (2006). The developmental neurotoxicity of fipronil: notochord degeneration and locomotor defects in zebrafish embryos and larvae. *Toxicol. Sci.* 92, 270–278.

Storm, J. E. (2012). *Organophosphorus Pesticides. Patty's Toxicology.* John Wiley & Sons.

Stueckle, T. A., Griffin, K. and Foran, C. M. (2008). No acute toxicity to Uca pugnax, the mud fiddler crab, following a 96-h exposure to sediment-bound permethrin. *Environ. Toxicol.* 23, 530–538.

Sturm, A., Wogram, J., Segner, H. and Liess, M. (2000). Different sensitivity to organophosphates of acetylcholinesterase and butrycholineterase from three-spined stickleback (*Gasterosteus aculeatus*) application in biomonitoring. *Environ. Toxicol. Chem.* 19, 1607–1615.

Sumith, J. A., Hansani, P. L. C., Weeraratne, T. C. and Munkittrick, K. R. (2012). Seasonal exposure of fish to neurotoxic pesticides in an intensive agricultural catchment, Uma-oya, Sri Lanka: Linking contamination and acetylcholinesterase inhibition. *Environ. Toxicol. Chem.* 31, 1501–1510.

Tanner, D. K. and Knuth, M. L. (1996). Effects of esfenvalerate on the reproductive success of the bluegill sunfish, *Lepomis macrochirus* in littoral enclosures. *Arch. Environ. Contam. Toxicol.* 31, 244–251.

Tingle, C. C., Rother, J. A., Dewhurst, C. F., Lauer, S. and King, W. J. (2003). Fipronil: environmental fate, ecotoxicology, and human health concerns. *Rev. Environ. Contam. Toxicol.* 176, 1–66.

Tišler, T., Jemec, A., Mozetič, B. and Trebše, P. (2009). Hazard identification of imidacloprid to aquatic environment. *Chemosphere* 76, 907–914.

Todd, N. E. and Van Leeuwen, M. (2002). Effects of sevin (carbaryl insecticide) on early life stages of zebrafish (*Danio rerio*). *Ecotox. Environ. Saf.* 53, 267–272.

U.S. EPA (1979). U.S. Environmental Protection Agency, Reviews of the Environmental Effects of Pollutants: I. Mirex and Kepone. Cincinnati, OH: Health Effects Research Laboratory, Office of Research and Development, p. 252.

U.S. EPA (1991). U.S. Environmental Protection Agency Registration and Eligibility Decision Facts: Methoprene, Washington, DC Office of Pesticides and Toxic Substances, pp. Report 21T–1003.

U.S. EPA (1998). U.S. Environmental Protection Agency Reregistration Eligibility Decision (RED) *Bacillus thuringiensis*. Washington, DC: Office of Prevention, Pesticides and Toxic Substances, p. 170.

U.S. EPA (1999). U.S. Environmental Protection Agency Fipronil Environmental Fate and Effects Assessment, p. 19. Washington, DC.

U.S. EPA (2000). U.S. Environmental Protection Agency, Pesticide Ecotoxicology Database. Washington DC.

U.S. EPA (2002). U.S. Environmental Protection Agency. Lindane, Reregistration Eligibility Decision on Lindane, p. 135. Washington, DC.

U.S. EPA (2005). U.S. Environmental Protection Agency, Permethrin, EFED revised risk assessment for the reregistration eligibility decision on permethrin, p. 93. Washington, DC.

U.S. EPA (2006a). U.S. Environmental Protection Agency, Lindane, Addendum to the 2002 Lindane Reregistration Eligibility Decision (RED), p. 19. Washington, DC.

U.S. EPA (2006b). U.S. Environmental Protection Agency Reregistration Eligibility Decision for Cypermethrin, p. 117. Washington, DC.

U.S. EPA (2007). U.S. Environmental Protection Agency Pesticides Registration Review Etofenprox Summary Document, p. 42. Washington, DC.

U.S. EPA (2008). U.S. Environmental Protection Agency reregistration eligibility decision for d-phenothrin. Office of Pesticide Programs, p. 54. Washington, DC.

U.S. EPA (2011a). U.S. Environmental Protection Agency Pesticides Industry Sales and Usage 2006 and 2007 Market Estimates. Washington, DC: Office of Chemical Safety and Pollution Prevention, p. 41.

U.S. EPA (2011b). U.S. Environmental Protection Agency Registration Review—Preliminary Problem Formulation for Ecological Risk and Environmental Fate, Endangered Species, and Drinking Water assessments for Spinosad and Spinetoram. Washington, DC: Office of Chemical Safety and Pollution Prevention, p. 80.

U.S. EPA (2012a) U.S. Environmental Protection Agency, Endosulfan Phase-out. Retrieved October 29, 2012 from http://www.epa.gov/pesticides/reregistration/endosulfan/endosulfan-agreement.html#agreement.

U.S. EPA (2012b) U.S. Environmental Protection Agency, Pesticide Chemical Search. Retrieved October 23, 2012 from http://iaspub.epa.gov/apex/pesticides/f?p=chemical-search:1).

U.S. FWS (2003). *Effects of Larvicides on Non-Target Organisms*. Washington, DC: U.S. Fish and Wildlife Service, p. 95.

van den Berg, H. (2009). Global status of DDT and its alternatives for use in vector control to prevent disease. *Environ. Health Perspect.* 117, 1656–1663.

Van Dolah, R., Maier, P., Fulton, M. and Scott, G. (1997). Comparison of azinphosmethyl toxicity to juvenile red drum (*Sciaenops ocellatus*) and the mummichog (*Fundulus heteroclitus*). *Environ. Toxicol. Chem.* 16, 1488–1493.

Velisek, J., Svobodova, Z. and Machova, J. (2009). Effects of bifenthrin on some haematological, biochemical and histopathological parameters of common carp (*Cyprinus carpio* L.). *Fish Physiol. Biochem.* 35, 583–590.

Wacksman, M. N., Maul, J. D. and Lydy, M. J. (2006). Impact of atrazine on chlorpyrifos toxicity in four aquatic vertebrates. *Arch. Environ. Contam. Toxicol.* 51, 681–689.

Wang, C., Lu, G. and Cui, J. (2010). Responses of AChE and GST activities to insecticide coexposure in *Carassius auratus*. *Environ. Toxicol.* 27, 50–57.

Ware, G. (1989). *The Pesticide Book*. Fresno, CA: Thomson Publications.

Weis, P. and Weis, J. S. (1976). Abnormal locomotion associated with skeletal malformations in the sheepshead minnow, *Cyprinodon variegatus*, exposed to malathion. *Environ. Res.* 12, 196–200.

Weiss, C. M. (1958). The determination of cholinesterase in the brain tissue of three species of fresh water fish and its inactivation *in vivo*. *Ecology* 39, 194–199.

Werner, I., Geist, J., Okihiro, M., Rosenkranz, P. and Hinton, D. E. (2002). Effects of dietary exposure to the pyrethroid pesticide esfenvalerate on medaka (*Oryzias latipes*). *Mar. Environ. Res.* 54, 609–614.

Werner, I. and Moran, K. (2008). Effects of pyrethroid insecticides on aquatic organisms. In *Synthetic Pyrethroids: Occurrence and Behavior in Aquatic Environments* (eds. J. Gan, F. Spurlock, P. Hendley and D. Weston), Washington, DC: American Chemical Society.

Wheelock, C. E., Miller, J. L., Miller, M. J., Gee, S. J., Shan, G. and Hammock, B. D. (2004). Development of toxicity identification evaluation procedures for pyrethroid detection using esterase activity. *Environ. Toxicol. Chem.* 23, 2699–2708.

Wheelock, C. E., Eder, K. J., Werner, I., Huang, H., Jones, P. D., Brammell, B. F., Elskus, A. A. and Hammock, B. D. (2005). Individual variability in esterase activity and CYP1A levels in Chinook salmon (*Oncorhynchus tshawytscha*) exposed to esfenvalerate and chlorpyrifos. *Aquat. Toxicol.* 74, 172–192.

Whitehead, A., Kuivila, K. M., Orlando, J. L., Kotelevtsev, S. and Anderson, S. L. (2004). Genotoxicity in native fish associated with agricultural runoff events. *Environ. Toxicol. Chem.* 23, 2868–2877.

Wirth, M. C., Georghiou, G. P. and Federici, B. A. (1997). CytA enables CryIV endotoxins of *Bacillus thuringiensis* to overcome high levels of CryIV resistance in the mosquito, Culex quinquefasciatus. *Proc. Natl Acad. Sci.* 94, 10536–10540.

Yameogo, L., Traore, K., Back, C., Hougard, J.-M. and Calamari, D. (2001). Risk assessment of etofenprox (vectron) on non-target aquatic fauna compared with other pesticides used as Simulium larvicide in a tropical environment. *Chemosphere* 42, 965–974.

Yang, D., Lauridsen, H., Buels, K., Chi, L.-H., La Du, J., Bruun, D. A., Olson, J. R., Tanguay, R. L. and Lein, P. J. (2011). Chlorpyrifos-oxon disrupts zebrafish axonal growth and motor behavior. *Toxicol. Sci.* 121, 146–159.

Yi, M. Q., Liu, H. X., Shi, X. Y., Liang, P. and Gao, X. W. (2006). Inhibitory effects of four carbamate insecticides on acetylcholinesterase of male and female *Carassius auratus in vitro*. *Comp. Biochem. Physio. Part C: Toxicol. Pharm.* 143, 113–116.

Zelikoff, J. T., Raymond, A., Carlson, E., Li, Y., Beaman, J. R. and Anderson, M. (2000). Biomarkers of immunotoxicity in fish: From the lab to the ocean. *Toxicol. Lett.* 112–113, 325–331.

Zhang, Z.-Y., Yu, X.-Y., Wang, D.-L. and Liu, X.-J. (2010). Acute toxicity to zebrafish of two organophosphates and four pyrethroids and their binary mixtures. *Pest. Manag. Sci.* 66, 84–89.

Zhou, S., Dong, Q., Li, S., Guo, J., Wang, X. and Zhu, G. (2009). Developmental toxicity of cartap on zebrafish embryos. *Aquat. Toxicol.* 95, 339–346.

DISCLAIMER

7

EFFECTS OF HERBICIDES ON FISH

KEITH R. SOLOMON
KRISTOFFER DALHOFF
DAVID VOLZ
GLEN VAN DER KRAAK

Organic Chemical Toxicology of Fishes: Volume 33
FISH PHYSIOLOGY

1. INTRODUCTION

Herbicides represent the largest proportion of pesticides used in agriculture. Worldwide use of pesticides in 2007 was estimated at 2.4 billion kg, of which the largest proportion, 40% or 950 million kg, was herbicides (U.S. EPA, 2012a). Herbicides are used in agriculture, forestry, and for other functions such as the control of vegetation on rights of way, industrial, and urban sites but agricultural uses dominate the market. With the number and variety of crops that have been genetically modified (GM) to be tolerant to herbicides, the use of some of these products will increase in the future. Until now, the herbicide most widely used on GM crops has been glyphosate (Duke and Powles, 2008), but other crops have been modified for resistance to other classes of pesticides such as the auxin mimic, 2,4-D, and related products (Mortensen et al., 2012). Increased uses on GM crops will result in increased inputs to agro-ecosystems with concomitant increases in the potential for exposures in the environment, particularly in aquatic organisms.

1.1. Specificity of Mode of Action

With the exception of biocides, such as chlorine gas for disinfection of water or the wood-preservative, pentachlorophenol, most pesticides used in agriculture are toxic to some groups of organisms and not others and their target sites are often specific to pests, which leads to specificity of mode of action. This is particularly the case for the herbicides that are designed to control plants. For example, photosynthetic inhibitors such as the urea and triazine herbicides specifically target components of the photosynthetic apparatus (Devine et al., 1993), which are not found in animals. As a result, animals are usually less sensitive to herbicides than plants. However, all chemicals are toxic at large enough concentrations, and, at sufficiently large concentrations, herbicides can be toxic to animals, but this is through a different mechanism—that of baseline narcosis.

In fish, baseline narcosis is characterized by progressive lethargy, unconsciousness, and death without any specific sustained symptoms such as hyperventilation, erratic or convulsive swimming, or hemorrhage (Veith and Broderius, 1990). All nonpolar organic chemicals have a generalized mode of action that targets the cell membrane, a feature of cells that is present in most life forms. This mode of action, narcosis or baseline toxicity, will affect any organism if the exposure concentration is large enough and the properties of the chemical (molecular weight and octanol-water partition coefficient (K_{OW})) allow partitioning into the lipid bilayer of the cell membrane (McCarty et al., 1992). If the combination of the partitioning

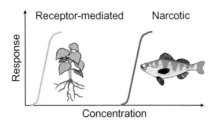

Figure 7.1. Illustration of the difference in toxicity of a herbicide to a plant and a fish and the difference in mode of action in these two groups of organisms. See color plate at the back of the book.

properties and exposure concentration causes enough of the chemical to partition into the membranes of the cells to cause disruption of function, toxicity will result. This is particularly the case for fish because of the large surface area of the membranes of the gill and their crucial role in respiration. Thus, although herbicides are usually less toxic to animals than plants, they will, at large enough concentrations, cause toxicity in animals (Figure 7.1). However, it is also important to recognize that all nonpolar chemicals can cause narcosis in fish and that these other chemicals represent a greater proportion of pollutants in surface waters than do herbicides.

Some herbicides with generalized modes of action can also be relevant to fish and other animals. These include substances that interact with metabolic pathways to result in the production of reactive oxygen species (ROS) and other free radicals, which cause damage to cells. However, this mode of action is not exclusive to herbicides and would mostly be observed in scenarios of high exposures.

Commercial pesticides as sold and used in the environment are usually mixed with other products to aid in the application to the target organisms. These substances are known as formulants and serve the same purpose as excipients that are added to some pharmaceuticals; that is, they help disperse the pesticides in the spray solution or facilitate the penetration of the pesticide into the target organism. Because most herbicides must be translocated through the xylem and phloem of the target plant, they are more soluble in water than other pesticides such as insecticides. Thus, most herbicides do not need the addition of dispersants to the formulation for the purpose of ease of application. However, some herbicides, such as glyphosate, are so polar that they will not penetrate through the waxy cuticle of plants. In these cases, surfactants are added to the formulation to increase penetration through the cuticle and into the plant tissues (Solomon and Thompson, 2003).

Surfactants are surface active agents and can interfere with the membranes of the gills of fish and cause toxicity (Könnecker et al., 2011).

As is illustrated in the case of glyphosate below, the toxicity of the surfactants may be greater than the active ingredient. However, surfactants are widely used in domestic and industrial uses, and the amounts in formulations of herbicides represent a small but unquantified fraction of the total annual use of 13 billion kg of surfactants used globally (Levinson, 2008). In a general sense, these other uses of surfactants are more likely to present risks to fish, but in some specific scenarios of use of herbicide formulants may be important for fish.

The above modes of action are related to direct toxicity where the herbicide directly affects the fish. However, herbicides may harm fish indirectly by removing plants that are either required as a source of food or as a habitat. Alternatively, the biological oxygen demand from the decaying plants may deplete oxygen to concentrations too low to sustain fish. For these indirect effects to occur, there first must be direct effects on the aquatic plants. These direct effects are most likely to occur when concentrations of herbicides are large, such as when they are deliberately used to control unwanted plants in surface waters.

1.2. Exposures of Fish to Herbicides

Herbicides may be directly applied to surface water for the control of aquatic weeds, or they may be transported to surface water via a number of pathways (Figure 7.2). Deposition from direct overspray or drift of spray from nearby fields is a potential route of exposure. Deliberate application to water for the control of nuisance aquatic weeds or weeds in rice paddies is probably the route of greatest exposure for fish. However, these uses are part of pest management operations, and nontarget effects are considered either by the agriculturalist or the provincial, state, or local agencies that regulate application of pesticides directly to surface waters. Applications to crops may result in some contamination of surface waters, but this effect is often mitigated by the use of buffers and setbacks for applications from the air or ground sprayers in agricultural uses of herbicides. Direct overspray of trees can be a source of inputs to surface waters in applications of herbicides to forests where small pools may not be visible and protectable by buffers or setbacks (Thompson, 2011). However, interception by the forest canopy will reduce directly deposited amounts and, from that, exposure concentrations (Linders et al., 2000; Thompson, 2011).

Once applied, herbicides may dissipate or degrade (Figure 7.2). These two terms describe different processes and are not interchangeable. Dissipation describes movement of the herbicide from one location to another. This could be via drift of droplets, volatilization, runoff, leaching, adsorption, and/or desorption to and from sediments and/or soils, or uptake

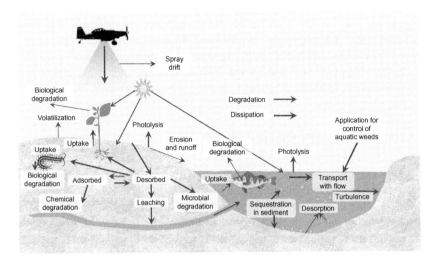

Figure 7.2. Graphical illustration of sources, routes of dissipation, and degradation of herbicides in the surface waters. See color plate at the back of the book.

of the active ingredient by organisms. Degradation describes the conversion of the herbicide from one molecular form to another and can be driven by chemical processes (e.g., hydrolysis and oxidation), physical processes (e.g., photolysis), or biological processes that occur in microbiota, plants, and animals. The degradates from these processes are normally less toxic than the parent product, and the pathway eventually leads to complete mineralization of the herbicide to simple compounds such as CO_2, $NaCl$, phosphate, and water.

Concentrations of herbicides have been extensively measured in some jurisdictions such as the United States. Programs such as the National Water-Quality Assessment Program (NAWQA) have sampled and analyzed surface and groundwaters for many pesticides, including herbicides, since the 1980s (USGS, 2012). In other cases, simple models have been used to estimate concentrations where analyses are unavailable or too few to properly characterize exposures. These models range from Tier-1 screening models such as GENEEC2 (U.S. EPA, 2001) used by the United States Environmental Protection Agency (U.S. EPA) to determine the need for more refined assessments of risk to more sophisticated models that provide more realistic estimates of concentrations under various scenarios of use, such as MUSCRAT (Mangels, 2001). Similar models are used in other jurisdictions such as the EU (FOCUS, 2003). A complete and detailed analysis of concentrations of herbicides in surface waters is beyond the scope of this chapter; however, concentrations of herbicides in surface wasters are, in general, low in comparison to their toxicity (see below). Even for pesticides that are relatively mobile and relatively persistent in surface waters, such as atrazine, 90th centiles of measured and predicted concentrations in surface waters are <20 µg/L (Giddings et al., 2005), well below the acute lethality or chronic toxicity of this herbicide for fish.

2. CLASSIFICATION OF HERBICIDES

There are a large number of herbicidal active ingredients. From the point of view of classifying these ingredients into categories relevant to the potential effects on nontarget organisms, the most appropriate groupings are those based on mode of action. Mode of action is specific to a target process in target and nontarget organisms and allows for a logical grouping of pesticides relevant to their use and effects in nontarget organisms (Stephenson and Solomon, 2007). We have used the classification system of the Herbicide Resistance Action Committee (HRAC, 2012) for grouping herbicides and for examining their acute toxicity to fish.

3. ACUTE LETHALITY OF HERBICIDES TO FISH

All major regulatory agencies through which herbicides are registered require data on toxicity to nontarget organisms. These data are provided by the registrant and, for the required data on fish (one warm-water and cold-water), the studies must be conducted according to guidelines such as OECD test method 203 (OECD, 1992), under good laboratory practice (GLP), and with quality assurance and quality control (QA/QC). Studies conducted under GLP are well documented, and the quality control is generally much more rigorous than papers published in peer-reviewed journals; these studies provide the most reliable toxicity data (McCarty et al., 2012). We have used these and other data from the open literature to characterize acute toxicity values for herbicides in fish.

Acute toxicity data for fish were obtained from the ECOTOX database (U.S. EPA, 2011). Data included all values from papers listed in the database from 1915 to the present and were selected as follows: Only LC50 and EC50 data where the end-point was mortality were used. These are the most commonly available data and provide the most robust data for characterizing toxicity. Subacute exposures (<48 h) and exposures >5 d were excluded. Where multiple acute data were available for a single species, the greatest duration up to 5 d was retained. The material tested was selected in order of preference for technical material > formulated product. Measured concentrations were selected over nominal (unmeasured). Flowthrough studies were selected > static renewal > static. If, after this selection, there was more than one toxicity value for a single species, the geometric mean was calculated as the most representative value for the species. As most of the fish used in these tests were young stages (fingerling) and few acute toxicity data were available for adults, it was not possible to segregate toxicity values by stage of development. Thus, where multiple stages were tested, these data were treated equally. All toxicity values are expressed as μg active ingredient/L. Toxicity values > the greatest concentration tested were retained if there were no other data for the tested species. These were included in the ranking but not in the regression.

The data were then sorted from smallest to greatest and assigned a rank number. It was assumed that the data were log-normally distributed (Burmaster and Hull, 1997). If there were >10 data points, plotting positions were calculated using the Weibull equation $P = i/(n+1) \times 100$ (from Parkhurst et al., 1996), where i is the rank number of the datum point and n is the total number of data points in the set. If there were ≥5 and <10 data points, the Hazen equation, $P = (I - 0.5)/n \times 100$ (Cunnane, 1978) was used. If there were <5 data points, the data were not plotted and only the smallest toxicity value (most sensitive) was listed in the table. The data were plotted on a log-probability scale using SigmaPlot (Systat Software Inc., 2010) and the 5th centile was calculated from the regression equation. An example of a graph is shown in Figure 7.3 for the herbicide dinitramine. The values for the regression equation (transformed to log-probability) are presented in the tables. The point estimate of the 5th centile is an approximation of the HC5 (hazard concentration 5%), which frequently is used for setting Canadian environmental criteria (CCME, 2007).

3.1. Inhibitors of Acetyl CoA Carboxylase (ACCase)

These herbicides are in HRAC Group A and include the aryloxyphen-oxypropionates (FOPs), cyclohexanediones (DIMs), and phenylpyrazoline (DEN) (HRAC, 2012). The mode of action of this group is inhibition of the enzyme acetyl CoA carboxylase, which is involved in fatty acid synthesis (Stephenson and Solomon, 2007); differences in this enzyme between plants and animals result in lower toxicity in animals. The most toxic herbicide in group A was clodinafop-propargyl with a lowest LC50 of 240 μg/L (Table 7.1). Some uncertainty is associated with this value as there were no publicly available data for several chemicals; however, the specificity of the mode of action makes it unlikely that other products in this group will be highly toxic to fish.

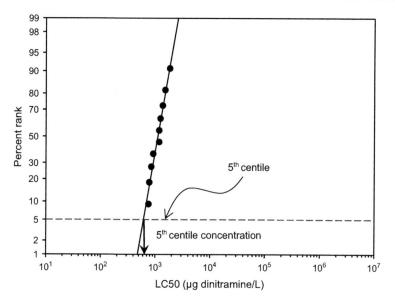

Figure 7.3. An exceedence plot for toxicity values of the herbicide dinitramine to fish. The 5th centile is shown but the horizontal dashed line.

Table 7.1
The most sensitive LC50s in fish for the inhibitors of acetyl CoA carboxylase (ACCase)

Herbicide[a]	n[b]	Smallest LC50 (μg/L)
Clodinafop-propargyl	2	240
Diclofop-methyl	2	273
Sethoxydim	3	3,500
Clethodim	2	19,000

[a]No toxicity data were available in the U.S. EPA database or the open literature for the following herbicides: Alloxydim; butroxydim; cycloxydim; cyhalofop-butyl; fenoxaprop-p-ethyl; fluazifop-P-butyl; haloxyfop-R-methyl; pinoxaden; profoxydim; propaquizafop; quizalofop-P-ethyl; tepraloxydin; and tralkoxydim.
[b]Number of toxicity values.

3.2. Inhibitors of Acetolactate Synthase (ALS) and Acetohydroxyacid Synthase (AHAS)

These herbicides are in HRAC Group B and include the sulfonylureas, imidazolinones, triazolopyrimidine, pyrimidinyl(thio)benzoate, and sulfonylaminocarbonyltriazolinones (HRAC, 2012). These herbicides inhibit the enzyme, acetohydroxyacid synthase (AHAS), which is involved in the biosynthesis of branched chain amino acids and is only found in plants. Thus, this confers strong selectivity to plants, while animals are relatively insensitive to Group B products (Stephenson and Solomon, 2007). The most toxic of the products in Group B was metsulfuron-methyl with a 5th centile of LC50s of 30,000 μg/L (see Table 7.2). Even though there is some uncertainty since there were no publicly available data for several chemicals, the specificity of the mode of action makes it very unlikely that other products in this group will be highly toxic to fish.

Table 7.2
The 5th centile or most sensitive LC50 for the inhibitors of acetolactate synthase (ALS) and acetohydroxyacid synthase (AHAS)

Herbicide[a]	n[b]	r^2	$y = ax + b$ a	b	5th centile (μg/L)
Metsulfuron-methyl	5	1	0.70	−4.79	30,351
Chlorsulforun	6	1	7.31	−35.03	36,833

Herbicide	n[b]				Smallest LC50
Imazethapyr	3				240,000
Imazaquin	3				280,000
Sulfosulfuron	3				>101
Primisulfuron-methyl	3				>13,000
Bensulfuron-methyl	6				>63,000
Imazamethabenz-methyl	2				>100,000
Imazapyr	4				>100,000
Thifensulfuron-methyl	2				>100,000
Triasulfuron	3				>100,000
Diclosulam	3				>110,000
Rimsulforun	3				>110,000
Halosulfuron-methyl	3				>1180,00
Imazamox	2				>119,000
Pyrithiobac-sodium	4				>145,000
Prosulfuron	3				>155,000
Flumetsulam	4				>293,000
Nicosulfuron	2				>1,000,000
Tribenuron-methyl	2				>1,000,000

[a]No toxicity data were available in the U.S. EPA database or the open literature for the following herbicides: Amidosulfuron; azimsulfuron; bispyribac-sodium; chlorimuron-ethyl; cinosulforun; cloransulam-methyl; cyclosulfamuron; ethametsulfuron-methyl; ethoxysulforun; flazasulfuron; florasulam; flucarbazone-sodium; flupyrsulfuron-methyl-Na; foramsulfuron; imazapic; imazosulfuron; iodosulfuron; mesosulfuron; metosulam; oxasulfuron; penoxsulam; propoxycarbazone-sodium; pyrazosulfuron-ethyl; pyribenzoxim; pyriftalid; pyriminobac-methyl; sulfometuron-methyl; trifloxysulfuron; triflusulfuron-methyl; and tritosulfuron.
[b]Number of toxicity values.

3.3. Inhibition of Photosynthesis at Photosystem II

These herbicides are in HRAC Group C and include the triazines, triazinones, triazolinones, uracils, pyridazinones, phenyl-carbamates, ureas, amides, nitriles, benzothiadiazinones, and phenyl-pyridazines (HRAC, 2012). These herbicides inhibit the transfer of electrons in photosystem II in the chloroplast, and the mechanism of action is specific to plants (Stephenson and Solomon, 2007). Thus, animals are less sensitive. The most toxic of the products in Group C was bromofenoxim with a lowest LC50 of 180 µg/L in fish (Table 7.3). The toxicity of one of the most widely used products in this group, atrazine, was low with a 5th centile of toxicity of 1185 µg/L. Again, because of the specificity of mode of action, herbicides in this group are not expected to be highly toxic to fish.

3.4. Diverters of Electrons in Photosystem I

These herbicides are in HRAC Group D, which includes only paraquat and diquat (HRAC, 2012). These herbicides divert electrons from photosystem I in plants and generate free radicals which cause general damage in the cell. When the paraquat^{++} ion intercepts an electron, it becomes a reduced paraquat free radical, which can react with oxygen to regenerate paraquat^{++} cations and superoxide O_2* (Stephenson and Solomon, 2007). Some of the superoxide can react with H^+ cations to produce hydrogen peroxide (H_2O_2), which interacts with paraquat to form •OH, the highly reactive hydroxyl radical. The •OH attacks fatty acids in the thylakoid membranes and the membranes of the cell, resulting in lipid peroxidation, destruction of membranes, and rapid death of the cell. Paraquat itself is not directly toxic. However, because its cation is quickly regenerated by reaction of the paraquat free radical with oxygen, it acts in a catalytic manner to produce toxic products. A few molecules of paraquat can cause the production of large quantities of the oxidants, superoxide, hydrogen peroxide, and hydroxyl radicals. Paraquat is toxic to plants by direct contact with leaves and to mammals via the oral route but exhibits low toxicity to fish, with a 5th centile of LC50s of 634 µg/L (Table 7.4). The only other product in this group is diquat, which is less toxic to fish than paraquat (Table 7.4). Lack of high toxicity in fish compared to mammals is likely due to the polarity of these molecules (they are totally ionized at physiologically relevant pH), large solubility (620–700 g/L), and the lack of partitioning into the tissues of the fish.

3.5. Inhibitors of Protoporphyrinogen Oxidase (PPO)

These herbicides are in HRAC Group E and include the diphenylethers, phenylpyrazoles, n-phenylphthalimides, thiadiazoles, oxadiazoles, triazolinone,

Table 7.3
The 5th centile or most sensitive LC50 for the inhibitors of photosystem II

Herbicide[a]	n	r^2	y = ax + b a	b	5th centile (μg/L)
Ametryne	12	0.85	1.27	−4.93	384
Diuron	12	0.81	1.22	−4.93	496
Propanil	14	0.76	1.81	−6.78	690
Terbuthylazine	6	0.86	1.70	−6.59	793
Atrazine	22	0.93	1.52	−6.32	1,185
Prometryne	8	0.90	3.12	−11.26	1,212
Simazine	28	0.94	1.02	−4.84	1,394
Terbutryne	7	0.93	6.43	−22.57	1,795
Cyanazine	11	0.95	2.08	−8.78	2,744
Chlorobromuron	9	0.90	5.28	−20.07	3,108
Simetryne	6	0.76	2.37	−9.98	3,299
Prometon	9	0.95	2.40	−10.85	6,910
Chlorotoluron	7	0.92	6.15	−29.44	33,053
Hexazinone	10	0.86	3.77	−21.31	164,780

Herbicide	n	Smallest LC50
Bromofenoxim	5	180
Neburon	1	320
Ioxynil	1	350
Phenmedipham	4	1,410
Desmedipham	2	1,700
Bromoxynil[b]	3	2,090
Dimethametryn	5	3,000
Terbumeton	5	10,000
Desmetryne	2	11,500
Siduron	2	13,000
Chloridazon	1	35,000
Bromacil	4	36,000
Metribuzin	5	42,000
Terbacil	3	54,000
Tebuthiuron	4	106,000
Bentazon	5	190,000

[a]No toxicity data were available in the U.S. EPA database or the open literature for the following herbicides: Amicarbazone; dimefuron; ethidimuron; fenuron; isoproturon; isouron; lenacil; linuron; metamitron; methabenzthiazuron; pentanochlor; pyridate; and trietazine.
[b]Bromoxynil is also an uncoupler of oxidative phosphorylation.

oxazolidinediones, and pyrimidindiones (HRAC, 2012). These herbicides inhibit the protoporphyrinogen IX oxidase (PROTOX), which is the last common enzyme in the pathway to the synthesis of chlorophyll and heme (Stephenson and Solomon, 2007). Porphyrin biosynthesis is an important pathway in most living organisms. Products of the pathway include the

Table 7.4
The 5th centile or most sensitive LC50 for the diverters of electrons in photosystem I

Herbicide	n	r^2	$y = ax + b$ a	b	5th centile (µg/L)
Paraquat	10	0.75	1.16	−4.88	634
Diquat	19	0.93	1.48	−6.94	3832

Table 7.5
The 5th centile or most sensitive LC50 for the inhibitors of protoporphyrinogen oxidase (PPO)

Herbicide[a]	n	r^2	$y = ax + b$ a	b	5th centile (µg/L) 5%
Oxadiazon	4	0.74	2.82	−8.37	241
Acifluorfen-Na	4	0.79	2.42	−11.48	11,667
Bifenox	5	0.95	2.59	−13.32	32,763
Herbicide	n				Smallest LC50
Fluthiacet-methyl	3				43
Oxyfluorfen	2				400
Flumiclorac-pentyl	1				1,100
Lactofen	2				3,700
Sulfentrazone	2				93,800
Fomesafen	3				680,000

[a]No toxicity data were available in the U.S. EPA database or the open literature for the following herbicides: Azafenidin; carfentrazone-ethyl; chlomethoxyfen; cinidon-ethyl; fluazolate; flumioxazin; fluoroglycofen-ethyl; halosafen; oxadiargyl; pyraflufen-ethyl; and thidiazimin.

chlorophyll pigments in plants as well as hemoglobin and myoglobin in animals and cytochrome enzymes in most organisms. The product most toxic to fish in this group is fluthiacet-methyl with a smallest LC50 of 43 µg/L (Table 7.5). The selectivity for plants is likely the result of lack of metabolism and/or excretion in plants as compared to animals. The range of sensitivity in Table 7.5 is likely due to differences between products in absorption, distribution, metabolism, and excretion (ADME).

3.6. Herbicides that Cause Bleaching

These herbicides are in HRAC Group F and inhibit carotenoid biosynthesis at the phytoene desaturase step (PDS), and 4-hydroxyphenyl-pyruvatedioxygenase (4-HPPD) and include the pyridazinones, pyridine-carboxamide, triketones, isoxazoles, pyrazoles, triazoles, isoxazolidinones, and diphenylethers (HRAC, 2012). HPPD is an important enzyme in the biosynthesis of plastoquinones in plants. Plastoquinones are important

Table 7.6
The 5th centile or most sensitive LC50 for herbicides that cause bleaching

Herbicide[a]	n	r^2	$y = ax + b$ a	b	5th centile (µg/L)
Fluridone	7	0.84	2.49	−9.88	2,040
Amitrole	8	0.96	1.07	−6.67	49,927

Herbicide	n				Smallest LC50
Isoxaflutole	3				6,400
Clomazone	4				7,320
Norflurazon	3				8,100
Mesotrione	1				532,000

[a]No toxicity data were available in the U.S. EPA database or the open literature for the following herbicides: Aclonifen; beflubutamid; benzobicyclon; benzofenap; bilanafos; diflufenican; flurochloridone; flurtamone; picolinafen; pyrazolynate; pyrazoxyfen; and sulcotrione.

cofactors for phytoene desaturase, and reduced activity of this enzyme decreases concentrations of carotenoids, which protect chlorophyll from photolysis. This is the cause of the bleaching and also reduces photosynthesis (Stephenson and Solomon, 2007). This mechanism is specific to plants, which accounts for the low toxicity of these products for mammals (BCPC, 2003) and to fish (Table 7.6). The most toxic product in this group of herbicides is fluridone with a 5th centile of toxicity values of 2040 µg/L (Table 7.6). Although toxicity data for fish were sparse, it is unlikely that herbicides in this group present a hazard to fish.

3.7. Herbicides that Inhibit EPSP Synthase and Glutamine Synthase

These herbicides are in HRAC Groups G and H and inhibit 5-enolpyruvylshikimate acid-3-phosphatase synthase (EPSPS) and glutamine synthase and include the glycines and the phosphinic acids (HRAC, 2012). Glyphosate and sulfosate are inhibitors of EPSPS, an important enzyme in the biosynthesis pathway for aromatic amino acids (the shikimic acid pathway) in plants. Biochemical symptoms in plants include decreased concentrations of the aromatic amino acids, tryptophan, phenylalanine, and tyrosine, as well as decreased rates of synthesis of protein, indole acetic acid, and chlorophyll and buildup of shikimate or shikimate-3-phosphate (Stephenson and Solomon, 2007). Glufosinate-ammonium inhibits glutamine synthase, an enzyme that is specific to plants. It is of low toxicity to fish (Table 7.7).

The active ingredient glyphosate is of low toxicity to fish (Table 7.7). Glyphosate is unusual in that some formulations (e.g., Roundup® and Vision®) include a surfactant (POEA) that is more toxic to aquatic

Table 7.7
The 5th centile or most sensitive LC50 for the inhibitors of EPSP synthase
and glutamine synthase

Herbicide	n	r^2	$y = ax + b$ a	b	5th centile (µg/L)
Glyphosate[a] formulated (Roundup®) and similar formulations)	8	0.80	1.74	−7.40	2,041
Glyphosate[a] AI	7	0.78	1.11	−5.80	5,689

Herbicide	n				Smallest LC50
Sulfosate (the trimethylsulfonium salt of glyphosate)	2				4,900
Glufosinate-ammonium	3				13,100

[a]Concentrations of glyphosate are expressed in acid equivalents (a.e.).

organisms than the parent material (Giesy et al., 2000; Solomon and Thompson, 2003). This accounts for the greater toxicity of the formulated product to fish than the active ingredient (Table 7.7). The surfactant facilitates uptake of the active ingredient across the waxy cuticles of leaves of plants and is particularly important in controlling some of the more tolerant species of plants (Solomon et al., 2007). Surfactants also affect membranes in aquatic animals, and formulated products may show relatively greater toxicity to nontarget organisms than the active ingredient (Wang et al., 2005). Agricultural formulations of glyphosate are not registered for use over water (Solomon and Thompson, 2003), and the only source of exposures to surfactants such as POEA in a glyphosate formulation is via drift of spray. Like glyphosate, POEA adsorbs strongly to sediments and has a short half-life for dissipation in natural aquatic systems (Wang et al., 2005). As a result, POEA and glyphosate would be expected to partition rapidly out of the water column and thus reduce exposures to fish. Also, because of strong adsorption to soil, toxicologically relevant concentrations of POEA are not expected to result from runoff into surface waters, thus reducing risks from normal agricultural uses. The interested reader is referred to an earlier review of glyphosate-based products and associated surfactants intended for direct application to water (Solomon and Thompson, 2003).

3.8. Herbicides that Interfere with Cell Division

These herbicides are in HRAC Groups K and L and inhibit assembly or organization of microtubules, the synthesis of very long-chain fatty acids (VLCFAs), or the synthesis of cellulose (HRAC, 2012). Although functioning via different mechanisms of action, these herbicides all inhibit division of cells, either through preventing the separation of the chromosomes during meiosis or interfering with the synthesis of new cell membranes or the cellulose component of the cell wall. These include the dinitroanilines, phosphoroamidates, pyridines, benzamides, benzoic acids, carbamates, chloroacetamides, acetamides, oxyacetamides, tetrazolinones, nitriles, benzamides, triazolocarboxamides, and quinolinecarboxylic acids. Toxicity values for herbicides that interfere with cell division cover a large range (Table 7.8). The most toxic of these compounds are trifluralin with a 5th centile LC50 of 2 μg/L and pretilachlor with a single LC50 of 2 μg/L.

The relatively high toxicity of some members of these classes of herbicides to fish (Table 7.8) may be due to their physical properties. Trifluralin, benfluralin, and ethalfluralin have log K_{OW}s of 4.83, 5.29, and 5.11, respectively (BCPC, 2003), suggesting that they would partition into cell membranes to a greater extent than most herbicides and could disrupt these by the mechanism of general narcosis. The effect of trifluralin on sodium transport in the gill (McBride and Richards, 1975) supports this suggestion. Pretilachlor, with a LC50 of 2 μg/L for the most sensitive fish tested (Table 7.8) and with a log K_{OW} of 4.08 (BCPC, 2003), may also act by the same mechanism. Other members of the class do not present a large hazard to fish.

3.9. Herbicides that Uncouple Electron Transport

These herbicides are in HRAC Group M, which represents uncouplers of electron transport and includes the dinitrophenols as the only chemical class (HRAC, 2012). Electron transport in the mitochondrion is an important process in plants and animals. In mitochondria exposed to uncouplers, the process of oxidation continues, but it is uncoupled from the process of phosphorylation or ATP production. The Krebs cycle continues, CO_2 is evolved, NADH and $FADH_2$ are produced, electron transport through the cytochrome system occurs, oxygen is taken up, and water is produced. However, phosphorylation or ATP production occurs at a reduced rate, and energy is not available for other essential functions in the cell (Stephenson and Solomon, 2007). These herbicides are not selective other than via processes related to ADME and are relatively hazardous to fish (Table 7.9).

3.10. Inhibitors of Synthesis of Lipids (Non-ACCase)

These herbicides are in HRAC Group N and inhibit synthesis of lipids (very long-chain fatty acids) by a mode of action that does not involve ACCase. Included in this group are the thiocarbamates,

Table 7.8
The 5th centile or most sensitive LC50 for herbicides that interfere with cell division

Herbicide[a]	n	r^2	$y = ax + b$ a	b	5th centile (µg/L)
Trifluralin	25	0.75	0.66	−1.90	2
Benfluralin	5	0.88	1.14	−3.11	19
Ethalfluralin	4	0.91	4.75	−10.61	77
Butachlor	9	0.94	2.38	−6.50	109
Alachlor	5	0.83	1.42	−5.18	305
Dinitramine	10	0.96	6.31	−19.23	614
Pendimethalin	4	0.98	6.02	−18.81	710
Piperophos	4	0.94	12.24	−45.36	3,736
Dichlobenil	12	0.95	4.74	−18.69	3,950
Metolachlor	7	0.85	5.25	−20.56	4,010
Diphenamid	4	0.92	6.93	−33.33	37,161
Propyzamide	5	0.98	1.85	−10.23	42,684

Herbicide	n	Smallest LC50
Pretilachlor	1	2
Propachlor	3	254
Butralin	3	370
Dithiopyr	2	465
Acetochlor	3	576
Isoxaben	3	>870
Oryzalin	2	2,880
Thiazopyr	3	2,900
Chlorpropham	3	3,200
Flufenacet	1	5,840
Chlorthal-dimethyl	2	>6,700
Napropamide	3	10,836
Propham	3	29,000
Chlorthiamid	1	33,000
Carbetamide	1	165,000

[a]No toxicity data were available in the U.S. EPA database or the open literature for the following herbicides: Amiprophos-methyl; anilofos; butamiphos; cafenstrole; dimethachlor; dimethanamid; fentrazamide; mefenacet; metazachlor; naproanilide; pethoxamid; propisochlor; tebutam; and thenylchlor.

phosphorodithioates, benzofurans, and chloro-carbonic acids (HRAC, 2012). These mechanisms are somewhat specific to plants. The three most toxic members of this group are bensulide, thiobencarb, and triallate (Table 7.10), which have log K_{OW}s of 4.2, 3.42, and 4.6, respectively, suggesting that they may be toxic to fish via partitioning into cell membranes and general narcosis. Butylate has a log K_{OW} of 4.1 but is not

Table 7.9
The 5th centile or most sensitive LC50 for herbicides that uncouple oxidative phosphorylation

Herbicide[a]	n	r^2	y = ax + b A	b	5th centile (μg/L)
Dinoseb	8	0.72	0.93	−1.89	2

Herbicide	n				Smallest LC50
DNOC	3				66

[a]No toxicity data were available in the U.S. EPA database or the open literature for dinoterb, the other member of HRAC class M.

Table 7.10
The 5th centile or most sensitive LC50 for the inhibitors of the synthesis of lipids

Herbicide[a]	n	r^2	y = ax + b a	b	5th centile (μg/L)
Bensulide	6	0.91	4.06	−11.12	216
Thiobencarb	19	0.96	2.18	−6.85	241
Triallate	5	0.75	2.77	−8.98	446
Butylate	4	0.88	2.86	−10.67	1,419
Vernolate	4	0.90	3.29	−12.48	1,972
EPTC	7	0.72	1.58	−7.18	3,228
Ethofumesate	5	0.93	2.16	−9.28	3,409
Dalapon	13	0.81	0.53	−3.54	3,678
Pebulate	5	0.93	12.89	−50.16	5,796
Molinate	14	0.93	3.34	−14.30	6,142

Herbicide	n				Smallest LC50
Cycloate	3				5,327
Trichloracetic acid-sodium salt	1				>2,000,000

[a]No toxicity data were available in the U.S. EPA database or the open literature for the following herbicides: Benthiocarb;dimepiperate; esprocarb; orbencarb; prosulfocarb; flupropanate;tiocarbazil; and benfuresate.

as toxic as bensulide, so this mechanism may be modified by ADME. Overall, this group of herbicides does not present a large hazard to fish.

3.11. Herbicides that Interfere with the Action of Auxins in Plants

These herbicides are in HRAC Groups O and P and interfere with the action of auxins in plants. Included in this group are the phenoxy-carboxylic acids, benzoic acids, pyridine carboxylic acid, quinoline carboxylic acids,

386

KEITH R. SOLOMON ET AL.

Table 7.11
The 5th centile or most sensitive LC50 for herbicides that interfere with the action
of plant auxins

Herbicide[a]	n	r^2	$y = ax + b$ a	b	5th centile (µg/L)
Picloram	10	0.95	1.23	−5.18	738
Triclopyr	7	0.75	1.40	−5.94	1,167
2,4-D	28	0.89	0.83	−4.31	1,691
MCPB	4	0.92	3.34	−12.71	2,085
Quinclorac	4	1.00	0.76	−4.56	7,039
Fluroxypyr	4	1.00	1.86	−8.89	7,758
MCPA	9	0.93	1.42	−7.82	22,141
Dicamba	4	1.00	3.26	−17.33	63,889

Herbicide	n				SmallestLC50
2,4-DB	3				4,323
2,3,6-TBA	3				8,500
Naptalam	2				76,100
Diflufenzopyr-Na	3				106,000
Clopyralid	2				700,000

[a]No toxicity data were available in the U.S. EPA database or the open literature for the following herbicides: Clomeprop; dichlorprop; mecoprop; chloramben; quinmerac; benazolin-ethyl; tiocarbazil; and benfuresate.

phthalamates, and semicarbazones (HRAC, 2012). The auxin hormones are specific to plants, and because animals do not possess the target for these herbicides, they are less toxic to fish. The most toxic of these herbicides is picloram with a 5th centile LC50 value of 738 µg/L (Table 7.11). Overall these herbicides do not present a hazard to fish.

3.12. Herbicides with Unknown Modes of Action

These herbicides are in HRAC Group Z, for which the mode of action is unknown. Included in this group are the arylaminopropionic acids, pyrazoliums, and organoarsenicals (HRAC, 2012). Toxicity values cover a wide range of values (Table 7.12), with the most toxic herbicide being dazomet. In the environment, dazomet degrades to an isothiocyanate (Stephenson and Solomon, 2007) which is reactive and biocidal, possibly accounting for the high toxicity of this herbicide. The other tested products in this group are all less toxic and do not present a hazard to fish.

Table 7.12
The 5th centile or most sensitive LC50 for the inhibitors for herbicides of unknown mode of action

Herbicide[a]	n	r^2	$y = ax + b$		5th centile (μg/L)
			a	b	
Dazomet	5	0.93	0.88	−3.06	41
MSMA	8	0.88	1.29	−6.93	12,701
Herbicide	n				Smallest LC50
Chlorflurenol	3				1160
Pelargonic acid	3				91,000
DSMA	3				>112,000
Oleic acid	1				205,000

[a]No toxicity data were available in the U.S. EPA database or the open literature for the following herbicides: Bromobutide; cinmethylin; cumyluron; difenzoquat; dymron; etobenzanid; flamprop-m-methyl; fosamine; indanofan; metam; methyldymron; oxaziclomefone; and pyributicarb.

4. SUBLETHAL EFFECTS OF HERBICIDES IN FISH

Herbicides may have sublethal effects on fish that express themselves in a number of responses related to development, growth, reproduction, and behavior. The following sections address these responses.

4.1. Effects of Herbicides on Early Development, Growth, and Reproduction

Normal early development, growth, and reproduction are critical for long-term maintenance of fish population health. As such, chronic exposure to single or multiple herbicides—if potent and at sufficient concentrations for extended durations—has the potential to adversely impact population stability and ecosystem health. This may be a particular concern within small freshwater streams throughout use-intensive agricultural regions such as the U.S. Corn Belt. Similar to acute toxicity data summarized in Section 3 above, the U.S. EPA's ECOTOX database (U.S. EPA, 2011) was mined in order to (1) identify whether sublethal fish toxicity data based on development, growth, and reproduction as end-points were available for herbicides classified by the Herbicide Resistance Action Committee (HRAC), and (2) identify the range of no observed effect concentrations (NOECs) and corresponding exposure durations for herbicides with available sublethal toxicity data. In order to identify these data, variables listed within Table 7.13 were selected prior to mining the ECOTOX

Table 7.13
Input variables selected using the Advanced Database Query within the
U.S. EPA's ECOTOX database

Database Tab	Variable Category	Variable Selected
Taxonomic	Taxonomic Name Entry	Animals (species box blank)
	Predefined Taxonomic Groups	Fish
Chemical	Chemical Entry	Herbicide 1, 2, 3…n
	Predefined Chemical Groups	None selected
Test results	Endpoints	NOEC
	Effect Measurements	Developmental
		Growth
		Reproduction
Test conditions	Test Locations	Lab
	Exposure Media	Freshwater
		Saltwater
	Exposure Types	Flow-through
		Renewal
	Chemical Analysis	None selected
Publications/Updates	Publication Years	1915–2012
	Reference Numbers	None inserted
	Independently Compiled Data	None selected
	Recent Modifications/Additions	None selected

database using the Advanced Database Query. These variables were selected in order to focus on continuous exposures (static renewal or flowthrough conditions) and sublethal end-points useful for ecological risk assessment (i.e., development, growth, and reproduction). As this minimally restrictive variable set occasionally identified other taxonomic groups (e.g., amphibians), all non-fish-specific data were eliminated prior to summarizing and analyzing the resulting toxicity data.

Using the ECOTOX database search variables listed in Table 7.13, a total of 107 results (or data points) were identified from peer-reviewed publications and/or registrant-submitted reports published from 1983–2012 (no results were identified prior to 1983). Moreover, these data were distributed across 17 different herbicides (Table 7.14), indicating that, despite 30 years of available data, fish-specific toxicity data based on development, growth, and reproduction as end-points were only available for approximately 6% of 274 HRAC herbicides queried. Within these results, end-points based on development, growth, and reproduction were the basis for 15% (16), 81% (87), and 4% (4) of the total number of data points, respectively, indicating that growth was the most prevalent end-point used for NOEC determinations. In addition, data derived from fathead minnows (*Pimephales promelas*)—one of the most commonly used species

Table 7.14
Summary of sublethal herbicide toxicity data based on development, growth, and reproduction as end-points

Chemical Family	Active Ingredient	Data Points (n)	Min.	Avg.	Max.
				NOEC (μg/L)	
Amide	Propanil	6	0.4	0.9	2.4
Chloroacetamide	Alachlor	7	500	879	2,000
Dinitroaniline	Pendimethalin	2	0.2	0.2	0.2
	Trifluralin	12	3	21.5	74
Dinitrophenol	DNOC	8	183	602	2,000
	Dinoseb	12	15	4,124	48,500
Nitrile	Bromoxynil	4	400	2,358	3,010
Phenoxy-carboxylic acid	2,4-D	2	27,200	28,600	30,000
Pyridine carboxylic acid	Clopyralid	4	68,000	170,500	273,000
	Picloram	4	600	890	1,180
	Cyanazine	1	400	n/a	n/a
Triazine	Simazine	1	4,500	n/a	n/a
	Atrazine	22	8.4	600.6	5,000
Triazinone	Hexazinone	12	7.9	21.8	85.5
Uracil	Bromacil	2	12,000	20,500	29,000
Urea	Tebuthiuron	1	9,300	n/a	n/a
	Diuron	7	14.5	510.4	3,400

n/a if N=1

for early life-stage and full life-cycle pesticide toxicity tests with fish—accounted for approximately 47% (50) of the data points, while Atlantic salmon (*Salmo salar*) and rainbow trout (*Oncorhynchus mykiss*) accounted for approximately 22% (24) and 9% (10) of the data points, respectively (Figure 7.4). The remaining 22% (23) of the data points was represented by 15 different species (Figure 7.4).

Across 107 data points, there was a broad range of reported exposure durations (3 to 396 days) and NOECs (0.2 to 273,000 μg/L) (Table 7.14), suggesting that non-target sublethal toxicity within fish is strongly dependent on the mode of action of the herbicide and metabolism, despite a common target pest (i.e., annual grasses and broadleaf weeds). Indeed, based on the intraherbicide variation in toxicity data (represented by the minimum, average, and maximum across data points), two of the most potent herbicides across multiple species and end-points were pendimethalin ($NOEC_{min}$ = 0.2 μg/L) and trifluralin ($NOEC_{min}$ = 3 μg/L) (Table 7.14)—both dinitroanilines that inhibit cellular microtubule assembly and are predominantly used as preplant/preemergence herbicides for soybean production (Pike et al., 2008). In addition, although inhibition of photosynthesis at photosystem II is selective to plants, the $NOEC_{min}$ for propanil (an amide-based herbicide) was 0.4 μg/L, suggesting that fish may

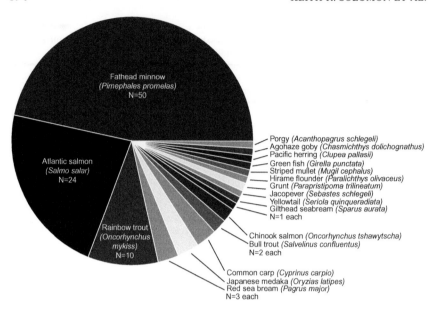

Figure 7.4. Number of sublethal herbicide toxicity data points by fish species. See color plate at the back of the book.

also be uniquely sensitive to this herbicide. The mechanism of toxicity is not understood, but the bioconcentration factor for propanil is minimal (1.6) and it is readily metabolized in fish (Call et al., 1983).

Given the broad range of exposure durations, NOECs across all data points and herbicide MoAs were plotted against exposure duration to (1) identify the extent of within-duration NOEC variation and (2) determine whether NOECs generally decreased with increasing duration of exposure. After plotting NOECs with available exposure durations (100 out of 107 data points; Figure 7.5), a Pearson's product moment correlation analysis was performed within SigmaPlot v12.0 to determine the goodness-of-fit (r^2) and statistical significance of association ($\alpha = 0.05$). Based on this analysis, there was strong within-duration variation in NOEC—in some cases spanning six orders of magnitude—across all exposure durations ($r^2 = 0.0002$), resulting in no significant negative correlation ($p = 0.691$) between NOEC and exposure duration. As such, this supports the conclusion that sublethal herbicide toxicity within fish is strongly dependent on herbicide MoA rather than concentration × duration alone.

In summary, an analysis of available sublethal herbicide toxicity data within the EPA's ECOTOX database revealed two major findings. First, although a large number of the HRAC herbicides have been registered and used for years, based on publicly available data there appear to be significant

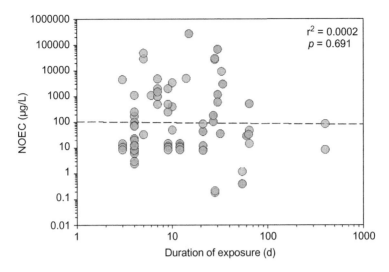

Figure 7.5. No observed effect concentrations (NOECs) and corresponding exposure duration for data on sublethal toxicity of pesticides.

data gaps and uncertainties about the potential effects of the majority (>95%) of these herbicides on fish development, growth, and reproduction. These gaps are surprising as chronic (early life stage) tests are normally required of all registered pesticides (U.S. EPA, 2012b). Second, although toxicity is generally dependent on Haber's law (toxicity α concentration × duration, ten Berge et al., 1986), there was no significant correlation between NOECs and duration of exposure. While this lack of association may be due to uncertainties in the "true" NOEC, these findings suggest that sublethal toxicity of herbicides in fish is strongly dependent on the mode of action and potential for off-target (nonselective) toxicity within fish. For example, the dinitroanilines (pendimethalin and trifluralin) were identified as two of the most potent herbicides within fish, presumably because microtubule inhibition may occur in plant and fish cells. On the contrary, herbicide classes with plant-specific MoAs (e.g., synthetic auxins such as 2,4-D) tend to exhibit very low toxicity despite continuous exposure across long durations. This characteristic suggests that these types of selective (e.g., plant-specific MoA) herbicides offer the ideal safety profile for fish.

4.1.1. SENSITIVITY OF DIFFERENT DEVELOPMENTAL STAGES TO HERBICIDES

It is generally believed that younger stages of fish are more sensitive to toxicants in general than older stages. For example, in a study on the toxicity of three metals (Cu, Zn, Ni) and pentachlorophenol, alevins (the stages from hatching until absorption of the yolk) of Coho salmon (*O. kisutch*) were consistently more sensitive than older and larger stages of fish (Hedtke et al., 1982).

A few studies have investigated the sensitivity of different developmental stages of fish to herbicides. A study on the toxicity of the herbicide thiobencarb (Table 7.10) in California grunion (*Leuresthes tenuis*), Atlantic silverside (*Menida menida*), and Tidewater silverside (*Menida peninsulae*) showed similar results (Borthwick et al., 1985). There was greater sensitivity in younger fish (0–7 d posthatch) than in older fish (28 d posthatch), and the differences in sensitivity ranged from < two- to threefold in these species. A study with the same herbicide on developmental end-points in medaka (*Oryzias latipes*) showed that thiobencarb was more toxic to stage-10 blastulas (EC50 of 3600 µg/L) than stage 23 fish (beating heart, EC50 of 4100 µg/L) (Villalobos et al., 2000). Differences in sensitivity were not large.

Other authors have reported the opposite. According to data provided in Mayes et al. (1987), the sensitivity of bluegill (*Lepomis macrochirus*) to the herbicide picloram (Table 7.11) increased with increasing weight. The 96 h LC50 for fish weighing 160 mg was 44,500 µg/L, which decreased to 14,500 to 23,000 µg/L in fish weighing 830 to 900 mg. The toxicity of dinoseb (Table 7.9) to eyed eggs and alevins of Chinook salmon (*Oncorhynchus tshawytscha*) was reported to be 335 and 71 µg/L, respectively, showing that the alevins were more sensitive (Viant et al., 2006).

The general assumption that younger stages of fish are more sensitive than older stages is not always the case for the few herbicides that have been tested. The reasons for this are not clear, but the differences in sensitivity were not large and would be included in uncertainty factors normally applied in risk assessments.

4.1.2. EFFECTS OF HERBICIDES ON SMOLTIFICATION OF FISH

Smoltification includes the behavioral, developmental, and physiological changes that accompany the transformation from parr to smolt in salmonids and the ability of the fish to initiate their downstream migration and the successful transition from life in freshwater to seawater (Hoar, 1988). The development of seawater tolerance results from a reorganization of the major osmoregulatory organs, including the gill, gut, and kidney, and is accompanied by an increase in gill Na+/K+-ATPase activity and the ability to regulate plasma ions following seawater transfer. A number of hormones, including thyroid hormones, prolactin, corticosteroids, and growth hormone, play pivotal roles in smoltification (Barron, 1986; McCormick, 2001; Ojima et al., 2009).

Recently, attention has focused on the possibility that exposure to anthropogenic chemicals may impair smoltification and contribute to declines in the numbers of Atlantic salmon (*S. salar*) in both Europe and North America. Few studies have been published on herbicides in general, but several studies have considered the responses to the triazine herbicide, atrazine. For example, Waring and Moore, (2004) reported that exposure of Atlantic salmon smolts to atrazine (Table 7.3) at concentrations of 2–22.7 μg/L for up to seven days led to elevated mortality upon transfer to seawater and contributed to increases in osmolarity and the concentrations of sodium, chloride, cortisol in plasma, and a significant decrease in activity of gill Na+/K+-ATPase. In one study, 100% mortality was reported upon transfer to seawater after exposure to 0.1 μg atrazine/L for 72 h in freshwater (Moore et al., 2008). In sharp contrast, several other studies have reported that Atlantic salmon smolts were far less sensitive to atrazine at similar or greater concentrations than those reported above (e.g., Moore et al., 2003, 2007; Nieves-Puigdoller et al., 2007; Matsumoto et al., 2010). These studies showed that substantially greater concentrations of atrazine were required to cause mortality, perturbations of ion homeostasis in freshwater, some loss of salinity tolerance, and a transient decrease in growth in seawater. For example, Nieves-Puigdoller et al. (2007) reported low mortality (9%) of smolts exposed to 100 μg atrazine/L for 21 d in freshwater and found that this treatment did not influence survival following a challenge with seawater for 24 h. In addition, there were no mortalities in controls or those exposed to 10 μg atrazine/L. There were also no differences in the weight or length of the control or atrazine-treated fish after a three-month grow-out period in seawater (Nieves-Puigdoller et al., 2007). Studies by Matsumoto et al. (2010) showed that atrazine at concentrations of up to 100 μg/L had no effect on the survival of Atlantic salmon following a four-day exposure in freshwater and then seven days in seawater.

The unanswered question is whether the effects of atrazine on survival and measures of iono-regulatory performance would be seen at concentrations encountered in the freshwater environment. Some studies raise significant concerns (Waring and Moore, 2004; Moore et al., 2008), while other research (Moore et al., 2003, 2007; Nieves-Puigdoller et al., 2007; Matsumoto et al., 2010) suggests that the concentrations at which responses to atrazine were observed would rarely be experienced in the environment, especially in lotic systems inhabited by salmonids. Overall, these studies point to the considerable uncertainty about the risks posed by atrazine to Atlantic salmon during smoltification.

4.1.3. EFFECTS OF HERBICIDES ON REPRODUCTION IN FISH

Relatively few herbicides have been assessed for effects on reproduction per se. Full life-cycle tests are resource-intensive, and the alternative short-term reproduction assay using fathead minnows (P. promelas) (Ankley et al.,

2001) has only recently been developed and validated. The effects of oryzalin (a dinitroaniline herbicide, Table 7.8) on reproduction and development of medaka (*O. latipes*) were reported by Hall et al. (2007). Treatment of either males and females, or both, resulted in an increase in the number of nonfertilized eggs produced. Compared to the control, male and female fish treated with 250, 500, but not 1000 μg oryzalin/L produced significantly more unfertilized eggs per day. The numbers of infertile eggs laid per day were 1.7, 4.7, 4.8, and 5.1 for the control and three exposure concentrations, respectively. There were no consistent effects on the number of eggs produced, but time-to-hatch of the eggs from exposed females, but not males, was increased in a concentration-dependent manner. Although no effects on the number of oocytes in the testes were reported, a significant increase in necrosis of spermatogonia was reported in fish exposed to 250 and 500 but not 1000 μg/L. The reason for the lack of concentration response is not clear.

In other studies, Xie et al. (2005) determined the estrogenic potency of triclopyr, 2,4-D (Table 7.11), diquat dibromide (Table 7.4), and glyphosate (Table 7.7) in an *in vivo* rainbow trout vitellogenin assay. Juvenile rainbow trout exposed to a large concentration (1640 μg/L) of 2,4-D for 7 days had a 93-fold increase in plasma concentrations of vitellogenin compared with untreated fish, while rainbow trout exposed to other pesticides alone did not show elevated concentrations of vitellogenin compared to the control fish. These results are consistent with the low estrogenic potencies of herbicides in fish.

A number of studies have been carried out to assess the effects of triazine herbicides (Table 7.3) on reproduction in fish. The effect of prometon (Table 7.3) on reproduction of fish was characterized using the short-term reproduction assay in fathead minnows (Villeneuve et al., 2006). No significant effects on cumulative production of eggs over the 21-d exposure were observed. In addition, other markers of endocrine disruption, such as concentrations of vitellogenin and estradiol in plasma or aromatase activity in the brain and ovary, and number of tubercles in males were also not significantly affected. Exposure to 20 μg prometon/L significantly increased concentrations of testosterone in the plasma of females, but the effect was not observed at 250 or 500 μg/L. Relative to body weight, the mass of the fat pad, an androgen-responsive tissue, was significantly reduced at all concentrations. The concentrations tested were greater than those measured in the surface waters of the United States (USGS, 2012). Of the 31,780 analyses of surface water for prometon conducted by the U.S. Geological Survey since 1991, the greatest value was 5.6 μg/L (USGS, 2012). The lack of response on the apical end-point of reproduction suggests that prometon does not cause adverse effects on reproduction of fish at realistic concentrations in the environment.

The effects of atrazine on reproduction and sexual development have been tested in a number of studies in fish. Exposure of Atlantic salmon (*S. salar*) to simazine (Table 7.3) at 0.1, 0.5, 1, and 2 μg/L and atrazine (Table 7.3) at 0.5 and 2 μg/L was reported to decrease the amount of milt expressible from males also exposed to priming pheromone (Moore and Lower, 2001); however, the statistical comparisons used in the data analysis were incorrect and the significance of this response could not be characterized. The authors compared the response of fish exposed to herbicide and priming pheromone to a water control instead of pheromone-exposed animals. There was no apparent concentration response or additivity between atrazine and simazine. The relevance of these apparent changes in production of sperm to reproductive function was not characterized.

In a study on zebrafish (*Danio rerio*) exposed from 17 d postfertilization to six months to a DMSO control or 10^{-7}, 10^{-6}, or 10^{-5} M atrazine (21.7, 217, and 2170 μg atrazine/L), it was reported that sex ratios of the resulting offspring were altered and that more females were produced by the exposed fish (Suzawa and Ingraham, 2008). The method was not clearly described, and the calculation of the sex ratios in the fish did not make sense in that the number of males decreased by 50% relative to the control, but the number of females increased by 400% relative to the control. As raw data were not provided, it was not possible to recalculate these ratios. Sex ratios were also characterized in a more robust study on zebrafish exposed to atrazine (Corvi et al., 2012). This study was conducted twice using the same design. Ten wild-type zebrafish (strain tropical 5D) per tank were exposed for 113 days (17 dpf through to maturation) to concentrations of 0, 21.6, 216, 2160 μg technical grade atrazine/L. Each treatment included eight replicate tanks. Estradiol (E2) was employed as positive control, and E2 and atrazine solutions were renewed every three days. Stock solutions were analyzed, and measured concentrations across treatments in Study 1 equaled 88% to 89% of nominal and for Study 2, 92% to 96% of nominal. Concentrations of dissolved oxygen (DO), pH, temperature, ammonia, and nitrite levels were within acceptable levels throughout the study, as demonstrated by lack of mortality. There were no significant differences ($p > 0.05$) in the ratio of male and female zebrafish compared to negative controls in either Study 1 or Study 2. In both studies, E2 produced significantly ($p < 0.01$) greater numbers of females than males compared to negative controls. Based on the results of the Corvi et al. study, atrazine does not appear to alter the sex ratios of fish at concentrations ranging from realistic worst-case environmental concentrations to 100-fold greater values.

Three studies using the 21-d reproduction protocol (Ankley et al., 2001) have evaluated the effects of atrazine on reproduction in fathead minnows

(Bringolf et al., 2004; U.S. EPA, 2005; Tillitt et al., 2010). A study conducted under Good Laboratory Practice (GLP) with full quality control showed that exposures to 25 and 250 μg atrazine/L had no effects on numbers of eggs laid, fertilization, hatchability, and development of the larvae (U.S. EPA, 2005). In addition, body weight, gonadosomatic index (GSI), concentration of steroid hormones in the plasma, and concentrations of vitellogenin in plasma of male fatheads were unaffected. A similar study by Bringolf et al. (2004) also used the 21-d reproduction protocol to assess the effects of exposures to 5 and 50 μg atrazine/L on reproduction in fathead minnows. No significant effects were observed on cumulative number of eggs laid, fertilization, hatching, and survival of the embryos (Bringolf et al., 2004). In addition, no effects were observed on GSI, development of the gonads, and secondary sexual characteristics. The third study (Tillitt et al., 2010) followed a different protocol from that of the 21-d reproduction study used above. The study exposed fish to 0.5, 5, and 50 μg atrazine/L, and no significant effects were observed on any responses related to survival of the adults, reproductive hormones, aromatase, GSI, and secondary sexual characteristics. However, there were significant decreases in the number of spawning events and total number of eggs laid between 14 d, when half the replicates were removed, and the end of the study (30 d instead of 21 d as recommended for the standard protocol). The reason for the difference in results between this study and the two described earlier that did not deviate from the standard protocol was not given; however, it is probably due to the use of only one male in each tank instead of two, and the increase in variance of the performance of the males as well as the reduced number of replicates after 14 d.

Although few studies of the effects of herbicides on reproduction of fish have been undertaken, these studies were conducted on herbicides such as atrazine and prometon, which are widely applied and found in surface waters near areas of use. Given the selectivity of most herbicides for processes that are only found in plants (atrazine and prometon are inhibitors of photosynthesis), effects on reproduction in fish would not be expected at the relatively small concentrations generally found in surface waters.

4.2. Effects of Herbicides on Stress Response in Fish

Stress is a common feature of life, and fish, like all vertebrates, have evolved a suite of defense reactions to protect themselves against stimuli that pose a challenge to the maintenance of the homeostatic equilibrium. One major component of the stress response is the endocrine stress response, which results in an increase in cortisol, the major glucocorticoid in fish. Cortisol acts to increase the availability of energy by reallocating energy away from nonessential physiological functions such as reproduction and toward activities that contribute to the restoration of homeostasis. Although this prioritizing of energy allocation is an integral component of allostasis, which represents part of the adaptive process for actively maintaining stability through change (McEwen and Wingfield, 2003), chronic inhibition of investment activities can be maladaptive. It is in this context that concern has been raised that exposure to herbicides may function as endocrine-disrupting chemicals and influence the stress response, although to date surprisingly few studies have addressed this issue.

By far the most extensive studies of herbicide effects on the stress response in fish have focused on the actions of atrazine (reviewed in Solomon et al., 2008). Common carp (*Cyprinus carpio*) held at 12–22°C and exposed to 100 µg atrazine/L for 72 h exhibited 2.5- to 5-fold increases in the plasma concentrations of cortisol (Gluth and Hanke, 1983, 1985). In the freshwater fish *Procilodus lineatus*, exposure to atrazine at 10 µg/L for 24 h had no effect on basal concentrations of cortisol, but stress-induced increases in cortisol observed in control animals 1 h and 3 h after a 3-min exposure to air were completely blocked in the atrazine-exposed fish (Nascimento et al., 2012). In studies with the Brazilian fish *Rhamdia quelen*, exposure to large concentrations of atrazine (1700, 3400 or 5100 µg/l) for 96 h led to a sixfold elevation of concentrations of cortisol in plasma (Cericato et al., 2008). In the same study, concentrations of cortisol were not increased further when the atrazine-treated fish were exposed to an acute stress by chasing them with a net. Atlantic salmon smolts exposed to atrazine in freshwater for seven days at concentrations between 6.5 and 22.7 µg/L and then tested directly or following transfer to seawater for 23 h exhibited a 2- to 2.5-fold increase in concentrations of cortisol in plasma (Waring and Moore, 2004). In other studies, concentrations of cortisol in plasma were significantly elevated in Atlantic salmon smolts treated for 21 d in freshwater with atrazine at 100 µg/L and after being held for a further 24 h in seawater (Nieves-Puigdoller et al., 2007). In the same study, Atlantic salmon exposed to 10 µg/L atrazine exhibited no change in concentrations of cortisol in plasma in either freshwater or seawater.

Other studies have tested the direct effects of atrazine on corticosteroid secretion by interrenal tissues from fish. For example, atrazine at concentrations up to (500 µM ≈ 100,000 µg/L) had no effect on the viability of head-kidney cells of *O. mykiss* incubated *in vitro* (Bisson and Hontela, 2002). However, atrazine at concentrations between 0.005 and 5 µM (≈ 1 to 1000 µg/L) caused a modest reduction in ACTH-stimulated cortisol secretion and at larger concentrations (500 µM atrazine ≈ 100,000 µg/L) caused a slight increase in cortisol secretion. In the same studies, atrazine at concentrations up to 500 µM had no effect on dibutryl cAMP-stimulated cortisol secretion.

Collectively, the results of these various studies suggest that there may be some subtle differences in the manner in which the HPI axis of different fishes respond to atrazine. In general, atrazine exposure seems to be associated with both a modest increase in cortisol levels but also a reduced ability to mount a cortisol response to an external stressor. Given that a stress response is adaptive in terms of maintaining homeostasis but can also be maladaptive in terms of effects on growth, reproduction, and survival, understanding the effects of atrazine and other herbicides on the stress response remains a priority. An important consideration in the future studies will be to test atrazine and other herbicides at environmentally relevant concentrations as many of the studies conducted to date have examined responses induced by concentrations that exceed the levels typically seen in the environment.

4.3. Effects of Herbicides on Olfactory Responses in Fish

The response of fish to olfactory clues has been the subject of a number of studies and is often also studied in relation to resulting behaviors related to feeding (discussed below). Most of these studies have focused on insecticides (e.g., Jarrard et al., 2004) as these are generally more toxic to fish and often affect the nervous system directly. A few studies have been carried out with herbicides and are reviewed here.

Olfactory response to benthiocarb (a thiocarbamate; Table 7.10) was measured in the olfactory bulb of carp (*Cyprinus carpio*) (Ishida et al., 1996). The threshold of olfactory response to benthiocarb was reported to be 0.17 µg/L, but responses measured via taste and the pit organ occurred only at ca. 100,000-fold greater concentrations, indicating olfaction as the primary sense for environmental stimuli. Exposure to dichlobenil (a benzonitrile; Table 7.8) caused changes in the G-protein(s), $G_{\alpha q}/G_{\alpha 11}$ proteins, in the olfactory neuroepithelium of the channel catfish (*Ictalurus punctatus*) at a concentration of 20,000 µg/L (Andreini et al., 1997). This was observed at a rather large and unrealistic concentration and, since no electrophysiological measurements were conducted, these results are not easily interpreted. Simazine and atrazine (Table 7.3 and Table 7.4) were reported to increase the response of the olfactory epithelium in Atlantic salmon (*S. salar*) at concentrations of 1 and 2 µg/L (Moore and Lower, 2001). Atrazine reduced the olfactory response to the female priming pheromone, prostaglandin F2, at a concentration of 1 µg/L. In studies on goldfish (*Carassius auratus*), Saglio et al. (2001) reported that bentazon and nicosulfuron (a benzothiadiazinone, Table 7.3) and a urea herbicide (Table 7.2), respectively, did not induce depolarization of the olfactory epithelium at concentrations of 1000 µg/L, although there were some behavioral responses at greater concentrations (see below). The effects of exposures of Coho salmon (*Oncorhynchus kisutch*) to technical glyphosate acid (99%), trifluralin (99.5%), and 2,4-D (98%) on electrical activity in the olfactory epithelium were reported by Tierney et al. (2006). Only glyphosate (an inhibitor of EPSP synthase; Table 7.7) and 2,4-D (Table 7.11) evoked L-serine-like olfactory responses over a range of concentrations from 10 to 100,000 µg/L. Despite being one of the more toxic herbicides to fish (Table 7.8), trifluralin did not evoke a response at 30 or 300 µg/L. When exposures to glyphosate and 2,4-D were removed, rapid recovery of the olfactory epithelium occurred. It is uncertain from the methods in the paper as to the pH of the solutions of glyphosate and 2,4-D. Both of these substances are acids, and this could confound the response unless the pH was buffered to neutral prior to exposure. Olfactory responses to linuron (a urea herbicide; Table 7.3) were measured in Coho and sockeye salmon, and rainbow trout (*O. kisutch, O. nerka*, and *O. mykiss*) (Tierney et al., 2007a). Unlike serine, the response to linuron did not follow a monotonic concentration response and evoked a greater response in sockeye salmon than in the other two salmonids. Linuron (10 µg/L) caused a decrease in olfactory response to L-serine (a feeding trigger) but had no effect on the response to taurocholic acid (bile salt). Similar olfactory responses were measured in rainbow trout exposed to atrazine (Table 7.3) (technical 97.4%) and Roundup®, a formulation of glyphosate containing glyphosate isopropylamine salt (IPA), and a surfactant POEA (Tierney et al., 2007b). Atrazine evoked no olfactory response at concentrations of 10 and 100 µg/L and Roundup® evoked responses at 100 and 1000 µg glyphosate IPA/L. Whether this response is to the glyphosate or the surfactant (POEA) in the formulation was not investigated. Atrazine evoked olfactory responses at 10 and 100 µg/L. When exposures were removed, recovery of the olfactory response was rapid for both herbicides. Atrazine reduced olfactory response to L-histidine at 10 and 100 µg/L and Roundup® at 10, 100, and 1000 µg glyphosate IPA/L. Olfactory responses were not strongly or proportionally related to behavioral responses (see below). However, as for behavioral responses, it is not clear what effect perception of a chemical or the interference of the chemical with perception of feeding and other environmental signals has on survival, growth, development, or reproduction. For herbicides, these apical responses have not been tested alongside those for olfaction.

4.4. Effects of Herbicides on Behavior of Fish

Many studies report stress-induced behaviors as observations in toxicity tests (e.g., Sarikaya and Yilmaz, 2003; Farah et al., 2004; Viant et al., 2006; Uyanikgil et al., 2009), but these are seldom quantified and data are usually qualitative. The behavior reported is most likely caused by stress on the fish and is a generalized response to escape from the stressor (e.g., increased swimming speeds) or "coughing" to clear the gills. These behaviors have been omitted from this review, and our focus is on more specific responses.

Several studies have been published on the behavior of fish exposed to herbicides. Most of these studies have assessed avoidance of the chemicals in experiments where fish were offered a choice between control and contaminated water. Sheepshead minnows (*Cyprinodon variegatus*) (Hansen, 1969) and mosquitofish (*Gambusia affinis*) (Hansen et al., 1972) avoided the auxinmimic herbicide, 2,4-D (Table 7.11; technical butoxyethanol ester, 70% acid equivalent) at concentrations of 10,000 and 1000 μg/L, but not 100 μg/L when exposed to solutions in an avoidance-preference chamber. The largest concentration of 2,4-D was greater than the LC50 in mosquitofish. Similar studies on avoidance of the commercial formulation of glyphosate (Table 7.7) (Roundup®) with fingerling rainbow trout (*O. mykiss*) showed that the lowest concentration avoided was 40,000 μg/L of formulated product, less than the LC50 for the product (52,000 μg/L) (Hildebrand et al., 1982). The authors suggested that this may provide protection to fish when this herbicide formulation is used in forestry. Coho salmon (*O. kisutch*) responded to the ethylene glycol butyl ether ester formulation of the auxin-mimic herbicide triclopyr (Table 7.11; Garlon®4 emulsifiable concentrate) by exhibiting signs of distress at nonlethal concentrations of 400 and 560 μg/L (Johansen and Geen, 1990). No behavioral responses were observed at 100 and 30 μg/L.

Studies on avoidance of the commercial formulation of glyphosate (Roundup®) with fingerling rainbow trout (*O. mykiss*) showed that the lowest concentration avoided was 40,000 μg/L of formulated product, less than the LC50 for the product (52,000 μg/L) (Hildebrand et al., 1982). Concentrations of 10,000 and 100,000 μg glyphosate (isopropylamine salt, IPA)/L (commercial Roundup®) were reported to evoke avoidance, but these concentrations were above the lethal level of the formulation (Tierney et al., 2007b). Exposure to 1000 μg glyphosate-IPA/L (in the formulation) did not cause a response; however, response to L-histidine was inhibited at 100 and 1000 μg/L. In a study on zebrafish (*Danio rerio*), animals were exposed to a mixture of pesticides that was representative of those found in urban runoff (Tierney et al., 2011). The mixture of technical-grade active ingredients was in the nominal ratio of 100 ng glyphosate/L, 2.5 ng dicamba/L, 15 ng mecroprop/L, and 25 ng 2,4-D/L (Table 7.7 and Table 7.11), and the mixture was tested at 1-X, 10-X, and 100-X. Measured concentrations in the exposures systems were less than the nominal values, and the ratios were different. A short-term attraction at 1 min and then an aversion at 5 min were observed only to the 100-X mixture in naïve fish. These responses were not observed in fish preexposed to the mixture for 96 h and then exposed to a doubled exposure (2-X) for 5 min. When compared to the control, preexposure to the mixture (1-, 10-, and 100-X) appeared to increase attraction to the L-alanine (a food attractant). Because a mixture was used, it was not possible to assign causality to a particular herbicide or class of herbicides.

Several studies have been conducted on the effects of atrazine (Table 7.3) on the behavior of fish. In experiments on behavior related to swimming and social activities in goldfish, technical atrazine (97.9% purity) was reported to increase burst swimming at exposures of 0.5 and 50 μg/L but not 5 μg/L (Saglio and Trijasse, 1998). Sheltering behavior was unaffected at all concentrations tested (0.5, 5, and 50 μg/L), but grouping decreased at 5 and 50 μg/L. Surfacing was increased at 5 but not 0.5 and 50 μg/L. Consistent concentration responses were not observed across all behaviors at these concentrations, and no adverse behavioral responses were observed after fish were moved to freshwater after 24-h exposure to atrazine at concentrations up to 50 μg/L. This suggests that the effects were rapidly reversible. Exposure to atrazine did not influence responses to extracts of goldfish skin except that grouping was decreased at 5 but not at 0.5 and 50 μg atrazine/L. At greater concentrations of 100, 1000, and 10,000 μg atrazine/L, no effects were observed on sheltering, grouping, surfacing, or attraction, but there were significant increases in burst swimming activity relative to the control. The relevance of these responses to survival was not investigated, and the lack of response at the larger concentrations and lack of concentration response was not mechanistically explained. In a study on larvae of red drum (*Sciaenops ocellatus*), effects on behavior of exposures to technical atrazine (98%) at 40 and 80 μg atrazine/L for 96 h were tested (del Carmen Alvarez and Fuiman, 2005). Larvae exposed to atrazine swam faster, were hyperactive, and swam considerably more convoluted paths. Monotonic concentration responses for these measures were not consistently observed. No effects of atrazine were observed for responsiveness, magnitude of response, and speed of response to visual stimuli. Similarly, there were no effects on responsiveness to a predator, gulf killifish (*Fundulus grandis*), suggesting that these behavioral responses were of little consequence to predation in the wild, although this was not specifically tested. The effects of atrazine on olfactory-mediated behavioral responses to L-histidine in rainbow trout were investigated (Tierney et al., 2007b). This was part of a study on olfaction as described above. Technical atrazine (97.4%) was used at two concentrations, 10 and 100 μg/L. Atrazine alone did not cause a preference or avoidance response at 10 and 100 μg/L; however, the tested exposures did inhibit the preference for L-histidine and, at 10 but not 100 μg/L, caused an avoidance of L-histidine. The significance of these responses to survival, growth, and development were not specifically tested but may be relevant as amino acids are one of the feeding triggers used by fish. The relative importance of the olfactory response compared to other stimuli, such as vision, is unknown.

A few other herbicides have been evaluated for effects on behavior. No behavioral effects (locomotion, respiration, or pigmentation) were reported in zebrafish (*D. rerio*) when these fish were exposed to paraquat (Table 7.4) at a concentration of 5000 μg/L for 4 weeks (Bretaud et al., 2004). The effects of diuron (a urea herbicide; Table 7.3) on goldfish (*C. auratus*) exposed to 0.5, 5, and 50 μg/L was reported (Saglio and Trijasse, 1998). Behaviors related to sheltering, grouping, surfacing, and burst swimming were evaluated. Only grouping activity was decreased, but only at 5 μg diuron/L for 24 h. None of these responses were observed after removal of fish to uncontaminated water. Exposure to 5 but not 0.5 and 50 μg diuron/L and a skin extract from goldfish resulted in a decrease in grouping but an increase in attraction. Studies were conducted on the effects of nicosulfuron (a urea herbicide; Table 7.3) and bentazon (a benzothiadiazinone; Table 7.3) on behavior in goldfish (Saglio et al., 2001). Attraction to flowing solutions of nicosulfuron (1000 and 10,000 μg/L) and bentazon (10 and 10,000 μg/L) was reported. No attraction to nicosulfuron was reported at 10 or 100 μg/L and to bentazon at 100 and 1000 μg/L. This lack of a concentration response suggests that these responses may have been spurious, which is consistent with a lack of stimulation of the olfactory epithelium (see above). Nicosulfuron caused several behavioral responses in goldfish exposed at 25, 50, and 100 μg/L for various times (2–8 h) (Saglio et al., 2003). No effects were observed on horizontal displacement, sheltering, antagonistic interaction, or buccal movements. Increases in burst swimming and grouping were observed, but with no consistent concentration response and only at exposures for 2, 4, and 6 h (not 8 h), suggesting habituation to the herbicide. Nicosulfuron had no effect on response to a mixture of the amino acids glycine, alanine, valine, and taurine (Saglio et al., 2003).

Although a number of studies have reported behavioral responses to herbicides, the significance of these responses to survival, growth, development, and reproduction or the sustainability of the populations of fish has not been quantified except in a speculative or modeling sense. These behaviors may be adaptive, and habituation may occur. The same is likely true for behavioral responses in other animals unless these are as a result of direct neurotoxicity or developmental effects on mammals (Allen, 2010).

5. INDIRECT EFFECTS OF HERBICIDES ON FISH

Herbicides are used to control plants that are designated as weeds. Most of the use of herbicides is in the terrestrial environment, but some herbicides are used to control macrophytes in aquatic systems. When macrophytes are killed by herbicides, two outcomes may adversely affect fish (and other aquatic animals). If the biomass of macrophytes is large and the herbicide acts rapidly, the death of the macrophytes may result in a greater biological demand for oxygen that deprives fish and other organisms of oxygen. Where herbicides are applied directly to water for this purpose, the areas treated are generally small, which reduces the risk of high biological oxygen demand. In some cases, slow-acting herbicides such as fluridone (Table 7.6) are used. The slow death of the macrophytes reduces the risk that biological oxygen demand will cause adverse effects in fish (Stephenson and Solomon, 2007).

Obviously, removal of plants will result in changes in habitat that may affect fish through lack of food for fish (or their prey) or physical habitat for shelter. No currently used herbicide has been documented to have effects via this mechanism, but chlorate (which was used historically as a herbicide in terrestrial situations) from pulp mill effluent was shown to adversely affect fish in the Baltic by removal of shelter for larval fish (Lehtinen et al., 1988). This resulted in increased predation and changes in the structure of the community. In an extensive review of the effects of herbicides in controlled experimental systems, Brock et al. (2000) noted only two instances of indirect effects of herbicides on fish. One was a reduction in growth of fish exposed to $\geq 20\,\mu g/L$ atrazine in large mesocosms that also contained macrophytes (Kettle et al., 1987). These effects may have been confounded by the use of small restrictive cages that limited the range of the fish and restricted access to food. No effects were reported in other studies on the same herbicide at larger concentrations or other herbicides (Brock et al., 2000). As would be expected, these indirect effects only occur after a direct effect on plants in the system and propagation of effects via the food chain.

6. CONCLUSIONS

Herbicides are used to control plants and are usually targeted to processes and target sites that are specific to plants. As a result, the active ingredients of most herbicides are not highly acutely toxic to fish. There are several exceptions to this general rule. Some herbicides, such as the uncouplers of oxidative phosphorylation (Section 3.9), are relatively toxic to fish because the target system is common to plants and animals. Other exceptions to this low toxicity include herbicides that interfere with cell division (Section 3.8). Three members of this class—trifluralin, benfluralin, and ethalfluralin—and a chloroacetamide, pretilachlor (Table 7.8) have greater toxicity, which does not appear to be conferred by processes related to ADME, and the precise mechanism is unknown at this time. Chronic and sublethal effects have been studied for some herbicides, but fewer data are available than for acute effects. Several sublethal effects of herbicides have been studied and include reproduction, stress, olfaction, and behavior. Although some of these responses have been observed in fish exposed to herbicides, these responses have either been observed at large concentrations that would be rarely found in surface waters inhabited by fish or, as in the case of behavior, have not been linked to ecologically relevant responses on survival, growth, development, and reproduction.

As with all pesticides, herbicides may have indirect effects in fish. These effects are mediated by herbicide-induced changes in food webs or in the physical environment. Indirect effects can only occur if direct effects occur first, and they would be mediated by removal of plants by herbicides. While this may occur when plants are controlled by direct treatment of surface waters, they appear to be unlikely to result from use of herbicides in terrestrial systems as concentrations in surface waters are, for the most part, very much less than those that could directly affect plants.

As is obvious from data presented in Tables 7.1 to 7.12, toxicity data are not publicly available for a large number of herbicides. Lack of data for these classes presents some uncertainty, but for many of the groups of herbicides there are sufficient data to support the argument that even the untested products are unlikely to be acutely toxic to the extent that they present a hazard to fish. This argument is supported by the absence of the target systems for these herbicides in fish and the fact that baseline narcosis is unlikely to be a significant factor in the toxicity of these mostly polar chemicals with small K_{OWS}.

The future may bring new herbicidal products onto the market that have modes of action relevant to fish. The routinely required testing of all new products for registration will identify these products, and their toxicity will be considered in the process of registration. In the short term, particularly as the use of genetically modified crops with resistance to herbicides increases, patterns of herbicide use will likely change. This may change exposures of fish to herbicides, but again is unlikely to change the loads of herbicides to surface waters to the extent that exposures will reach concentrations that will cause acute toxicity.

ACKNOWLEDGMENTS

The authors wish to thank Dr. John Sprague for his constructive comments on the MS.

REFERENCES

Allen, S. L. (2010). Regulatory Aspects of Acute Neurotoxicity Assessment. In *Handbook of Pesticide Toxicology* (eds. R. I. Krieger, J. Doull, J. J. van Hemmen, E. Hodgson, H. I. Maibach, L. Ritter, J. Ross and W. Slikker), vol. 1, pp. 587–602. Burlington, MA: Elsevier.

Andreini, I., DellaCorte, C., Johnson, L. C., Hughes, S. and Kalinoski, D. L. (1997). G-protein (S), G alpha q/G alpha 11, in the olfactory neuroepithelium of the channel catfish (*Ictalurus punctatus*) is altered by the herbicide, dichlobenil. *Toxicology* 117, 111–122.

Ankley, G. T., Jensen, K. M., Kahl, M. D., Korte, J. J. and Makynen, E. A. (2001). Description and evaluation of a short-term reproduction test with the fathead minnow (*Pimephales promelas*). *Environ. Toxicol. Chem.* 20, 1276–1290.

Barron, M. G. (1986). Endocrine control of smoltification in anadromous salmonids. *J. Endocrinol.* 108, 313–319.

BCPC (2003). *The e-Pesticide Manual.* Farnham, Surrey, UK: British Crop Protection Council.

Bisson, M. and Hontela, A. (2002). Cytotoxic and endocrine-disrupting potential of atrazine, diazinon, endosulfan, and mancozeb in adrenocortical steroidogenic cells of rainbow trout exposed in vitro. *Toxicol. Appl. Pharmacol.* 180, 110–117.

Borthwick, P. W., Patrick, J. M., Jr. and Middaugh, D. P. (1985). Comparative acute sensitivities of early life stages of atherinid fishes to chlorpyrifos and thiobencarb. *Arch. Environ. Contam. Toxicol.* 14, 465–473.

Bretaud, S., Lee, S. and Guo, S. (2004). Sensitivity of zebrafish to environmental toxins implicated in Parkinson's disease. *Neurotox. Teratolog.* 26, 857–864.

Bringolf, R. B., Belden, J. B. and Summerfelt, R. C. (2004). Effects of atrazine on fathead minnow in a short-term reproduction assay. *Environ. Toxicol. Chem.* 23, 1019–1025.

Brock, T. C. M., Lahr, J. and van den Brink, P. J. (2000). Ecological Risks of Pesticides in Freshwater Ecosystems. *Part 1: Herbicides.* Wageningen, The Netherlands: Alterra, p. 124.

Burmaster, D. E. and Hull, D. A. (1997). Using lognormal distributions and lognormal probability plots in probabilistic risk assessments. *Human Ecol. Risk Assess.* 3, 235–255.

Call, D. J., Brooke, L. T., Kent, R. J., Knuth, M. L., Anderson, C. and Moriarity, C. (1983). Toxicity, bioconcentration, and metabolism of the herbicide propanil (3′,4′-dichloropropionanilide) in freshwater fish. *Arch. Environ. Contam. Toxicol.* 12, 175–182.

CCME (2007). *Canadian Water Quality Guidelines for the Protection of Aquatic Life. A Protocol for the Derivation of Water Quality Guidelines for the Protection of Aquatic Life 2007.* Ottawa, ON, Canada: Canadian Council of Ministers of the Environment, p. 37.

Cericato, L., Neto, J. G., Fagundes, M., Kreutz, L. C., Quevedo, R. M., Finco, J., da Rosa, J. G., Koakoski, G., Centenaro, L., Pottker, E., et al. (2008). Cortisol response to acute stress in jundia *Rhamdia quelen* acutely exposed to sub-lethal concentrations of agrichemicals. *Comp. Biochem. Physiol. C.* 148, 281–286.

Corvi, M., Stanley, K., Peterson, T., Kent, M., Feist, S., La Du, J., Volz, D., Hosmer, A. and Tanguay, R. (2012). Investigating the impact of chronic atrazine exposure on sexual development in zebrafish. *Birth. Def. Res. B* 95, 276–288.

Cunnane, C. (1978). Unbiased plotting positions. A review. *J. Hydrol.* 37, 205–222.

del Carmen Alvarez, M. and Fuiman, L. A. (2005). Environmental levels of atrazine and its degradation products impair survival skills and growth of red drum larvae. *Aquat. Toxicol.* 74, 229–241.

Devine, M. D., Duke, S. O. and Fedtke, C. (1993). *Physiology of Herbicide Action.* Englewood Cliffs, NJ: Prentice Hall.

Duke, S. O. and Powles, S. B. (2008). Glyphosate: A once-in-a-century herbicide. *Pest Manag. Sci.* 64, 319–325.

Farah, M. A., Ateeq, B., Ali, M. N., Sabir, R. and Ahmad, W. (2004). Studies on lethal concentrations and toxicity stress of some xenobiotics on aquatic organisms. *Chemosphere* 55, 257–265.

FOCUS (2003). FOCUS Surface Water Scenarios in the EU Evaluation Process under 91/414/ EEC. Report of the FOCUS Working Group on Surface Water Scenarios. Rev 2, pp. 245. Brussels, Belgium: European Commission.

Giddings, J. M., Anderson, T. A., Hall, L. W., Jr, Kendall, R. J., Richards, R. P., Solomon, K. R. and Williams, W. M. (2005). *A Probabilistic Aquatic Ecological Risk Assessment of Atrazine in North American Surface Waters.* Pensacola, FL: SETAC Press.

Giesy, J. P., Dobson, S. and Solomon, K. R. (2000). Ecotoxicological risk assessment for Roundup® herbicide. *Rev. Environ. Contam. Toxicol.* 167, 35–120.

Gluth, G. and Hanke, W. A. (1983). The effect of temperature on physiological changes in carp, *Cyprinus carpio* L., induced by phenol. *Ecotoxicol. Environ. Safety* 7, 313–389.

Gluth, G. and Hanke, W. A. (1985). Comparison of physiological changes in carp, *Cyprinus carpio*, induced by several pollutants at sublethal concentrations. I. The dependency on exposure time. *Ecotoxicol. Environ. Safety* 9, 179–188.

Hall, L. C., Okihiro, M., Johnson, M. L. and Teh, S. J. (2007). Surflan and oryzalin impair reproduction in the teleost medaka (*Oryzias latipes*). *Mar. Environ. Res.* 63, 115–131.

Hansen, B. H. (1969). Avoidance of pesticides by untrained sheepshead minnows. *Trans. Am. Fish. Soc.* 98, 426–429.

Hansen, D. J., Matthews, E., Nall, S. L. and Dumas, D. P. (1972). Avoidance of pesticides by untrained mosquitofish. Gambusia affinis. *Bull. Environ. Contam. Toxicol.* 8, 46–51.

Hedtke, J. L., Robinson-Wilson, E. and Weber, L. J. (1982). Influence of body size and developmental stage of coho salmon (*Oncorhynchus kisutch*) on lethality of several toxicants. *Fundamental and Applied Toxicology: Official Journal of the Society of Toxicology* 2, 67–72.

Hildebrand, L. D., Sullivan, D. S. and Sullivan, T. P. (1982). Experimental studies of rainbow trout populations exposed to field applications of Roundup herbicide. *Arch. Environ. Contam. Toxicol.* 11, 93–98.

Hoar, W. S. (1988). The physiology of smolting salmonids. In *Fish Physiology* (eds. W. S. Hoar and D. Randall), vol. XIB, pp. 275–343. New York: Academic Press.

HRAC (2012). *Classification of Herbicides According to Site of Action*, vol. 2012. Frankfurt: Herbicide Resistance Action Committee.

Ishida, Y., Yoshikawa, H. and Kobayashi, H. (1996). Electrophysiological responses of three chemosensory systems in the carp to pesticides. *Physiol. Behav.* 60, 633–638.

Jarrard, H. E., Delaney, K. R. and Kennedy, C. J. (2004). Impacts of carbamate pesticides on olfactory neurophysiology and cholinesterase activity in coho salmon (*Oncorhynchus kisutch*). *Aquat. Toxicol.* 69, 133–148.

Johansen, J. A. and Geen, G. H. (1990). Sublethal and acute toxicity of the ethylene glycol butyl ether ester formulation of triclopyr to juvenile coho salmon (*Oncorhynchus kisutch*). *Arch. Environ. Contam. Toxicol* 19, 610–616.

Kettle, W. D., deNoyelles, F., Jr., Heacock, B. D. and Kadoum, A. M. (1987). Diet and reproductive success of bluegill recovered from experimental ponds treated with atrazine. *Bull. Environ. Contam. Toxicol.* 38, 47–52.

Könnecker, G., Regelmann, J., Belanger, S., Gamon, K. and Sedlak, R. (2011). Environmental properties and aquatic hazard assessment of anionic surfactants: physico-chemical, environmental fate and ecotoxicity properties. *Ecotoxicol. Environ. Safety* 74, 1445–1460.

Lehtinen, K.-J., Notini, M., Mattsson, J. and Landner, L. (1988). Disappearance of bladder-wrack (*Fucus vesiculosis* L.) in the Baltic Sea; Relation to pulp-mill chlorate. *Ambio* 17, 387–393.

Levinson, M. I. (2008). Handbook of Detergents, *Part F*. Boca Raton, FL: CRC Press/Taylor and Francis.

Linders, J., Mensink, H., Stephenson, G. R., Wauchope, D. and Racke, K. D. (2000). Foliar interception and retention values after pesticide application. A proposal for standardized values for environmental risk assessment. *Pure Appl. Chem.* 72, 2199–2218.

Mangels, G. (2001). *The Development of MUSCRAT (Multiple Scenario Risk Assessment Tool): A Software Tool for Conducting Surface Water Exposure Assessments*, vol. 2007. Waterborne Environmental, Inc.

Matsumoto, J., Hosmer, A. J. and Van Der Kraak, G. (2010). Survival and iono-regulatory performance in Atlantic salmon smolts is not affected by atrazine exposure. *Comp. Biochem. Physiol. C* 152, 379–384.

Mayes, M. A., Hopkins, D. I. and Dill, D. C. (1987). Toxicity of picloram (4-amino-3,5,6-trichloropicolinic acid) to life stages of the rainbow trout. *Bull. Environ. Contam. Toxicol.* 38, 653–660.

McBride, R. K. and Richards, B. D. (1975). The effects of some herbicides and pesticides on sodium uptake by isolated perfused gills from the carp *Cyprinus carpio. Comp. Biochem. Physiol. C* 51, 105–109.

McCarty, L. S., Borgert, C. J. and Mihaich, E. M. (2012). Information quality in regulatory decision-making: Peer review versus good laboratory practice. *Environ. Health Perspect.* 120, 927–934.

McCarty, L. S., Mackay, D., Smith, A. D., Ozburn, G. W. and Dixon, G. D. (1992). Residue-based interpretation of toxicity and bioconcentration QSARs from aquatic bioassays: Neutral narcotic organic. *Environ. Toxicol. Chem.* 11, 917–930.

McCormick, S. D. (2001). Endocrine control of osmoregulation in teleost. *Am. Zool.* 41, 781–794.

McEwen, B. S. and Wingfield, J. C. (2003). The concept of allostasis in biology and biomedicine. *Homones. Behav.* 43, 2–15.

Moore, A. and Lower, N. (2001). The impact of two pesticides on olfactory-mediated endocrine function in mature male Atlantic salmon (*Salmo salar* L.) parr. *Comp. Biochem. Physiol. B* 129, 269–276.

Moore, A., Lower, N., Mayer, I. and Greenwood, L. (2007). The impact of a pesticide on migratory activity and olfactory function in Atlantic salmon (*Salmo salar* L.) smolts. *Aquaculture* 273, 350–359.

Moore, A., Scott, A. P., Lower, N., Katsiadaki, I. and Greenwood, L. (2003). The effect of 4-nonylphenol and atrazine on Atlantic salmon (*Salmo salar* L.) smolts. *Aquaculture* 222, 253–263.

Moore, A., Cotter, D., Quayle, V., Rogan, G., Poole, R., Lower, N. and Privitera, L. (2008). The impact of a pesticide on the physiology and behaviour of hatchery-reared Atlantic salmon, *Salmo salar*, smolts during the transition from fresh water to the marine environment. *Fish. Manage. Ecol.* 15, 385–392.

Mortensen, D. A., Egan, J. F., Maxwell, B. D., Ryan, M. R. and Smith, R. G. (2012). Navigating a critical juncture for sustainable weed management. *Bioscience* 62, 75–84.

Nascimento, C. R., Souza, M. M. and Martinez, C. B. (2012). Copper and the herbicide atrazine impair the stress response of the freshwater fish *Prochilodus lineatus. Comp. Biochem. Physiol. C* 155, 456–461.

Nieves-Puigdoller, K., Björnsson, B. T. and McCormick, S. D. (2007). Effects of hexazinone and atrazine on the physiology and endocrinology of smolt development in Atlantic salmon. *Aquat. Toxicol.* 84, 27–37.

OECD (1992). *Test No. 203: Fish, Acute Toxicity Test, OECD Guidelines for the Testing of Chemicals, Section 2: Effects on Biotic Systems*. Paris, France: OECD.

Ojima, D., Peterson, R. J., Wolkers, H. K., Johnsen, J. and Jörgensen, E. H. (2009). Growth hormone and cortisol treatment stimulate seawater tolerance in both anadromous and landlocked Arctic charr. *Comp. Biochem. Physiol. A.* 153, 378–385.

Parkhurst, B. R., Warren-Hicks, W. J., Cardwell, R. D., Volosin, J. S., Etchison, T., Butcher, J. B. and Covington, S. M. (1996). *Aquatic ecological risk assessment: A multi-tiered approach to risk assessment.* Alexandria, VA: Water Environment Research Foundation.

Pike, D. R., Knake, E. L. and McGlamery, M. D. (2008). Weed Control Practices in North America. In *The Triazine Herbicides: 50 Years Revolutionizing Agriculture* (eds. H. M. LeBaron, J. E. McFarland and O. C. Burnside), pp. 45–56. San Diego, CA: Elsevier.

Saglio, P. and Trijasse, S. (1998). Behavioral responses to atrazine and diuron in goldfish. *Arch. Environ. Contam. Toxicol.* 35, 484–491.

Saglio, P., Olsen, K. H. and Bretaud, S. (2001). Behavioral and olfactory responses to prochloraz, bentazone, and nicosulfuron-contaminated flows in goldfish. *Arch. Environ. Contam. Toxicol.* 41, 192–200.

Saglio, P., Bretaud, S., Rivot, E. and Olsen, K. H. (2003). Chemobehavioral changes induced by short-term exposures to prochloraz, nicosulfuron, and carbofuran in goldfish. *Arch. Environ. Contam. Toxicol.* 45, 515–524.

Sarikaya, R. and Yilmaz, M. (2003). Investigation of acute toxicity and the effect of 2,4-D (2,4-dichlorophenoxyacetic acid) herbicide on the behavior of the common carp (*Cyprinus carpio* L., 1758; Pisces, Cyprinidae). *Chemosphere* 52, 195–201.

Solomon, K. R. and Thompson, D. G. (2003). Ecological risk assessment for aquatic organisms from over-water uses of glyphosate. *J. Toxicol. Environ. Hlth. B* 6, 211–246.

Solomon, K. R., Anadón, A., Carrasquilla, G., Cerdeira, A., Marshall, J. and Sanin, L.-H. (2007). Coca and poppy eradication in Colombia: Environmental and human health assessment of aerially applied glyphosate. *Rev. Environ. Contam. Toxicol.* 190, 43–125.

Solomon, K. R., Carr, J. A., Du Preez, L. H., Kendall, R. J., Smith, E. E. and Van Der Kraak, G. J. (2008). Effects of atrazine on fish, amphibians, and aquatic reptiles: A critical review. *Crit. Rev. Toxicol.* 38, 721–772.

Stephenson, G. R. and Solomon, K. R. (2007). *Pesticides and the Environment.* Guelph, Ontario, Canada: Canadian Network of Toxicology Centres Press.

Suzawa, M. and Ingraham, H. A. (2008). The herbicide atrazine activates endocrine gene networks via non-steroidal NR5A nuclear receptors in fish and mammalian cells. *PLoS ONE* 3, e2117.

Systat Software Inc. (2010). *SigmaPlot for Windows.* Chicago: Systat Software Inc.

ten Berge, W. F., Zwart, A. and Appelman, L. M. (1986). Concentration-time mortality response relationship of irritant and systemically acting vapours and gases. *J. Haz. Matter* 13, 301–309.

Thompson, D. G. (2011). Ecological impacts of major forest-use pesticides. In *Ecological Impacts of Toxic Chemicals* (eds. F. Sánchez-Bayo, P. van den Brink and R. M. Mann), Oak Park, IL: Bentham Science Publishers, (pp. 88–110).

Tierney, K. B., Ross, P. S. and Kennedy, C. J. (2007a). Linuron and carbaryl differentially impair baseline amino acid and bile salt olfactory responses in three salmonids. *Toxicology* 231, 175–187.

Tierney, K. B., Singh, C. R., Ross, P. S. and Kennedy, C. J. (2007b). Relating olfactory neurotoxicity to altered olfactory-mediated behaviors in rainbow trout exposed to three currently-used pesticides. *Aquat. Toxicol.* 81, 55–64.

Tierney, K. B., Ross, P. S., Jarrard, H. E., Delaney, K. R. and Kennedy, C. J. (2006). Changes in juvenile coho salmon electro-olfactogram during and after short-term exposure to current-use pesticides. *Environ. Toxicol. Chem.* 25, 2809–2817.

Tierney, K. B., Sekela, M. A., Cobbler, C. E., Xhabija, B., Gledhill, M., Ananvoranich, S. and Zielinski, B. S. (2011). Evidence for behavioral preference toward environmental concentrations of urban-use herbicides in a model adult fish. *Environ. Toxicol. Chem.* 30, 2046–2054.

Tillitt, D. E., Papoulias, D. M., Whyte, J. J. and Richter, C. A. (2010). Atrazine reduces reproduction in fathead minnow (*Pimephales promelas*). *Aquat. Toxicol.* 99, 149–159.

U.S. EPA (2001). GENEEC—(GEN)eric (E)stimated (E)nvironmental (C)oncentration model. Washington, DC: Environmental Fate and Effects Division, Office of Pesticide Programs, U.S. EPA.

U.S. EPA (2005). Draft Final Report on Multi-Chemical Evaluation of the Short-Term Reproduction Assay with the Fathead Minnow. pp. 165. Washington, DC: U.S. Environmental Protection Agency.

U.S. EPA. (2011). ECOTOXicology Database System. Version 4.0. United States Environmental Protection Agency, Office of Pesticide Programs, Environmental Fate and Effects Division, United States EPA, Washington, DC, http://www.epa.gov/ecotox/, Accessed December 2012.

U.S. EPA (2012a). 2006–2007 Pesticide Market Estimates: Usage, vol. 2012. Washington, DC: U.D. Environmental Protection Agency.

U.S. EPA (2012b). Data Requirements for Pesticide Registration, vol. 2012. Washington, DC: U.S. Environmental Protection Agency.

USGS. (2012). NAWQA Database. United States Geological Survey, http://infotrek.er.usgs.gov/servlet/page?_pageid=543&_dad=portal30&_schema=PORTAL30, Accessed, December 2012.

Uyanikgil, Y., Yalcinkaya, M., Ates, U., Baka, M. and Karakisi, H. (2009). Effects of 2,4-dichlorophenoxyacetic acid formulation on medulla spinalis of *Poecilia reticulata*: A histopathological study. *Chemosphere* 76, 1386–1391.

Veith, G. D. and Broderius, S. J. (1990). Rules for distinguishing toxicants that cause Type I and Type II narcosis syndromes. *Environ. Health Perspect.* 87, 207–211.

Viant, M. R., Pincetich, C. A. and Tjeerdema, R. S. (2006). Metabolic effects of dinoseb, diazinon and esfenvalerate in eyed eggs and alevins of Chinook salmon (*Oncorhynchus tshawytscha*) determined by 1H NMR metabolomics. *Aquat. Toxicol.* 77, 359–371.

Villalobos, S. A., Hamm, J. T., Teh, S. J. and Hinton, D. E. (2000). Thiobencarb-induced embryotoxicity in medaka (*Oryzias latipes*): Stage-specific toxicity and the protective role of chorion. *Aquat. Toxicol.* 48, 309–326.

Villeneuve, D. L., Murphy, M. B., Kahl, M. D., Jensen, K. M., Butterworth, B. C., Makynen, E. A., Durhan, E. J., Linnum, A., Leino, R. L., Curtis, L. R., et al. (2006). Evaluation of the methoxytriazine herbicide prometon using a short-term fathead minnow reproduction test and a suite of in vitro bioassays. *Environ. Toxicol. Chem.* 25, 2143–2153.

Wang, N., Besser, J. M., Buckler, D. R., Honegger, J. L., Ingersoll, C. G., Johnson, B. T., Kurtzweil, M. L., MacGregor, J. and McKee, M. J. (2005). Influence of sediment on the fate and toxicity of a polyethoxylated tallowamine surfactant system (MON 0818) in aquatic microcosms. *Chemosphere* 59, 545–551.

Waring, C. P. and Moore, A. (2004). The effect of atrazine on Atlantic salmon (*Salmo salar*) smolts in fresh water and after sea water transfer. *Aquat. Toxicol.* 66, 93–104.

Xie, L., Thrippleton, K., Irwin, M. A., Siemering, G. S., Mekebri, A., Crane, D., Berry, K. and Schlenk, D. (2005). Evaluation of estrogenic activities of aquatic herbicides and surfactants using an rainbow trout vitellogenin assay. *Toxicol. Sci.* 87, 391–398.

8

PERSONAL CARE PRODUCTS IN THE AQUATIC ENVIRONMENT: A CASE STUDY ON THE EFFECTS OF TRICLOSAN IN FISH

ALICE HONTELA

HAMID R. HABIBI

Organic Chemical Toxicology of Fishes: Volume 33
FISH PHYSIOLOGY

1. INTRODUCTION

The number of pharmaceutical and personal care products (PPCPs) has increased exponentially during the past few decades, and while products made from natural plant and animal materials once dominated the market, today most are synthetic compounds to which fish may not have evolved tolerance. Synthetic compounds can mimic natural ones, as discussed in the chapter on estrogenic compounds (Chapter 5), and so detoxification systems in fishes can, to some extent, cope (detoxify and excrete) and reduce the toxicity of many synthetic chemicals (Chapter 1). However, as shown repeatedly throughout this book, minor differences in chemical structure can significantly alter toxicity. Thus, the large number of PPCPs and their diversity in chemical structures and toxicology currently pose a significant challenge in understanding the adverse environmental and health impact of these products on animals and humans. Moreover, in the environment organisms are exposed to mixtures of chemicals which, as discussed in a separate chapter on mixtures (Chapter 10), exert their effects through complex chemical, toxicological, and physiological interactions. Consequently, it is beyond the scope of this chapter to consider all the environmentally relevant PPCPs, many of which have not even been tested in fish. Instead, this chapter focuses on a single product, triclosan (TCS), an approach that is intended as a case study for other personal care products that enter the hydrosphere often via wastewater treatment plant (WWTP) effluent.

Triclosan is a broad-spectrum antimicrobial used in personal care products, including deodorants, disinfectants, soap, shampoo, toothpaste and mouthwash, in fabrics used for sport clothing, in plastic additives, and in other products. The antimicrobial action of TCS, a chemical that permeates the bacterial cell wall, targets multiple cytoplasmic and membrane sites, including the synthesis of fatty acids and macromolecules such as RNA (McMurray et al., 1998; Levy et al., 1999). TCS enters the hydrosphere mainly via WWTP effluent. At the present time, it is not technologically possible to remove all contaminants at municipal WWTP, and some compounds are of special concern as they may bioaccumulate in aquatic organisms to concentrations that exert adverse physiological effects (discussed in Chapter 1).

The use of TCS increased over the past 25 years as the public became concerned about transmission of disease and personal hygiene (Russell, 2004). A majority of the 700 antibacterial products available to consumers in recent years contained TCS as an active ingredient (Schweizer, 2001). Among brand-name products containing TCS are Irgasan DP300,

Aquasept, Sapoderm, and Ster-Zac, while fabrics with fibers impregnated with TCS may be labeled as Ultra-Fresh, Amicor, Microban, Monolith, Bactonix, and Sanitized (Adolfsson-Erici et al., 2002).

Even though TCS is a halogenated compound, its use is not highly regulated because the antimicrobial has a low acute toxicity to mammals and is well tolerated and considered safe to humans (Jones et al., 2000; Rodricks et al., 2010). However, given the volumes of personal care products used and their concentrations of TCS (0.1–0.3% of product weight) (Sabaliunas et al., 2003), significant amounts of TCS enter wastewater treatment facilities and surface waters. Concerns are emerging that the product might be harmful to nontarget species and the environment. TCS has been detected in the aquatic environment and effects in aquatic species, including fish, have been reported (Reiss et al., 2002; Servos et al., 2007; Torres-Duarte et al., 2012). The objective of this chapter is to review the literature on the occurrence of TCS in aquatic environments, exposure levels in fish, its toxicity, and its endocrine disrupting potential.

2. PHYSICAL AND CHEMICAL PROPERTIES OF TRICLOSAN, AND ITS USE AS A PERSONAL CARE PRODUCT

2.1. Physical and Chemical Properties

Triclosan (TCS) is a chlorinated aromatic hydrocarbon containing phenol and diphenyl ether (Figure 8.1), with a structure similar to

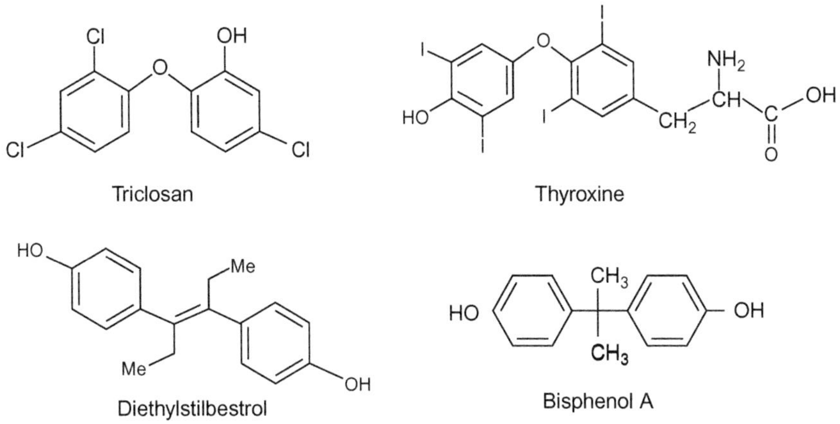

Figure 8.1. Chemical structure of triclosan, thyroxine, diethylstilbestrol, and bisphenol A.

polychlorinated biphenyls, polybrominated diphenyl ethers, bisphenol A, dioxins, and thyroid hormones (Crofton et al., 2007; Ahn et al., 2008). Its molecular formula is C_{12}-H_7-Cl_3-O_2 (CAS registration number of 3380-34-5, molecular weight 289.55), and it is known as either 5-chloro-2-(2,4-dichlorophenoxy)phenol or 2,4,4'-trichloro-2'-hydroxydiphenyl ether. It is a thermally stable compound, available as a whitish powder, with a boiling point between 280 and 290°C, and melting point between 54 and 57°C. TCS is easily dissolved in organic solvents, and even though it is not readily soluble in water, the solubility increases at alkaline pH (Bhargava and Leonard, 1996). Mostly present in the ionized form in the aquatic environment, the half-life of TCS in surface water is approximately 41 min, with 2,4-dichlorophenol as the main degradation product (Reiss et al., 2002; Lyndall et al., 2010). In the synthesis of TCS, trace amounts of other chlorinated chemicals, including dibenzodioxins and furans, can be formed (Beck et al. 1989). The concentrations of these microcontaminants in commercial TCS depend on the purity of the materials used for the synthesis, as well as on temperature during the process (Ni et al., 2005). Although the concentrations of dioxins and furans detected in commercial TCS are well below levels of concern, the potential for contamination is considered significant enough for some countries, including EU, Canada, and the United States to be setting standards for maximum permissible levels of impurities in TCS.

2.2. Uses of Triclosan

TCS is widely used as an antimicrobial, most commonly in personal care products, such as toothpaste, shampoos, deodorants, liquid soaps, and disinfectants. The thermal stability of TCS is a property that makes this antimicrobial suitable for incorporation into plastics and fibers such as sporting clothes impregnated with TCS. As public concern about transmission of disease and disease outbreaks increases, over 700 antibacterial products, the majority of which contained TCS, have entered the consumer market in recent years. These products and the concentrations of TCS in them (typically, 0.1% to 0.3%) are regulated by the European Community Cosmetic Directive or the U.S. Food and Drug Administration (USFDA) (Sabaliunas et al., 2003; Rodricks et al., 2010).

3. EXPOSURE IN THE AQUATIC ENVIRONMENT

3.1. Concentrations of Triclosan and Its Degradation Products in Water and Sediments

3.1.1. TRICLOSAN

The antimicrobial is commonly detected in surface waters at concentrations ranging from 0.01 to 2.3 µg l^{-1} (Dann and Hontela, 2011). The main source of TCS is wastewater effluent that has collected wastewater from houses, hospitals, and other sources of this personal care product. Effluent concentrations of TCS entering WWTPs range from 1.9 to 26.8 µg l^{-1} (Chalew and Halden, 2009; Lyndall et al., 2010). Even though WWTPs have the capacity to remove TCS from the effluent, this capacity varies greatly depending on the treatment process used (Heidler and Halden, 2007), with TCS concentrations in the treated effluent ranging from 0.027 to 2.7 µg l^{-1} (Chalew and Halden, 2009; Nakada et al. 2010). Aerobic sludge digestion is more efficient in removing TCS compared to anaerobic digestion (McAvoy et al., 2002).

TCS discharged from the WWTPs enters the aquatic environment, usually river systems (Nakada et al. 2008). Kolpin et al. (2002) detected TCS at concentrations ranging from below the detection limit up to 2.3 µg l^{-1} in 57.6% of U.S. streams and rivers sampled between 1999 and 2002. Also, TCS can enter the aquatic systems from land applications of biosolids contaminated with residues of TCS and other organic contaminants (Chu and Metcalfe, 2007; Mackay and Barnthouse, 2010). Concentrations of TCS in activated sludge range from 580 to 14,700 µg kg^{-1} of dry weight, with a median concentration of 5000 µg kg^{-1} of dry weight using data from WWTPs in diverse U.S. states (Reiss et al., 2002). Moreover, there is evidence that TCS can persist in sediments. TCS is detected in sediment at concentrations from 800 to 53,000 µg kg^{-1} (Miller et al., 2008; Chalew and Halden, 2009; Maruya et al., 2012). Singer et al. (2002) and also Miller et al. (2008) measured TCS and its degradation products in 30- to 40-year-old sediment and were able to associate increased sediment concentrations with the temporal patterns of TCS use over this extended time period.

Given its potential longevity in the aquatic environment, there is concern about bioaccumulation of TCS in marine mammals. Top-level predators such as bottlenose dolphins (*Tursiops truncates*) and killer whales (*Orcinus orca*) may be especially susceptible to contamination by TCS and its transformation products. Plasma TCS concentrations in these aquatic mammals correlated with environmental concentrations in water or prey (Fair et al., 2009; Bennett et al., 2009).

3.1.2. DEGRADATION PRODUCTS OF TRICLOSAN

Methyltriclosan (MTCS, 5-chloro-2-[2,4 dichlorophenoxy)anisole; CAS No. 4640-01-1]) is generated by biological methylation of TCS during the wastewater treatment in the WWTPs (Boehmer et al., 2004; Bester, 2005). This transformation product, released in treated effluent, is more lipophilic than TCS and has been in fact proposed for use as a marker of exposure to lipophilic WWTP contaminants (Balmer et al., 2004). It is also more resistant to degradation than TCS and more persistent in the aquatic environment (Lindström et al., 2002).

Dioxins—During the process of degradation of TCS, of particular concern is the occurrence and persistence of chlorinated transformation products, specifically dioxins. There is some evidence that photolysis of TCS generates chlorinated derivates, including dichlorodibenzo-*p*-dioxin (2,8-DCDD), a highly toxic chemical (Buth et al., 2009). Solar irradiation of these by-products leads to formation of higher level chlorinated dioxins. The evidence that photodegradation of TCS in the aquatic environment has the potential to generate dioxins, with yields up to 2.5%, has been provided by several studies and raised concerns about continued use of this antimicrobial personal care product (Latch et al., 2005; Aranami and Readman, 2007; Buth et al., 2009). The importance of pH, irradiation wavelengths, and presence of organic matter in the transformation of TCS into dioxins in experimental waters as well as natural waters has been investigated in these studies.

Chlorophenols and chloroform are other transformation products of concern, generated by TCS in the presence of free chlorine or chloramine, including under drinking water conditions (Rule et al., 2005; Greyshock and Vikesland, 2006; Fiss et al., 2007). Moreover, Rule et al. (2005) reported the formation of chloroform, albeit at small concentrations ($49 \, \mu g \, l^{-1}$ after 120 min) when dish soap containing TCS was added to chlorinated water. These results indicate that the daily use of TCS containing household products has the potential for production of chloroform and chlorophenols, and warrants further risk assessment for TCS.

3.2. Exposure to Triclosan and Its Degradation Products in Fish

Chronic exposure of aquatic biota to TCS released from the WWTPs leads to accumulation of this lipophilic antimicrobial and its degradation products in tissues of organisms. TCS and its methylated metabolite MTCS were reported at concentrations ranging from 100 to 150 $\mu g \, kg^{-1}$ and 50 to 89 $\mu g \, kg^{-1}$, respectively, in algae (*Cladophora* spp.) sampled in a stream receiving treated effluent (Capdeveille et al., 2008; Coogan et al., 2007).

Freshwater snails (*Helisoma trivolvis*) also accumulated TCS and MTCS, with relatively high bioaccumulation factors of >500, similar to the algae (bioaccumulation factors >1000) (Coogan and La Point, 2008). The bioaccumulation of TCS in snails and other invertebrates, including estuarine grass shrimp (DeLorenzo et al., 2008) makes these compounds likely to be available to higher consumers, including fish.

Numerous studies detected TCS and its transformation products in fish. Miyazaki et al. (1984) detected MTCS in freshwater fish (1 to 38 µg kg^{-1} whole body) sampled in the Tama River, Japan. A later study detected TCS in the bile of rainbow trout (*Oncorhynchus mykiss*) caged downstream from a WWTP in Sweden, as well as in wild fish collected downstream from the plant and rainbow trout exposed to treated water in tanks (Adolfsson-Erici et al., 2002). Bream (*Abramis brama*) sampled in Dutch waters also had high concentrations of TCS in the bile (Houtman et al., 2004). MTCS was detected in muscle or bile of numerous fish species sampled at sites in Switzerland (Balmer et al., 2004) and Germany (Boehmer et al. 2004) impacted by treated wastewater effluents. Interestingly, in the large monitoring study in Germany (Boehmer et al., 2004), MTCS was detected in all of the fish muscle samples analyzed, while TCS was only detected in a small number of samples. A relatively small number of studies characterized the exposure to TCS or MTCS in fish in North America. Both compounds were detected in plasma of 13 fish species sampled in the Detroit River (Valters et al., 2005), at concentrations ranging from 1.87 to 10.26 µg/g ww. Common carp (*Cyprinus carpio*), a species with a sediment foraging behavior, sampled in Las Vegas Bay, Nevada, had MTCS at relatively high mean whole-body concentrations (0.596 µg/g ww) compared to other fish (Leiker et al., 2009). In contrast, a large monitoring study in the United States reported only trace concentrations of TCS in muscle tissue of fish sampled near WWTPs and reference sites, while norfluoxetine and carbamazepine were measured in ng/g concentrations (Ramirez et al., 2009).

4. KINETICS AND METABOLISM OF TRICLOSAN

The absorption and pharmacokinetics of TCS following dermal, subcutaneous, oral, and intravenous exposure have been characterized in laboratory rodents as well as human subjects (Queckenberg et al., 2010) but the understanding of these processes in fish and other organisms is very limited. TCS induces rat liver cytochrome P450 enzymes (Hanioka et al., 1997; Jinno et al., 1997), but neither ethoxyresorufin O-deethylase

(EROD) nor 7-pentoxyresorufin-O-dealkylation (PROD) was induced by TCS in Japanese medaka, *Oryzias latipes* (Ishibashi et al., 2004). In contrast, TCS stimulated EROD activity and mRNA expression of CYP1A and CYP3A in swordtail fish *Xiphophorus helleri* (Liang et al., 2013). Recently, James et al. (2012) reported demethylation of MTCS to TCS, as well as its sulfonation and glucuronidation in liver and intestine of catfish *Ictalarus punctatus*. Additional studies are needed to determine if these differences are caused by species-specific enzymatic activities, or simply differences in dose and exposure time used in studies with mammals and different species of fish.

5. TOXICITY OF TRICLOSAN

5.1. Acute Toxicity

The toxicity of TCS, primarily a waterborne pollutant, has been characterized in many aquatic organisms, including algae, daphnids, and fish (Orvos et al., 2002). Algae are highly sensitive to TCS, with a median effective concentration (96 h EC_{50}) for growth inhibition of 1.4 µg l^{-1} and a no observed effect concentration (NOEC) of 0.69 µg l^{-1} in *Scenedesmus subspicatus* (Orvos et al., 2002). Toxicity data for other species of algae or phytoplankton are similar (Tatarazako et al., 2004; DeLorenzo et al., 2008), with the lowest NOEC for algae at less than 1 µg l^{-1}. Considering that concentrations of TCS in WWTP effluents range from 0.2 to 2.7 µg l^{-1} (Reiss et al., 2002), current exposures to TCS in rivers and streams (predicted environmental concentration [PEC]) may surpass the NOEC for algae, as indicated by the HQ value (hazard quotient, PEC divided by NOEC) greater than 1 (see Chapter 10 for HQ calculations). There is some evidence that TCS bioaccumulates in algae (Coogan et al., 2007), although detailed studies of toxicokinetics of TCS in algae are lacking. Aquatic invertebrates are less sensitive to TCS than algae; the LC_{50} values in a 10-day exposure toxicity test for *Daphnia magna* and the amphipod *Hyalella azteca* were 0.4 and 0.2 mg l^{-1}, respectively (Orvos et al., 2002). Recent evidence suggests that marine benthic organisms are more sensitive to TCS than freshwater species (Perron et al., 2012).

Acute toxicity of TCS has been investigated extensively in several fish species (Table 8.1). The LC_{50} (24 h and 96 h) of TCS in medaka, *Oryzias latipes*, was 0.60 mg l^{-1} and 0.40 mg l^{-1}, respectively (Tatarazako et al., 2004; Kim et al., 2009). Nassef et al. (2009) reported an LC_{50} (96 h) of TCS for adult medaka of 1.7 mg l^{-1}, with the NOEC estimated at 1.7 µg l^{-1}, a concentration 12 times higher than the PEC for the chemical. Even though

Table 8.1

Physiological and endocrine-disrupting effects of triclosan (TCS) in fish

Fish Species	Life Stage	Route of Exposure	Test Duration	TCS Exposure	End-point	Reference
Rainbow trout (*Oncorhynchus mykiss*)	embryo adult	water (flowthrough)	35 d 61 d acute (96-h)	71.3 µg l^{-1} 390 µg l^{-1}	delayed swim-up; ↓35-dph survival; erratic swimming, locked jaw LC$_{50}$	Orvos et al. (2002) CIBA (1998)
Medaka (*Oryzias latipes*)	eggs	water (renewal)	14 d	313 µg l^{-1}	↓ hatching; delayed hatching	Ishibashi et al. (2004)
	eggs	water *in ovo* injection	14 d post-fertilization in SW	400 µg l^{-1} 4.2 ng egg^{-1}	IC$_{50}$ (hatching) EC$_{50}$ (survival)	Tatarazako et al. (2004) Nassef et al. (2010)
	embryos	water	14 days	100 µg l^{-1}	weak androgenic (or anti-estrogenic) effect (↑male fin size, slight male bias sex ratio)	Foran et al. (2000)
	larvae (24-h old)	water	acute (96-h)	602 µg l^{-1}	LC$_{50}$	Ishibashi et al. (2004)
	fry	water	acute (48-h)	350 µg l^{-1}	LC$_{50}$	Foran et al. (2000)
	larvae	water	acute (96-h)	600 µg l^{-1}	LC$_{50}$	Kim et al. (2009)
	adult	water	14–21 d	20 µg l^{-1}	weak estrogenic activity; ↑Vtg in male fish; activity in yeast assay	Ishibashi et al. (2004)
	adult	water (renewal)	acute (96 h) in SW	1700 µg l^{-1}	LC$_{50}$	Nassef et al. (2009)
Bluegill sunfish (*Lepomis macrochirus*)	adult	water (renewal)	acute (96 h)	370 µg l^{-1}	LC$_{50}$	Orvos et al. (2002)

(Continued)

Table 8.1 (continued)

Fish Species	Life Stage	Route of Exposure	Test Duration	TCS Exposure	End-point	Reference
Fathead minnow (*Pimephales promelas*)	adult	water (renewal)	acute (96 h)	260 $\mu g \, l^{-1}$	LC_{50} (at pH 7.5)	Orvos et al. (2002)
	full life cycle	water (TCS in mixture)		0.1 and 0.3 $\mu g \, l^{-1}$ mixture of products	no effects F_0; ↑ larval deformities in F_1	Parrott and Bennie (2009)
	fry, adult	water	12 d	$ng/ng \, l^{-1}$	no effects; effect on behavior, no effect on Vtg or physiological endpoints	Schultz et al. (2012)
	larvae	water	7 d	up to 0.52 μM	↓ swim performance	Cherednichenko et al. (2012)
	larvae	water	7 d	10 $\mu g/ng \, l^{-1}$	↓ swim activity	Fritsch et al. (2013)
Zebrafish (*Danio rerio*)	eggs	water (renewal)	9 d	220 $\mu g \, l^{-1}$	IC_{50} (hatching)	Tatarazako et al. (2004)
	embryo	24-well micro plates	acute (96-h)	420 $\mu g \, l^{-1}$	LC_{50}; teratogenic effects	Oliveira et al. (2009)
	43 dpf	semi-static	7 d	19.3 μM	IC_{50} binding to estrogen receptor	Torres-Duarte et al. (2012)
	adult	water semi-static	acute (96-h)	340 $\mu g \, l^{-1}$	LC_{50}	Oliveira et al. (2009)
	adult	diet	21 d	100 $\mu g \, g^{-1}$ fish per day	hyperplasia of thyroid tissue, ↑number of follicles, ↑TSH and sodium/iodide transporter transcription	Pinto et al. (2013)

Mosquitofish, (*Gambusia affinis*)	male fish	water	35 d	101.3 µg l^{-1}	↑vitellogenin, ↓ sperm count in males	Raut and Angus (2010)
Bream (*Abramis brama*)	bile of male fish	field sites Netherlands		No activity up to 0.1 mM	no estrogenic activity detected in ER-CALUX assay	Houtman et al. (2004)
Cell-based assays						
MCF37 breast cancer cells		In vitro		10 µM	estrogenic and androgenic effects	Gee et al. (2008)
Cell-based nuclear-receptor-responsive and calcium signaling bioassays (AhR, ER, AR, RyR)		In vitro		110 µM (for ER- and AR-responsive gene expression; 0.110 µM (for RyR response)	weak AhR activity; antagonistic activity in ER- and AR-dependent gene expression; interaction with RyR1, ↑Ca2+ mobilization in skeletal myotubes	Ahn et al. (2008)
HuH7 cells (human hepatoma cell line) transfected with human pregnane X receptor (hPXR)		In vitro		>10 µM	activation of hPXR	Jacobs et al. (2005)
Induced rat liver microsomes		In vitro		3.1 µM (IC$_{50}$)	↓diiodothyronine (T2) sulfotransferase activity	Schuur et al. (1998)
2933Y cells (human)		In vitro		1.0 µM and 10 µM	↓testosterone-induced transcriptional activity	Chen et al. (2007)
T4-human transthyretin assay (T4-hTTR assay)		In vitro		treated wastewater samples	competitive binding of TCS	Metcalfe et al. (2013)

EC50— Effective concentration; LC50—Lethal concentration; NOEC—No Observed Effect Concentration

early life stages are more sensitive than adults to some toxicants, at least in medaka, the toxicity of TCS to fry (LC_{50} 48-h 0.35 mg l^{-1}, Foran et al. 2000; LC_{50} 96 h 0.6 mg l^{-1}, Ishibashi et al. 2004) is similar to that in adults. The embryonic toxicity of TCS was investigated in medaka by Nassef et al. (2010). TCS was administered *in ovo* by nanoinjection, and EC_{50} (effective concentration) based on survival and embryonic development was set at 4.2 ng egg^{-1}.

Acute toxicity data for TCS are remarkably similar across fish species. Toxicity testing of TCS in zebrafish, *Danio rerio*, at different life stages, using the OECD guidelines on Fish Embryo Toxicity, demonstrated that at concentrations above 0.7 mg l^{-1}, TCS had teratogenic effects, delaying embryo development and causing mortality within 48 h (Oliveira et al., 2009). The toxicity of TCS was slightly higher in embryo/larvae of zebrafish, with LC_{50} (96 h) of 0.42 mg l^{-1}. These toxicity tests led to the conclusion that TCS at concentrations equal to or above 0.3 mg l^{-1} constitute a hazard for aquatic ecosystems. In fathead minnow, *Pimephales promelas*, and sunfish, *Lepomis macrochirus*, LC_{50} values for TCS were similar to those for zebrafish and medaka, 0.26 and 0.37 mg l^{-1}, respectively (Orvos et al., 2002). There is evidence that fish, and also algae and invertebrates, are sensitive to TCS; however, the NOEC for fish range between 0.034 and 0.200 mg l^{-1} (Orvos et al., 2002; Ishibashi et al., 2004; Capdevielle et al., 2008), concentrations that exceed the PECs (0.01–0.14 µg l^{-1}) for TCS. Collectively, these observations suggest that TCS at predicted environmental concentrations may not be acutely toxic to fish.

5.2. Subacute/Subchronic and Chronic Toxicity of Triclosan

Few studies have characterized the long-term toxicity of TCS in aquatic species. Again, algae were more sensitive than fish, including rainbow trout exposed for up to 21 days (Orvos et al., 2002; Table 8.1). Interestingly, the survival and growth of adult fathead minnow did not seem adversely affected by a chronic exposure to TCS at environmentally relevant concentrations in life-cycle experiments (Parrott and Bennie, 2009). TCS at concentrations up to 115 ng l^{-1}, administered in a mixture with five other pharmaceuticals and personal care products, did not alter survival, growth, development or reproduction in adults. An effect on morphological development of larvae was observed in the F1 generation. The genotoxic, mutagenic, or carcinogenic potential has not been tested in fish or fish cell lines. The mammalian data indicate that TCS does not have this type of activity (Rodricks et al., 2010). Yet, other studies have linked environmental levels of TCS to abnormal feeding, swim performance, and behavior, with possible implications for reproductive success and survival in wild fish population (see below).

6. REPRODUCTIVE AND DEVELOPMENTAL EFFECTS

Reports released by government environmental protection agencies in North America and other OECD countries indicate that TCS can be detected in surface water, particularly in the regions that receive inputs from wastewater systems (U.S. EPA 739-RO-8009, 2008; U.S. Environment Canada Chemical Abstracts Service Registry Number 3380-34-5, 2012). Both Environment Canada and the United States Environmental Protection Agency (U.S. EPA) concluded that environmental TCS may not be at concentrations that constitute danger to human health and most aquatic organisms, although there is a concern, expressed in some recent studies detailed below, that TCS may be a contributing factor in disrupting normal reproduction and development in wild fish inhabiting contaminated water.

An additional concern is that health hazard assessments are predominantly based on data from testing chemicals individually, whereas the potency and mode of action of contaminants are increasingly recognized as being potentially different in complex mixtures (see Chapter 10). This is relevant when considering the effects of TCS in fishes. Also, TCS exposure may be higher than predicted from water or sediment concentrations measured because TCS can adsorb to suspended solids and can bioaccumulate (Coogan et al., 2007; DeLorenzo et al., 2008). Furthermore, TCS can be converted into dioxins at the environmental levels known to have antibacterial activity and other biological effects (Buth et al., 2009; Ricart et al., 2010). Therefore, it is important and relevant to consider the results of all studies that tested the effects of TCS in fish and other vertebrates. The focus here is on reviewing information on the potential adverse effects of TCS on reproduction and development of fish (Table 8.1).

The reproductive biology of fishes varies considerably due to the plastic nature of gonadal development and differentiation. Fishes can be hermaphroditic or gonochoristic species and reproduce either once a year, once in a lifetime (semelparous), or several times per year (iteroparous). Thus, physiological mechanisms of reproduction and control of gonadal differentiation vary significantly. Compounds that interact with the hormones of the brain-pituitary-gonadal axis can have different disruptive effects ranging from no detectable change in reproduction to abnormal development of gonads and even complete sex reversal in case of prolonged exposure to contaminants with estrogen-like activity.

A number of studies have demonstrated the adverse effects of TCS on fish reproduction. Acute toxicity of TCS to embryos and larvae, and disruptive effects of TCS on the behavior of adults were observed in zebrafish (Oliveira et al., 2009). Exposure to TCS (20 and 100 μg/l) for 21 days increased the production of vitellogenin in male medaka, indicating the estrogen-like activity of TCS in this species (Ishibashi et al., 2004). Furthermore, exposure of medaka fertilized eggs to 313 μg/l TCS for 14 days decreased the hatchability and delayed time to hatching (Ishibashi et al., 2004). In mature male western mosquitofish, *Gambusia affinis*, exposure to TCS (350 nM) for 35 days decreased sperm counts and increased liver size and vitellogenin mRNA (Raut and Angus, 2010).

In fathead minnow, life-cycle exposure to a mixture of environmentally relevant concentrations of TCS (115 ng/l) with five other common pharmaceuticals at concentrations ranging from 0.01 to1.0 μg/l caused no significant changes in growth and development, external sexual characteristics, and egg production in adults, but increased deformities in larvae from F1 (Parrott and Bennie, 2009). In another study with fathead minnow exposed to TCS alone at concentrations in the ng/l range for 21 days, decreased aggressive behavior was observed in adult males, despite the lack of significant morphological and physiological changes in adults and newly hatched larvae (Schultz et al., 2012). It was suggested that decreased aggression might have reproductive consequences and decrease the ability to defend and hold a nest site necessary for spawning (Schultz et al., 2012). However, plasma vitellogenin level, expression of secondary sexual characteristics, and relative size and morphology of liver and gonads were unaffected in adult fathead minnow (Schultz et al., 2012). This finding suggests that this species, in contrast to medaka and mosquitofish (Ishibashi et al., 2004; Raut and Angus, 2010), might be relatively tolerant to TCS. The variation in sensitivity and level of TCS estrogenicity in different species of fish could be related to differences in the characteristics of the estrogen receptors. TCS is known to have a low affinity for estrogen receptors in fish, which might explain the lack of effect on vitellogenin production in some fish species

In zebrafish, estrogen receptor-α (ER-α) has a low affinity for TCS (inhibitory concentration $IC_{50} = 19.3$ μM), which is 2000-fold lower than estradiol (Torres-Duarte et al., 2012). Accordingly, exposure to relatively high concentrations of TCS (1 μM) is needed to significantly increase vitellogenin1 mRNA levels (Torres-Duarte et al., 2012). Thus, TCS likely has a weak estrogen-like activity. However, the specificity of ER-α for TCS may not be the same for all vertebrate species since TCS failed to bind to human ER-α in the same study (Torres-Duarte et al., 2012). Similarly exposure to TCS failed to elicit an estrogenic response in human ovarian cancer cells stably transfected with a human ER reporter system (Ahn et al., 2008). Studies in rodents, however, reported the disruptive actions of TCS on rat reproduction. In this context, TCS exposure was found to alter the uterine response to low concentrations of ethinyl estradiol in the weanling rat, but failed to activate the ER reporter system (Louis et al., 2013). The latter study concluded that TCS is neither an agonist nor an antagonist of ER. Also in the intact rats, exposure to TCS alone was without effect on uterine weight, morphology, and gene expression, but enhanced ethinyl estradiol response (Louis et al., 2013). Comparative studies designed to characterize the affinity of the estrogen receptors to TCS in different species, including fish, would provide important mechanistic data to predict the potential impact of TCS on the reproductive system.

7. EFFECTS OF TRICLOSAN ON THE THYROID AXIS

A number of phenolic compounds are known to have endocrine disruptive properties in mammalian and nonmammalian vertebrates. The most extensively investigated phenolic compound is bisphenol A (BPA), which is widely used in plastic manufacturing. It is not surprising that TCS with its structural similarity to thyroid hormones (TH), BPA, and other phenolic endocrine disruptors, can interfere with normal endocrine function, depending on the concentration and duration of exposure.

There is evidence that TCS can influence TH-mediated functions and potentially disrupt growth, development, and metabolism, which are regulated by THs. One of the best known examples of the developmental action of TH is control of metamorphosis in amphibians, and there is evidence that TCS may disrupt this function. Exposure to low environmental concentrations of TCS affected TH-mediated metamorphosis of the North American bullfrog, *Rana catesbeiana*, and altered the expression profile of TH receptors, basic transcription element binding protein, and proliferating nuclear cell antigen (PCNA) (Veldhoen et al., 2006). Exposure of tadpoles to low concentrations of TCS (0.15 μg/l) for four days resulted in earlier onset of metamorphosis (Veldhoen et al., 2006); this suggests the possible agonistic action of TCS as a TH mimic. In the same study, TCS decreased expression of thyroid receptor-β indicating activation of negative feedback. This is consistent with an earlier report that TCS activates the pregnane-X receptor (PXR)-reporter system in human hepatocarcinoma cells (Jacobs et al., 2005), indicating that TCS can potentially disrupt TH homeostasis. Similarly, in weanling female rats oral administration of TCS resulted in dose-dependent decreases in the total thyroxine (T4) level, further supporting the view that TCS can disrupt thyroid hormone homeostasis (Crofton et al, 2007). Changes in normal TH levels can potentially disrupt amphibian metamorphosis and interfere with growth, and bone and neuronal development in mammals and other vertebrates (Zoeller and Crofton, 2000; Koibuchi and Iwasaki, 2006). In weanling rats, oral administration of TCS (0, 3, 30, 100, 200, or 300 mg/kg) from postnatal day 23 to 53, reduced serum testosterone, T4, and triiodothyronine (T3) (Zorrilla et al., 2009). In this study, the authors also report increased hepatosomatic index (HSI) and suggested that the induction of hepatic enzymes may have contributed to the altered T4 and T3 concentrations (Zorrilla et al., 2009).

A recent study with adult zebrafish (Pinto et al., 2013) demonstrated that TCS, administered through diet for 21 days, interfered with the thyroid axis by modifying the morphology of the thyroid tissue, and upregulating thyroid-stimulating hormone and sodium-iodide transporter. Further evidence for the capacity of TCS to disrupt the thyroid axis was also provided by a multi-assay screening study with EDCs present in treated wastewater (Metcalfe et al., 2013). An *in vitro* competitive binding assay using T4 and the recombinant human thyroid hormone transport protein, transthyretin, revealed a positive relationship between the concentrations of TCS in the wastewater and the response in the binding assay. The consequences of thyroid impairment caused by TCS on basic physiological processes such as neural development and metabolism in fish have not been investigated yet. It is interesting that some of the observed effects of TCS on thyroid hormone action could not be replicated when tested *in vitro*, indicating indirect actions, possibly involving the metabolism of THs (Hinther et al., 2011). In *Rana catesbeiana*, treatment of tadpole tail fin biopsy cultures with TCS *in vitro* resulted in altered transcript levels of the heat shock protein 30 (HSP30) and catalase (CAT) (Hinther et al., 2011). A possible implication of this response for an intact animal would be an increased stress response due to TCS exposure, and possible disruption of frog metamorphosis. Similarly, in rat pituitary GH3 cells, TCS increased Hsp70 transcript levels at low doses, suggesting that induction of a cellular stress response may be a common mechanism (Hinther et al., 2011).

8. INTERACTIONS OF TRICLOSAN WITH ARYL HYDROCARBON AND RYANODINE RECEPTORS

Interaction of TCS with the aryl hydrocarbon receptor (AhR) has been suggested by the activation of cytochrome P450, including EROD and

CYP1A expression in the rat and in swordtail fish (Hanioka et al., 1997; Jinno et al., 1997; Liang et al., 2013). This view is consistent with the finding that TCS exerts weak agonistic and/or antagonistic activity in the AhR-responsive bioassay, using a mouse skeletal myotube preparation (Ahn et al., 2008). In the same study, treatment with TCS (0.1–10 μM) was found to significantly increase the binding of type 1 ryanodine receptors (RyR1) and increase resting cytosolic [Ca^{2+}] in primary skeletal myotubes (Ahn et al., 2008). TCS also impaired the excitation–contraction coupling of cardiac and skeletal muscle in murine cardiac and skeletal muscle *in vitro* and *in vivo* (Cherednichenko et al., 2012). An implication of the *in vivo* observations was depressed hemodynamics and grip strength in mice.

The effects of TCS on muscle cells and swim performance were also reported in fish species (Table 8.1). In Japanese medaka, *in ovo* nanoinjection of TCS into 8-h postfertilization eggs resulted in delays in embryonic development (eye, body size, and internal organs), increased heart beat rate, and failure in upward swimming of larvae (Nassef et al., 2010). The effects of TCS on swim performance were also assessed in fathead minnow, a species relatively insensitive to the reproductive effects of TCS, as reported by Parrott and Bennie (2009). In larval fathead minnow, exposure to TCS (0.035, 0.26, or 0.52 μM) for seven days resulted in significant reduction in swim performance assessed by measuring unprovoked and forced distance traveled (Cherednichenko et al., 2012). The implications of this decline in swimming ability in wild fish would be potentially negative, adversely affecting predator avoidance, feeding success, courtship behavior, and migration ability. In this context, exposure of fathead minnow to low environmental levels of TCS (10 μg/l) for seven days resulted in significant decreases in swim activity as measured by total distance traveled (Fritsch et al., 2013). In the same study, TCS exposure in fathead minnow adversely affected to a different degree the expression and activity of ryanodine receptor isoforms, dihydropyridine receptor subunit1alpha-S, N-methly-D-aspartate receptor (NMDAR), Selenoprotein N1 (SEPN1), Junctophilin 1 (JPH1), Homer 1 (h1), Sarco/Endoplasmic reticulum ATPase (SERCA), FK-Binding Protein 12 (FKBP12), and Creatine Kinase (CK) (Fritsch et al., 2013). Furthermore, the adverse effects of TCS on the feeding ability of fathead minnow was observed after just one day, and was reduced after four and seven days of exposure (Fritsch et al., 2013).

These findings collectively indicate that TCS has the potential to impact fish physiology and health, with widespread adverse effects, ranging from reproduction and metabolism to behavior. Additional studies with TCS at environmentally realistic exposures and in mixtures, mimicking concentrations detected in treated wastewaters, are needed to provide data for risk assessment.

9. FUTURE USE OF TRICLOSAN AND RELATED ENVIRONMENTAL ISSUES

The efficacy of the broad-spectrum antimicrobial action of TCS and the concerns about the spread of disease in human populations, especially in high-density urban settings, are factors that make future use of TCS very likely. However, the uses of some TCS products are under debate, and several issues, including the impact of TCS on the aquatic environment, have been raised.

9.1. Use of TCS in Medical Facilities

Even though use of TCS-containing consumer products such as shampoos, deodorants, and liquid soaps has been criticized and there is evidence that these products are not significantly more efficient in preventing growth of bacteria than regular soap (Aiello et al., 2007), TCS use in medical settings is still favored. Use of TCS for hand washing or bathing patients in hospitals and clinics has been reviewed (Jones et al., 2000), concluding that disinfection with TCS in such settings is appropriate because these soaps are mild and do not cause dermal irritation or photosensitization, two side effects linked to daily use of other soap-like products, leading to noncompliance by medical personnel. Another use of TCS in a health care setting is incorporation of the antimicrobial into polymers and plastics, including medical polyethylene and polyvinyl chloride. The efficacy of TCS incorporated into these materials to prevent or reduce bio-film formation on medical devices has been investigated, and the results are promising (Ji and Zhang, 2009).

9.2. Domestic Uses

Use of personal care products containing TCS in a domestic setting is more controversial, since the benefit/risk consideration is different from the hospital setting. The number of TCS products is enormous, ranging from antimicrobial soaps, shampoos, deodorants, and mouthwash to sport clothing (Sabaliunas et al., 2003; Gao and Cranston, 2008). The concentrations of TCS in these products are not subject to strict regulations and therefore vary, generally within the range of 0.1 to 0.45% (w/v) (Sabaliunas et al., 2003). The efficacy of TCS-based soaps compared to regular soaps has been questioned (Aiello et al., 2007), and even though TCS-impregnated textiles designed for sport clothing are widely available in North America, these fabrics have been banned in Europe, mainly because of issues of antibiotic resistance and generation of toxic by-products such as dioxins.

9.3. Food Industry

Incorporation of TCS into packaging materials or surfaces in contact with food has been considered (Canosa et al., 2008) but since the efficacy of such practices in preventing bacterial proliferation in food has been inconclusive (Chung et al., 2003; Camilloto et al., 2009), TCS was deleted from the EU list of provisional additives for use in plastic food-contact materials in March 2010.

9.4. Antibiotic Cross Resistance

Even though the effects of TCS in nontarget organisms, especially aquatic organisms including fish, are considered important to warrant a reevaluation of the risks associated with contamination of the aquatic environment caused by large-scale use of TCS in personal care products, it is the concern about development of bacterial resistance to TCS that is at the forefront of the debate about TCS uses. Broad-action antimicrobials that target multiple sites within the bacteria are traditionally considered unlikely to lead to development of resistance. Recent studies, however, provided evidence that, although TCS does target multiple sites, the bacterial enzyme enoyl-acyl carrier protein reductase is highly vulnerable to the antimicrobial, especially at lower concentrations (Russell, 2004). These studies are important because they provide evidence to suggest that TCS use could lead to bacterial resistance and resistance to other antimicrobials, including antibiotics (Aiello et al., 2007). Moreover, a recent Norwegian study provided evidence for a link between TCS exposure in children and allergic sensitization (Bertelsen et al., 2013).

10. CONCLUSIONS AND KNOWLEDGE GAPS

The antimicrobial TCS is currently being used in many personal care products, in domestic and hospital settings, to reduce bacterial proliferation and disease. While use of TCS in the hospital setting can to some extent be justified, because TCS-based soaps and products are low irritants in humans, there are concerns regarding extensive use of TCS-containing personal care products at home. Notwithstanding the view that TCS soaps may not be more effective than normal soap, there is evidence that environmental TCS may pose a significant risk to aquatic organisms in locations downstream of municipal WWTPs. Moreover, there is serious concern that use of TCS may lead to or facilitate development of resistance to this and possibly other antimicrobial products, including antibiotics. The Canadian Medical Association has issued a call for a ban on domestic use of TCS-based products in 2009 while maintaining its use in hospital settings. European countries, such as Sweden, have actively discouraged consumers from using antimicrobial products for many years. To protect the health of the aquatic organisms as well as human health, the permissible concentrations of TCS in surface waters should be reevaluated and additional data need to be provided on the effects of TCS on biota.

The physiological and endocrine effects of TCS on aquatic species, including fish, have been characterized to some extent in the laboratory. There is substantial evidence that TCS has the potential to impair normal physiological processes in fish and also amphibians. However, the current understanding of the mechanisms of action of TCS in fish is limited, and there is an urgent need to assess the effects of environmentally realistic exposures to TCS. Adverse effects of environmental contaminants are caused by disruption of physiological and developmental processes resulting largely from alteration of normal cellular and molecular pathways, homeostasis, metabolism, and stress. As PPCPs are largely synthetic, their activity and mode of action are often nonspecific and dependent on their active molecular conformation and relative concentration which determines which cellular or receptor-mediated responses are affected due to overlap of specificity. Therefore, it is difficult to clearly define the nature of the response as we can do for specific endogenous hormones. In case of TCS, there is evidence for its weak estrogenicity in fish but not in mammals, possibly due to more stringent specificity of ERs in mammals. This is not unusual as hormone receptors are often more promiscuous in lower nonmammalian vertebrates such as fish compared to mammals and humans, in particular. A number of well-documented studies demonstrate, for example, the increased specificity of hormones such as growth hormone and peptides in mammals compared to fish. Therefore there should be less generalization of data, as the effects of a contaminant may be different depending on species, concentration, season, sex, age, and developmental stage. As end-points are variable and complex, a system approach should be used where possible to investigate multiple end-points simultaneously as well as studying targeted end-points previously validated. Use of the "omics" approach may prove to be very helpful as it identifies pathways as potential targets that were not considered previously. Furthermore, the effects and potency of an environmental contaminant may be different, depending on its action individually or as a complex mixture, as demonstrated in a recent study investigating the effects of low concentrations of contaminants on fish metabolism (Jordan et al., 2012).

11. GENERAL LESSONS LEARNED FROM THE CASE STUDY OF TRICLOSAN

The case study of TCS demonstrates that the toxicity of a PPCP can be species-specific. Even though TCS is considered relatively safe at current environmental concentrations to mammals, including humans, the toxicity to lower vertebrates such as fish and amphibians may be significant. Therefore, extrapolation of toxicity data across species, and especially across vertebrate classes, may not be appropriate for this chemical. In contrast, the structural similarity of TCS to bisphenol A, thyroid hormones, and other endocrine disruptors can to some extent predict endocrine-disrupting activity, as shown by experimental data for fish and amphibians. Thus, toxicity studies designed to evaluate the risk in the receiving environment, specifically rivers impacted by WWTP effluents, must include key species representing the potential targets in the environment exposed to environmentally relevant concentrations, at different time points *in vivo*. Furthermore, exposure tests, especially *in vitro* tests, must be validated with toxicokinetic data for physiologically relevant half-life, absorption, and excretion data. Toxicity and adverse effects on physiology, morphology, metabolism, and receptor-mediated interactions need to be evaluated under exposures to chemicals both individually and in complex mixtures.

The major issue involved in evaluating the impact of PPCPs on the receiving environment is the unpredictability of PPCP use by humans, in contrast to, for example, industrial pollutants generated on predictable temporal patterns linked to production schedules. PPCPs enter the receiving environment in variable concentrations and temporal patterns, depending on human use. The toxicity to humans is low since these products are used for beneficial effects in humans, whether related to health and other personal benefits. The amounts entering the receiving aquatic environment can be substantial, as shown in the TCS case study. Thus the challenge of providing risk assessments for the large number of PPCPs used at the present time and in the future is daunting. However, mechanistic studies designed to discover the quantitative structure–activity relationships (QSARs), in a comparative across-species approach and in carefully controlled laboratory studies, complemented by epidemiological field studies, are likely to provide data that risk assessors can use to set acceptable NOEC for TCS and similar PPCPs.

This approach, complemented by a clear understanding of basic comparative physiology and toxicology, will help us to predict the toxicity and to set protective guidelines for large numbers of new chemicals, specifically PPCPs.

ACKNOWLEDGMENT

The authors are grateful to Deborah MacLatchy (Wilfrid Laurier University) for comments on an earlier version of this manuscript.

REFERENCES

Adolfsson-Erici, M., Pettersson, M., Parkkonen, J. and Sturve, J. (2002). Triclosan, a commonly used bactericide found in human milk and in the aquatic environment in Sweden. *Chemosphere* 46, 1485–1489.

Ahn, K. C., Zhao, B., Chen, J., Cherednichenko, G., Sanmarti, E., Denison, M. S., Lasley, B., Pessah, I. N., Kültz, D., Chang, D. P. Y., Gee, S. J. and Hammock, B. D. (2008). *In vitro* biologic activities of the antimicrobials Triclocarban, its analogs, and Triclosan in bioassay screens: Receptor-based bioassay screens. *Environ. Health Perspect.* 116, 1203–1210.

Aiello, A. E., Larson, E. L. and Levy, S. B. (2007). Consumer antibacterial soaps: Effective or just risky? *Clin. Infectious Diseases* 45, S137–S147.

Aranami, K. and Readman, J. W. (2007). Photolytic degradation of triclosan in freshwater and seawater. *Chemosphere* 66, 1052–1056.

Balmer, M. E., Poiger, T., Droz, C., Romanin, K., Bergqvist, P. A., Muller, M. D. and Buser, H. (2004). Occurrence of methyl triclosan, a transformation product of the bactericide triclosan, in fish from various lakes in Switzerland. *Environ. Sci. Technol.* 38, 390–395.

Beck, H., Droß, A., Eckart, K., Mathar, W. and Wittkowski, R. (1989). Determination of PCDDs and PCDFs in Irgasan DP 300. *Chemosphere* 19, 167–170.

Bennett, E. R., Ross, P. S., Huff, D., Alaee, M. and Letcher, R. J. (2009). Chlorinated and brominated organic contaminants and metabolites in the plasma and diet of a captive killer whale (*Orcinus orca*). *Mar. Pollut. Bull.* 58, 1078–1083.

Bertelsen, R. J., Longnecker, M. P., Løvik, M., Calafat, A. M., Carlsen, K. H., London, S. J. and Lødrup Carlsen, K. C. (2013). Triclosan exposure and allergic sensitization in Norwegian children. *Allergy* 68, 84–91.

Bester, K. (2005). Fate of triclosan and triclosan-methyl in sewage treatment plants and surface waters. *Arch. Environ. Contam.Toxicol.* 49, 9–17.

Bhargava, H. N. and Leonard, P. A. (1996). Triclosan: Applications and safety. *Am. J. Infect. Control* 24, 209–218.

Boehmer, W., Ruedel, H., Weinzel, A. and Schroeter-Kerman, C. (2004). Retrospective monitoring of triclosan and methyl-triclosan in fish: results from the German environmental specimen bank. *Organohalogen compd.* 66, 1516–1521.

Buth, J. M., Grandbois, M., Vikesland, P. J., McNeill, K. and Arnold, W. A. (2009). Aquatic photochemistry of chlorinated triclosan derivatives: potential source of polychlorodibenzo-*p*-dioxins. *Environ. Toxicol. Chem.* 28, 2555–2563.

Camilloto, G. P., Soares, N. D. F., Pires, A. C. D. and de Paula, F. S. (2009). Preservation of sliced ham through triclosan active film. *Packaging Technol. Sci.* 22, 471–477.

Canosa, P., Rodriguez, I., Rubi, E., Ramil, M. and Cela, R. (2008). Simplified sample preparation method for triclosan and methyltriclosan determination in biota and foodstuff samples. *J. Chromatogr., A.* 1188, 132–139.

Capdevielle, M., Egmond, R. V., Whelan, M., Versteeg, D., Hofmann-Kamensky, M., Inauen, J., Cunningham, V. and Woltering, D. (2008). Consideration of exposure and species sensitivity of triclosan in the freshwater environment. *Integr. Environ. Assess. Manage.* 4, 15–23.

Chalew, T. E. A. and Halden, R. U. (2009). Environmental exposure of aquatic and terrestrial biota to triclosan and triclocarban. *J. Am. Water Works Assoc.* 45, 4–13.

Chen, J. G., Ahn, K. C., Gee, N. A., Gee, S. J., Hammock, B. D. and Lasley, B. L. (2007). Antiandrogenic properties of parabens and other phenolic containing molecules in personal care products. *Toxicol. Appl. Pharmacol.* 221, 278–284.

Cherednichenko, G., Zhang, R., Bannister, R. A., Timofeyev, V., Li, N., Fritsch, E. B., Feng, W., Barrientos, G. C., Schebb, N. H., Hammock, B. D., Beam, K. G., Chiamvimonvat, N. and Pessah, N. I. (2012). Triclosan impairs excitation−contraction coupling and Ca2+ dynamics in striated muscle. *Proc. Natl. Acad. Sci. U. S. A.* 109, 14158–14163.

Chu, S. and Metcalfe, C. D. (2007). Simultaneous determination of triclocarban and triclosan in municipal biosolids by liquid chromatography tandem mass spectrometry. *J. Chromatogr.* 1164, 212–218.

Chung, D., Papadakis, S. E. and Yam, K. L. (2003). Evaluation of a polymer coating containing triclosan as the antimicrobial layer for packaging materials. *Int. J. Food Sci. Technol.* 38, 165–169.

CIBA (1998). Brochure no. 2521: IRGASAN DP 300/IRGA-CARE MP: Toxicological and ecological data/Official registration.

Coogan, M. A., Edziyie, R. E., La Point, T. W. and Venables, B. J. (2007). Algal bioaccumulation of triclocarban, triclosan, and methyl-triclosan in a North Texas wastewater treatment plant receiving stream. *Chemosphere* 67, 1911–1918.

Coogan, M. A. and La Point, T. W. (2008). Snail bioaccumulation of triclocarban, triclosan, and methyltriclosan in a North Texas, USA, stream affected by wastewater treatment plant runoff. *Environ. Toxicol. Chem.* 27, 1788–1793.

Crofton, K. M., Paul, K. B., De Vito, M. J. and Hedge, J. M. (2007). Short-term *in vivo* exposure to the water contaminant triclosan: Evidence for disruption of thyroxine. *Environ. Toxicol. Pharmacol.* 24, 194–197.

Dann, A. B. and Hontela, A. (2011). Triclosan: environmental exposure, toxicity and mechansims of action. *J. Appl. Toxicol.* 31, 285–311.

DeLorenzo, M. E., Keller, J. M., Arthur, C. D., Finnegan, M. C., Harper, H. E., Winder, V. L. and Zdankiewicz, D. L. (2008). Toxicity of the antimicrobial compound triclosan and formation of the metabolite methyl-triclosan in estuarine systems. *Environ. Toxicol.* 23, 224–232.

Fair, P. A., Lee, H. B., Adams, J., Darling, C., Pacepavicius, G., Alaee, M., Bossart, G. D., Henry, N. and Muir, D. (2009). Occurrence of triclosan in plasma of wild Atlantic bottlenose dolphins (*Tursiops truncatus*) and in their environment. *Environ. Poll.* 157, 2248–2254.

Fiss, E. M., Rule, K. L. and Vikesland, P. J. (2007). Formation of chloroform and other chlorinated byproducts by chlorination of triclosan-containing antibacterial products. *Environ. Sci. Technol.* 41, 2387–2394.

Foran, C. M., Bennett, E. R. and Benson, W. H. (2000). Developmental evaluation of a potential non-steroidal estrogen: triclosan. *Mar. Environ. Res.* 50, 153–156.

Fritsch, E. B., Connon, R. E., Werner, I., Davies, R. E., Beggel, S., Feng, W. and Pessah, I. N. (2013). Triclosan impairs swimming behavior and alters expression of excitation-contraction coupling proteins in fathead minnow (*Pimephales promelas*). *Environ. Sci. Technol.* 47, 2008–2017.

Gao, Y. and Cranston, R. (2008). Recent advances in antimicrobial treatment of textiles. *Text. Res. J.* 78, 60–72.

Gee, R. H., Charles, A., Taylor, N. and Darbre, P. D. (2008). Oestrogenic and androgenic activity of triclosan in breast cancer cells. *J. Appl. Toxicol.* 28, 78–91.

Greyshock, A. E. and Vikesland, P. J. (2006). Triclosan reactivity in chloraminated waters. *Environ. Sci. Technol.* 40, 2615–2622.

Hanioka, N., Jinno, H., Nishimura, T. and Ando, M. (1997). Effect of 2,4,4′-trichloro-2′-hydroxydiphenyl ether on cytochrome p450 enzymes in the rat liver. *Chemosphere* 34, 719–730.

Heidler, J. and Halden, R. U. (2007). Mass balance assessment of triclosan removal during conventional sewage treatments. *Chemosphere* 66, 362–369.

Hinther, A., Bromba, C. M., Wulff, J. E. and Helbing, C. C. (2011). Effects of triclocarban, triclosan, and methyl triclosan on thyroid hormone action and stress in frog and mammalian culture systems. *Environ. Sci. Technol.* 45, 5395–5402.

Houtman, C. J., Van Oostveen, A. M., Brouwer, A., Lamoree, M. H. and Legler, J. (2004). Identification of estrogenic compounds in fish bile using bioassay-directed fractionation. *Environ. Sci. Technol.* 38, 6415–6423.

Ishibashi, H., Matsumura, N., Hirano, M., Matsuoka, M., Shiratsuchi, H., Ishibashi, Y., Takao, Y. and Arizono, K. (2004). Effects of triclosan on the early life stages and reproduction of medaka *Oryzias latipes* and induction of hepatic vitellogenin. *Aquat. Toxicol.* 67, 167–179.

Jacobs, M. N., Nolan, G. T. and Hood, S. R. (2005). Lignans, bacteriocides and organochlorine compounds activate the human pregnane X receptor (PXR). *Toxicol. Appl. Pharmacol.* 209, 123–133.

James, M. O., Marth, C. J. and Rowland-Faux, L. (2012). Slow O-demethylation of methyl triclosan to triclosan, which is rapidly glucuronidated and sulfonated in channel catfish liver and intestine. *Aquat. Toxicol.* 124–125, 72–82.

Ji, J. H. and Zhang, W. (2009). Bacterial behaviors on polymer surfaces with organic and inorganic antimicrobial compounds. *J. Biomed. Mater. Res., Part A* 88, 448–453.

Jinno, H., Hanioka, N., Onodera, S., Nishimura, T. and Ando, M. (1997). Irgasan(R) DP 300 (5-chloro-2-(2,4-dichlorophenoxy)phenol) induces cytochrome P450s and inhibits haem biosynthesis in rat hepatocytes cultured on Matrigel. *Xenobiotica* 27, 681–692.

Jones, R. D., Jampani, H. B., Newman, J. L. and Lee, A. S. (2000). Triclosan: A review of effectiveness and safety in health care settings. *Am. J. Infect. Control* 28, 184–196.

Jordan, J., Zare, A., Jackson, L. J., Habibi, H. R. and Weljie, A. M. (2012). Exposure to low concentrations of waterborne contaminants, individually and in a mixture, causes disruption in metabolism in fish. *J. Proteome Res.* 11 (2), 1133–1143.

Kim, J. W., Ishibashi, H., Yamauchi, R., Ichikawa, N., Takao, Y., Hirano, M., Koga, M. and Arizono, K. (2009). Acute toxicity of pharmaceutical and personal care products on freshwater crustacean (*Thamnocephalus platyurus*) and fish (*Oryzias latipes*). *J. Toxicol. Sci.* 34, 227–232.

Koibuchi, N. and Iwasaki, T. (2006). Regulation of brain development by thyroid hormone and its modulation by environmental chemicals. *Endocr. J.* 53, 295–303.

Kolpin, D. W., Furlong, E. T., Meyer, M. T., Thurman, E. M., Zaugg, S. D., Barber, L. B. and Buxton, H. T. (2002). Pharmaceuticals, hormones, and other organic wastewater contaminants in US streams, 1999-2000: A national reconnaissance. *Environ. Sci. Technol.* 36, 1202–1211.

Latch, D. E., Packer, J. L., Stender, B. L., VanOverbeke, J., Arnold, W. A. and McNeill, K. (2005). Aqueous photochemistry of triclosan: Formation of 2,4-dichlorophenol, 2,8-dichlorodibenzo-p-dioxin, and oligomerization products. *Environ. Toxicol. Chem.* 24, 517–525.

Leiker, T. J., Abney, S. R., Goodbred, S. L. and Rosen, M. R. (2009). Identification of methyl triclosan and halogenated analogues in male common carp (*Cyprinus carpio*) from Las Vegas Bay and semipermeable membrane devices from Las Vegas Wash, Nevada. *Sci. Total Environ.* 407, 2102–2114.

Levy, C. W., Roujeinikova, A., Sedelnikova, S., Baker, P. J., Stuitje, A. R., Slabas, A. R., Rice, D. W. and Rafferty, J. B. (1999). Molecular basis of triclosan activity. *Nature* 398, 383–384.

Liang, X., Nie, X., Ying, G., An, T. and Li, K. (2013). Assessment of toxic effects of triclosan on the swordtail fish (*Xiphophorus helleri*) by a multi-biomarker approach. *Chemosphere* 90, 1281–1288.

Lindström, A., Buerge, I. J., Poiger, T., Bergqvist, P. A., Müller, M. D. and Buser, H. R. (2002). Occurrence and environmental behavior of the bactericide triclosan and its methyl derivative in surface waters and in wastewater. *Environ. Sci. Technol.* 36, 2322–2329.

Louis, G. W., Hallinger, D. R. and Stoker, T. E. (2013). The effect of triclosan on the uterotrophic response to extended doses of ethinyl estradiol in the weanling rat. *Reprod. Toxicol.* 36, 71–77.

Lyndall, J., Fuchsman, P., Bock, M., Barber, T., Lauren, D., Leigh, K., Perruchon, E. and Capdevielle, M. (2010). Probabilistic risk evaluation for triclosan in surface water, sediments, and aquatic biota tissues. *Integr. Environ. Assess. Manage.* 6, 419–440.

Mackay, D. and Barnthouse, L. (2010). Integrated risk assessment of household chemicals and consumer products: addressing concerns about triclosan. *Integr. Environ. Assess. Manage.* 6, 390–392.

Maruya, K. A., Vidal-Dorsch, D. E., Bay, S. M., Kwon, J. W., Xia, K. and Armbrust, K. L. (2012). Organic contaminanats of emerging concern in sediments and flatfish collected near outfalls discharging treated wastewater effluent to the Southern California Bight. *Environ. Toxicol. Chem.* 31, 2683–2688.

McAvoy, D. C., Schatowitz, B., Jacob, M., Hauk, A. and Eckhoff, W. S. (2002). Measurement of triclosan in wastewater treatment systems. *Environ. Toxicol. Chem.* 21, 1323–1329.

McMurray, L. M., Oethinger, M. and Levy, S. B. (1998). Triclosan targets lipid synthesis. *Nature* 394, 531–532.

Metcalfe, C. D., Kleywegt, S., Letcher, R. J., Topp, E., Wagh, P., Trudeau, V. L. and Moon, T. W. (2013). A multi-assay screening approach for assessment of endocrine-active contaminants in wastewater effluent samples. *Sci. Total Environ.* 454, 132–140.

Miller, T. R., Heidler, J., Chillrud, S. N., Delaquil, A., Ritchie, J. C., Mihalic, J. N., Bopp, R. and Halden, R. U. (2008). Fate of triclosan and evidence for reductive dechlorination of triclocarban in estuarine sediments. *Environ. Sci. Technol.* 42, 4570–4576.

Miyazaki, T., Yamagishi, T. and Matsumoto, M. (1984). Residues of 4-chloro-1(2,4-dichlorophenoxy)-2-methoxybenzene(triclosanmethyl) in aquatic biota. *Bull. Environ. Contam. Toxicol.* 32, 227–232.

Nakada, N., Kiri, K., Shinohara, H., Harada, A., Kuroda, K., Takizawa, S. and Takada, H. (2008). Evaluation of pharmaceuticals and personal care products as water-soluble molecular markers of sewage. *Environ. Sci. Technol.* 42, 6347–6353.

Nakada, N., Yasojima, M., Okayasu, Y., Komori, K. and Suzuki, Y. (2010). Mass balance analysis of triclosan, diethyltoluamide, crotamiton and carbamazepine in sewage treatment plants. *Water Sci. Technol.* 61, 1739–1747.

Nassef, M., Matsumoto, S., Seki, M., Kang, I. J., Moroishi, J., Shimasaki, Y. and Oshima, Y. (2009). Pharmaceuticals and personal care products toxicity to Japanese medaka fish (*Oryzias latipes*). *J. Fac. Agric., Kyushu Univ.* 54, 407–411.

Nassef, M., Kim, S. G., Seki, M., Kang, I. J., Hano, T., Shimasaki, Y. and Oshima, Y. (2010). *In ovo* nanoinjection of triclosan, diclofenac and carbamazepine affects embryonic development of medaka fish (*Oryzias latipes*). *Chemosphere* 79, 966–973.

Ni, Y., Zhang, Z., Zhang, Q., Chen, J., Wu, Y. and Liang, X. (2005). Distribution patterns of PCDD/Fs in chlorinated chemicals. *Chemosphere* 60, 779–784.

Oliveira, R., Domingues, I., Grisolia, C. K. and Soares, A. M. (2009). Effects of triclosan on zebrafish early-life stages and adults. *Environ. Sci. Pollut. Res.* 16, 679–688.

Orvos, D. R., Versteeg, D. J., Inauen, J., Capdevielle, M., Rothenstein, A. and Cunningham, V. (2002). Aquatic toxicity of triclosan. *Environ. Toxicol. Chem.* 21, 1338–1349.

Parrott, J. L. and Bennie, D. T. (2009). Life-cycle exposure of fathead minnows to a mixture of six common pharmaceuticals and triclosan. *J. Toxicol. Environ. Health, Part A* 72, 633–641.

Perron, M. M., Ho, K. T., Cantwell, M. G., Burgess, R. M. and Pelletier, M. C. (2012). Effects of triclosan on marine benthic and epibenthic organisms. *Environ. Toxicol. Chem.* 31, 1861–1866.

Pinto, P. I., Guerreiro, E. M. and Power, D. M. (2013). Triclosan interferes with the thyroid axis in the zebrafish (*Danio rerio*). *Toxicol. Res.* 2, 60–69.

Queckenberg, C., Meins, J., Wachall, B., Doroshyenko, O., Tomalik-Scharte, D., Bastian, B., Abdel-Tawab, M. and Fuhr, U. (2010). Absorption, pharmacokinetics, and safety of triclosan after dermal administration. *Antimicrob. Agents Chemother.* 54, 570–572.

Ramirez, A. J., Brain, R. A., Usenko, S., Mottaleb, M. A., O'Donnell, J. G., Stahl, L. L., Wathen, J. B., Snyder, B. D., Pitt, J. L., Perez-Hurtado, P., Dobbins, L. L., Brooks, B. W. and Chambliss, C. K. (2009). Occurrence of pharmaceuticals and personal care products in fish: results of a national pilot study in the United States. *Environ. Toxicol. Chem.* 28, 2587–2597.

Raut, S. A. and Angus, R. A. (2010). Triclosan has endocrine-disrupting effects in male western mosquitofish, *Gambusia affinis*. *Environ. Toxicol Chem.* 29, 1287–1291.

Reiss, R., Lewis, G. and Griffin, J. (2009). An ecological risk assessment for triclosan in the terrestrial environment. *Environ.Toxicol. Chem.* 28, 1546–1556.

Reiss, R., Mackay, N., Habig, C. and Griffin, J. (2002). An ecological risk assessment for triclosan in lotic systems following discharge from wastewater treatment plants in the United States. *Environ. Toxicol. Chem.* 21, 2483–2492.

Ricart, M., Guasch, H., Alberch, M., Barceló, D., Bonnineau, C., Geiszinger, A., Farré, M., Ferrer, J., Ricciardi, F., Romani, A. M., Morin, S., Proia, L., Sala, L., Sureda, D. and Sabater, S. (2010). Triclosan persistence through wastewater treatment plants and its potential toxic effects on river biofilms. *Aquat. Toxicol.* 100, 346–353.

Rodricks, J. V., Swenberg, J. A., Borzelleca, J. F., Maronpot, R. R. and Shipp, A. M. (2010). Triclosan: A critical review of the experimental data and development of margins of safety for consumer products. *Crit. Rev. Toxicol.* 40, 422–484.

Rule, K. L., Ebbett, V. R. and Vikesland, P. J. (2005). Formation of chloroform and chlorinated organics by free-chlorine-mediated oxidation of triclosan. *Environ. Sci. Technol.* 39, 3176–3185.

Russell, A. D. (2004). Whither triclosan? *J. Antimicrob. Chemother.* 53, 693–695.

Sabaliunas, D., Webb, S. F., Hauk, A., Jacob, M. and Eckhoff, W. S. (2003). Environmental fate of triclosan in the river Aire basin, UK. *Water Res.* 37, 3145–3154.

Schultz, M. M., Bartell, S. E. and Schoenfuss, H. L. (2012). Effects of Triclosan and Triclocarban, two ubiquitous environmental contaminants, on anatomy, physiology, and behavior of the fathead minnow (*Pimephales promelas*). *Arch. Environ. Contam. Toxicol.* 63, 114–124.

Schuur, A. G., Legger, F. F., van Meeteren, M. E., Moonenm, M. J. H., van Leeuwen-Bol, I., Bergman, A., Visser, T. J. and Brouwer, A. (1998). In vitro inhibition of thyroid hormone sulfation by hydroxylated metabolites of halogenated aromatic hydrocarbons. *Chem. Res. Toxicol.* 11, 1075–1081.

Schweizer, H. P. (2001). Triclosan: A widely used biocide and its link to antibiotics. *FEMS Microbiol. Lett.* 202, 1–7.

Servos, M. R., Smith, M., McInnis, R., Burnison, B. K., Lee, B. H., Seto, P. and Backus, S. (2007). The presence of selected pharmaceuticals and the antimicrobial triclosan in drinking water in Ontario, Canada. *Water Qual. Res. J. Can.* 42, 130–137.

Singer, H., Muller, S., Tixier, C. and Pillonel, L. (2002). Triclosan: Occurrence and fate of a widely used biocide in the aquatic environment: Field measurements in wastewater treatment plants, surface waters, and lake sediments. *Environ. Sci. Technol.* 36, 4998–5004.

Tatarazako, N., Ishibashi, H., Teshima, K., Kishi, K. and Arizono, K. (2004). Effects of triclosan on various aquatic organisms. *Environ. Sci.* 11, 133–140.

Torres-Duarte, C., Viana, M. T. and Vazquez-Duhalt, R. (2012). Laccase-mediated transformations of endocrine disrupting chemicals abolish binding affinities to estrogen receptors and their estrogenic activity in zebrafish. *Appl. Biochem. Biotechnol.* 168, 864–876.

U.S. EPA 739-R0-8009, Reregistration eligibility decision for Triclosan, 2008.

U.S. Environment Canada Chemical Abstracts Registry Number 3380-34-5, Triclosa, 2012.

Valters, K., Li, H. X., Alaee, M., D'Sa, I., Marsh, G., Bergman, A. and Letcher, R. J. (2005). Polybrominated diphenyl ethers and hydroxylated and methoxylated brominated and chlorinated analogues in the plasma of fish from the Detroit River. *Environ. Sci. Technol.* 39, 5612–5619.

Veldhoen, N., Skirrow, R. C., Osachoff, H., Wigmore, H., Clapson, D. J., Gunderson, M. P., Van Aggelen, G. and Helbing, C. C. (2006). The bactericidal agent triclosan modulates thyroid hormone-associated gene expression and disrupts postembryonic anuran development. *Aquat. Toxicol.* 80, 217–227.

Zoeller, R. T. and Crofton, K. M. (2000). Thyroid hormone action in fetal brain development and potential for disruption by environmental chemicals. *Neurotoxicology* 21, 935–945.

Zorrilla, L. M., Gibson, E. K., Jeffay, S. C., Crofton, K. M., Setzer, W. R., Cooper, R. L. and Stoker, T. E. (2009). The effects of triclosan on puberty and thyroid hormones in male Wistar rats. *Toxicol. Sci.* 107, 56–64.

9

EMERGING THREATS TO FISHES: ENGINEERED
ORGANIC NANOMATERIALS

TYSON J. MACCORMACK
GREG G. GOSS
RICHARD D. HANDY

1. INTRODUCTION

Engineered nanomaterials (ENMs) are purposely manufactured to have novel material properties at the nanoscale, that is, either at least one dimension < 100 nm (Masciangioli and Zhang, 2003; Roco, 2003), or a primary size in the 1–100 nm range (SCENIHR, 2007). However, the current debate on a precise definition of an ENM focuses on the fraction of

439

Organic Chemical Toxicology of Fishes: Volume 33
FISH PHYSIOLOGY

the material with a particle size range of less than 100 nm (Lövestam et al., 2010).

ENMs can be produced in many different forms with numerous possible core materials and surface chemistries. One approach to categorizing ENMs has been to adopt a chemical classification system, as is the case for more conventional, noncolloidal chemicals (Stone et al., 2010). The main types of ENMs include inorganic metals and metal oxides ("nano metals"; Shaw and Handy, 2011), organic carbon-based products (see review by Petersen and Henry, 2012) such as single- or multiwalled carbon nanotubes (SWCNT/ MWCNTs) and carbon spheres (fullerenes), organic ENMs such as dendrimers (Astuc et al., 2010), organic polymers designed to encapsulate organic chemicals (Pérez-de-Luque and Rubiales, 2008), and lipid micelles (Igbal et al., 2012).

ENMs also can be made from more than one substance. These composite ENMs include a variety of nanoscale ceramics, clays, and quantum rods and dots (Stone et al., 2010). An added complexity is when the surface of ENMs is covalently functionalized to derive numerous chemical reactivities on the surface of a particle, even within one major class of ENMs. In addition ENMs can have a less permanent surface coating, usually derived from the adsorption of chemicals onto the surface of the particle. These include anionic surfactants and detergents, which are often used as dispersing agents when commercially available ENMs are supplied as liquids. Coating ENMs using "biologically relevant" substances such as citrate, humic substances (humic or fulvic acids), proteins, and cell membrane lipids (e.g., phosphatidylcholine) are now being applied. These functionalizations and coatings, generally referred to as capping agents, impart the surface of the ENM with different properties; these are especially important to toxicity and bioavailability, since it is arguably the surface that is first presented to the organism. When considering the ecotoxicity of organic ENMs, it is therefore necessary to consider inorganic ENMs with organic capping agents, as the bioactivity of a specific ENM formulation may be determined largely by the properties of its surface.

From the viewpoint of ecotoxicity, multiple factors must be considered when assessing the hazard of ENMs. Toxicity may be associated with the primary ENM (i.e., single particles <100 nm in size) as well as aggregates or clusters of particles that may be several hundred nanometers or larger (Handy et al., 2008a). Chemicals associated with ENMs, such as those used to improve particle dispersion or prevent breakdown, may pose their own set of hazards to fish.

ENMs can be made into a plethora of different shapes and/or sizes, raising some concern regarding how shape and size effects of ENMs can be included in environmental risk assessment strategies (Klaine et al., 2012). Shapes currently include spherical, or rod-shaped, nanoparticles (NPs) (Nowack and Bucheli, 2007) and various types of hollow spheres, nanowires, and nanotubes. Of particular concern for respiratory health in humans are high-aspect-ratio (long and thin) ENMs. New forms of high-aspect-ratio materials include nanocrystalline cellulose, nanofibrillar cellulose, and nano needles, which are long, thin crystals often with very sharp points. These appear to invoke inflammatory responses as have been described for mice exposed to various types of carbon nanotubes (CNTs) (Poland et al., 2008). Similarly, high-aspect-ratio materials like CNTs have been shown to be toxic to the respiratory systems of fishes (Smith et al., 2007) but spherical ENMs can also affect gill function (e.g., Schultz et al., 2012). The composition and properties of "first-generation" ENMs were relatively simple and generally consisted of a single-core material (e.g., TiO_2 or Ag) with one functionality. In contrast, some "second-"generation ENMs can have very complex shapes (e.g., rosette nanotubes; Borzsonyi et al., 2010), or are engineered with new crystal structures to give application-specific chemical reactivity (e.g., tin-doped nano titania for catalytic remediation of diesel pollution in seawater; Yu et al., 2011).

The ability to decorate the surface of ENMs to carry out certain functions enabled the explosion of product applications that engage ENMs in electronics, fuel additives, building materials, textiles, paints, food and food packaging, nanomedicine, medical devices, bioremediation, wastewater treatment technology, and personal care products (Aitken et al., 2006; Chaudhry et al., 2008; Hansen et al., 2008; Sozer and Kokini, 2009). Some of these applications are clearly of direct relevance as potential routes of exposure of aquatic species (e.g., in water treatment technology), but there may also be many potential benefits of ENMs to cultured fish health, such as the potential for new fish foods, veterinary medicines, and materials for aquacultural engineering (Handy, 2011).

The existing and rapidly emerging ENMs present a grand challenge to toxicologists. The ecotoxicity of ENMs to wildlife has been documented in a number of recent reviews (Moore, 2006; Handy et al., 2008a; Klaine et al., 2008; Pérez et al., 2009; Kahru and Savolainen, 2010; Scown et al., 2010). The heavy bias toward the ecotoxicology of metal-containing ENMs is primarily because researchers were able to verify total metal concentrations during exposures and sometimes in the tissues by existing spectroscopic methods, such as inductively coupled plasma emission spectrophotometry (ICP-OES, e.g., Cu; Shaw et al., 2013). Metal-containing ENMs are also being produced in large quantities (e.g., Piccino et al., 2012) and therefore represent an immediate environmental risk. There is a reasonable body of work on fullerenes and CNTs (Petersen and Henry, 2012). However, predicting the toxicity of many other types of purely organic ENMs has been largely neglected because of the difficulties in assigning exposure dose due to an inability to detect the materials quantitatively. Most of the research that has occurred has focused on the end-points used in regulatory toxicology (e.g., survival, growth), with less attention given to resulting aberrant animal physiology (see Handy et al., 2011 for a critique of ecotoxicity tests). A recent review of the effects of ENMs on the different body systems of fishes (Handy et al., 2012) identified numerous knowledge gaps in our understanding of fish physiology and also on the species of fish investigated, with most of the limited data being on the few freshwater teleosts typically used in regulatory testing.

Considerations relevant to the physiological responses of fishes to ENMs, like other toxicants, are that they are dynamic, with bioavailability highly dependent on the environmental chemistry, and that uptake mechanisms exist on barriers like the gills or gut (see Chapter 1). A special concern is the interactions of ENMs with other chemicals already in the environment. The environmental chemistry of ENMs is complex, involving both solution chemistry and colloid theory (see reviews, Lead and Wilkinson, 2006; Handy et al., 2008a; Ju-Nam and Lead, 2008; Klaine et al., 2008; von der Kammer et al., 2012). Furthermore, some ENMs are very good at adsorbing hydrophobic chemicals, and other ENMs aggregate on the surfaces of organisms (e.g., Handy et al., 2008a), both of which could lead to increased bioavailability of the organic chemical in the environment. Hydrophobic ENMs do not readily disperse in aqueous solutions and may exhibit bioaccumulative properties similar to persistent organic pollutants. As with hydrophobic organic compound adsorption to sediments and organic carbon, this "delivery vehicle" or "Trojan horse" effect is a significant area of uncertainty for environmental risk assessment. Whether or not the aspects of the hazard of organic and/or hydrophobic ENMs can be regulated in the same way (i.e., using similar rules or logic) as persistent organic pollutants is unclear. To make these judgments, a more detailed understanding of the organismal biology is needed.

This chapter outlines what is known about the physiological effects of ENMs on fishes, with a comparison to established knowledge on organic chemicals. Example organic ENMs—those having available data regarding toxicity to fish (CNTs, NCCs and polymer nanocapsules)—are reviewed with the goal of illustrating key principles concerning bioavailability and physiological effects. To close, we address the important issue of particle surface chemistry by exploring the effects of different capping agents and their influence on ENM toxicity to fishes.

2. THE ENVIRONMENTAL CHEMISTRY OF ENGINEERED NANOMATERIALS AND BIOAVAILABILY TO FISHES

The environmental chemistry of ENMs is detailed elsewhere (Lead and Wilkinson, 2006; Ju-Nam and Lead, 2008; von der Kammer et al., 2012), including the context of ecotoxicity (Handy et al., 2008a; Klaine et al., 2008). The particle chemistry issues relating to fishes have also been discussed (Handy et al., 2012). The key points from these reviews are briefly summarized here.

First, fishes have evolved in a geochemical environment that is rich in natural nanoscale-sized materials. Natural aquatic colloids include particulate matter in the 1 nm to 1 μm size range (natural nanomaterials; NMs) primarily comprised of macromolecular organic material (e.g., humic and fulvic acids, peptides, protein) as well as colloidal inorganic species (hydrous iron and manganese oxides; see Buffle, 2006). Soils and sediments also contain nanoscale clay particles, iron oxide particles (hematite), and many other mineral particles; some of which will inevitably wash into aquatic systems during rainfall. However, even though they evolve in a nano-rich world, fishes are not likely to be physiologically adapted to ENMs because the ENMs have combinations of structure, physicochemical properties, and chemical reactivities that are not found in nature. Therefore, from the perspective of environmental hazard assessment ENMs toward fishes should therefore be regarded as "new chemicals" with unintended and incidental releases (e.g., Crane et al., 2008).

Until recently, environmental concentrations of ENMs were predicted concentrations from exposure modeling that estimated ENM concentrations in surface waters in the range of $ng\,l^{-1}$ to low $\mu g\,l^{-1}$ (Biswas and Wu, 2005; Boxall et al., 2007; Nowack and Bucheli, 2007; Gottschalk et al., 2009). However, single-particle inductively coupled mass spectroscopy (SP-ICP-MS) allows detection of some metallic particles in defined dispersions in water (e.g., silver particles; Mitrano et al., 2012), opening the door for prospecting for metal ENM concentrations in the environment. Still lacking are national monitoring programs for ENMs in aquatic systems, but these may be available in the near future. A very large challenge will be the ability to detect ENMs against the substantial background of natural colloids already in the water (von der Kammer et al., 2012).

Abiotic aquatic factors are well known to affect the bioavailability of organic chemicals. Key factors are the uptake across fish gills, including the hydrophobicity of the substance (water–octanol partition coefficients; see Chapter 1), the pH of the water and the tendency of the chemical substance to show an electrical dipole or not at acidic pH values, the dissolved oxygen concentration, and the presence of complexing dissolved natural organic matter (DOM) in the water (Erickson et al., 2006; Nichols et al., 1990). Whether or not these ideas apply to hydrophobic and/or organic ENMs remains unclear. However, some fundamental features of colloid chemistry theory (DLVO theory; see Derjaguin and Landau, 1941; Verwey et al., 1948) likely apply to many ENMs (see Handy et al., 2008a). These include that (i) ENMs form emulsions and dispersions in liquids—they do not usually form *solutions* and are not "dissolved" in the water; (ii) ENMs can form aggregates (particles sticking to each other), or agglomerates (particles loosely joined together, or sometimes tangled up with DOM or other bridging agents); and (iii) the aggregation behavior of ENMs is influenced by water pH, the presence of divalent ions, ionic strength, and likely water temperature and the presence/type of DOM in the water (Ryman-Rasmussen et al., 2006; Handy et al., 2008a; Zhang and Monteiro-Riviere, 2009).

The ability of ENMs to form dispersions, or to aggregate, seems to be particularly critical to bioavailability, assuming that well-dispersed particles may be more bioavailable. One could also argue that the aggregation and settling of particles would increase the ENM exposure of benthic species. In freshwater, the presence of humic and fulvic acids (from rotting leaf litter; peaty water) may stabilize ENM suspensions in the water column by slowing particle aggregation and precipitation (Lead and Wilkinson, 2006), although this is not always the case (Furman et al., 2013). Such waters also tend to be soft waters, and so one might expect particles to disperse more easily in waters such as those found in the Great Lakes area in North America, Scandinavia, and Scotland. On the other hand, very hard water, such as the chalky streams in southern England may tend to aggregate particles. In seawater, the high ionic strength may dominate particle behavior. Small increases in salinity above that of freshwater, such as 2.5 parts per thousand, can result in particle aggregation, precipitation, and subsequent large decreases in the particle number (or mass concentration) in the water (Stolpe and Hassellöv, 2007). Thus, there are considerable differences in particle dispersion in freshwater compared to even very dilute seawater, with particles likely to be rapidly removed from the water column at the freshwater–seawater interface (Klaine et al., 2008). From an ecological perspective, therefore, the aggregation and settling of ENMs at the top of an estuary might present a special hazard to flatfish. However, the water chemistry effects on particle aggregation need to be considered in the context of the surface charge of the ENM. For example, a particle with a net negative surface charge can be titrated with hydrogen ions to the point of zero charge to promote aggregation (e.g., iron oxyhydroxide particles; Gilbert et al., 2007); in this case, water pH may be a critical factor. For an ENM designed as an electroneutral particle, agglomeration with DOM may be more important to particle behavior in the water.

The topography and hydrology of the water catchment may also be important. Particle aggregation kinetics are affected by stirring and sonication, and in general, the more energy added to the water, the better the dispersion. This implies that water flow rate and turbulence will be important in natural systems, with a fast-moving stream giving better dispersion than a pond or lake. Thus, ENMs accumulate in slow-moving water, or due to the viscous forces in biofilms (e.g., on rocks in the riverbed), or in the dead spaces between gravel/sand on the riverbed (e.g., for natural iron oxides; Hasellöv and von der Kammer, 2008) may present a risk to grazing organisms. Similar concerns have been raised for the sea surface microlayer (see Klainc et al., 2008), which is a thin film of unstirred water that is home to many planktonic and larval stages of marine organisms.

3. ACUTE TOXICITY OF ORGANIC ENMs

The existing literature suggests that the lethal concentrations (LC_{50} values) of ENMs in fishes are on the order of mg l^{-1}, which far exceeds predicted environmental concentrations (Handy et al., 2012). However, this dataset is still small compared to the wealth of literature on other chemicals and is biased with more information on metal-containing particles (reviewed by Shaw and Handy, 2011). Information on marine fishes and the sensitivity of different life stages within a species is scarce. For organic ENMs, most of the data relate to either C_{60} fullerenes or CNTs.

A significant problem at the bench is finding dispersing agents that work well with ENMs, but also have reasonable biocompatibility. Early work on the toxicity of C_{60} fullerenes in fishes (Oberdörster, 2004; Zhu et al., 2006; 2007) were confounded by solvent toxicity. Decomposition products of the tetrahydrofuran (THF), used in early experiments to disperse C_{60}, are the likely cause of toxicity in some experimental conditions (see discussion in Henry et al., 2007; Spohn et al., 2009), as subsequent experiments showed limited waterborne toxicity of C_{60} to fishes (Shinohara et al., 2009). Thus, proper solvent controls are essential (see Handy et al., 2012).

4. UPTAKE ROUTES AND TARGET ORGANS FOR ORGANIC ENMs

The application of toxicokinetics (absorption, distribution, metabolism, excretion; ADME; see Chapter 1) to ENMs in fishes has been reviewed

(Handy et al., 2008b), and the major target organs have been identified (Handy et al., 2012). Confirmation of toxicant concentration in target organs is the traditional approach to confirm direct toxicity, which for pesticides and industrial organics have usually involved either lengthy extraction procedures to recover the test substance from the tissue (McCarthy and Jimenez, 1985) or radiolabeled chemicals (Black and McCarthy, 1988). For the carbon-based ENMs, the challenge of this approach is the high carbon background already present in the organism. In a few cases, ^{14}C labeled fullerenes were shown to accumulate mainly in the liver and lungs of rats following intravenous injection (Sumner et al., 2010). Near-infrared fluorescence spectroscopy has also been used to detect CNTs in biological samples (Wild and Jones, 2009), but the technique is limited to "unfunctionalized" single-walled CNTs (SWCNTs), so it is of limited value in determining the toxicity of otherwise functionalized SWCNTs.

Some experiments have demonstrated potential target organs for some organic ENMs, especially the carbon-based materials (Table 9.1). An external target organ is the gill, although gill uptake mechanisms have yet to be unequivocally demonstrated for most organic ENMs. Internal target organs for organic ENMs include liver, brain, and immune organs. SWCNTs potentially affect the brain and vasculature (Smith et al., 2007), while carbon-based ENMs affect reproduction and embryonic development (Table 9.1). Such effects are well known for organic chemicals (Kime, 1995). In addition, secondary effects (without appreciable accumulation of the organic ENM) may occur. For example, Smith et al. (2007) showed evidence of oxidative gill damage in trout during exposures to 0.1–$0.5 \, \text{mg} \, \text{l}^{-1}$ of SWCNTs. One might therefore argue that observed changes in the brain in the same fish were related to systemic hypoxia rather than direct SWCNT toxicity in the nervous system.

For traditional organic chemicals, the lipid solubility drives diffusional uptake across all regions of the gut (not requiring a specific carrier; see Chapter 1). However, information on dietary exposure to organic ENMs in fishes is very limited (a handful of studies), and the rules for gastrointestinal absorption of organic ENMs are unknown. For example, it was unclear whether or not rainbow trout (*Oncorhynchus mykiss*) fed food containing carbon-based ENMs ($500 \, \text{mg} \, \text{kg}^{-1}$ C_{60} or SWCNT diets, Fraser et al., 2010) internalized the ENMs. More studies are needed on the physiological effects of both dietary and waterborne exposures to fishes with organic ENMs.

Table 9.1

Example sublethal effects of organic engineered and carbon-based nanomaterials in fishes, modified from Handy et al. (2011)

Nanomaterial/Characteristics	Concentration and Exposure Time	Exposure Method	Species	Toxic Effects	Authors
Single-walled carbon nanotubes (SWCNTs). ~1–2 nm diameter, 5–30 μm length, suspended in SDS and sonicated.	0.1–0.5 mg l^{-1} for 10 d	Waterborne exposure	*Oncorhynchus mykiss* (juveniles)	Respiratory distress evidenced by increased ventilation rates and mucus production. Gill pathologies observed in exposed fish, and vascular injury in the brain.	Smith et al. (2007)
Fluorescently labeled multiwalled CNTs 20 nm diameter, length of 800 mm, suspended in ultrapure water.	2 ng single dose at 1 or 72 hpf	Microinjection into embryo	*Danio rerio* (embryos)	No developmental abnormalities or toxic effects seen in the exposed embryos.	Cheng et al. (2009)
C$_{60}$ fullerenes (prepared with THF vehicle), 30–100 nm aggregates suspended in ultrapure water.	0.5 or 1 mg l^{-1} for 48 h	Waterborne exposure	*Micropterus salmoides* (juveniles)	Significant increase in lipid peroxidation products reported in the brain and gill with the 0.5 mg l^{-1} treatment. The authors did not conduct a solvent control for the THF vehicle.	Oberdörster (2004)
C$_{60}$ fullerenes, 10–200 nm aggregates formed by stirring in ultrapure water for at least two months.	0.5 mg l^{-1} for 96 h	Waterborne exposure	*Pimephales promelas* (adult males)	Downregulation in expression of the peroxisomal membrane protein PMP70.	Oberdörster et al. (2006)
C$_{60}$ fullerenes prepared either with or without use of THF vehicle, suspended in ultrapure water. No sizes given.	0.5 mg l^{-1} of either water-stirred or THF prepared C$_{60}$ for 48 h	Waterborne exposure	*Pimephales promelas* (adult males)	100% mortality in fish exposed to THF-prepared C$_{60}$ within 18 h. Water-stirred C$_{60}$ was less toxic and small (< twofold induction), but statistically significant increases in CYP-like genes were observed in the liver.	Zhu et al. (2006)
C$_{60}$, C$_{70}$ and C$_{60}$(OH)$_{24}$ fullerenes suspended in DMSO. No sizes given.	100–500 μg l^{-1} for C$_{60}$ and C$_{70}$, 500–5000 μg l^{-1} for C$_{60}$(OH)$_{24}$. Exposed at 24 hpf until 96 hpf.	Waterborne exposure	*Danio rerio* (dechorionated embryos)	Delayed development and abnormalities in embryos exposed to C$_{60}$ and C$_{70}$. Functionalized C$_{60}$(OH)$_{24}$ was significantly less toxic, causing similar injuries at 2500 μg l^{-1}, compared to 200 μg l^{-1} for the same injuries with other fullerenes.	Usenko et al. (2007)

(Continued)

Table 9.1 (continued)

Nanomaterial/Characteristics	Concentration and Exposure Time	Exposure Method	Species	Toxic Effects	Authors
C_{60} fullerenes suspended in DMSO. No sizes given.	$0-500\ \mu g\ l^{-1}$, exposed at 24 hpf for 5 d	Waterborne exposure	*Danio rerio* (dechorionated embryos)	Experiments were conducted with or without antioxidants, and with or without reduced light levels. Decreased light showed significantly less mortality and malformations compared to the same dose in the normal light regime, except for highest $500\ \mu g\ l^{-1}$ concentration, which still produced mortalities. Antioxidants generally protected from toxicity.	Usenko et al. (2008)
Carbon: C_{60} fullerenes (aggregates between 50 and 300 nm) dispersed in water by either stirring for 7 days with sonication or by using THF.	$0-25\%$ (v/v) for 72 h	Waterborne exposure	*Danio rerio* (larvae)	Survival reduced in THF C_{60} and THF control treatments, but not in the water-stirred treatment. Minimal gene expression changes in the latter treatment compared to control, while changes seen in THF C_{60} and THF control fish were deemed to be linked to a THF degradation product (γ-butyrolactone).	Henry et al. (2007)
Carbon: C_{60} fullerene aggregates in suspension.	$0-1.5\ mg\ l^{-1}$ for 96 hpf	Waterborne exposure	*Danio rerio* (dechorionated embryos)	Embryonic development delayed and reduced hatching success.	Zhu et al. (2007)
Carbon: C_{60} fullerenes suspended in H_2O (diameter unknown).	Nominal $0-10\ mg\ l^{-1}$ for 96 h	Waterborne exposure (20 ppt seawater)	*Fundulus heteroclitus* (adults)	No significant mortality. Dose-dependent increase in total GSH in liver tissue. No effect on total GSH in gill tissue. No significant lipid peroxidation.	Blickley and McClellan-Green (2008)
Carbon: C_{60} fullerenes suspended in H_2O (diameter unknown).	Nominal $0-10\ mg\ l^{-1}$ for 12 d	Waterborne exposure (20 ppt seawater)	*Fundulus heteroclitus* (larvae)	No significant mortality. Dose-dependent increase in total GSH. No significant lipid peroxidation.	Blickley and McClellan-Green (2008)
Carbon: C_{60} fullerenes suspended in H_2O (diameter unknown).	Nominal $0-10.2\ mg\ l^{-1}$	Waterborne exposure (20 ppt seawater)	*Fundulus heteroclitus* (embryos)	No significant mortality or developmental abnormalities. C_{60} adhered to chorion	Blickley and McClellan-Green (2008)

Material	Dose	Exposure	Species	Effects	Reference
Carbon: C_{60} fullerenes, hydroxylated (409 nm diameter).	0.2 μg NP g^{-1} bodyweight	Microinjected into embryo	*Danio rerio* (embryos)	313 mRNA probes were differentially expressed by ±2 fold on a 15617 probe microarray. Changes in genes related to circadian rhythm, kinase activity, vesicular transport, immune response.	Jovanović et al. (2011)
Carbon: C_{60} fullerenes (toluene functionalized and hydroxylated, average 226–245 nm diameter, respectively).	0.5–2 mg l^{-1} for 4 d	Waterborne exposure	*Oryzias latipes* (dechorionated embryos)	Delayed hatch in toluene-C_{60}. 25–30% mortality at 2 mg l^{-1} exposure. Toxicity generally reduced with addition of NOM.	Kim et al. (2012)
Carbon: CNT (multiwalled, acid functionalized, 10–25 nm diameter).	0.5–2 mg l^{-1} for 4 d	Waterborne exposure	*Oryzias latipes* (dechorionated embryos)	Developmental abnormalities. 40% mortality at 2 mg l^{-1} exposure. Toxicity generally reduced with addition of NOM.	Kim et al. (2012)
Carbon: CNT (single-walled, 0.7–1.2 nm diameter).	1–100 mg l^{-1} for 96 hpf	Waterborne exposure	*Danio rerio* (embryos)	No effect on hatch or morphology post-hatch.	Ong et al. (2013)
Carbon: CNT (single-walled, same batch as in Smith et al., 2007). and C_{60} fullerenes (as used in Henry et al., 2007).	0 or 500 mg SWCNT or C_{60} kg^{-1} diet for 6 weeks	Dietary exposure	*Oncorhynchus mykiss* (juveniles)	Transient elevation (week 4 only) in brain TBARS from fish exposed to SWCNT.	Fraser et al. (2010)
Carbon: CNT (various forms, with or without capping agents) of single- and multiwalled, diameters 10–20 nm, and single-walled nanohorns, diameter 2–3 nm).	0–10 μg ml^{-1} for 6 or 24 h	*In vitro* exposure	*Oncorhynchus mykiss* (macrophage cells)	Dose-dependent increases in inflammatory gene expression (IL-1β) in all treatments, with capped nanotubes more stimulatory than uncapped.	Klaper et al. (2010)
Cellulose: Nanocrystalline, 75 nm average diameter.	1 or 10 g l^{-1}	Waterborne exposure	*Oncorhynchus mykiss* (adults)	No overt toxicity.	Kovacs et al. (2010)
Cellulose: Nanocrystalline, 75 nm average diameter.	16–2000 mg l^{-1} for 48 h	*In vitro* exposure	*Oncorhynchus mykiss* (primary hepatocytes)	Cell swelling at 16 mg l^{-1}, increased lipid peroxidation and cell death at doses >80 mg l^{-1}.	Kovacs et al. (2010)
Cellulose: Nanocrystalline, 75 nm average diameter.	1–6 g l^{-1} for 96 hpf	Waterborne exposure	*Danio rerio* embryos	No morphological abnormalities, no overt toxicity.	Kovacs et al. (2010)
Cellulose: Nanocrystalline, 75 nm average diameter.	30–480 mg l^{-1} for 10 days	Waterborne exposure	*Pimephales promelas* (adult breeding pairs)	Decreased egg production at 480 mg l^{-1} exposure dose.	Kovacs et al. (2010)

(Continued)

Table 9.1 (continued)

Nanomaterial/Characteristics	Concentration and Exposure Time	Exposure Method	Species	Toxic Effects	Authors
Latex (polystyrene): Fluorescent monodispersed NPs (39.4, 474, 932,18,600, and 42,000 nm diameters).	$1–30$ mg l^{-1} for 3 d	Waterborne exposure	*Oryzias latipes* (eggs and juveniles)	No egg mortality after 1 mg l^{-1} exposure for 3 days. Fluorescence was detected in whole egg and chorion. In the juveniles, fluorescence was detected in gills, gut, blood, and possibly other internal organs.	Kashiwada (2006)
Polystyrene: (24–28 nm diameter in H$_2$O).	Fish were fed *Daphnia magna* raised on NP-exposed algae for 29 d	Dietary exposure	*Carassius carassius* (adults)	Inhibition of normal feeding behavior. Decreased weight loss relative to controls, changes in triglyceride:cholesterol ratio and distribution of cholesterol between muscle and liver.	Cedervall et al. (2012)
Dendrimers: PAMAM (various forms).	$0–20$ μM from 6–120 hpf	Waterborne exposure	*Danio rerio* (embryos)	G3.5 carboxylic acid and ligand-capped dedrimers showed no mortality or sublethal effects. G4 amino-capped dendrimers showed dose- and time-dependent mortality above 0.2 μM and sublethal toxicity was observed at low exposure doses. G4 ligand-capped dendrimers exhibited increased mortality only at 20 μM exposures.	Heiden et al. (2007)

DMSO, dimethyl sulfoxide; SDS, sodium dodecyl sulfate; hpf, hours postfertilization; THF, tetrahydrofuran; PAMAM, polyamidoamine; TBARS, thiobarbituric acid reactive substances.

5. EFFECTS OF ORGANIC ENMs ON THE PHYSIOLOGICAL SYSTEMS OF FISHES

The effects of organic ENMs on each of the major physiological systems in mainly teleost fishes are summarized below for ENMs (reviewed by Handy et al., 2012).

5.1. Respiratory System

Rainbow trout exposed to SWCNTs showed increased ventilation rate (hyperventilation) and gill irritation, which led to mucus secretion, edema, and other gill pathologies (Smith et al., 2007). The hyperventilation could be a secondary respiratory response to systemic hypoxia, but blood gases in ENM-exposed fishes were not measured. These effects on fish are also well known for many other organic chemicals (e.g., Mallat, 1985). However, there may also be some nano-specific gill injury. For example, exposure to SWCNTs led to epithelial cell hyperplasia in trout gills (Smith et al., 2007), and this has been observed with other ENMs (TiO_2 NPs, Federici et al., 2007; nano iron, Li et al., 2009). Lung pathology, especially inflammation, is reasonably well known for the mammalian respiratory system exposed to organic ENMs (e.g., Kreyling et al., 2002; Bermudez et al., 2004; Lam et al., 2004; Warheit et al., 2006).

A key question for physiology is whether or not the functions of the gills in respiration and osmoregulation are compromised. Exposure to polyvinyl pyrrolidone (PVP) capped Ag ENMs significantly increased critical oxygen tension (*P*crit) in Eurasian perch (Bilberg et al., 2010), suggesting an impairment of respiratory gas exchange. The effects on osmotic and ionic homeostasis in fish exposed to organic ENMs are largely unknown. Citrate-capped Ag ENMs decreased ^{22}Na uptake in juvenile rainbow trout (Shultz et al., 2012), but exposure to SWCNTs did not elicit an overt osmotic stress (Smith et al., 2007). Nothing is known regarding the effects of organic ENMs on gill acid/base homeostasis (e.g., Dymowska et al., 2012) or nitrogenous waste excretion (e.g., Wright and Wood, 2009) in fishes. Again, more work is required in this area.

5.2. Gastrointestinal Tract

The gut is also an important site of uptake for traditional organic chemicals, and the environmental concerns include direct oral toxicity to fishes, as well as food chain transfer of organic chemicals and biomagnification of the contaminants to higher trophic levels (Veith et al., 1979). Uptake depends partly on the hydrophobicity and dipole moment of the substance, and simple diffusion into the lipid bilayers of the mucosal membrane, or by the formation of lipid micelles (Sanborn et al., 1977) (see Chapter 1). Similar concerns arise for hydrophobic organic ENMs. Biomagnification of inorganic ENMs occur from bacteria to protozoans (Werlin et al., 2011). Proof-of-principle studies have established that organic ENMs can be transferred through the food chain from algae to zooplankton to fish and accumulate in tissues to detectable levels (Cedervall et al., 2012).

Rainbow trout that consumed diets containing 100 mg kg^{-1} of SWCNTs or C_{60} (Fraser et al., 2010) for over 8 weeks did not exhibit any loss of appetite or changes in growth rate. However, whether or not dietary exposure to organic ENMs has adverse effects on the animal's physiology, and especially on the major nutritional functions of the gut (motility, secretion, digestion, absorption), is less clear. There are too few studies to reach any consensus. Dietary exposure of crucian carp (*Carassius carassius*) to polystyrene ENMs changed fatty acid metabolism, including alterations in serum triglyceride: cholesterol ratio and the distribution of cholesterol between muscle and liver (Cedervall et al., 2012). Eating SWCNTs or C_{60} by rainbow trout (Fraser et al., 2010) resulted in a normal carcass proximate composition (protein, fat, carbohydrate, and ash content), suggesting no deficiency in the major categories of nutrients. In rainbow trout given waterborne SWCNT exposure via drinking, histological evidence suggested damage to the gut epithelium with erosion of the gut epithelium and fusion of villi and vacuole formation (Smith et al., 2007). However, the same SWCNT added to the food caused only minor effects on the histological integrity of the gut (Fraser et al., 2010). Thus, the form of SWCNT delivery, as with other chemicals (Handy et al., 2005), is important to bioavailability and toxicity, perhaps with materials incorporated into food pellets being less toxic than in water. Similar observations have been made for metallic ENMs comparing waterborne (Federici et al., 2007) and dietary (Ramsden et al., 2009) exposure to TiO_2 ENMs.

Many data gaps exist on the nutritional effects of organic ENMs in fishes and mammals. Information is lacking on the effects of any type of ENMs on gut motility or the secretion mechanisms that control the release of fluids into the digestive system (e.g., secretion of HCl in the stomach for acid digestion), the absorption of specific nutrients (amino acids, sugars, etc.), and the osmoregulatory or trace element nutrition functions of the gut. Metallic ENMs have been demonstrated to alter the gut microbial biodiversity of trout (Merrifield et al., 2013), and a recent study has demonstrated nystatin-sensitive (an endocytosis inhibitor) uptake of titanium for inorganic TiO_2 ENMs in the perfused intestine of rainbow trout (Al-Jubory and Handy, 2012). However, mechanistic details and uptake rates are unknown for most organic ENMs, partly because of the technical problems of detecting the particles in either the tissues or the exposure medium.

5.3. Circulatory System

There are few data on the body distribution of organic ENMs or how they are carried in fish blood to arrive at internal target organs. The effects of ENMs on the cardiovascular functions are mostly unknown. Colloid chemistry theory predicts that ENMs will aggregate or agglomerate in the saline conditions found in the blood supply, unless they adhere to either lymphocytes (perhaps altering their defense function) or macromolecules (albumins, lipids, etc.) (Hellstrand et al., 2009). These macromolecules coating the ENM could change their toxicological properties as well as their dispersal and body distribution. Another immunological concern is that such phenomena might "hide" potentially immunogenic ENMs from the immune system, or indeed make them more inflammatory if antibodies are adsorbed to the surface. The effects of this adsorption chemistry and particle aggregation on the bioavailability of nutrients, blood clotting, the circulating components of the immune system, circulating hormones, and so on, are unknown.

The ability of some organic ENMs to generate reactive oxygen species (ROS) is a concern because mammalian inhalation of airborne organic ENMs can rapidly trigger oxidative stress, resulting in vascular inflammation and cardiovascular dysfunction (Duffin et al., 2007). Systemic inflammation is also indicated from oxidative stress parameters and pathologies in rainbow trout during SWCNT exposures (Smith et al., 2007), perhaps impacting blood pressure regulation and autonomic control of the heart and creating long-term cardiovascular problems in fish with chronic exposures to organic ENMs. Indeed, preliminary studies with metallic ENMs on white sucker (*Catostomus commersonii*) suggest that heart rate is significantly decreased in fish exposed to 1 mg l^{-1} ZnO ENMs with an organic capping agent (KMA Butler and T. J. MacCormack, unpublished). Acute cardiotoxicity may not be obvious at low ENM exposure concentrations but may manifest itself as decreases in swimming performance and aerobic scope (see below on fish behavior and Boyle et al., 2013). Studies with metallic ENMs so far suggest that the effects on hematology are limited and that circulating blood cells are maintained (Federici et al., 2007; Shaw et al., 2012), and this is the case for SWCNTs (Smith et al., 2007).

5.4. Liver and Kidney

The liver is a critical organ for the organic xenobiotic metabolism and biliary excretion of their metabolites (Lefebvre et al., 2007) (see Chapter 1).

Soluble products from phase II liver metabolism are also eliminated in the urine via the kidneys. However, it remains uncertain if these fundamental principles apply to organic ENMs. The ability of the renal corpuscle to filter organic ENMs in fishes is questioned because the filter (circa 60 kDa molecules) is simply too small to filter out ENMs from the blood directly, and there are also some concerns about the processing of aggregates in the vesicular pathways in liver cells (see Handy et al., 2008b). In stenohaline marine fishes glomerular filtration rate (GRF) is absent or very low, suggesting that the liver would be the main route for excretion.

For traditional organic chemicals, sublethal exposure to substances such as pesticides (e.g., Glover et al., 2007) can cause fatty change and lipidosis in the livers of fish (see Chapter 6). Provided these changes are moderate, there may not be a pathological loss of function (e.g., the flow of bile is not compromised). Currently, most of the data on livers from ENM-exposed fish come from histopathological studies focused on metallic particles (TiO_2, Federici et al., 2007; Hao et al., 2009; Cu-ENMs, Griffitt et al., 2007; Al-Bairuty et al., 2013; Ag-ENMs, Gagné et al., 2012). These studies show similar lesions such as fatty change and foci of necrosis. This result has also been reported with waterborne exposure to SWCNTs (Smith et al., 2007). No clear treatment-dependent lesions were observed in rainbow trout fed SWCNTs (500 mg kg^{-1} food; Fraser et al., 2010), but in the same study a hepatitis-like injury was observed in two fish exposed to 500 mg kg^{-1} food of C_{60} for six weeks. However, these associations are difficult to relate to exposure dose because carbon-based ENMs are difficult to detect in tissues. In addition, there is little information on the effects of any type of ENMs on liver enzymes and the nutritional functions of the liver.

Studies on renal function in fishes following exposure to organic ENMs are lacking. Scown et al. (2009) investigated renal function in trout infused with TiO_2 ENMs and found no effects on creatinine clearance or total plasma protein concentrations, suggesting that GFR was unaffected. Al-Bairuty et al. (2013) reported some mild renal tubule pathology, changes in the Bowman's space, and increased melanoma-crophage activity in trout exposed to Cu NPs, without affecting normal plasma electrolytes and hematology (Shaw et al., 2012). The melanoma-crophage deposits may indicate an immune response in the kidney to particulate or other waste materials. Scown et al. (2009) also report particulate material in electron micrographs from the kidney tissue of TiO_2-infused fish. However, this is within the hematopoietic tissue of the kidney and is not part of renal function in terms of the production of urine or osmotic regulation.

5.5. Hematopoietic System and Immunology

The physiological effects of ENMs on the hematopoetic system of fishes have been discussed at length in Handy et al. (2012), and suggestions for research on most aspects of fish immunology have been made (Jovanović and Palić, 2012).

The mammalian literature clearly shows the inflammatory effects of organic and inorganic ENMs and raises concerns that immunotoxicity is a major mechanism of toxicity for ENMs (see reviews, Nel et al., 2006; Dobrovolskaia and McNeil, 2007). High-aspect-ratio (long, thin) ENMs are especially worrisome as they can cause frustrated phagocytosis (Brown et al., 2007), leading to a prolonged respiratory burst with excessive release of ROS and therefore inflammation. Indeed, an asbestos-like inflammation paradigm is suggested for CNTs in the lung (Poland et al., 2008; Mühlfeld et al., 2012).

The immunotoxicity of ordinary organic chemicals, with the possible exception of pesticides (Galloway and Handy, 2003) and PAHs like benzo-α-pyrene (Haggar and Galloway, 2009), is poorly understood, and ENMs even less so. Most of the literature is on metal and metal oxide ENMs (Handy et al., 2011). The spleen is important in removing damaged blood cells from the circulation, and an increase in the red pulp (mostly sinusoids containing red blood cells) would suggest that the spleen is working to remove or contain damaged red blood cells, while an increase in white pulp (mostly immune cells) would suggest an immunological stress response (see Handy et al., 2012). Boyle et al. (2013) found a small decrease in the proportion of red pulp in trout following exposure to 1 mg l^{-1} of either TiO$_2$ ENMs or the bulk material compared to controls that was likely associated with respiratory distress (hypoxia) but with no changes in the white pulp. The circulating blood cells, including total white blood cell counts, were unaffected by exposure to TiO$_2$ or SWCNTs in trout (Federici et al., 2007; Smith et al., 2007).

5.6. Brain and Behavior

The brain is a relatively fatty tissue with a rich blood supply and a propensity to accumulate lipophilic chemicals that diffuse across the blood–brain barrier. Consequently, the effects of lipophilic substances (e.g., methylmercury; Berntssen et al., 2003, see Chapter 5; and pesticides; Galloway and Handy, 2003, see Chapter 6) on the central nervous system have been an ongoing concern. Thus, direct neurotoxicity concerns for organic ENMs exist, especially metallic ENMs with surface coatings or capping agents that are lipophilic. Experiments need to be conducted to assess this and other direct toxicity hazards from ENMs.

Early research on the toxicity of organic ENMs in the brain of fishes has been confounded by solvent toxicity. Lipid peroxidation was reported in brains (but not gills or liver) of juvenile largemouth bass (*Micropterus salmoides*) exposed to C_{60} (Oberdörster, 2004). However, solvent controls were not included, and subsequently the toxicity has been attributed to the tetrahydrofuran solvent used rather than the C_{60} (Henry et al., 2007; Shinohara et al., 2009). Data confirming direct target organ toxicity (actual measurements of ENMs inside the brain) is lacking for all the studies to date reporting biochemical disturbances and histopathological changes (e.g., TiO_2, Ramsden et al., 2009; Boyle et al., 2013; SWCNTs, Smith et al., 2007; Cu NPs, Al-Bairuty et al., 2013). Indeed, Kashiwada (2006) found only background fluorescence in the brain in the internal organs of Japanese medaka (*O. latipes*) exposed to latex ENMs. Also, waterborne metal ENMs pathology was observed without appreciable metal accumulation in the brain (Boyle et al., 2013; Al-Bairuty et al., 2013), suggesting a secondary mode of toxicity. Reported brain pathologies have largely been subtle histological changes, including a low incidence of necrotic cell bodies and small foci of vacuoles in different regions of the brain for metal ENMs (Federici et al., 2007; Ramsden et al., 2009; Boyle et al., 2013; Al-Bairuty et al., 2013) and carbon-based materials (SWCNTs; Smith et al., 2007). The brain is very intolerant of hypoxia and is sensitive to oxidative stress, and secondary toxicity associated with respiratory distress due to gill injury is possible (Boyle et al., 2013). Evidence for a secondary toxicity due to changes in the vasculature of the brain (potential blood pressure and/or hypoxia effects) comes from at least three studies that have reported the appearance of swollen blood vessels in the brain. Smith et al. (2007) showed swelling of the blood vessels on the ventral surface of the cerebellum (consistent with aneurysm) of trout during SWCNT exposures. The fish showed increased ventilation, and the observation might be explained by a change in blood pressure associated with altering ventilation: perfusion ratios to cope with damaged gills. Similar vascular swellings have now been reported in trout brain for TiO_2 ENMs (Boyle et al., 2013) and Cu ENMs (Al-Bairuty et al., 2013). Notably, Al-Bairuty et al. (2013) also found a thickening of the *stratum periventriculare* layer in the mesencephalon, suggestive of osmotic disturbances to the cerebral spinal fluid and/or the presence of inflammatory factors in the fluid.

One dietary study (with no gill injury) has found elevated total Ti concentrations in the brains of rainbow trout following 8-week exposures to 10 or 100 mg kg^{-1} dietary TiO_2, which was associated with disturbances to brain Cu and Zn concentrations, and a 50% decrease in whole-brain

Na^+/K^+-ATPase activity (Ramsden et al., 2009). Although the form of the Ti was not established, direct metal toxicity derived from a metal oxide particle exposure is clearly possible. Smith et al. (2007) also found changes in brain Cu and Zn levels in rainbow trout exposed to SWCNTs through the water (0.1–0.5 mg l^{-1} for 10 days), and some changes in lipid peroxidation that were only partly attributed to the solvent control (Smith et al., 2007). Thus, ENMs can disturb brain biochemistry.

Whether or not these subtle biochemical disturbances and brain pathologies alter the neurological functions of fishes, such as sensory perception, motor output to muscles for locomotion, and maintaining the rhythm of vital processes such as respiration and tone of the heart, is a critical issue. Different ENMs at high mg l^{-1} concentrations had no direct effect on the action potentials in isolated crab nerve preparations (Windeatt and Handy, 2012), suggesting that peripheral nerves could continue to propagate action potentials. No data exist on the effects of any ENMs on sensory perception in fishes. A number of organic compounds are known to interact with and inhibit olfaction in teleosts (Tierney et al., 2010), in part because olfactory neurons are in direct contact with the environment. It is unclear, however, if this kind of phenomenon also applies to organic ENMs. Feeding behavior was impaired in crucian carp exposed to polystyrene ENMs via the diet (Cedervall et al., 2012), but the underlying mechanism of action was not investigated. Boyle et al. (2013) concluded that the decreases in the proportion of time spent at high swimming speeds in trout after waterborne exposure to 1 mg l^{-1} TiO_2 ENMs were more easily explained by respiratory distress and the animal's bioenergetics than by direct brain injury. There is a shortage of studies specifically on the effects of organic ENMs in the brain and on fish behavior.

5.7. Effects of Organic ENMs on Reproduction and Development

The effects of inorganic ENMs on the early life stages of fishes were summarized in Handy et al. (2012). For organic ENMs, like conventional organic pollutants, direct reproductive effects and endocrine disruption of sex hormones are a concern. But apart from Ramsden et al. (2012) recently showing poor-quality embryo production from adult zebrafish exposed to TiO_2 ENMs without overt pathology of the reproductive organs, almost no data exist for organic ENMs impacts on the embryonic development of fishes. This is a concern as acute lethal concentrations to embryos for metal and metal oxide ENMs are around a few mg l^{-1} (see Handy et al., 2011). Again, few data exist for organic ENMs.

Dose-dependent increases in embryo mortality and malformations were observed in Japanese medaka exposed to several formulations of carbon

ENMs (Kim et al., 2012). However, delayed hatching in zebrafish exposed to SWCNTs (Cheng et al., 2007) was attributed to metal catalysts used in ENM production that were found in the SWCNTs. Subsequently, SWCNT exposure did not delay hatching in zebrafish, but a number of metallic ENMs did, possibly by inhibiting the proteases responsible for breaking down the chorion (Ong et al., 2013). Indeed, exposing chorionic fluid to metallic ENMs significantly reduced gross protease activity and was highly correlated to the decrease in hatching at 72 hours (Ong et al., 2013). A similar delay in hatching was also demonstrated with very high concentrations (800 mg l^{-1}) of organic 5 nm polyacrylic acid (PAA) nanocapsules, but the correlation with protease activity was not demonstrated (Felix et al., 2013). Both acute and sublethal toxicities were observed in zebrafish embryos exposed to various formulations of organic dendrimers (Heiden et al., 2007).

ENMs have the ability to penetrate the chorion barrier and then penetrate the embryo itself to have direct toxic effects on development. Metallic ENMs can move through the chorionic pores in zebrafish embryos (Nallathamby et al., 2008). The chorionic pores of a zebrafish embryo are approximately 0.5–0.7 μm in diameter (Rawson et al., 2000), which is sufficiently large to allow passage of ENMs, even in aggregated forms. Only those ENMs with aggregates that were small enough to pass through the chorionic pores delayed zebrafish hatching (Ong et al., 2013). Also, dye-containing organic nanocapsules (primarily PAA) are able to transverse the chorion and interact with surface of zebrafish embryos. There is no reason to believe that smaller monodispersed organic ENMs cannot do the same thing, although rates of transfer across the chorion will likely depend highly on the physicochemical properties (e.g., agglomeration state, charge) of the specific ENM in the media.

6. STUDIES ON ORGANIC ENMs AND CAPPING AGENTS

The following examples are used to highlight two emerging priority issues that are different from the established hazards of traditional organic chemicals.

6.1. Organic Capping Agents

As mentioned earlier, a diverse array of capping agents have been employed for organic and inorganic ENMs to make them surface-functionalized and facilitate the formation of a stable colloidal suspension, to protect the particle core from dissolution or oxidation, and/or to alter the physicochemical properties of the ENM to better suit their intended application. The diverse array of capping agents includes, but is not limited to, peptides and proteins, fatty acids and fatty amines, carboxylic acids, surfactants, and polymers. Dendrimer ENMs can also be used as a capping agent for other organic and inorganic ENM cores.

The potential toxicity of any ENM formulation will be related to its component parts. Considerable effort has been focused on the development of biocompatible capping agents, particularly in relation to ENMs designed for clinical applications, where minimizing interactions between the ENM surface and biological materials (e.g., serum proteins) is highly desirable (Knop et al., 2010). Unbound cap material may be present as a contaminant in ENM stock solutions, or it can be released from particles under specific environmental conditions, either unintentionally or by design (Romberg et al., 2008). Alternatively, the capping or functionalizing agent may be stripped from the core ENM inside the animal by bacterial or other biochemical processes, resulting in intended or unintended effects on cell physiology. ENMs with similar cores and different capping agents frequently exhibit disparate bioactivity, and there are a number of potential explanations for this observation. Not all capping agents are equally effective at preventing ENM aggregation or dissolution of the core material, so the effective exposure dose and/or the mechanism of ENM toxicity can change quickly under test conditions. The relative hydrophobicity of the cap may influence the uptake and distribution kinetics of the particle or change the characteristics of the "protein corona" that is formed on its surface after introduction into a biological medium (e.g., Aggarwal et al., 2009). In addition, if the capping material itself is toxic, its release from the ENM over time or with changing environmental conditions could also have negative effects on the animal.

Given the large variety of core materials and organic capping agents currently in use in ENM synthesis, we chose to focus the discussion on inorganic Ag ENMs capped with either citrate (Ag ENM-C) or the organic polymer polyvinylpyrrilodone (Ag ENM-PVP). Relevant data on the toxicity of organic ENMs with similar core materials and different capping agents are unavailable, but information on Ag ENMs will illustrate some general considerations applicable to all types of ENMs.

Citrate and PVP are two of the most common capping agents used to stabilize Ag ENMs, and both impart a negative surface potential to the particle at pHs relevant to most fish studies (Tolaymat et al., 2010). Citrate is a small, highly charged, organic acid (Figure 9.1) linked to energy metabolism and biosynthetic intermediate production (Icard et al., 2012). Citrate forms relatively weak associations with the surface of Ag ENMs

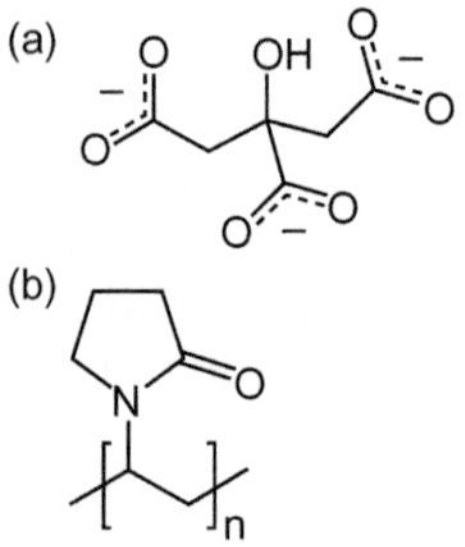

Figure 9.1. Chemical structures of common engineered nanomaterial organic capping agents. (a) Citrate. (b) Polyvinyl pyrrolidone (PVP).

through multiple hydrogen bonds (Kilin et al., 2008), so some level of release from the particle surface is likely, particularly at elevated Cl^- levels (Thio et al., 2012). Changes in solution pH associated with the degradation/dissolution of Ag ENM-C appear to be minor (Tejamaya et al., 2012) and are likely to be overshadowed by changes due to biological activity in the system. Citrate is not considered a hazardous compound but it is a chelating agent and its ability to bind ionic Ca^{2+} makes it an effective anticoagulant. In a clinical setting, citrate can lead to toxicity via hypocalcemia (Weinstein, 2001), but only with extremely high intravenous doses that are unlikely to be of any relevance to ENM toxicity. PVP is a comparatively large molecule (Figure 9.1) that strongly adsorbs to the surface of Ag ENMs via interactions with its carbonyl oxygen (Mdluli et al., 2011), reducing its likelihood to be released under typical exposure conditions. Brief exposures to high concentrations (>10% v/v) of PVP can affect hatching rate in sea bream (*Pagrus major*) embryos (Xiao et al., 2008), but at lower concentrations it showed no effects on zebrafish embryos (Powers et al., 2011). Information is unavailable regarding PVP toxicity in adult fish or under chronic exposure conditions. In general, the influence of specific capping agents on the physicochemical properties of the ENM will play a much larger role in determining bioactivity than will the free-capping agent alone (see below). Regardless, it is important to account for the potential bioactivity of freely dissociated capping agents in nanotoxicity tests, as their release from the ENM surface is a possibility.

Citrate and PVP stabilize ENM suspensions by different mechanisms and therefore have varying capacities for preventing aggregation of the particles and/or oxidation or dissolution of the ENM core material in test media. The latter point is particularly important for metal particles but is also relevant to organic ENMs. Electrostatic repulsion between the negatively charged surfaces of individual Ag ENM-C inhibits interactions between adjacent ENM cores, which would otherwise aggregate and precipitate. The effectiveness of a charge-stabilized suspension is strongly influenced by the ionic strength and composition of the media, and as mentioned earlier, relatively small changes in solution properties can lead to a rapid loss of colloidal stability. For example, Ag ENM-C (ca. 10 nm core diameter) aggregates immediately in standard Organisation for Economic Cooperation and Development (OECD) media for *Daphnia* sp. toxicity testing (Römer et al., 2011; Tejamaya et al., 2012) and is almost completely absent from suspension within three days (Tejamaya et al., 2012).

Unlike citrate, PVP mainly prevents the association of ENM core materials and stabilizes the suspension using its bulky structure (e.g., Tolaymat et al., 2010). Sterically stabilized ENM suspensions are generally less sensitive to changes in the ionic strength and composition of the media than charge-stabilized suspensions, but aggregation may still be observed. Ag ENM-PVP (ca. 10 nm core diameter) were stable in the same *Daphnia* sp. OECD test media mentioned above for substantially longer than Ag ENM-C and exhibited little aggregation over 21 days of observation (Tejamaya et al., 2012). In other studies, Ag ENM-PVP suspensions were more stable, but the differences were subtle, with both formulations remaining relatively well dispersed over several days (Farkas et al., 2011). In contrast, Powers et al. (2011) found that in water or Hanks's solution, Ag ENM-PVP aggregated much more quickly than Ag ENM-C, with less than 10% of the initial total Ag remaining in suspension after 24 hours versus 55% for Ag ENM-C. This type of variance in the efficacy of capping agents can lead to substantial changes in the effective ENM exposure dose over the time course of an experiment and alter the mechanism of ENM uptake by the fish (e.g., gill vs. gut). From the aforementioned examples, it is clear that ENM characteristics should be assessed on a case-by-case basis, as subtle changes in particle formulation or exposure conditions may greatly alter the effective exposure dose.

The potential for dissolution of ENM core material is a clear toxicological concern as numerous ENMs are synthesized from metals such as Ag, Cd, and Cu, which are known to be toxic to fish (Shaw and Handy, 2011; Borm et al., 2006). Capping agents that are effective at preventing aggregation are not always correspondingly effective at preventing dissolution of the ENM core, and gradual loss of core material over time appears to be inevitable (Thio et al., 2011). Ag ENM-PVP is usually more stable in suspension than Ag ENM-C, but the citrate capped particles release less ionic Ag^+ over time than PVP capped particles (Meyer et al., 2010). Schultz et al. (2012) also showed minimal release of ionic Ag^+ from Ag ENM-C in dechlorinated tap water used in rainbow trout ^{22}Na uptake experiments. Dissolution of Ag^+ can sometimes explain differential toxicity in similar

ENM cores with different capping agents. The toxicity of Ag ENM-PVP to the nematode, *Caenorhabditis elegans*, could be entirely explained by dissolution of ionic Ag^+ from the particle core, whereas toxicity of the citrate-capped Ag ENMs was largely unrelated to Ag^+ dissolution (Meyer et al., 2010). The effects of Ag ENM-C on ^{22}Na uptake in trout also appeared to be unrelated to the release of ionic Ag^+ (Schultz et al., 2012), at least not to the bulk solution. Small molecules like citrate may provide better particle surface coverage (i.e., more capping molecules per nm^2) than bulky polymers. This could explain their enhanced ability to minimize core dissolution. It is unclear at this point how uptake into the animal influences the behavior of ENMs with different capping agents.

The influence of capping agents on various indicators of toxicity in fish has been addressed in a number of studies (e.g., Powers et al., 2011; Farkas et al., 2011), but it is often difficult to define whether observed differences relate specifically to the capping agent or to its ability to influence particle properties during the exposure. Capping agents that produce a more monodispersed suspension of ENMs increase the potential for ENM–protein interactions and enhance subsequent bioactivity (e.g., MacCormack et al., 2012). Ag ENM-C that was more stable in suspension than equivalent Ag ENM-PVP exhibited correspondingly greater toxicity in embryonic and larval zebrafish (Powers et al., 2011). However, exposure to Ag ENM-PVP during development still caused behavioral abnormalities posthatch, despite evidence for substantial particle aggregation and loss from suspension (Powers et al., 2011). In a study by Farkas et al. (2011), both Ag ENM-C and Ag ENM-PVP remained relatively well dispersed during testing and caused similar levels of cytotoxicity to cultured rainbow trout gill cells. Clearly, more evidence is required to establish a firm relationship between ENM characteristics during exposure and their potential toxicity. Nevertheless, it is apparent that the choice of capping agent can potently influence ENM toxicity and that its effects on the physicochemical characteristics of the particle play an important role in this phenomenon.

6.2. Carbon-Based ENMs

Carbon-based nanomaterials such as single-walled carbon nanotubes (SWCNTs), multiwalled carbon nanotubes (MWCNTs), graphene sheets, and fullerenes are likely the most studied of all nanomaterials in terms of breadth and volume with >36,000 CNT and graphene publications and >3500 patents filed by the end of 2011 (De Volder et al., 2013). Despite this, the relative cytotoxic potential remains the topic of some discussion and debate. Most carbon-based nanomaterials developed to date are highly hydrophobic and are generally not thought to be present in high quantities in the water column. For this reason, the primary mechanisms of CNT toxicity in fishes are thought to be mediated through ingestion and therefore are more directed at benthic feeding fishes. Early studies on CNTs and fullerenes engaged dispersants/surfactants to induce exposure, and often the findings of these studies attributed toxicity to the CNTs without proper control for the surfactant or other contaminants in the material preparation. More recent and well-controlled studies in fish have found specific CNT-mediated effects such as excessive mucus production by the gills of rainbow trout after exposure to carbon-based nanoparticles (Smith et al., 2007), increased oxidative stress as measured in the gills, along with physical damage to the gills and other organs (Auffan et al., 2011). Moreover, recent developments in CNT technology have allowed for new side-chain functionalizations to significantly increase the use of these materials in commercial products (De Volder et al., 2013). Many of these newer functionalizations are specifically directed at increasing water solubilization for use in specific processes, and this may significantly alter both the kinetics and route of exposure, potentially increasing the toxicity of these materials. Further vigilance and study on these newer materials is required to ensure their safer application into commerce.

7. NANOCRYSTALLINE AND NANOFIBRILLAR CELLULOSIC MATERIALS

Cellulose is one of the most abundantly available materials, and with the advent of the enabling umbrella of "nanotechnology" the forestry industry around the world has actively promoted the development of high-value-added nanomaterials made from cellulose fibers. Nanocrystalline cellulose (NCC) and nanofibrillar cellulose (NFC) are the crystalline rodlike components of cellulosic fibers. NCC is an organic-based fiber that is obtained from material such as straw, wood, or other cellulosic biological materials. Based on the initial starting material used, its size can range from 3 to 10 nm in diameter to 140–220 nm in length (Ruiz et al., 2000). Through development of effective isolation technologies, NCC and NFC can be isolated from a large variety of feedstock materials in a cost-efficient manner. NCC and NFC have the advantages of being materials that are highly soluble in water and biological in origin. NCC and NFC, as organic ENMs, are currently being proposed for wide-scale use in a number of commercial products. One of the more important and attractive (from a commercial sense) aspects of NCC/NFC is the availability of a number of –OH side chains running along the length of the fiber. These available –OH groups allow for a tailored chemistry approach to be applied and these functionalized materials to be customized for use in many commercial products. Table 9.2 lists a number of proposed functionalizations as published in the literature, along with their proposed use in commerce. However, this list is far from comprehensive, and current increases in production will likely stimulate many different types of functionalized NCCs and NFCs. Despite proposals for large-scale production and extensive functionalization for use in a broad number of product classes, almost no testing for toxicity has been done in any of the above-mentioned studies. The reason is that NCC and NFC are generally considered to have low toxicity and likely pose little environmental risk. Toxicity testing of NCC consists of a single industry-sponsored study (Kovacs et al., 2010) in

Table 9.2

Demonstrated functionalizations and proposed commercial uses for nanocrystalline cellulosic materials

Functionalization	Proposed Use	Reference
Glucose oxidase conjugated on thiol-functionalized Au NPs	Enzyme immobilization for biosensors	Incani et al. (2012)
Cationic hydroxypropyltrimethyl-ammonium chloride (HPTMAC)	Reinforcing fillers in polymer composites	Hasani et al. (2008)
Carboxyl	Biosensing; enzyme immobilization; affinity chromatography; drug targeting/delivery	Leung et al. (2011)
Poly(ethylene oxide)	Nanocomposite materials with improved mechanical properties	Kloser and Gray (2010)
Fluorescein-5′-isothiocyanate (FITC)	Fluorescence bioassay and bioimaging	Dong and Roman (2007)
Waterborne polyurethane (WPU)	Reinforcing materials	Cao et al. (2009)
DNA oligomers	Nucleic acid research; templates for nucleotide-polymerizing enzymes	Mangalam et al. (2009)
Maleated polypropylene	Filler in polymeric matrices	Ljungberg et al. (2005)
Poly(styrene)	Smart windows; optical films; security papers; optical devices	Yi et al. (2008)
Ferrocene	Smart membranes; sensors; bioelectrochemical devices; 3D conducting systems	Eyley et al. (2012)

which the toxicity of the base material NCC was comprehensively evaluated using nine aquatic species and one fish cell line (Kovacs et al., 2010). As mentioned previously, one of the main advantages of NCC and NFC is the ability to perform chemistry on the available side groups. The industry-sponsored study by Kovacs et al. (2010) represents the first and only published toxicological assessment of NCC. It has been quoted in all the subsequently published studies as obviating the need for further testing. This generalization makes NCC appealing for use in applications such as personal care, food, and pharmaceuticals (Peng et al., 2011). However, caution must be exercised since many nanotoxicity studies have shown that functionalization or capping agents can alter the properties of uptake, agglomeration, and interaction, with organisms resulting in new mechanisms of toxicity. Further toxicity testing and life-cycle analyses are absolutely required to determine the potential toxic effects of these functionalized NCCs and NFCs prior to their incorporation into commercial products.

8. POLYMER-COATED ENMs IN THE AGRICULTURAL SECTOR

The agricultural sector has recognized the potential of using nanomaterials and is investing heavily in the development of new products to enhance productivity in this sector. Control of active systemic herbicides and pesticides using nanotechnology-enabled products holds promise for reductions in the use of pesticides and herbicides through controlled release and/or delivery of materials to the crop with reduced soil mobility and reduced nontarget organism toxicity. Furthermore, reduced runoff of fertilization and controlled delivery of micronutrients in the soil are long-sought properties that are being accomplished through the application of nanoscience-enabled technologies. One of the more popular methods for controlled delivery and release of herbicides, pesticides, and micronutrients is through the application of polymer-coated materials or development of encapsulating polymer-based coatings that provide apparently good biocompatibility, solubility, and excellent dispersion characteristics (Pérez-de-Luque and Rubiales, 2008). One of the more tested materials of this type to date is the PAA polymer-based ENMs manufactured by Vive Crop Protection Inc. (formerly Vive Nano Inc.).

Although zebrafish toxicity tests with the capsules and associated inorganic cores have shown a range of effects, these are only seen at relatively high concentrations, far above the scenarios for proposed application rates. These materials are highly monodispersed in water, are small ($\sim$7–10 nm), and have a high zeta potential (< -20 mV), suggesting that they will remain suspended in aqueous solutions for an extended period. In a recent paper, Felix et al. (2013) demonstrated that PAA-coated CeO_2 ENMs are indeed able to traverse the chorion of zebrafish embryos, thereby directly exposing the embryo during important developmental stages. This study also found high concentrations ($\geq$800 mg/L) of the PAA coating itself (nanocapsules) inhibit hatching of zebrafish embryos, while PAA-coated ZnO metal core ENMs were not shown to inhibit zebrafish hatch.

Further study of these ENMs demonstrated that the PAA particles themselves (along with some inorganic materials as well) inhibited the protease activity of hatching enzyme isolated from chorionic fluid of 48 hours postfertilization (hpf) zebrafish embryos (Ong et al., 2013). Inhibition of hatching is thought to occur when the polymer-encapsulated ENM, as a colloid in suspension, interacts with the hatching enzyme, thereby reducing its activity similar to the reduction in enzymatic activity of purified lactate dehydrogenase (LDH; EC1.1.1.27) in the presence of many different types of small (<10 nm) nanoparticles (MacCormack et al., 2012). Studies with PAA nanocapsules loaded with Nile red as a marker to follow their behavior have demonstrated that PAA capsules are indeed able to penetrate the chorionic membrane of zebrafish (Figure 9.2) and associate with the surface of the embryo preferentially. Presumably, this allows for interaction and inhibition of the hatching enzyme within the chorionic fluid. An important consideration is that these capsules may be loaded with systemic herbicides, pesticides, or other organic or inorganic agents, and selective partitioning to the embryo surface may facilitate the delivery of associated contaminants and increase their toxicity. Therefore, despite the low level of toxicity directly attributable to the nanocapsule itself, it will be critical to investigate how this method of delivery affects the toxicity of the active agent itself. Furthermore, these results point to the fact that hatching inhibition/hatching success, while not normally considered to be a toxicological end-point, should be more thoroughly investigated as a potential indicator of biological and ecological toxicity.

9. KEY NEEDS IN ENVIRONMENTAL RISK ASSESSMENT OF ORGANIC ENMs

The regulatory community would like to continue to use the existing hazard assessment framework for organic ENMs, albeit with modifications and revalidation of the test methods for ENMs. For risk assessment, the traditional triggers for concern at the population level will likely remain (i.e., survival, growth, and reproduction), but key issues in the environmental risk

200mg/L vnCAP-NR   Control

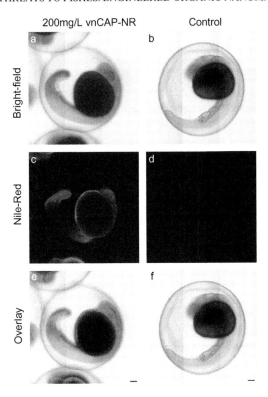

Fig. 9.2. Nile red nanocapsules are able to transverse the chorion of zebrafish. Bright field (a, b), Nile red nanocapsule fluorescence (c, d) and overlay (e, f) confocal laser scanning micrographs of 24-hour postfertilization zebrafish exposed to 200 mg/L dialyzed nanocapsules containing Nile red (vnCAP-NR; left) or dechlorinated tap water control (right). Scale bars are 100 μm. Photos courtesy of Lindsey Felix and James Ede. See color plate at the back of the book.

assessment of organic ENMs also include the environmental persistence and bioaccumulation potential. For example, an important parameter for understanding both the toxicity and fate/behavior of substances in the environment is the measurement of hydrophobicity. The octanol:water partition coefficient measurement has long since been correlated with bioconcentration factors in fishes (Veith et al., 1979; see Chapter 1), but the fundamental assumption that organic ENMs can bioconcentrate in fishes has not been verified. The fate and behavior of most ENMs have not been detailed in the environment, and data on the breakdown of the organic ENMs (persistence in the environment), as well as the ability of fishes to metabolize and excrete them (persistence in the organism), are unclear. These issues represent significant data gaps for environmental risk assessment. It is essential to conduct studies at lower concentrations and for longer times in order to support the risk assessment process with environmentally relevant chronic exposure data. Moreover, the basic assumptions of differential species sensitivity need to be established for ENMs. In essence, while limited laboratory data collected to date indicates that there are concerns for both lethal and sublethal effects of organic ENMs in fishes, it remains difficult to put this information into context for risk of effects to wild populations of fishes in a formal risk assessment.

The scientific community has been unable to agree on a procedure to let this information be used in an approved way for risk assessment that would be internationally recognized by bodies involved in standardization of testing such as the OECD; or the regulatory bodies with some responsibility for approving new products (European Chemicals Agency, the U.S. Environmental Protection Agency, etc.). The uncertainty in risk assessments, especially for organic ENMs, brings this issue sharply into focus. There are ongoing discussions on how to use nonstandard data in environmental risk assessment, such as agreeing on quality criteria for research papers (see Crane et al., 2008 on minimum characterization issues). Further uncertainty exists regarding the compatibility of standard laboratory and biochemical testing methods with ENMs (Handy et al., 2012; Ong et al., 2013a). Numerous ENM formulations interact with critical components of common toxicity test assays (Monteiro-Riviere et al., 2009), often in unpredictable ways (reviewed by Kroll et al., 2009). It is well established that ENMs bind to proteins and inhibit their function (Aggarwal et al., 2009; MacCormack et al., 2012); this also occurs in biochemical assay systems that rely on enzymatic activity to generate results. For example, ENMs inhibit LDH activity, sometimes completely, and this could lead to erroneous estimates of cytotoxicity based on traditional LDH release assays (Wohlleben et al., 2011; Han et al., 2011; MacCormack et al., 2012). ENMs also interact with colorimetric dyes such as MTT (3-(4,5-dimethylthiazole-2-yl)-2,5-diphenyl tetrazolium bromide), which is commonly used in cell viability assays (often linked to enzymatic activity). These interactions can generate artifacts at multiple steps in the assay (Worle-Knirsch et al., 2006; Davoren et al., 2007; Belyanskaya et al., 2007; Monteiro-Riviere et al., 2009; Ciofani et al., 2010; Ong et al., 2013b). Similar concerns exist with other colorimetric and fluorometric agents utilized in commercially available toxicity test assays (Monteiro-Riviere et al., 2009; Ong et al., 2013b). Such artifacts can lead to a substantial under- or overestimate of bioactivity; yet similar assays continue to be used in nanotoxicity testing without appropriate validation (Kroll et al., 2009; Ong et al., 2013b), particularly in fundamental research. In some instances, appropriate controls can account for ENM interference, but in other cases it may be necessary to consider alternative test protocols. Assessing data quality to be used in the risk assessment is central to the acceptance of the risk analysis (see Chapter 10).

10. CONCLUSIONS AND PERSPECTIVES

ENMs can interact with traditional organic and inorganic agents that may alter their characteristics of adsorption, distribution, metabolism, and excretion in fish. In addition, the growing interest in developing nano-enabled pharmaceuticals and chemicals could introduce even more effective "Trojan horses" into the environment, requiring a reassessment of ecotoxicity models for fate, transformation, and effects of these materials. The need for sufficient characterization of ENMs in test media has been emphasized frequently in the literature and throughout this chapter, and its importance in experimental design and in the interpretation of toxicity data cannot be overstated. As illustrated in the discussion on capping agents, ENMs with identical formulations can behave quite differently in different media, heavily influencing the nature and magnitude of observed bioactivity. ENMs bearing identical capping agents but with different core materials also exhibit differential bioactivity (MacCormack et al., 2012), illustrating that the properties of the particle are not dominated solely by the properties of its capping agent. The majority of nanotoxicity studies thus far have focused on model organisms, and there is a need to survey a broader range of animals to identify sentinel species and develop appropriate ecological monitoring procedures. There is also a pressing need for work on estuarine and marine fish, for which there is a conspicuous lack of information on potential ENM toxicity.

ACKNOWLEDGMENTS

We thank Thomas Moon and Anthony Farrell for comments on earlier versions of this chapter. T.J.M. was supported by the Natural Sciences and Engineering Research Council (NSERC) Discovery Grants program. G.G.G. was supported by Environment Canada, the NRC-NSERC BDC Nanotechology Initiative and Alberta Innovates Nanoworks grants. R.D.H. was supported by NERC (NE/G001812/1).

REFERENCES

Aggarwal, P., Hall, J. B., McLeland, C. B., Dobrovolskaia, M. A. and McNeil, S. E. (2009). Nanoparticle interaction with plasma proteins as it relates to particle biodistribution, biocompatibility and therapeutic efficacy. *Advanced Drug Delivery Reviews* **61**, 428–437.

Aitken, R. J., Chaudhry, M. Q., Boxall, A. B. A. and Hull, M. (2006). Manufacture and use of nanomaterials: current status in the UK and global trends. *Occupational Medicine (Oxford)* **56**, 300–306.

Al-Bairuty, G. A., Shaw, B. J., Handy, R. D. and Henry, T. B. (2013). Histopathological effects of waterborne copper nanoparticles and copper sulphate on the organs of rainbow trout (*Oncorhynchus mykiss*). *Aquatic Toxicology*. doi: 10.1016/j.aquatox.2012.10.005 (in press).

Al-Jubory, A. R. and Handy, R. D. (2012). Uptake of titanium from TiO_2 nanoparticle exposure in the isolated perfused intestine of rainbow trout: nystatin, vanadate and novel CO_2-sensitive components. *Nanotoxicology.* DOI:10.3109/17435390.2012.735268.

Astuc, D., Boisselier, E. and Ornelas, C. (2010). Dendrimers designed for functions: from physical, photophysical, and supramolecular properties to applications in sensing, catalysis, molecular electronics, photonics, and nanomedicine. *Chemical Reviews* **110**, 1857–1959.

Auffan, M., Flahaut, E., Thill, A., Mouchet, F., Carriere, M., Gauthier, L., Achouak, W., Rose, J., Weisner, M. R. and Bottero, J.-Y. (2011). Ecotoxicology: Nanoparticle reactivity and living Organisms. *Nanoethics and Nanotoxicology* 325–357.

Belyanskaya, L., Manser, P., Spohn, P., Bruinink, A. and Wick, P. (2007). The reliability and limits of the MTT reduction assay for carbon nanotubes–cell interaction. *Carbon* **45**, 2643–2648.

Bermudez, E., Mangum, J. B., Wong, B. A., Asgharian, B., Hext, P. M., Warheit, D. B. and Everitt, J. I. (2004). Pulmonary responses of mice, rats, and hamsters to subchronic inhalation of ultrafine titanium dioxide particles. *Toxicological Sciences* **77**, 347–357.

Berntssen, M. H., Aatland, A. and Handy, R. D. (2003). Chronic dietary mercury exposure causes oxidative stress, brain lesions, and altered behaviour in Atlantic salmon (*Salmo salar*) parr. *Aquatic Toxicology* **65**, 55–72.

Bilberg, K., Malte, H., Wang, T. and Baatrup, E. (2010). Silver nanoparticles and silver nitrate cause respiratory stress in Eurasian perch (*Perca fluviatilis*). *Aquatic Toxicology* **96**, 159–165.

Biswas, P. and Wu, P. (2005). Nanoparticles and the environment. *Journal of the Air and Waste Management Association* **55**, 708–746.

Black, M. C. and McCarthy, J. F. (1988). Dissolved organic macromolecules reduce the uptake of hydrophobic organic contaminants by the gills of rainbow trout (*Salmo gairdneri*). *Environmental Toxicology and Chemistry* **7**, 593–600.

Blickley, T. M. and McClellan-Green, P. (2008). Toxicity of aqueous fullerene in adult and larval *Fundulus heteroclitus*. *Environmental Toxicology and Chemistry* **27**, 1964–1971.

Borm, P., Klaessig, F. C., Landry, T. D., Moudgil, B., Pauluhn, J., Thomas, K., Trottier, R. and Wood, S. (2006). Research strategies for safety evaluation of nanomaterials, part V: role of dissolution in biological fate and effects of nanoscale particles. *Toxicological Sciences* **90**, 23–32.

Borzsonyi, G., Johnson, R. S., Myles, A. J., Cho, J.-Y., Yamazaki, T., Beingessner, R. L., Kovalenko, A. and Fenniri, H. (2010). Rosette nanotubes with 1.4 nm inner diameter from a tricyclic variant of the Lehn–Mascal G ^ C base. *Chemical Communication* **46**, 6527–6529.

Boxall, A.B.A., Chaundhry, Q., Sinclair, C., Jones, A., Aitken, R., Jefferson, B. and Watts, C. (2007). Current and future predicted environmental exposure to manufactured nanoparticles. Report by the Central Science Laboratory (CSL) York for the Department of the Environment and Rural Affairs (DEFRA), UK. Available at: http://www.defra.gov.uk/science/Project_Data/DocumentLibrary/CB01098/CB01098_6270_FRP.pdf

Boyle, D., Al-Bairuty, G. A., Ramsden, C. S., Sloman, K. A., Henry, T. B. and Handy, R. D. (2013). Subtle alterations in swimming speed distributions of rainbow trout exposed to titanium dioxide nanoparticles are associated with gill rather than brain injury. *Aquatic Toxicology* **126**, 116–127.

Brown, D. M., Kinloch, I. A., Bangert, U., Windle, A. H., Walter, D. M., Walker, G. S., Scotchford, C. A., Donaldson, K. and Stone, V. (2007). An *in vitro* study of the potential of carbon nanotubes and nanofibres to induce inflammatory mediators and frustrated phagocytosis. *Carbon* **45**, 1743–1756.

Buffle, J. (2006). The key role of environmental colloids/nanoparticales for the sustainability of life. *Environmental Chemistry* **3**, 155–158.

Cao, X., Habibi, Y. and Lucia, L. A. (2009). One-pot polymerization, surface grafting, and processing of waterborne polyurethane-cellulose nanocrystal nanocomposites. *Journal of Materials Chemistry* **19**, 7137–7145.

Cedervall, T., Hansson, L.-A., Lard, M., Frohm, B. and Linse, S. (2012). Food chain transport of nanoparticles affects behaviour and fat metabolism in fish. *PLoS ONE* **7**, e32254. DOI:10.1371/journal.pone.0032254.

Chaudhry, Q., Scotter, M., Blackburn, J., Ross, B., Boxall, A., Castle, L., Aitken, R. and Watkins, R. (2008). Applications and implications of nanotechnologies for the food sector. *Food Additives and Contaminants* **25**, 241–258.

Cheng, J., Chan, C. M., Veca, L. M., Poon, W. L., Chan, P. K., Qu, L., Sun, Y.-P. and Cheng, S. H. (2009). Acute and long-term effects after single loading of functionalized multi-walled carbon nanotubes into zebrafish (*Danio rerio*). *Toxicology and Applied Pharmacology* **235**, 216–225.

Ciofani, G., Danti, S., D'Alessandro, D., Moscato, S. and Menciassi, A. (2010). Assessing cytotoxicity of boron nitride nanotubes: interference with the MTT assay. *Biochemical and Biophysical Research Communications* **394**, 405–411.

Crane, M., Handy, R. D., Garrod, J. and Owen, R. (2008). Ecotoxicity test methods and environmental hazard assessment for manufactured nanoparticles. *Ecotoxicology* **17**, 421–437.

Davoren, M., Herzog, E., Casey, A., Cottineau, B., Chambers, G., Byrne, H. J. and Lyng, F. M. (2007). *In vitro* toxicity evaluation of single walled carbon nanotubes on human A549 lung cells. *Toxicology In Vitro* **21**, 438–448.

Derjaguin, B. V. and Landau, L. D. (1941). Theory of the stability of strongly charged lyophobic sols and of the adhesion of strongly charged particles in solutions of electrolytes. *Acta Physicochimica* **14**, 733–762.

De Volder, M. F. L., Tawfick, S. H., Baughman, R. H. and Hart, A. J. (2013). Carbon nanotubes: present and future commercial applications. *Science* **339**, 535–539.

Dobrovolskaia, M. A. and McNeil, S. E. (2007). Immunological properties of manufactured nanomaterials. *Nature Nanotechnology* **2**, 469–478.

Dong, S. and Roman, M. (2007). Fluorescently labeled cellulose nanocrystals for bioimaging applications. *Journal of the American Chemical Society* **129**, 13810–13811.

Duffin, R., Tran, L., Brown, D., Stone, V. and Donaldson, K. (2007). Proinflammogenic effects of low-toxicity and metal nanoparticles *in vivo* and *in vitro*: Highlighting the role of particle surface area and surface reactivity. *Inhalation Toxicology* **19**, 849–856.

Dymowska, A. K., Hwang, P. P. and Goss, G. G. (2012). Structure and function of ionocytes in the freshwater fish gill. *Respiratory Physiology and Neurobiology* **184**, 282–292.

Eyley, S., Shariki, S., Dale, S. E. C., Bending, S., Marken, F. and Thielemans, W. (2012). Ferrocene-decorated nanocrystalline cellulose with charge carrier mobility. *Langmuir* **28**, 6514–6519.

Farkas, J., Christian, P., Gallego-Urrea, J. A., Roos, N., Hassellöv, M., Tollefsen, K. E. and Thomas, K. V. (2011). Uptake and effects of manufactured silver nanoparticles in rainbow trout (*Oncorhynchus mykiss*) gill cells. *Aquatic Toxicology* **101**, 117–125.

Federici, G., Shaw, B. J. and Handy, R. D. (2007). Toxicity of titanium dioxide nanoparticles to rainbow trout (*Oncorhynchus mykiss*): Gill injury, oxidative stress, and other physiological effects. *Aquatic Toxicology* **84**, 415–430.

Felix, L. C., Ortega, V. A., Ede, J. D. and Goss, G. G. (2013). Physicochemical characteristics of polymer-coated metal-oxide nanoparticles and their toxicological effects on zebrafish (*Danio rerio*) Development. *Environmental Science and Technology*. dx.doi.org/10.1021/es401403p (in press).

Fraser, T. W. K., Reinardy, H. C., Shaw, B. J., Henry, T. B. and Handy, R. D. (2010). Dietary toxicity of single-walled carbon nanotubes and fullerenes (C_{60}) in rainbow trout (*Oncorhynchus mykiss*). *Nanotoxicology* **5**, 98–108.

Furman, O., Usenko, S. and Lau, B. L. T. (2013). Relative importance of the humic and fulvic fractions of natural organic matter in the aggregation and deposition of silver nanoparticles. *Environmental Science and Technology* **47**, 1349–1356.

Gagné, F., André, C., Skirrow, R., Gélinas, M., Auclair, J., van Aggelen, G., Turcotte, P. and Gagnon, C. (2012). Toxicity of silver nanoparticles to rainbow trout: A toxicogenomic approach. *Chemosphere* **89**, 615–622.

Galloway, T. and Handy, R. D. (2003). Immunotoxicity of organophosphorus pesticides. *Ecotoxicology* **12**, 345–363.

Gilbert, B., Lu, G. and Kim, C. S. (2007). Stable cluster formation in aqueous suspensions of iron oxyhydroxide nanoparticles. *Journal of Colloid and Interface Science* **313**, 152–159.

Glover, C. N., Petri, D., Tollefsen, K.-E., Jørum, N., Handy, R. D. and Berntssen, M. H. G. (2007). Assessing the sensitivity of Atlantic salmon (*Salmo salar*) to dietary endosulfan exposure using tissue biochemistry and histology. *Aquatic Toxicology* **84**, 346–355.

Gottschalk, F., Sonderer, T., Scholz, R. W. and Nowack, B. (2009). Modelled environmental concentrations of manufactured nanomaterials (TiO_2, ZnO, Ag, CNT, fullerenes) for different regions. *Environmental Science and Technology* **43**, 9216–9222.

Griffitt, R. J., Weil, R., Hyndman, K. A., Denslow, N. D., Powers, K., Taylor, D. and Barber, D. S. (2007). Exposure to copper nanoparticles caused gill injury and acute lethality in zebrafish (*Danio rerio*). *Environmental Science and Technology* **41**, 8178–8186.

Han, X., Gelein, R., Corson, N., Wade-Mercer, P., Jiang, J., Biswas, P., Finkelstein, J. N., Elder, A. and Oberdörster, G. (2011). Validation of an LDH assay for assessing nanoparticle toxicity. *Toxicology* **287**, 99–104.

Handy, R.D. (2011). FSBI briefing paper: Nanotechnology in fisheries and aquaculture. The Fisheries Society of the British Isles. Available at: http://www.fsbi.org.uk/assets/brief-nanotechnology-fisheriesaquaculture.pdf.

Handy, R. D., Al-Bairuty, G., Al-Jubory, A., Ramsden, C. S., Boyle, D., Shaw, B. J. and Henry, T. B. (2011). Effects of manufactured nanomaterials on fishes: a target organ and body systems physiology approach. *Journal of Fish Biology* **79**, 821–853.

Handy, R. D., Cornelis, G., Fernandes, T., Tsyusko, O., Decho, A., Sabo-Attwood, T., Metcalfe, C., Steevens, J. A., Klaine, S. J., Koelmans, A. A. and Horne, N. (2012). Ecotoxicity test methods for engineered nanomaterials: practical experiences and recommendations from the bench. *Environmental Toxicology and Chemistry* **31**, 15–31.

Handy, R. D., Henry, T. B., Scown, T. M., Johnston, B. D. and Tyler, C. R. (2008b). Manufactured nanoparticles: their uptake and effects on fish—a mechanistic analysis. *Ecotoxicology* **17**, 396–409.

Handy, R., Kammer, F. v., Lead, J., Hassellöv, M., Owen, R. and Crane, M. (2008a). The ecotoxicology and chemistry of manufactured nanoparticles. *Ecotoxicology* **17**, 287–314.

Handy, R. D., van den Brink, N., Chappell, M., Mühling, M., Behra, R., Dušinská, M., Simpson, P., Ahtiainen, J., Jha, A. N., Seiter, J., Bednar, A., Kennedy, A., Fernandes, T. F. and Riediker, M. (2012). Practical considerations for conducting ecotoxicity test methods with manufactured nanomaterials: what have we learnt so far? *Ecotoxicology* **21**, 933–972.

Hasani, M., Cranston, E. D., Westman, G. and Gray, D. G. (2008). Cationic surface functionalization of cellulose nanocrystals. *Soft Matter* **4**, 2238–2244.

Hansen, S. F., Michelson, E. S., Kamper, A., Borling, P., Stuer-Lauridsen, F. and Baun, A. (2008). Categorization framework to aid exposure assessment of nanomaterials in consumer products. *Ecotoxicology* **17**, 438–447.

Hao, L., Wang, Z. and Xing, B. (2009). Effect of sub-acute exposure to TiO_2 nanoparticles on oxidative stress and histopathological changes in Juvenile Carp (*Cyprinus carpio*). *Journal of Environmental Sciences* **21**, 1459–1466.

Hasellöv, M. and von der Kammer, F. (2008). Iron oxides as geochemical nanovectors for metal transport in soil-river systems. *Elements* **4**, 401–406.

Heiden, T. C. K., Dengler, E., Kao, W. J., Heideman, W. and Peterson, R. E. (2007). Developmental toxicity of low generation PAMAM dendrimers in zebrafish. *Toxicology and Applied Pharmacology* **225**, 70–79.

Hellstrand, E., Lynch, I., Andersson, A., Drakenberg, T., Dahlback, B., Dawson, K. A., Linse, S. and Cedervall, T. (2009). Complete high-density lipoproteins in nanoparticle corona. *FEBS Journal* **276**, 3372–3381.

Henry, T. B., Menn, F. M., Fleming, J. T., Wilgus, J., Compton, R. N. and Sayler, G. S. (2007). Attributing effects of aqueous C-60 nano-aggregates to tetrahydrofuran decomposition products in larval zebrafish by assessment of gene expression. *Environmental Health Perspectives* **115**, 1059–1065.

Icard, P., Poulain, L. and Lincet, H. (2012). Understanding the central role of citrate in the metabolism of cancer cells. *Biochimica et Biophysica Acta* **1825**, 111–116.

Igbal, M. A., Md, S., Sahni, J. K., Baboota, S., Dang, S. and Ali, J. (2012). Nanostructured lipid carriers system: recent advances in drug delivery. *Journal of Drug Targeting* **20**, 813–830.

Incani, V., Danumah, C. and Boluk, Y. (2012). Nanocomposites of nanocrystalline cellulose for enzyme immobilization. *Cellulose* **20**, 191–200.

Jovanović, B., Ji, T. and Palić, D. (2011). Gene expression of zebrafish embryos exposed to titanium dioxide nanoparticles and hydroxylated fullerenes. *Ecotoxicology and Environmental Safety* **74**, 1518–1525.

Jovanović, B. and Palić, D. (2012). Immunotoxicology of non-functionalized engineered nanoparticles in aquatic organisms with special emphasis on fish—Review of current knowledge, gap identification, and call for further research. *Aquatic Toxicology* **118–119**, 141–151.

Ju-Nam, Y. and Lead, J. R. (2008). Manufactured nanoparticles: An overview of their chemistry, interactions and potential environmental implications. *Science of the Total Environment* **400**, 396–414.

Kahru, A. and Savolainen, K. (2010). Potential hazard of nanoparticles: From properties to biological and environmental effects. *Toxicology* **269**, 89–91.

Kashiwada, S. (2006). Distribution of nanoparticles in the see-through medaka (*Oryzias latipes*). *Environmental Health Perspectives* **114**, 1697–1702.

Kilin, D. S., Prezhdo, O. V. and Xia, Y. (2008). Shape-controlled synthesis of silver nanoparticles: Ab initio study of preferential surface coordination with citric acid. *Chemical Physics Letters* **458**, 113–116.

Kim, K.-T., Jang, M.-H., Kim, J.-Y., Xing, B., Tanguay, R. L., Lee, B.-G. and Kim, S. D. (2012). Embryonic toxicity changes of organic nanomaterials in the presence of natural organic matter. *Science of the Total Environment* **426**, 423–429.

Kime, D. E. (1995). The effects of pollution on reproduction in fish. *Reviews in Fish Biology and Fisheries* **5** (1), 52–95.

Klaine, S. J., Alvarez, P. J. J., Batley, G. E., Fernandes, T. F., Handy, R. D., Lyon, D. Y., Mahendra, S., McLaughlin, M. J. and Lead, J. R. (2008). Nanomaterials in the environment behavior, fate, bioavailability, and effects. *Environmental Toxicology and Chemistry* **27**, 1825–1851.

Klaper, R., Arndt, D., Setyowati, K., Chen, J. and Goetz, F. (2010). Functionalization impacts the effects of carbon nanotubes on the immune system of rainbow trout, *Oncorhynchus mykiss*. *Aquatic Toxicology* **100**, 211–217.

Kloser, E. and Gray, D. G. (2010). Surface grafting of cellulose nanocrystals with poly(ethylene oxide) in aqueous media. *Langmuir* **26**, 13450–13456.

Knop, K., Hoogenboom, R., Fischer, D. and Schubert, U. S. (2010). Poly(ethylene glycol) in drug delivery: pros and cons as well as potential alternatives. *Angewandte Chemie International Edition* **49**, 6288–6308.

Kovacs, T., Naish, V., O'Connor, B., Blaise, C., Gagné, F., Hall, L., Trudeau, V. and Martel, P. (2010). An ecotoxicological characterization of nanocrystalline cellulose (NCC). *Nanotoxicology* **4**, 255–270.

Kreyling, W. G., Semmler, M., Erbe, F., Mayer, P., Takenaka, S., Schulz, H., Oberdörster, G. and Ziensenis, A. (2002). Translocation of ultrafine insoluble iridium particles from lung epithelium to extrapulmonary organs is size dependent but very low. *Journal of Toxicology and Environmental Health A* **65**, 1513–1530.

Kroll, A., Pillukat, M. H., Hahn, D. and Schnekenburger, J. (2009). Current *in vitro* methods in nanoparticle risk assessment: Limitations and challenges. *European Journal of Pharmaceutics and Biopharmaceutics* **72**, 370–377.

Lam, C. W., James, J. T., McCluskey, R. and Hunter, R. L. (2004). Pulmonary toxicity of single-wall carbon nanotubes in mice 7 and 90 days after intratracheal instillation. *Toxicological Sciences* **77**, 126–134.

Lead, J. R. and Wilkinson, K. J. (2006). Aquatic colloids and nanoparticles: Current knowledge and future trends. *Environmental Chemistry* **3**, 156–171.

Lefebvre, K. A., Noren, D. P., Schultz, I. R., Bogard, S. M., Wilson, J. and Eberhart, B. T. (2007). Uptake, tissue distribution and excretion of domoic acid after oral exposure in coho salmon (*Oncorhynchus kisutch*). *Aquatic Toxicology* **81**, 266–274.

Leung, A. C. W., Hrapovic, S., Lam, E., Liu, Y., Male, K. B., Mahmoud, K. A. and Luong, J. H. T. (2011). Characteristics and properties of carboxylated cellulose nanocrystals prepared from a novel one-step procedure. *Small* **7**, 302–305.

Li, H., Zhou, Q., Wu, Y., Fu, J., Wang, T. and Jiang, G. (2009). Effect of waterborne nano-iron on medaka (*Oryzias latipes*): antioxidant enzymatic activity, lipid peroxidation and histopathology. *Ecotoxicology and Environmental Safety* **72** (3), 684–692.

Ljungberg, N., Bonini, C., Bortolussi, F., Boisson, C., Heux, L. and Cavaillé, J. Y. (2005). New nanocomposite materials reinforced with cellulose whiskers in atactic polypropylene: Effect of surface and dispersion characteristics. *Biomacromolecules* **6**, 2732–2739.

Lövestam, G., Rauscher, H., Roebben, G., Klüttgen, B. S., Gibson, N., Putaud, J.-P. and Stamm, H. (2010). *Considerations on a definition of nanomaterial for regulatory purposes.* Ispra: European Commission: Joint Research Centre (JRC).

McCarthy, J. F. and Jimenez, B. D. (1985). Reduction in bioavailability to bluegills of polycyclic aromatic hydrocarbons bound to dissolved humic material. *Environmental Toxicology and Chemistry* **4**, 511–521.

MacCormack, T. J., Clark, R. J., Dang, M. K. M., Ma, G., Kelly, J. A., Veinot, J. G. C. and Goss, G. G. (2012). Inhibition of enzyme activity by nanomaterials: Potential mechanisms and implications for nanotoxicity testing. *Nanotoxicology* **6**, 514–525.

Mallat, J. (1985). Fish gill structural changes induced by toxicants and other irritants: A statistical review. *Canadian Journal of Fisheries and Aquatic Sciences* **42**, 630–648.

Mangalam, A. P., Simonsen, J. and Benight, A. S. (2009). Cellulose/DNA hybrid nanomaterials. *Biomacromolecules* **10**, 497–504.

Masciangioli, T. and Zhang, W. X. (2003). Environmental technologies at the nanoscale. *Environmental Science and Technology* **37**, 102A–108A.

Mdluli, P. S., Sosibo, N. M., Mashazi, P. N., Nyokong, T., Tshikhudo, R. T., Skepu, A. and van der Lingen, E. (2011). Selective adsorption of PVP on the surface of silver nanoparticles: A molecular dynamics study. *Journal of Molecular Structure* **1004**, 131–137.

Merrifield, D. L., Shaw, B. J., Harper, G. M., Saoud, I. P., Davies, S. J., Handy, R. D. and Henry, T. B. (2013). Ingestion of metal-nanoparticle contaminated food disrupts endogenous microbiota in zebrafish (*Danio rerio*). *Environmental Pollution* **174**, 157–163.

Meyer, J. N., Lord, C. J., Yang, X. Y., Turner, E. A., Badireddy, A. R., Marinakos, S. M., Chilkoti, A., Wiesner, M. R. and Auffan, M. (2010). Intracellular uptake and associated toxicity of silver nanoparticles in *Caenorhabditis elegans*. *Aquatic Toxicology* **100**, 140–150.

Mitrano, D. M., Lesher, E. K., Bednar, A., Monserud, J., Higgins, C. P. and Ranville, J. F. (2012). Detecting nanoparticulate silver using single-particle inductively coupled plasma–mass spectrometry. *Environmental Toxicology and Chemistry* **31**, 115–121.

Monteiro-Riviere, N. A., Inman, A. O. and Zhang, L. W. (2009). Limitations and relative utility of screening assays to assess engineered nanoparticle toxicity in a human cell line. *Toxicology and Applied Pharmacology* **234**, 222–235.

Moore, M. N. (2006). Do nanoparticals present ecotoxicological risk for the health of the aquatic environment? *Environment International* **32**, 967–976.

Mühlfeld, C., Poland, C. A., Duffin, R., Brandenberger, C., Murphy, F. A., Rothen-Rutishauser, B., Gehr, P. and Donaldson, K. (2012). Differential effects of long and short carbon nanotubes on the gas-exchange region of the mouse lung. *Nanotoxicology* **6**, 867–879.

Nallathamby, P. D., Lee, K. J. and Xu, X. N. (2008). Design of stable and uniform single nanoparticle photonics for *in vivo* dynamics imaging of nanoenvironments of zebrafish embryonic fluids. *ACS Nano* **2**, 1371–1380.

Nel, A., Xia, T., Madler, L. and Li, N. (2006). Toxic potential of materials at the nanolevel. *Science* **311**, 622–627.

Nichols, J. W., McKim, J. M., Andersen, M. E., Gargas, M. L., Clewell, H. J., 3rd and Erickson, R. J. (1990). A physiologically based toxicokinetic model for the uptake and disposition of waterborne organic chemicals in fish. *Toxicology and Applied Pharmacology* **106**, 433–447.

Nowack, B. and Bucheli, T. D. (2007). Occurrence, behavior and effects of nanoparticles in the environment. *Environmental Pollution* **150**, 5–22.

Oberdörster, E. (2004). Manufactured nanomaterials (fullerenes, C_{60}) induce oxidative stress in the brain of juvenile largemouth bass. *Environmental Health Perspectives* **112**, 1058–1062.

Oberdörster, E., Zhu, S., Blickley, T. M., McClellan-Green, P. and Haasch, M. L. (2006). Ecotoxicology of carbon-based manufactured nanoparticles: Effects of fullerene (C_{60}) on aquatic organisms. *Carbon* **44**, 1112–1120.

Ong, K.J., MacCormack, T.J., Clark, R.J., Ede, J.D., Ortega, V.A., Felix, L.C., Dang, M.K.M., Ma, G., Fenniri, H., Veinot, J.G.C. and Goss, G.G. (2013a) Widespread nanoparticle-assay interference: Implications for nanotoxicity testing. Unpublished.

Ong, K. J., Zhao, X., Thistle, M. E., MacCormack, T. J., Clark, R. J., Ma, G., Martinez-Rubi, Y., Simard, B., Loo, J. S., Veinot, J. G. and Goss, G. G. (2013b). Mechanistic insights into the effect of nanoparticles on zebrafish hatch. *Nanotoxicology.* doi: 10.3109/17435390.2013.778345 (in press).

Petersen, E. J. and Henry, T. B. (2012). Methodological considerations for testing the ecotoxicity of carbon nanotubes and fullerenes: Review. *Environmental Toxicology and Chemistry* **31**, 60–72.

Peng, B. L., Dhar, N., Liu, H. L. and Tam, K. C. (2011). Chemistry and applications of nanocrystalline cellulose and its derivatives: A nanotechnology perspective. *Canadian Journal of Chemical Engineering* 89, 1191–1206.

Pérez-de-Luque, A. and Rubiales, D. (2008). Nanotechnology for parasitic plant control. *Pest Management Science* 65 (5), 540–545.

Pérez, S., la Farré, M. and Barceló, D. (2009). Analysis, behavior and ecotoxicity of carbon-based nanomaterials in the aquatic environment. *Trends in Analytical Chemistry* 28, 820–832.

Piccino, F., Gottschalk, F., Seeger, S. and Nowack, B. (2012). Industrial production quantities and uses of ten engineered nanomaterials in Europe and the world. *Journal of Nanoparticle Research* 14, 1109–1119.

Poland, C. A., Duffin, R., Kinloch, I., Maynard, A., Wallace, W. A., Seaton, A., Stone, V., Brown, S., Macnee, W. and Donaldson, K. (2008). Carbon nanotubes introduced into the abdominal cavity of mice show asbestos-like pathogenicity in a pilot study. *Nature Nanotechnology* 3, 423–428.

Powers, C. M., Slotkin, T. A., Seidler, F. J., Badireddy, A. R. and Padilla, S. (2011). Silver nanoparticles alter zebrafish development and larval behavior: distinct roles for particle size, coating and composition. *Neurotoxicology and Teratology* 33, 708–714.

Ramsden, C. S., Smith, T. J., Shaw, B. S. and Handy, R. D. (2009). Dietary exposure to titanium dioxide nanoparticles in rainbow trout, (*Oncorhynchus mykiss*): No effect on growth, but subtle biochemical disturbances in the brain. *Ecotoxicology* 18, 939–951.

Rawson, D. M., Zhang, T., Kalicharan, D. and Jongebloed, W. L. (2000). Field emission scanning electron microscopy and transmission electron microscopy studies of the chorion, plasma membrane and syncytial layers of the gastrula-stage embryo of the zebrafish *Brachydanio rerio*: A consideration of the structural and functional relationships with respect to cryoprotectant penetration. *Aquaculture Research* 3, 325–336.

Roco, M. C. (2003). Nanotechnology: Convergence with modern biology and medicine. *Current Opinion in Biotechnology* 14, 337–346.

Romberg, B., Hennink, W. E. and Storm, G. (2008). Sheddable coatings for long-circulating nanoparticles. *Pharmaceutical Research* 25, 55–71.

Römer, I., White, T. A., Baalousha, M., Chipman, K., Viant, M. R. and Lead, J. R. (2011). Aggregation and dispersion of silver nanoparticles in exposure media for aquatic toxicity tests. *Journal of Chromatography A* 1218, 4226–4233.

Ruiz, M. M., Cavaille, J. Y., Dufresne, A., Gerard, J. F. and Graillat, C. (2000). Processing and characterization of new thermoset nanocomposites based on cellular whiskers. *Composite Interfaces* 7, 117–131.

Ryman-Rasmussen, J. P., Riviere, J. E. and Monteiro-Riviere, N. A. (2006). Surface coatings determine cytotoxicity and irritation potential of quantum dot nanoparticles inepidermal keratinocytes. *Journal of Investigative Dermatology* 127, 143–153.

Sanborn, J. R., Childers, W. F. and Hansen, L. G. (1977). Uptake and elimination of [^{14}C]-hexachloro-benzene (HCB) by the green sunfish, *Lepomis cyanellus* Raf., after feeding contaminated food. *Journal of Agriculture and Food Chemistry* 25, 551–553.

SCENIHR (2007). *Opinion on the appropriateness of the risk assessment methodology in accordance with technical guidance documents for new and existing substances for assessing the risks of nanomaterials.* Scientific Committee on Emerging and Newly Identified Health Risks (SCENIHR), European Commission.

Schultz, A., Ong, K. J., MacCormack, T. J., Ma, G., Veinot, J. G. C. and Goss, G. G. (2012). Silver nanoparticles inhibit sodium uptake in juvenile rainbow trout (*Oncorhynchus mykiss*). *Environmental Science and Technology* 46, 10295–10301.

Scown, T. M., van Aerle, R., Johnston, B. D., Cumberland, S., Lead, J. R., Owen, R. and Tyler, C. R. (2009). High doses of intravenously administered titanium dioxide nanoparticles accumulate in the kidneys of rainbow trout but with no observable impairment of renal function. Toxicological Sciences 109, 372–380.

Scown, T. M., van Aerle, R. and Tyler, C. R. (2010). Review: Do engineered nanoparticles pose a significant threat to the aquatic environment? Critical Reviews in Toxicology 40, 653–670.

Shaw, B. J. and Handy, R. D. (2011). Physiological effects of nanoparticles on fish: A comparison of nanometals versus metal ions. Environment International 37, 1083–1097.

Shaw, B. J., Ramsden, C. S., Turner, A. and Handy, R. D. (2013). A simplified method for determining titanium from TiO$_2$ nanoparticles in fish tissue with a concomitant multi-element analysis. Chemosphere. doi: 10.1016/j.chemosphere.2013.01.065 (in press).

Shinohara, N., Matsumoto, T., Gamo, M., Miyauchi, A., Endo, S., Yonezawa, Y. and Nakanishi, J. (2009). Is lipid peroxidation induced by the aqueous suspension of fullerene C$_{60}$ nanoparticles in the brains of Cyprinus carpio? Environmental Science and Technology 43, 948–953.

Smith, C. J., Shaw, B. J. and Handy, R. D. (2007). Toxicity of single walled carbon nanotubes to rainbow trout (Oncorhynchus mykiss): Respiratory toxicity, organ pathologies, and other physiological effects. Aquatic Toxicology 82, 94–109.

Sozer, N. and Kokini, J. L. (2009). Nanotechnology and its applications in the food sector. Trends in Biotechnology 27, 82–89.

Spohn, P., Hirsch, C., Hasler, F., Bruinink, A., Krug, H. F. and Wick, P. (2009). C$_{60}$ fullerene: A powerful antioxidant or a damaging agent? The importance of an in-depth material characterization prior to toxicity assays. Environmental Pollution 157, 1134–1139.

Stolpe, B. and Hassellöv, M. (2007). Changes in size distribution of freshwater nanoscale colloidal matter and associated elements on mixing with seawater. Geochimica et Cosmochimica Acta 71, 3292–3301.

Stone, V., Nowack, B., Baun, A., van den Brink, N., von der Kammer, F., Dusinska, M., Handy, R., Hankin, S., Hassellöv, M., Joner, E. and Fernandes, T. F. (2010). Nanomaterials for environmental studies: Classification, reference material issues, and strategies for physico-chemical characterisation. Science of the Total Environment 408, 1745–1754.

Sumner, S. C. J., Fennell, T. R., Snyder, R. W., Taylor, G. F. and Lewin, A. H. (2010). Distribution of carbon-14 labeled C60 ([^{14}C] C60) in the pregnant and in the lactating dam and the effect of C60 exposure on the biochemical profile of urine. Journal of Applied Toxicology 30, 354–360.

Tejamaya, M., Römer, I., Merrifield, R. C. and Lead, J. R. (2012). Stability of citrate, PVP, and PEG coated silver nanoparticles in ecotoxicology media. Environmental Science and Technology 46, 7011–7017.

Thio, B. J. R., Montes, M. O., Mahmoud, M. A., Lee, D., Zhou, D. and Keller, A. A. (2012). Mobility of capped silver nanoparticles under environmentally relevant conditions. Environmental Science and Technology 46, 6985–6991.

Tierney, K. B., Baldwin, D. H., Hara, T. J., Ross, P. S., Scholz, N. L. and Kennedy, C. J. (2010). Olfactory toxicity in fishes. Aquatic Toxicology 96, 2–26.

Tolaymat, T. M., El Badawy, A. M., Genaidy, A., Scheckel, K. G., Luxton, T. P. and Suidan, M. (2010). An evidence-based environmental perspective of manufactured silver nanoparticle in syntheses and applications: A systematic review and critical appraisal of peer-reviewed scientific papers. Science of the Total Environment 408, 999–1006.

Usenko, C. Y., Harper, S. L. and Tanguay, R. L. (2007). In vivo evaluation of carbon fullerene toxicity using embryonic zebrafish. Carbon 45, 1891–1898.

Usenko, C. Y., Harper, S. L. and Tanguay, R. L. (2008). Fullerene C60 exposure elicits an oxidative stress response in embryonic zebrafish. *Toxicology and Applied Pharmacology* **229**, 44–55.

Veith, G. D., DeFoe, D. L. and Bergstedt, B. V. (1979). Measuring and estimating the bioconcentration factor of chemicals in fish. *Journal of the Fisheries Research Board of Canada* **36**, 1040–1048.

Verwey, E. J. W., Overbeek, J. Th. G. and van Nes, K. (1948). *Theory of the Stability of Lyophobic Colloids: the Interaction of Sol Particles Having an Electric Double Layer*. New York: Elsevier Publishing.

von der Kammer, F., Ferguson, P. L., Holden, P. A., Masion, A., Rogers, K. R., Klaine, S. J., Koelmans, A. A., Horne, N. and Unrine, J. M. (2012). Analysis of engineered nanomaterials in complex matrices (environment and biota): General considerations and conceptual case studies. *Environmental Toxicology and Chemistry* **31**, 32–49.

Warheit, D. B., Webb, T. R., Sayes, C. M., Colvin, V. L. and Reed, K. L. (2006). Pulmonary instillation studies with nanoscale TiO_2 rods and dots in rats: toxicity is not dependent upon particle size and surface area. *Toxicological Sciences* **91**, 227–236.

Weinstein, R. (2001). Hypocalcemic toxicity and atypical reactions in therapeutic plasma exchange. *Journal of Clinical Apheresis* **16**, 210–211.

Werlin, R., Priester, J. H., Mielke, R. E., Kramer, S., Jackson, S., Stoimenov, P. K., Stucky, G. D., Cherr, G. N., Orias, E. and Holden, P. A. (2011). Biomagnification of cadmium selenide quantum dots in a simple experimental microbial food chain. *Nature Nanotechnology* **6**, 65–71.

Wild, E. and Jones, K. C. (2009). Novel method for the direct visualization of *in vivo* nanomaterials and chemical interactions in plants. *Environmental Science and Technology* **43**, 5290–5294.

Windeatt, K. M. and Handy, R. D. (2012). Effect of nanomaterials on the compound action potential of the shore crab, *Carcinus maenas*. *Nanotoxicology*. doi:10.3109/17435390.2012.663809 (in press).

Wright, P. A. and Wood, C. M. (2009). A new paradigm for ammonia excretion in aquatic animals: role of Rhesus (Rh) glycoproteins. *Journal of Experimental Biology* **212**, 2303–2312.

Wohlleben, W., Kolle, S. N., Hasenkamp, L.-C., Böser, A., Vogel, S., von Vacano, B., van Ravenzwaay, B. and Landsiedel, R. (2011). Artifacts by marker enzyme adsorption on nanomaterials in cytotoxicity assays with tissue cultures. *Journal of Physics: Conference Series* **304**, 012061.

Worle-Knirsch, J. M., Pulskamp, K. and Krug, H. F. (2006). Oops they did it again! Carbon nanotubes hoax scientists in viability assays. *Nano Letters* **6**, 1261–1268.

Xiao, Z. Z., Zhang, L. L., Xu, X. Z., Liu, Q. H., Li, J., Ma, D. Y., Xu, S. H., Xue, Y. P. and Xue, Q. Z. (2008). Effect of cryoprotectants on hatching rate of red seabream (*Pagrus major*) embryos. *Theriogenology* **70**, 1086–1092.

Yi, J., Xu, Q., Zhang, X. and Zhang, H. (2008). Chiral-nematic self-ordering of rodlike cellulose nanocrystals grafted with poly(styrene) in both thermotropic and lyotropic states. *Polymer* **49**, 4406–4412.

Yu, X., Du, Q., Zhu, P. F., Hu, D. D. and Yang, L. (2011). Study on the photocatalytic degradation of diesel pollutants in seawater by a stannum- doped nanometer titania. *Advanced Materials Research* **197/198**, 780–785.

Zhang, L. W. and Monteiro-Riviere, N. A. (2009). Mechanisms of quantum dot nanoparticle cellular uptake. *Toxicological Sciences* **110**, 138–155.

Zhu, S. Q., Oberdorster, E. and Haasch, M. L. (2006). Toxicity of an manufactured nanoparticle (fullerene, C-60) in two aquatic species, *Daphnia* and fathead minnow. *Marine Environmental Research* **62**, S5–S9.

Zhu, X., Zhu, L., Li, Y., Duan, Z., Chen, W. and Alvarez, P. J. (2007). Developmental toxicity in zebrafish (*Danio rerio*) embryos after exposure to manufactured nanomaterials: buckminsterfullerene aggregates (nC$_{60}$) and fullerol. *Environmental Toxicology and Chemistry* **26**, 976–979.

10

HANDLING FISH MIXTURE EXPOSURES IN RISK ASSESSMENT

DICK DE ZWART
LEO POSTHUMA

Organic Chemical Toxicology of Fishes: Volume 33
FISH PHYSIOLOGY

1. INTRODUCTION

Handling the organic toxicology of fish should encompass handling mixtures, as is evident from large-scale field inventories. Body residues of contaminants in captured wild fish are present, concern multiple compounds, and are highly variable in space and time and across species. Environmental policies generally aim to prevent or reduce exposures and impacts, and thus should also encompass mixtures. Environmental policies thereby apply the principles of risk assessment and subsequent risk management when needed. Risk assessment is a procedure that aims to characterize risks of contaminant exposures. Risk assessment starts with problem definition, then moves to exposure assessment, and lastly provides the effect that is to be expected from concentration-effect relationships.

In this chapter, single- and mixture-compound and risk assessment for the aquatic compartment are described in their various formats. These range from deterministic procedures that show whether the emissions of a compound can be considered safe in all realistic circumstances, commonly considered as conservative methods, to higher-tier and more refined methods that provide a quantitative estimate of the magnitude of expected impacts. Mixture risk assessment methods have been designed to address the issue of mixture exposures. In addition to methods for assessment of identified compounds, there are also methods to predict the cumulative impact of mixtures of unknown composition. Finally, there are eco-epidemiological methods, which are used to diagnose whether and to what extent mixture impacts natural species assemblages. Although data on mixture effects, including some for fish, have been collected for a few decades, basic approaches for mixture risk assessment have only recently been adopted for formal policies. There is latitude to further address mixture exposures in fish and other biota in relation to environmental policies in the future. It is concluded that mixture exposures occur, that mixture risk and impact methods exist, and that these methods unveil mixture issues in the natural environment. It is also concluded that the methods yield useful results for planning risk reduction strategies, though as yet these strategies have not necessarily been formulated specifically for fish.

1.1. (Policy) Problem Definition

Worldwide, the legislation on environmental chemical exposures is predominantly based on assessments carried out using individual substances. Protection from adverse chemical exposures in surface waters has long taken the shape of single-compound aquatic life criteria (water quality criteria, or WQC). These criteria help prevent the introduction and use of new, highly hazardous compounds and reduce the emissions of existing compounds, while also furthering the restoration of damaged ecosystems where those criteria were found to have been exceeded. As an easy day-to-day applicable approach, predicted or measured ambient concentrations of a compound are compared to a defined effect limit value (e.g., a maximum permissible concentration), which is a concentration chosen so that aquatic life is assumed to be fully or sufficiently protected. The judgment boils down then to calculating a simple ratio, commonly called the risk quotient (RQ), risk characterization ratio (RCR), or a hazard index (HI), between exposure concentration and limit concentration. An RCR value higher than 1 signals a risk, which then triggers preventive or remediative action.

The water quality criteria are often derived from single-species, single-compound tests, which not only involve fish tests but also tests with other species. In various regulatory contexts, criteria can be derived when data are collected on algae, daphnids, and fish, as a minimum data requirement for regulatory risk assessment. This common procedure neglects the fact that exposure to diverse mixtures of toxicants is the rule rather than the exception in the environment. Since animals and ecosystems are exposed to a wide variety of substances, there is increasing concern about the potential adverse effects due to the cumulative effects of all substances present simultaneously. Thereby, there is a large spatiotemporal variability in exposures to mixtures. The policy problem that arises is how to handle mixtures in regulatory risk assessment, given scientific evidence on the one hand and single-compound practices on the other?

1.2. Toxicological State of the Art

Only recently has consensus grown on the question of how to handle mixture issues in human health policies on contaminants. Theoretically, a suite of issues need be considered; for example, which compounds really do occur together? Do they have similar or dissimilar modes of action? That is, do compounds A and B toxicologically affect the same molecular target or different ones? Do they interact? That is, does the presence of compound A influence the environmental behavior of compound B, or its uptake, or its toxicodynamics? Based on an analysis of the available scientific literature, the nonfood Scientific Committees of the European Commission reached the following conclusions regarding chemical mixture exposure and risk assessment for humans (see EC 2012):

1. Under certain conditions, chemicals will act jointly in a way that the overall level of toxicity is affected.

DICK DE ZWART AND LEO POSTHUMA

2. Chemicals with similar modes of action (MoA, e.g., two compounds affecting the nervous system in the same way) will act jointly to produce combinatorial effects that are larger than the effects of each mixture component individually. These effects can be described by dose/concentration addition.
3. For chemicals with different MoA (independently acting), no robust evidence is available that exposure to a mixture of such substances is of health or environmental concern if the individual chemicals are present at or below their zero-effect levels.
4. Interactions (including antagonism, potentiation, and synergies) usually occur at medium- or high-dose levels (relative to the lowest effect levels). At low-exposure levels, they are either unlikely to occur or are toxicologically insignificant.
5. In view of the almost infinite number of possible combinations of chemicals to which species are exposed, some form of a screen is necessary to focus on mixtures of potential concern.
6. With regard to the assessment of chemical mixtures, a major knowledge gap is the lack of exposure information and the rather limited number of chemicals for which there is sufficient MoA information. Currently, there is neither an agreed MoA inventory nor a defined set of criteria of how to characterize or predict an MoA for data-poor chemicals.
7. If no MoA information is available, the dose-concentration addition method should be preferred to the independent action approach. Prediction of possible interaction requires expert judgment and hence needs to be considered on a case-by-case basis.

Toxicologically, mixtures matter (1), modes of actions matter (2 and 3), and interactions among compounds may matter, especially at higher exposures (4). Practically, focus is relevant (5), addressing mixtures is still complex, (6) but a conservative stance is advised when data are lacking (assume concentration additivity) (7). There is no reason to believe that the response patterns of fish exposed to mixtures of chemicals would in this toxicological way deviate from the overall picture generated in the preceding seven points.

1.3. Ecotoxicological Issues

Environmental mixture issues partly differ from toxicological issues. Current mixture toxicity approaches for the environment should account for the following interaction options.

1. Chemical interactions between the mixture constituents—*reactions*

2. Environmental interactions between mixture constituents and the exposure matrix—*reactions and partitioning*
3. Toxicological interactions between mixture constituents and the receptors in the exposed organism's tissues—*selectivity and receptor saturation*
4. Ecological interactions between the mixture toxicity effects and other types of stress—*multiple stressors (e.g., climatic change)*
5. Ecological interactions between exposed organisms—*indirect effects*

In many ecotoxicological experiments and observations, there is no systematical approach to deal with the various levels of interactions that can be distinguished in the context of a contaminated environment. Misinterpretation of experimental data of ecotoxicological mixture studies may result from disregarding the interactions that were mentioned. For example, when the net impact of a mixture of toxicants in an experiment is lower than expected from a current mixture model, it can be characterized as a case of "antagonism" (toxicological) when in fact it may be something else (e.g., altered bioavailability). A good approach in handling mixtures should account for all levels of abiotic interaction, toxicological principles, as well as biological interactions in species assemblages.

1.4. Aims and Readers Guide

This chapter aims to

show whether and how fish are exposed to mixtures.
show whether and how mixture responses differ from single-compound responses.
provide an overview of scientific principles regarding single-compound and mixture impact assessments.
put these in a risk assessment and management context.
provide evidence for mixture and multistress impacts.
describe pathways to advances in exposure, effect, and risk assessment and management of mixtures

2. FISH AND MIXTURES

The relevance of mixture assessment for fish specifically requires attention for two key aspects: Does exposure to mixtures occur? And, if so, is there evidence that the aforementioned toxicological conclusions apply to fish too? This section summarizes the key evidence on field mixture exposures of fish and fish mixture studies.

2.1. Fish Mixture Exposures in the Field

There is ample evidence for the exposure of fish in natural water bodies to suites of mixtures of chemical compounds. A comprehensive recent example is the United States National Lake Fish Tissue Study (U.S. EPA, 2009), which may represent the variability in space and time present in current-day waters for a suite of compounds and across a large range of water bodies.

The study investigated body residues of the largest known subselection of chemical compounds in fish tissue. The measured compounds were selected because they had detailed information available for interpretation, they were known to accumulate, and they were identified as important in one or more of the EPA study programs. The list of monitored compounds consisted of 268 chemicals, including mercury, five forms of arsenic, 17 dioxins and furans, 159 PCB congeners, 46 pesticides, and 40 semivolatile organic compounds. Five hundred stations were sampled in a statistically valid way, so that the findings could be extrapolated to nearly 80,000 lakes for predatory fish species and more than 45,000 lakes for bottom-dwelling species. The key conclusions, cited here, clearly indicate that mixture exposures of fish are common in our present-day landscapes.

Results from the National Lake Fish Tissue Study indicate that mercury, PCBs, dioxins and furans are widely distributed in lakes and reservoirs in the lower 48 states. Mercury and PCBs were detected in all the fish samples collected from the 500 sampling sites. Dioxins and furans were detected in 81% of the predator samples (fillet composites) and 99% of the bottom-dweller samples (whole-fish composites). In contrast, there were a number of chemicals that were not detected in any of the fish samples collected during the study. Forty-three of the 268 target chemicals were not detected in any samples, including all nine organophosphate pesticides (e.g., chlorpyriphos and diazinon), one PCB congener (PCB-161), and 16 of the 17 polycyclic aromatic hydrocarbons (PAHs) analyzed as semi-volatile organic chemicals. There were also seventeen other semi-volatile organic chemicals that were not detected.

Although the cited report clearly shows multiple exposures of fish by body residue data, the evidence for mixture exposure is still not complete. The study may underestimate true exposures, since a multitude of chemicals and possible situations of exposure may be underrepresented.

Nonpersistent, nonaccumulating compounds may be absent in the screening, but exposure to them may still affect fish assemblages. Local factors may temporarily reduce the bioavailability of the compounds measured, so that false negatives may be present in the inventory. The situation elsewhere may also differ from that in U.S. lakes. Many water bodies may be cleaner due to less intense land use, but they also may be more heavily contaminated due to more intense land use, specific industrial sources, high population densities, and less rigorous regulatory activities than in the United States. Nonetheless, this comprehensive study suggests that mixture exposures of fish may widely occur and vary.

2.2. Laboratory Fish Mixture Effect Studies

That fish appear to be exposed to mixtures in the field (Section 2.1) triggers the question of what this means in terms of toxicological impacts. Can impacts be cumulative, that is, are the effects of the mixture always larger than the effects of the most toxic compound in the mixture? If only the most toxic compound would cause the effect (e.g., mortality), then the other compounds would not matter (a fish can die only once). The concentration-effect curve of the mixture would be similar to the concentration-effect curve of the most toxic compound in the mixture. Or if it is different, do various compounds together cause impacts that are higher than those of the most toxic compound? If so, is this cumulative response predictable by certain rules, which can be derived from toxicological principles, like (dis)similarity in modes of action?

Many laboratory mixture effect studies have been carried out. The studies have been done over a long time period and pertain in part to mixture studies with fish. Investigations have been done on whether and how mixture impacts can be understood. A recent overview of such mixture studies was presented by Kortenkamp et al. (2009), but various earlier summary reports have been published—for example, EIFAC (1987). Early studies done by Könemann (1980, 1981) clearly showed that mixture effects were not similar to the effects of the most toxic compound. Based on LC50-tests with guppies (*Poecilia reticulata*) exposed to various mixtures of organic compounds, Könemann concluded that "in our experiments the toxicity of these mixtures [of chemicals with a simple similar action] follows the concentration addition procedure, which makes no-effect levels for separate chemical inapplicable for mixtures." That is, chemicals that affect similar molecular target sites in the organism act together in determining the net, aggregated mixture effect. The concept of no-effect level for a compound, which is a key concept for risk assessment of chemicals, is thus taken not to be a constant, due to mixtures. Already by the early 1980s, clear conclusions were drawn on the necessity of handling mixtures in risk assessment and on the complexity of doing so (see Sections 3, 4 and 5).

A suite of studies has been made since, with mixtures of chemicals with assumed simple similar actions and with mixtures with assumed independent joint actions, both for mixtures of metals, of organic compounds, and of mixtures containing both compound groups. Recent overview studies showed, that the predicted impact of mixture exposure often resembles a concentration-additive type of response, independent of the combination of modes of action. That is, the observed mixture responses differ by a factor of often less than two from the predicted responses under the assumption of simple similar action (concentration addition), though the real responses appeared to vary widely among studies. In the review presented by Coors and Frische (2011), for example, distributions of mixture responses across a large number of mixture studies were described and analyzed. The median of the mixture responses across a range of studies was near the one expected from concentration addition, while the distribution of responses around this median was wide and bell-shaped. That is, there are nearly equal numbers of under- and overpredictions of mixture impacts, given the concentration–addition situation as null-model. These authors reflected on this variance across mixture studies and suggested that the exact prediction of mixture impacts in individual cases is difficult: On average, a mixture response is well predicted via concentration addition modeling (frequently within a factor of two), but there is a large uncertainty regarding the effects to be expected for individual mixtures.

2.3. Implications for Handling Mixtures

Fish in the field are exposed to mixtures (Section 2.1), and mixtures matter in determining an aggregate response (Section 2.2). Chemical management policies and water quality management policies should therefore consider mixtures. This topic is covered in the following section, which is introduced by an overview of risk assessment principles for single compounds.

3. PRINCIPLES OF RISK AND IMPACT ASSESSMENT
OF CHEMICALS AND MIXTURES

3.1. Risk Assessment and Management

Regulations concerning chemical compounds in the environment are often based on risk assessments. Risk is a combination of the probability and magnitude of exposure and severity of the hazard (toxic potency of a chemical). Risk assessment is the technical support for decision making with uncertainty (Suter, Barnthouse et al., 1993).

Risk assessments are often used to underpin risk management actions, either focusing on preventing or limiting emissions of a compound (preventive chemical management policies), or focusing on environmental (water) quality restoration. Broadly adopted guidelines on procedures providing the technical support for ecological risk assessment have been formulated over the last decennia, but—by virtue of the problem of risk itself—some uncertainties often remain unsolved at the moment a decision is taken.

This section describes how scientific facts are used in contaminant policies and environmental management policies, starting from the broad principles and single-compound assessment, progressing toward mixture exposures, and finally attaining multiple stress conditions. When possible, specific attention is paid to aquatic exposures and specifically fish. Note, however, that, while fish mixture studies exist, the chapter describes the position in which handling mixture data in risk assessment has evolved only very recently.

3.2. General Principles of Risk Assessment and Management

This section describes general features of risk assessment and management, as needed, to understand the current state of art in mixture risk assessment.

3.2.1. PRINCIPLES, TIERING, AND REFINEMENT

Ecological risk assessment (ERA) proceeds according to adopted principles (Figure 10.1, left; Suter et al., 1993) and in clear steps (hazard definition, exposure assessment, effect assessment, risk characterization, risk management). The steps ask for a refined assessment loop when the risk assessment does not deliver final results considered sufficient for decision making. Loops are made by refining the hazard definition (via, e.g., more specific information on the environment and more data on exposure and/or effects), until the results are sufficient for risk management. Given this commonly adopted process, Solomon et al. (2008) explored and sketched the consequences of this so-called tiered approach (Figure 10.1, right; see Solomon et al., 2008). Tiering implies simple risk assessment (exposure and effects assessment models) when possible, but more complex and more precise assessment when needed for more complex hazard definitions and/or costly remediation activities. The latter would not likely be undertaken when costs are high, and the signal is highly uncertain. The Solomon et al.

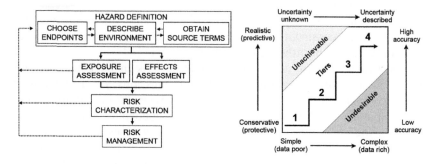

Fig. 10.1. Ecological risk assessment (ERA) is a pragmatic science-policy approach. Left (Suter, Barnthouse et al., 1993): concept of single-stressor (compound) risk assessment and management, including the loop that implies refined and tiered assessment approaches. Right (Solomon, Brock et al., 2008): the implications of this loop in the format of tiering.

approach shows that risk assessment processes have trade-offs: It is unachievable to be precise at low investment in the assessment, and it is undesirable to do costly assessments of low accuracy.

3.2.2. STEPS

The steps in ecological risk assessment are hazard definition, exposure assessment, effect assessment, risk characterization, and risk management. These steps have been mainly applied to single compounds.

Hazard identification is the trigger for a risk assessment. This step considers the inherent toxicity of a chemical or a mixture. As shown in Section 2, there is ample evidence that fish are exposed to multiple chemicals in natural water bodies, most of which are considered intrinsically hazardous. The general hazard identification for wild fish is that fish are exposed not only to chemical mixtures, but also simultaneously to other stressors. This warrants attention not only for mixtures but also for other stressors (e.g., temperature, low dissolved oxygen).

Exposure assessment is based on either measured environmental concentrations or predicted environmental concentration (PEC) modeling. PEC modeling can be done according to the European Union Technical Guidance Document (EU-TGD, EC, 2003).

The measured environmental concentrations generally only concern total concentrations, though these values are sometimes corrected for bioavailability. The bioavailability differences of a compound among water bodies will exist due to processes such as chemical sorption to the matrix or due to the effect of water chemistry on metal speciation. Bioavailability correction can be carried out by applying formulas in which the relationship between total concentration and the portion available for uptake is described as an empirical function of environmental variables, such as pH and organic matter (De Zwart et al., 2008). Mechanistic modeling may also be applied, like biotic ligand models (BLM, Niyogia and Wood 2004).

The first step in the calculation of a predicted environmental concentration is to estimate the substance's release rate based on its use pattern. All potential emission sources need to be analyzed, and the releases and the receiving environmental compartment(s) identified. After assessing releases, the fate of the substance once released to the environment must be considered. This is estimated by considering likely routes of exposure and biotic and abiotic transformation processes and bioavailability. The quantification of emission, distribution, and degradation of the substance as a function of time and space leads to estimates of local and regional concentration values. Like measured concentrations, a predicted total concentration can also be refined using models for the bioavailability correction.

Effects assessment considers the concentration-effect relationship between a contaminant's concentration and some kind of adverse effect. Usually, adverse effects are chosen such that they are of proven ecological relevance, or they are considered as such. Effects assessments are usually based on total concentrations, though sometimes the bioavailability of the toxicant is considered in the assessment (see Section 3.5.3). It is a common practice to derive certain ecotoxicity end-points from concentration–effect tests executed under laboratory conditions, such as a no observed effect concentration (NOEC) or a median effect concentration (EC50). According to adopted principles, a set of such data for different species can be used to derive a predicted no effect concentration (PNEC) (see the discussion of risk characterization below).

The set of ecotoxicity (effects assessment) data for aquatic organisms is not restricted to fish. However, for aquatic toxicity evaluation, OECD directs testing of at least one species each of algae, crustaceans (often *Daphnia*), and fish, therefore including the autotrophs, herbivores, and carnivores (OECD, 1981). Many chemicals have been tested with a much larger array of species. This is, in part, due to the desire of ecotoxicologists to identify and protect the "most sensitive species," as well as the wish to test and protect species of local or regional importance. In principle, the PNEC is determined by applying an assessment factor to available toxicity data. The PNEC is then calculated by dividing the lowest median lethal concentration (LC50), median effective concentration (EC50), or no observed effect concentration (NOEC) value by an appropriate assessment factor in accordance with adopted guidance, such as the EU-TGD (EC, 2003). The assessment factor is applied to extrapolate from laboratory single-species toxicity test data to multispecies ecosystem effects. The assessment factor is used to address a number of uncertainties: (1) interspecies variation in susceptibility; (2) short-term to long-term toxicity extrapolation, and (3) laboratory data to field impact extrapolation, so that protective water quality criteria are indeed protective for aquatic life.

Risk characterization often involves the calculation of the PEC/PNEC ratio, also called the risk characterization ratio (RCR and the indices of similar kind). When the value of this ratio exceeds unity, we can come to the scientific conclusion that exposure is potentially larger than the predicted no-effect concentration, so that the regulatory conclusion follows that there is a potential for impacts. A value lower than unity is interpreted as a zero or negligible probability of the occurrence of any effect in an exposed assemblage of species. In short, the ecological community is considered to be sufficiently protected when PEC remains below PNEC (RCR < 1).

Risk characterization may also be undertaken with higher-tier methods (Figure 10.1, right) when needed. This involves a quantitative estimate of risk (probability and magnitude of impact) rather than a discrimination between "protected" (RCR < 1) and "unprotected" (RCR > 1), as discussed in the next section.

3.3. Quantitative Single-Compound Toxic Pressure Assessment

3.3.1. General Principle

The effect assessment performed with assessment factors is the most often used approach in chemical risk assessment, but—according to the principle of tiering—this approach can be refined by a statistical extrapolation method. This is called the species sensitivity distribution (SSD) approach, and it can be applied when the database of ecotoxicity test data is sufficient for its application (Posthuma, Traas et al., 2002; Posthuma and Suter, 2011). This model is used in two ways: to derive water quality criteria and to estimate the fraction of species affected. Both formats give opportunities to address mixture issues (Section 4).

If a large dataset from long-term tests for different taxonomic groups is available, statistical extrapolation methods may be used to derive the aforementioned PNEC (OECD, 1992), or a quantitative estimate of risk (see below). There is no scientific problem when these methods are applied to ecotoxicity data for fish only, though it is common in single-chemicals regulations to follow established guidance documents that consider multiple species groups. The main underlying assumptions of the statistical extrapolation methods are as follows:

- The distribution of species sensitivities follows a theoretical distribution function.

- The group of species tested in the laboratory is a random sample of this distribution.

In the framework of deriving a PNEC, guidance documentation (EC 2003) prescribes a minimum dataset. For example, in order to derive a valid SSD in the EC-context, chronic NOEC values are needed for a minimum of 10 species selected from at least eight different taxonomic groups (e.g., fish, insect larvae, crustaceans, etc.) (EC, 2003).

In general, the method works as follows (Figure 10.2, left graph). The ecotoxicity data are collected from literature or existing databases, log transformed and fitted according to the chosen distribution function and a prescribed percentile is used as criterion. To determine the PNEC, generally the 5th percentile of this model is chosen to derive the hazardous concentration for 5% of the exposed species (HC5-NOEC); this value is then adopted as PNEC, unless there are reasons to take another value (Van Straalen and Denneman, 1989). Aldenberg and Slob (1991) refined the way to estimate the uncertainty of the 5th percentile by introducing the lower 5% confidence margin of the HC5 as the proposed PNEC. Several distribution functions have been proposed. The U.S. EPA (Stephan, Mount et al., 1985) assumes a log-triangular function, Kooijman (1987) and Van Straalen and Denneman (1989) assume a log-logistic function, and Wagner and Løkke (1991) assume a log-normal function.

Van Straalen and Denneman (1989), and various authors such as Hamers et al. (1996) and Posthuma and Suter (2011), have proposed to use the SSD distribution curve in an inverse way. This is done to infer which

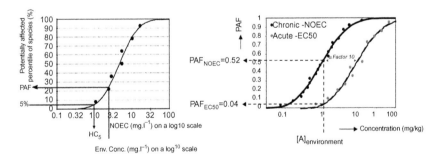

Figure 10.2. Left: The dual use of ecotoxicity data and species sensitivity distributions (SSDs) for single compounds (1) to derive the hazard quotient at 5% (HC5) as a policy-defined maximum allowable concentration, and (2) to derive the potentially affected fraction of species (PAF) from the ambient exposure concentration. Right: Various SSDs can be derived for a single compound: (1) from a set of chronic NOEC data, or (2) from a set of acute EC50 data. The resulting SSDs are shifted relative to each other and will result in two estimates of PAF (a higher PAF-NOEC and a lower PAF-EC50).

fraction of species is exposed above the NOEC at any given ambient concentration (Figure 10.2, left graph). This so-called toxic pressure is expressed as the potentially affected fraction (PAF) of species. The toxic pressure of an environmental sample is interpreted as the ecological risk of a chemical to the local ecosystem and quantifies the fraction of species, which would be affected in the given sample given the samples' contaminant load. For example, when the toxic pressure of sample B is higher than that of sample A, then the fraction of test species that will be affected in B will likely be higher in B than in A. By virtue of assuming a similarity between the sensitivities of tested species and the sensitivities of all species in the contaminated system, the fraction of field species that will be affected in B will also be higher than in A.

3.3.2. CHRONIC AND ACUTE ASSESSMENTS

The procedures described above can also be used with L(E)C50 data or other toxicity end-points (Figure 10.2, right graph). For acute ecotoxicity data, this would then yield the acute toxic pressure. For derivation of conservative water quality criteria values, the legally binding choice for toxicity data to construct SSDs is the NOEC after chronic exposure (Scott-Fordsmand and Jensen, 2002; Sijm, Van Wezel et al., 2002; Stephan, 2002; EC, 2003). For various risk assessment problems, the use of possibly other test end-points is advocated (Solomon and Takacs, 2002; Traas, Van de Meent et al., 2002; Warren-Hicks, Parkhurst et al., 2002; Posthuma and Suter, 2011). There might even be preference for acute data over chronic data because of higher data availability (see, e.g., De Zwart, 2002; De Zwart and Posthuma, 2005) and a possibly higher resolution capacity in case of comparative risk assessment at highly contaminated sites. For assessment of contaminated sites, the issue is "select the data so that they fit the problem": At very high exposures, it is wise to consider acute SSDs. NOEC data are relatively scarce as compared to acute data. Therefore, if chronic data are used, the resulting SSDs may be less precise in estimating the sensitivity distribution of natural species assemblages (Forbes and Calow, 2002).

3.3.3. CONSIDERING MODES OF ACTION AND SPECIFIC SPECIES GROUPS

For the derivation of water quality criteria, the SSDs should be based on the toxicity data of as many species as possible from different taxonomical groups, the data of which are then lumped into a single SSD model.

For the determination of any local toxic pressure, a focus on the problem definition and pragmatic considerations should be considered to determine which part of the dataset is of interest, and whether to model the dataset as a whole or subsets of the data. For example, a bimodal curve as presented in Figure 10.3 (top right panel) (Posthuma, Traas et al., 2002) could be a

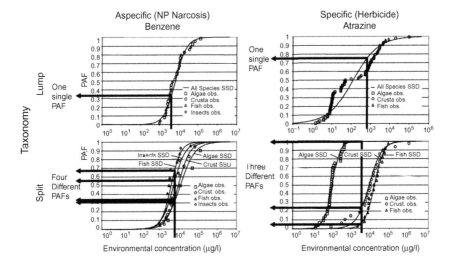

Figure 10.3. An illustration of the assignment of one SSD to all compiled aquatic toxicity data for single compounds (top panels) versus the assignment of various SSDs to subselections of data (lower panels) for a compound with nonspecific and specific toxic modes of action, respectively (benzene, left panels; atrazine, right panels) (Posthuma, Traas et al., 2002).

pragmatically good choice to derive a robust estimate of the HC5. However, for a different assessment with the same dataset, such as a risk assessment for a contaminated environment, it can be a poor choice, as demonstrated in Figure 10.3, lower right curve. It is also possible to generate SSD curves for one compound, but separately for different taxonomical groups, and thus also separately for fish. This approach may provide information on the most vulnerable groups of organisms, which may give a clue on potential indirect ecological effects as shown in Table 10.1. If, for example, algae and cyanobacteria were the only taxa experiencing a high risk of extinction, then it is not difficult to imagine the ecological consequences for a range of taxa operating on higher trophic levels, even without formal food web modeling. Such reasoning is impossible when a single risk estimate is produced from one overall SSD for each compound. As a rule of thumb, tailoring the SSD to the assessment problem would imply a consideration of known modes of action of the compounds, which might reveal a specific lack of fit of the SSD model for some groups of species. For example, a neurotoxin will have a specific impact on the subset of fish (all relatively sensitive), while the effect on species without a nervous system will be completely different (probably much less sensitive). If this happens, it is unwise to discard this knowledge in calculating toxic pressures.

Table 10.1
Ecological risk assessment for a hypothetical freshwater pond with a realistic exposure to a mixture of toxicants with different modes of action (MoA). The values in bold identify the cases where the probability of effects exceeds 5% (De Zwart and Posthuma, 2005)

Compounds	Benzene and Naphthalene	Zinc	Cadmium	Atrazine	Malathion
MoA	NP Narcosis	Zn action	Cd action	Photos. inh.	AChE inh.
Algae	0.010	0.013	**0.058**	**0.143**	ND
Cyanobacteria	ND	ND	ND	**0.381**	ND
Crustaceans	0.005	**0.062**	0.014	0.001	**0.283**
Insect (larvae)	0.010	0.026	0.000	0.000	0.009
Mollusks	0.006	**0.104**	0.007	0.000	0.000
Worms/leeches	0.003	0.012	0.001	ND	0.000
Amphibia	0.004	ND	0.000	0.000	0.000
Fish	0.007	0.045	0.017	0.000	0.000

NP Narcosis = non-polar narcosis; Photos. inh. = Photosynthesis inhibitor; AChE inh. = Acetyl cholinesterase inhibitor; ND = No toxicity data available.

3.4. Validation of Risk Assessment Approaches

Risk assessments are used in environmental management and thus need to be validated. Where probabilistic toxic effect predictions are matched with biological field observations, it has been noted that effects in the field are rarely observed at concentrations equivalent to lower centiles from chronic NOEC distributions such as the aforementioned PNEC-level, for example, the HC5 derived from an NOEC-SSD (Giddings, Solomon et al., 2001; Van den Brink, Blake et al., 2006; Mebane, 2010). This suggests that, in general, the lower-tier method results in water quality criteria for single chemicals that are sufficiently protective for aquatic life.

As discussed earlier, risk assessments can proceed according to subsequent tiers (see Figure 10.1). Apparently, higher-tier approaches may be developed and executed for both theoretical and practical reasons. There are several theoretical reasons why higher-tier probabilistic ecological risk assessment (PERA) derived from SSD information can conceptually be preferred over lower tier RCR (HQ, HI) point estimates of risk (Solomon and Takacs, 2002).

1. There is a consensus that the worst-case scenarios applied to the HQ approach in combination with safety factors for either or both exposure and effects assessment may overestimate exposure and effects and thus perceived risk.

2. The use of the HQ approach in comparative studies assumes a linear relationship between concentration and effects, while in reality the real response pattern is sigmoidal, as in an SSD (Figure 10.2).
3. Direct comparison of HQs for different compounds incorrectly assumes that the concentration range (SSD slope) with partial responses is similar.
4. Probabilistic methods give more usable information to environmental managers and regulators by providing an educated estimate on the order of effective restoration measures.
5. Adopting more realistic and less conservative criteria derived from the probabilistic approach will reduce the societal cost of implementation and risk mitigating measures.

In addition, RCR methods generate data that have no upper boundary, while the SSD-based methods do. In the case of the SSD methods, the maximum response relates to a toxic pressure on 100% of the species.

The major advantage of PERA is that it uses all relevant single species toxicity data and, when combined with exposure distributions, allows for the derivation of a quantitative estimate of risk. In addition, the data may be revisited and the decision criteria become more robust with additional data, while the method is fully reproducible. The quantitative probabilistic method also has some disadvantages. More data are usually needed, although these are mostly from low-cost studies. Also, it is not easily applied to highly bioaccumulative substances where exposure occurs via the food chain as well as the environmental matrix.

For the inverse use of SSDs, these arguments are important, since they suggest a second level of validation, namely: Do field responses in contaminated water systems increase when toxic pressure increases? This validation is more complex, since the field responses are determined via mixtures (see Section 2.1) and other stressors. The distinction of exposure effects from control (multiple-stress) variation in field studies is often possible only given very distinct population responses, which can best be described with "substantial response" end-points (e.g., acute EC50 on abundance or biomass) rather than with nonobserved response end-points, such as a chronic NOEC (Van den Brink, Brock et al., 2002). Despite this problem, it has been shown that (mixture) toxic pressure is a meaningful estimate of impact magnitude in the field, since increasing ecologically significant effects have been observed in relation to increases in predicted toxic pressures, when the latter are based on SSDs derived from laboratory-based acute EC50 and LC50 data (Posthuma and De Zwart, 2006, 2012).

3.5. Technical Steps of Risk Assessment

3.5.1. DATA FOR QUANTITATIVE RISK ASSESSMENT

For the construction of SSD curves, ample data are freely available on the Internet. For instance, the ecotoxicology database (ECOTOX) is a source for locating single chemical toxicity data for aquatic life, terrestrial plants, and wildlife (U.S. EPA, 2007). ECOTOX was created and is maintained by the U.S. EPA, Office of Research and Development (ORD), and the National Health and Environmental Effects Research Laboratory's (NHEERL's) Mid-Continent Ecology Division (MED). On September 25, 2012, the aquatic ECOTOX database contained the results of a total of 345,495 tests covering 5625 different species and 8165 different chemicals, of which 94% were organics. The number of toxicity tests with fish species was 152,745 (44%), with a total number of 886 fish species and 5849 different compounds. The top 10 fish species tested cover 55% of all the fish tests performed (Table 10.2). Thus, even though some species are overrepresented, it will be possible to collect fish ecotoxicity data from an existing source, to the end of deriving both hazardous concentrations (of fish specifically, or of all species including fish) and toxic pressures for species assemblages (for fish specifically or for all species including fish).

3.5.2. METHODS FOR QUANTITATIVE RISK ASSESSMENT

The construction of SSD curves is fully described in all its details in the 2002 review book (Posthuma, Traas et al., 2002). An SSD is often a cumulative log-normal distribution fully characterized by the average, calculated from an appropriate series of log-transformed laboratory-derived

Table 10.2

Top 10 fish species used in toxicity tests represented in the ECOTOX database (U.S. EPA, 2007). The top 10 species cover 55% of all 152,745 fish toxicity tests performed

Species Latin Name	Species Common Name	Number of Tests	Cumulative Percentage
Oncorhynchus mykiss	Rainbow Trout	27,027	18%
Pimephales promelas	Fathead Minnow	15,416	28%
Lepomis macrochirus	Bluegill	9,700	34%
Cyprinus carpio	Common Carp	7,371	39%
Danio rerio	Zebra Danio	4,919	42%
Oryzias latipes	Japanese Medaka	4,591	45%
Carassius auratus	Goldfish	4,121	48%
Ictalurus punctatus	Channel Catfish	3,945	50%
Oncorhynchus kisutch	Coho Salmon, Silver Salmon	3,570	53%
Poecilia reticulata	Guppy	3,288	55%

toxicity data for different species as the midpoint, in combination with the standard deviation of the same log-transformed toxicity data as the slope. SSDs can be constructed by using standard Microsoft Excel functionality (see De Zwart and Posthuma, 2005, for formulas). Alternative methods applying probit transformation and linear regression are also possible using Excel spreadsheet functionality (Giddings, 2011). However, to derive SSDs with more specific and useful output for practical risk assessments, a number of computer programs are freely available on the Internet, including the SSD program ETX2.0 (http://www.rivm.nl/rvs/Risicobeoordeling/Modellen_voor_risicobeoordeling/ETX_2_0 as accessed September 27, 2012) and the Species Sensitivity Distribution Generator V1 (http://water.rutgers.edu/TMDLs/default.htm as accessed February 27, 2012).

3.5.3. METHODS FOR BIOAVAILABILITY CORRECTION

It is well known that site-specific impacts are lower when exposure levels are modified by processes such as contaminant sorption to the matrix. Unlike the concentrations of toxicants that are applied in more or less standardized and controlled laboratory toxicity tests, matrix interactions in natural ecosystems may interfere with bioavailability. Generally speaking, physicochemical processes (ionization, dissolution, precipitation, complexation, sorption, and partitioning) reduce the concentration of toxicants actually experienced by the biota. These processes are dependent on the properties of the toxicants and simultaneously on the local abiotic characteristics of the exposed ecosystem. Uptake, body burden, and reallocation to target sites of action in specific tissues also depend on the physiological properties of the exposed organism. Studies monitoring chemicals in the environment mostly report total concentrations, regardless of the form, binding, and availability of the toxicants. In such cases, model calculations can be used to estimate the bioavailable fraction from measured total concentrations, in combination with ecosystem qualifier data and physicochemical toxicant properties.

A wide variety of speciation models with different levels of complexity are available that are either mechanistic or empirical in nature. A compact overview of the available models to include considerations on bioavailability is given in De Zwart et al. (2008). Field studies have shown that bioavailability in ecosystems is a major variable determining exposure and thus impacts. So far, equilibrium partitioning and speciation have served to derive quality criteria for soil and sediments from quality criteria for water. However, risk assessments for specific water bodies based on risk ratio (RCR) currently often lack considerations of speciation and bioavailability (thus, in the lower, standardized tier). When, however, local toxic pressures are derived in a higher tier, it is more common to correct for bioavailability differences among samples.

4. CUMULATIVE RISK ASSESSMENT FOR MIXTURES OF TOXICANTS

4.1. General Steps

Much progress has been made in chemical and water quality policies through the traditional risk assessment approach of studying the effects of exposures to individual chemicals (Section 3). Such an approach does not, however, reflect the current reality of chemical exposures where organisms are exposed to thousands of chemicals simultaneously (Section 2). Cumulative risk assessment is an approach that focuses on the combined effects of multiple chemicals, though the concept can also be expanded to cover mixtures and other stressors as well. This section describes how the single-chemical approaches expand when mixture issues are taken into account.

Hazard identification for mixtures in relation to fish has been addressed in Section 2. Apart from incidents (due, for example, to high emissions caused by a spill), however, the body of evidence suggests that the current ambient concentrations are not to be characterized as peak exposures with acute lethal effects on fish. The current level of exposures to hazardous compounds does have an implication for the way the exposure assessment proceeds (with or without chemical-to-chemical or chemical-to-substrate interactions as potential major co-drivers of mixture effects). There are two fundamentally different cases when there is a need to consider mixture impacts on biota. These two approaches are based on known mixture compositions (compounds and concentrations). Below, we separately address mixtures of known and unknown composition.

Exposure assessment for mixtures of compounds commonly reveals that at the generally low ambient concentrations of toxic compounds, many are unlikely to chemically interfere with each other in a risk-relevant manner. That is, it is unlikely that chemical A, due to processes such as competition for sorption sites in the matrix, would cause a relevant extra release of compound B, which would have remained sorbed in the absence of A. Such processes would induce "higher responses than expected" for the mixture of A and B when compound B is more toxic than A. The current exposure levels in the field imply that the bioavailable fractions of individual toxicants may often be simply used as input for the mixture effects assessment. In some cases, however, combinations of toxicant complexes or undissolved precipitates may form; these may alter exposure and uptake, and therefore any ensuing toxicity. In such cases, the mixture would cause lower impacts than expected because either would reduce exposures and thus lower their mixture effects.

Effects assessment for mixtures of compounds generally uses either of two distinct reference models: the concentration addition (CA) model (Hewlett and Plackett, 1959), or the response addition or independent action (IA) model (Bliss, 1939). The CA and response addition models have been developed in the context of pharmacology and toxicology, for a species exposed to multiple compounds. The mathematical models that were derived are useful to describe the effects of mixtures of components having similar and dissimilar modes of action.

A mode of action is assigned via known or assumed intoxication processes, so that compounds with the same molecular target are placed into a single group of compounds. For example, "acetyl cholinesterase inhibitors" are compounds that inhibit the breakdown of acetylcholine, thereby increasing the level and duration of action of the neurotransmitter acetylcholine in nervous systems. The assignment of modes of actions to compounds is under debate, and one may follow gross and refined classifications (see, e.g., the discussion in Escher et al., 2002). A gross classification considers that compounds might specifically affect any of the three main highly conserved target domains in biological organisms, which are membranes, proteins, or genetic material. This would define three major modes of action types. In the most refined classification, each chemical has its own specific (set of) modes of action in each biological species. Since it is typical for environmental sciences not only to try to unravel real mechanisms of action, but to develop descriptive and predictive models useful for risk assessment and management (Escher et al., 2002), risk assessment of mixtures could pragmatically proceed given the aforementioned facts and models, while making practical choices on mode of action issues.

In the context of risk analyses based on SSDs, the mathematical concepts as derived for single-species conditions have also been applied and evaluated in a pragmatic way in the context of assessing risks for species assemblages. De Zwart and Posthuma (2005) proposed this as a methodology and suggested a mixed-model approach for multiple compounds with various modes of action. Within a group of compounds with the same mode of action the toxic pressure per compound is aggregated using the CA model, and aggregation over the different groups of compounds with different modes of action proceeds through the response addition model. This yields the mixture toxic pressure for a water, soil, or sediment sample, expressed as the fraction of the species exposed beyond a certain effect level, given the ambient mixture.

Simple similar action (SSA) is a special restricted case of CA, assuming that the individual curves of the components are parallel. The theory of SSA is based on comparing mixtures where the constituents are considered to represent a dilution (less potent variant) of itself (Berenbaum, 1989). The broader interpretation of CA can also be used for chemicals that have a similar mode of action. CA and response addition both exclude interactions between toxicants.

Simple interaction (SI) is occurring when one of the toxicants influences the expression of toxicity of the other toxicant through an indirect mechanism, such as synergism or antagonism. Although specific examples of toxicant interactions have been well studied, models to predict interactions are not available. Most observed examples of toxicant interactions can be explained by changes in toxicokinetics (Haddad and Krishnan, 1998).

Many mixture toxicity evaluations from chemical monitoring and fate modeling studies provide evidence that the number of individual chemicals accounting for the majority of the toxic potency of the mixture is very low (EC, 2012). Assuming concentration additivity, usually only a few chemicals explain more than 90% of the total toxicity of the local environmental mixture (De Zwart, 2005; Harbers et al., 2006; Junghans et al., 2006), though individual sites also have site-specific subsets of a few major-impact compounds. Among sites, these subsets of most hazardous compounds are often different, and there is no rule of thumb suggesting that only few typical mixtures need attention.

Many experimental mixture studies provide evidence that both CA and response addition predict a similar magnitude of effects, often with an observed difference of less than a factor of two (Belden, Gilliom et al., 2007). Specific experimental mixture studies (e.g., Altenburger, Backhaus et al., 2000; Backhaus, Altenburger et al., 2000) demonstrate that the difference between the two models can be validated with experimental data, as depicted in Figure 10.4. In this figure, mixtures of similarly and dissimilarly acting compounds are tested at two exposure levels (EC10 and EC50, respectively), and observed responses are compared to the two null models CA and response addition. CA was observed to be the more conservative model of the two at both exposure levels, predicting more severe effects. The big difference between CA and response addition is that in CA a concentration of a compound not producing any effect by itself may still add to the effect of the mixture, while in response addition the possible effects of very low concentrations of toxicants are excluded (many times "no response" remains "no response").

Mixture studies can be conducted using either simple binary or tertiary mixtures, or more complex mixtures with many components. The complex type of mixture toxicity experiments more closely resembles real-world

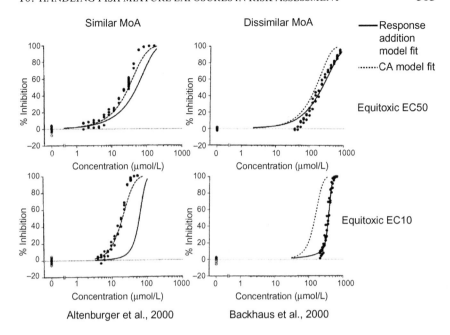

Figure 10.4. Single-species ecotoxicity data show that mixture impacts often resemble two null models, CA and response addition that can be used to predict mixture impacts (Altenburger, Backhaus et al., 2000; Backhaus, Altenburger et al., 2000).

situations where the environment is nearly always contaminated with small concentrations of many toxic chemicals (e.g., De Lange, van der Pol et al., 2006). The limitations of the available scientific data for chemical mixtures challenge decision makers. The majority of mixture studies in the available literature are experiments testing high doses of a few constituents, using completely artificial experimental schemes, such as equitoxic mixtures (a study is called equitoxic when all mixture constituents are added at, e.g., their EC10 level). Most real-world human and environmental exposures, however, are concerned with low concentrations and with a complex range of chemicals. Warne (1991) observed that mixtures containing only a few compounds had highly variable toxicities as compared to CA. Complex mixtures with more than 10 compounds generally exhibited a toxicity level well predicted by the CA model. Warne and Hawker (1995) proposed an explanation for this observation. All chemicals, regardless of whether they have a specific mode of action, also exert a nonspecific toxic effect (baseline toxicity; Verhaar, Van Leeuwen et al., 1992). The observed phenomenon can be explained if there are threshold concentrations below which the specific modes of action would not occur and above which they do. At the concentrations leading to acute toxicity, the toxicity will be predominantly due to the specific mode of action. At the low concentrations that occur in environmental mixtures, the relative contribution of the specific mode of action decreases, while that of the nonspecific concentration additive mode remains. The so-called funnel hypothesis (Warne and Hawker, 1995) states that as the number of components in mixtures increases, there is an increased tendency for the toxicity to act similar to CA.

Risk Characterization. According to recent scientific reviews of mixture concepts and data, and current agreement regarding the science-policy interpretation of the scientific state of art, all the above observations have led to the recent adoption of concentration addition as an agreed, possibly conservative reference model. This should be used as the starting point for practical mixture risk assessment, even for mixtures of unrelated toxicants producing similar effects in both ecotoxicity and human effect studies (NRC 2008; Kortenkamp, Backhaus et al., 2009).

The Kortenkamp et al. (2009) review for the EC explicitly posed and answered a series of questions as presented in Table 10.3. They conclude that where chemicals in a mixture have diverse modes of action, theory suggests that such independent actions should not yield a combination effect. However, concepts of dose additivity suggest that the toxic effects of a mixture could be seen even when individual components are below (presumably only slightly below) their individual no effect level. The authors propose that a dose addition approach is likely to provide a more realistic estimate of mixture toxicity than assumption of an independent action model.

4.2. Deterministic Mixture Risk Assessment for Known Mixtures

For the mixtures of known composition, several deterministic indicators for mixture assessment have been formulated, all based on the concept of concentration addition (Kortenkamp et al., 2009). All methods are, in fact, ratio-based approaches, and they have different names and details.

4.2.1. HAZARD INDEX

The hazard index (HI) (Teuschler and Hertzberg, 1995) is a regulatory approach to component-based mixture risk assessment, which is based on the concept of CA and which can be generally defined by the formula:

$$HI = \sum_{i=1}^{n} \frac{EL_i}{AL_i},$$

where EL_i is the exposure level and AL_i is the acceptable level for compound i (e. g., the aforementioned PNEC), and n is the number of chemicals in the mixture.

Table 10.3
Questions asked and answered by Kortenkamp et al. (2009)

Questions	Answers
Is an assessment of the effects of chemical mixtures necessary from a scientific viewpoint?	Yes, empirical evidence showed that the joint action of chemicals is typically larger than the effect of the most toxic compound.
Is there not sufficient protection against mixture effects if we make sure that each chemical is present individually at exposures unlikely to pose risks?	Assuming concentration additivity, any concentration of any compound needs to be considered because it adds to the mixture concentration and thus may invoke effects.
Is it necessary to test every conceivable combination of chemicals, or is it possible to predict the effects of a mixture?	There is strong evidence that it is possible to predict the toxicity of chemical mixtures with reasonable accuracy and precision by assuming concentration additivity or independent action for individual species.
Which of the two assessments and prediction concepts—concentration additivity or independent action—should be utilized in practice?	The practical relevance of independent action has been questioned on the basis of considerations of biological organization. Concentration addition has been deemed more broadly applicable and has even been termed the "general solution" for mixture toxicity assessment.
Which chemicals should be subjected to mixtures risk assessment?	Ecotoxicological studies often employ broad, integrating end-points such as mortality or reproduction. The consensus is that if a compound affects such end-points, it is considered to be of relevance from a mixture perspective and dose (concentration) addition, independent action, or a combination of both is applied.
How should mixture effect assessment concepts be applied in practice?	To deal with data gaps and to take account of differing data quality, tiered approaches to mixture risk assessment have been proposed, such as those advocated in the following references (De Zwart and Posthuma, 2005; Posthuma, Richards et al., 2008; Ragas, Teuschler et al., 2010).
	Tier 1: Questioning whether combined exposures are in fact likely may reveal that the situation to be evaluated does not in fact present an issue for mixtures risk assessment.
	Tier 2: Evaluate whether adoption of a specific mixtures assessment factor may safeguard against the possibility of joint effects.
	Tier 3: Apply the assumption of concentration addition for all mixture constituent semiquantitative approaches like the toxic equivalent factor/quantity approach (TEF/TEQ), the toxic unit summation approach (TUS), the hazard index approach (HI), the summation of relative potency factors (RPF), or the point of departure index approach (PODI) can be applied (see Section 4.1).

(Continued)

Table 10.3 (continued)

Questions	Answers
	Tier 4: If information on modes of action is available, mixed mixture assessment models can be applied with concentration addition within groups of compounds causing similar types of effects, followed by independent action across the groups of compounds with different types of effects.
	Finally, the highest tier might be used to address both issues of modes of action and differences in the vulnerability of various groups of species.
What knowledge gaps hamper the consideration of mixture toxicology and ecotoxicology in chemical risk assessment?	Whether or not ecological risks arise from combined exposures of toxicants can only be decided on the basis of better information about relevant combined exposures in "real-world" exposure settings. That information is currently often incomplete, presenting a major challenge to risk assessment.
	Clear cases of synergism are observed. Such cases are very specific for the type of mixture, exposed organisms, and toxicity end-points. These cases cannot be evaluated with a general risk assessment scheme, but must be treated on a case-by-case basis.
	Mixture evaluations require that the composition of the mixture of interest is known. In reality, this is almost never the case. For all practical purposes, mixtures will usually not be known to their very last compound. Criteria are therefore needed to define the "relevant" components of a mixture.

Various measures for exposure levels and expectable levels may be applied. The only constraint is that EL and AL must be expressed in the same unit. If $HI > 1$, the total concentration (or dose) of mixture components exceeds the level considered to be acceptable. The method offers flexibility in applying different uncertainty factors when defining AL for the individual substances.

4.2.2. TOXIC UNIT SUMMATION

The method of toxic unit summation (TUS) (Sprague, 1970) is a direct application of the CA concept and defined by the formula:

$$TUS = \sum_{i=1}^{n} TU_i = \sum_{i=1}^{n} \frac{c_i}{ECx_i}$$

where c_i are the actual concentrations of the individual substances in a mixture and ECx_i denote equi-effective concentrations of these substances if present singly (e.g., $EC50_i$). The quotients c_i/ECx_i are termed toxic units (TU). Toxic units rescale absolute concentrations (or doses) of substances to their different individual toxic potencies. They express the concentrations of mixture components as fractions of equi-effective individual concentrations (or doses) ECx_i. Typically, $x = 50\%$ ($EC50_i$) is chosen as the reference level, but TUS can also be calculated for any other effect level x. If TUS $= 1$, the mixture is expected to elicit the total effect x. If the sum of toxic units is smaller or larger than 1, the mixture is expected to elicit effects smaller or larger than x, respectively.

4.2.3. POINT OF DEPARTURE INDEX

The point of departure index (PODI) is an approach to component-based mixture risk assessment that is similar to the HI and TUS and also based on the CA concept. In contrast to the HI, however, exposure levels (ELs) of chemicals in a mixture are not expressed as fractions of individually acceptable levels (ALs) but as fractions of their respective points of departure (PODs) such as a predicted no effect concentration (PNEC) or a benchmark concentration. In this way, different uncertainty factors that may be included in AL values (see HI) are removed from the calculation (Wilkinson, Christoph et al., 2000):

$$PODI = \sum_{i=1}^{n} \frac{EL_i}{POD_i}$$

A PODI can be used to estimate margins of exposure for the mixture of interest.

4.2.4. RELATIVE POTENCY FACTORS

The relative potency factor (RPF) approach is a practical regulatory application of the CA concept for mixtures of chemical substances that are assumed to be toxicologically similar (U.S. EPA, 2000). The concentrations (or doses) of mixture components are scaled relative to the concentration of an index compound and then summed up. The scaling factor is called RPF. The total toxicity of the mixture is assessed in terms of the toxicity of an equivalent concentration of the index compound. In general, the mixture concentration C_m expressed in terms of the index compound for n compounds is:

$$C_m = \sum_{i=1}^{n} (C_i * RPF_i)$$

where c_i is the concentration of the ith mixture component, and $RPF_i = 1$, as $i = 1$ indicates the index chemical.

4.2.5. TOXIC EQUIVALENCY FACTORS

The toxic equivalence factor (TEF) is a specific type of RPF formed through a scientific consensus procedure (U.S. EPA, 2000). Based on the assumptions of a similar mechanism of action of structurally related chemicals and parallel concentration response curves, they were first developed for dioxins. The total toxicity of the mixture is assessed in terms of the toxicity of an equivalent concentration of an index compound. The total equivalent quantity (TEQ) is estimated by summation of the concentrations of mixture components c_i multiplied by the respective TEF_i:

$$TEQ = \sum_{i=1}^{n}(c_i * TEF_i)$$

4.3. Probabilistic Mixture Risk Assessment for Known Mixtures

4.3.1. GENERAL

The risk of exposure to individual chemicals as calculated with the SSD method is based on the same mathematical principles as used in the derivation of concentration–response curves in single-species toxicity evaluation. As for individual species, both the concentration- and response-addition models can conceptually be applied in ecological risk assessment for species assemblages exposed to mixtures of toxicants, now being formulated probabilistically (Traas et al., 2002).

Species assemblages of fish are exposed to mixtures (see Section 2), and the fraction of species that is potentially affected by a given mixture can be quantified using SSDs and the mixed-model approach (De Zwart and Posthuma, 2005). The mixture toxic pressure expresses the fraction of species that is potentially exposed beyond a certain toxicity end-point, such as an NOEC or EC50. Using ambient concentrations per compound, the quantification of the local toxic pressure of each compound proceeds as described above. Thereafter, the single-substance PAF (ssPAF) values are aggregated within each of the assigned subgroups of compounds with assumed similar modes of action and then, among these groups, by CA and response addition modeling, respectively.

4.3.2. PROTOCOL FOR THE MULTIPLE SPECIES RISK MODEL FOR CA

Toxicity data are scaled into dimensionless hazard units (HUs), defined as the concentration where the effect criterion (e.g., EC50) is exceeded for 50% of all species tested, that is, the median of the toxicity data of the whole dataset:

$$HU_i^j = \frac{EC50_i^j}{\overline{EC50_i}}$$

for $i = 1$ to n compounds and for $j = 1$ to m species, HU_i^j is the scaled EC50s in dimensionless hazard units (mg/L/mg/L), and $\overline{EC50}_i$ is the median NOEC for substance i.

The SSDs for each compound are obtained by fitting a log-normal model to the log toxicity data in hazard units (HU). For the log-normal procedure, the SSDs are fully characterized by the slope of the CDF that equals the standard deviation (σ) of log(HU_i^j), the median of the distributions is zero by definition. For the CA model to be applicable, it is essential to verify that the slopes for the different mixture constituents are about equal (we suggest a maximum deviation of $\pm 10\%$), otherwise the compounds with outlying SSD slopes should be attributed to a different MoA, and their risk should be added by the response addition model. For each compound present in an environmental sample, the bioavailable exposure concentration is again recalculated to hazard units:

$$HU_i = \frac{Bioavailable\ concentration_i}{\overline{NOEC}_i}$$

The HU values are added (non-log-transformed) for substances with corresponding TMoA and corresponding slope:

$$HU_{TmoA} = \sum_i HU_i$$

The log-normal CA model gives the toxic risk for mixture constituents with the same MoA by applying the Microsoft Excel function:

$$msPAF_{CA,MoA} = NORMDIST\left(\log\left(\sum HU_{MoA}\right), 0, \sigma, 1\right)$$

with $msPAF_{CA,MoA}$ representing the multisubstance potentially affected fraction of various (groups of) compounds calculated by response addition.

4.3.3. Protocol for the Multiple Species Risk Model for Response Addition

The $msPAF_{CA,MoA}$ values for the different MoA in the mixture are calculated according to the multiple species risk model for CA, even if an MoA is only represented by a single substance. The combination effect for compounds with different modes of action is again calculated analogous to the probability of two nonexcluding processes (Hewlett and Plackett, 1979). For the present use in SSDs, it is assumed that sensitivities are uncorrelated in response addition. For more than two chemicals or groups of chemicals with different TMoA, this leads to:

$$msPAF_{response\ addition} = 1 - \prod_{TMoA} (1 - msPAF_{TMoA})$$

510   DICK DE ZWART AND LEO POSTHUMA

for TMoA = 1 to n substances or modes of action, with $msPAF_{response\ addition}$ representing the multisubstance potentially affected fraction of various (groups of) compounds calculated by response addition.

4.3.4. MEANING OF THE MIXED-MODEL OUTPUT

The mixed-model approach yields a large set of ssPAF values (per compound), a smaller set of mode-of-action specific multisubstance $msPAF_{CA}$ values (per assigned subgroup of similarly acting compounds), and one $msPAF_{response\ addition}$ (over all compounds) per sample. Mixture toxic pressure can be of use in setting priorities in risk management, since the higher values imply higher impacts.

The meaning of the ssPAF and msPAF values is as follows. Imagine two water bodies with different mixture exposures (different compounds present at different concentrations), and assume that the overall toxic pressures derived using SSD models based on EC50 values are 20% and 55%, respectively. Under this scenario, the first water body would impose impacts larger than the EC50 in 20% of the tested species, while the second water body would impose similar impacts in 55% of the tested species. The hazard potential of the latter water body is larger, as summarized by a PAF of 20 and 55% of the tested species assemblages. When the tested species assemblage has the same sensitivity distribution as the natural assemblages in the water bodies, for example, of fish, the PAF of the tested species can be interpreted as the PAF for the natural assemblage of species.

Multisubstance toxic pressures can, in the same reasoning, be calculated for all sites mentioned in the fish body residue study in Section 2, yielding a GIS map delineating the spatial divergence in the potential of mixture effects to occur. In the derivation of SSDs, data for all groups of species can be used, or one may opt to perform an evaluation for fish only.

Note that, apart from measured compound concentrations, it is easy to envisage that predicted environmental concentrations can also be used to quantify mixture toxic pressures. In this setting, we can explore the impact of risk management scenarios on the potential for recovery of fish assemblages. If compound emissions get reduced, so would downstream mixture toxic pressures, and analyses of scenarios using predicted environmental concentrations and SSD modeling could show the magnitude of the predicted recoveries.

The magnitude of recovery upon taking measures against mixture exposures needs take into account other stressors too. The aforementioned procedure would only evaluate the limitations potentially imposed by mixtures. Removing mixture stress, however, need not always lead to recovery according to the predicted reduced toxic pressure, since other factors might remain limiting. Hence, multiple stress analyses are also warranted. An example of a statewide assessment of mixture toxic pressures for fish was made by De Zwart et al. (2006), where exposure levels were in part measured (metals) and in part modeled (household chemicals).

4.4. Risk Assessment of Whole Mixtures of Unknown Composition

Definitions of realistic environmental assessment problems may also consist of an assumed hazardous situation without knowledge of the mixture composition. In a river basin, a multitude of chemicals may reach the water, and each sampling site is likely to have a different chemical mixture composition. To enable assessment of hazards of this kind, exploratory whole-sample methods have been developed, and these also lead to estimates of the fraction of species potentially affected, which includes fish.

One such method is the pT method (toxic potency). According to this method, water samples are taken and thereafter used for ecotoxicity testing with high-throughput microtests. The sample of tests used for the pT evaluation (Struijs et al., 2010) consists of five bioassays, selected to represent three important trophic levels in an aquatic ecosystem (bacteria, algae, and invertebrates): (1) Microtox® (bacteria), (2) PAM (algae), (3) Thamnotoxkit F® (crustaceans), (4) Rotoxkit F® (rotifers), and (5) Daphnia IQ® (crustaceans).

Since most water bodies have become less contaminated over the last several decades, these tests do not often respond to the toxicants in the ambient samples. The pT procedure therefore first employs a concentration step, whereby the organic unknown compounds are concentrated by a factor of 1000. A serial stepwise dilution is then applied prior to applying the set of microtests. In effect, this is a concentration-response test for five microtest species, whereby the concentration factor causing 50% effect is of interest. The more contaminated the sample with unknown mixtures is, the lower the concentration factor that causes 50% of effect in the test. The concentration factor data are then summarized over the test species, and an SSD-pT is constructed (a log-normal distribution of concentration factors among the test species). The fraction of species that would be affected at a concentration factor of 1 (which represents the original water sample) can now be determined.

As in the toxic pressure (msPAF) approach for known mixtures, this pT-msPAF is a value that—at least relatively—shows the hazard potency of the unknown mixture of contaminants in a water sample to affect a species assemblage: the higher the pT, the higher the probability of impact on local species assemblages, including fish. The latter is based on the assumption that the responses of the microtest species are caused by the same toxicological processes that also occur in fish.

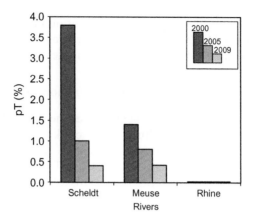

Figure 10.5. Variability in space (three rivers) and time (gray shades) of the pT value (%), redrawn from Struijs, Van der Grinten et al. (2010). The pT is derived from multiple microtoxicity tests executed with concentrates of water samples taken in the field. The unit of pT is again a multisubstance potentially affected fraction of species (msPAF, potentially affected fraction of species).

A broad study on the spatiotemporal variation of pT values in western European rivers has revealed relevant insights into the toxic potency of unknown mixtures. Figure 10.5 summarizes the observations for three major rivers (the Scheldt, Meuse, and Rhine).

The first finding is that the rivers have different characteristics regarding toxic potency: The pT values reduce in the order Scheldt, Meuse, and Rhine. The pT values were larger in the Scheldt and Meuse in the year 2000 than they are currently (time-related reduction), but the highest values were estimated as being below 10% of species potentially affected by the unknown mixtures. By considering not only the pT as average estimate of toxic potency but also the underlying five test data individually, it appeared that some outliers (high toxicity) were found for the microtests with algae. These individual-species adverse peak effects could, upon chemical analyses, be related to specific high emissions of photosynthesis inhibiting compounds.

5. EVIDENCE FOR MIXTURE IMPACTS IN THE FIELD

5.1. Overview

In Section 2 it was shown that fish are exposed to mixtures, and in Sections 3 and 4 the principles of single-compound and mixture impact assessment were discussed. The next step is to address the question of whether mixture impacts can indeed be observed under field conditions. Note that in Section 4 the issue was that predicted impacts were generated for either known mixtures (msPAF) or unknown ones (pT-msPAF), but that these values were not yet demonstrated to be accurate predictors of observed impacts. This section considers field-observed effects and relates these values when possible to the predicted values resulting from mixture assessment modeling.

5.2. Field Effects Assessment and Mixtures

The collection of monitoring data on a landscape scale is, in various jurisdictions, a legal obligation, for example, the EU Water Framework Directive. This means that over time datasets will evolve into massive datasets on the occurrence of species, chemical compounds, and ranges of other habitat data, like hydromorphological characteristics and a suite of in-stream variables. Such datasets have been used to investigate mixture impacts on fish under field conditions, so that mixture analyses were made in the context of multiple other potential stressors. This approach is aimed to generate a preliminary diagnosis of local impacts on a landscape scale so as to support optimizing river basin management planning. The term coined for this type of work in 1984 was *ecoepidemiology* (Bro-Rasmussen and Lokke, 1984). Assessment of chemical impacts is commonly difficult, since the sheer number of chemicals and other potential stressors in monitoring databases simply cause a lack of statistical power when all chemicals are considered individually. Uniquely, by first aggregating mixture risks through the models that were defined, it became possible to assess mixture impacts at current exposure levels.

De Zwart et al. (2006) and Kapo et al. (2008) designed two key studies of this kind on the ecoepidemiology of fish assemblages on the landscape scale. Fish monitoring data from the state of Ohio in the United States were collected and combined with monitoring data on a suite of other stressors, including a set of chemicals (metals, household chemicals, and ammonia). The chemical concentrations were either measured or modeled and expressed as mixture toxic pressures. The eco-epidemiological analysis of De Zwart et al. (2006) first aimed to derive direct Pearson correlations between species abundances and the toxic pressure of the local mixtures (Posthuma and De Zwart, 2006). This first analysis showed an absence of significant associations between mixture toxic pressure and species abundance for the majority of fish species (Figure 10.6). Few species exhibited a significant negative correlation between their abundance and toxic pressure, and few had a significant positive one. Most species would be characterized as "neutral <" without a significant correlation coefficient

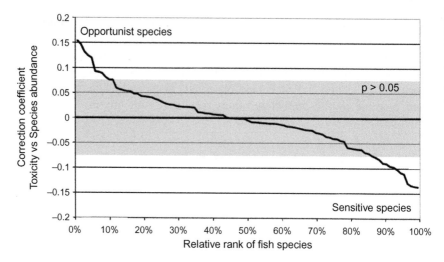

Figure 10.6. Example data on fish ecoepidemiology suggesting that field mixture impacts may go unnoticed, due to the multiple-stress influences of many other local stressors (see next figure for that evidence) (Posthuma and De Zwart, 2006).

between mixture exposure and abundance. However, it is well known that species abundances in the field are usually influenced by all local stress factors together, so that the correlations lead to the conclusion of almost absence of mixture impacts. Mixture impacts may be hidden by the effects of other stressor combinations.

Hence, in a second analytical step, the same data were reanalyzed with a technique that shows multiple-stress abundance effects, including the association between toxic pressure of mixtures and species abundances. This analysis resulted in a clear statistical association between the abundance of a majority of fish species and mixture exposure level. Table 10.4 summarizes the multiple, significant associations between the abundances of fish species and the suite of stressors. It became clear that almost all fish species showed an association between abundance and a suite of predictors, given the high percentages of predictor occurrences in species-abundance formulas (second column). Within this multitude of influences per species, it became clear that 55% of the fish species exhibited abundance changes when mixture toxic pressure variation occurred for household chemicals ($msPAF_{HH, NOEC}$), and that this value was 50% for metals and ammonia exposure. The table shows that toxic pressure variance in the field is reflected in abundance changes in the field, for more than half of the fish species.

Combining the U.S. studies on exposure and ecoepidemiology (Section 2 and this section), the body residue studies imply that fish are exposed to

Table 10.4

Examples of significant stressors, which are relevant in determining local species abundances. Highly significant influence of mixtures on most species is shown (Posthuma and De Zwart, 2006)

Predictor Variables	Predictor Occurs in Percentage of Species	Significance of Regression Terms						
		$P \leq 0.001$	$0.001 < P \leq 0.01$	$0.01 < P \leq 0.05$	$0.05 < P \leq 0.1$	$0.1 < P \leq 0.5$	$0.5 < P \leq 1.0$
Latitude	99%	71%	7%	3%	2%	11%	6%
Longitude	96%	64%	7%	3%	3%	23%	0%
Log(gradient)	94%	52%	7%	9%	6%	27%	0%
Log(drainage area)	98%	72%	3%	4%	2%	18%	0%
Dissolved oxygen (m)	45%	86%	5%	5%	0%	5%	0%
Hardness (m)	55%	85%	13%	0%	0%	2%	0%
% Cum. effluent (m)	65%	92%	6%	0%	0%	2%	0%
pH (m)	43%	95%	5%	0%	0%	0%	0%
Total susp. solids (m)	67%	89%	11%	0%	0%	0%	0%
Channel alterations	47%	96%	2%	0%	0%	2%	0%
Cover alterations	66%	92%	5%	3%	0%	0%	0%
Warm water attributes	48%	91%	9%	0%	0%	0%	0%
Pool alterations	46%	86%	14%	0%	0%	0%	0%
Riffle alterations	47%	87%	9%	2%	0%	2%	0%
Riparian alterations	44%	88%	12%	0%	0%	0%	0%
Substrate alterations	50%	88%	8%	4%	0%	0%	0%
msPAF$_{HH, NOEC}$	55%	91%	8%	0%	2%	0%	0%
msPAF$_{NH3\&Metals, NOEC}$	50%	92%	6%	0%	2%	0%	0%

mixtures, while the current analyses demonstrate that fish probably react to mixture exposures in terms of abundance impacts and eventually (at higher exposures) presence/absence effects.

The combination of Figure 10.6 with Table 10.4 suggests that the absence of a correlation between abundances of fish species and mixture toxic pressure need not imply absence of influences. It is likely that mixture impacts are masked by the impacts of other stressors.

Similar phenomena as presented here for fish in Ohio have been found for other species groups, other landscapes, and other stressor combinations, for example, in the River Scheldt for macro invertebrates (De Zwart, Posthuma et al., 2009) and in the Netherlands for macrofauna and macrophyte species (De Zwart, 2005).

5.3. Presenting Mixture and Multiple Stress Impacts

We have argued so far that field species are exposed to a variety of chemical mixtures, that these mixtures probably cause different responses, and that other stressors also matter. This results in an issue in risk characterization. Rather than presenting simple ratios per chemical, like the PEC/PNEC ratio, there is a need to help water quality management with clear risk communication. Based on the results obtained for fish in Ohio, the statistical associations between toxic pressure, other stressor variables, and species abundances were analyzed further and plotted on a map. This can be done by using pie diagrams (Figure 10.7) or by geographic information system (GIS) modeling as described in Kapo et al. (2008).

De Zwart et al. (2006) used statistical associations to delineate preliminary diagnostic graphs on the map of Ohio (Figure 10.7). They do so by deriving effect and probable cause (EPC) pie diagrams, for which pie size relates to species expected but absent (magnitude of impact) and the colors relate to the different stressor categories. Large slice sizes are interpreted as a high relative influence of the pertinent combination of stress factors. The map shows that the local fish assemblages of various regions in Ohio are apparently affected by different subsets of stressors, including chemical mixtures. In area A, an industrial area, water chemistry (pH) appeared to be the dominant stressor. In area B, characterized as a metropolitan area, the assessment identified water chemistry, habitat alteration, municipal effluent, and toxicity from metals and ammonia as likely causes. In area C, identified as a natural area, a marked influence of toxic pressure influences was indicated, whereby the toxic pressure contributions were mainly attributed to metals. Ammonia toxicity was identified as the main causal factor in area D and could be attributed to a malfunctioning sewage treatment plant.

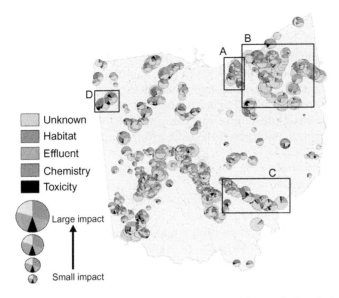

Figure 10.7. A GIS map of magnitudes of impacts (fractions of fish species lost depicted by pie size differences) and probable causes (colors); mixture ecotoxicity summarized in red (via msPAF) (De Zwart, Dyer et al., 2006). See color plate at the back of the book.

The EPCmethod (De Zwart, Dyer et al., 2006) and the methods applied by Kapo et al. (2008) take into account the fact that species assemblages also differ among sites and water bodies due to natural variability. The in-stream assemblage of a small river in the North is naturally different from that of a lake in the South. This fact is acknowledged and quantitatively addressed in the diagnostic methods by RIVPACS-type modeling (see Wright, Sutcliffe et al., 2000). This type of modeling yields, for all sites, the species expected to be present given the local natural attributes, and thus allows the derivation of which species are missing when local samples are taken.

5.4. Validation and Interpretation of Toxic Pressure

The four areas A, B, C, and D were identified based on double-blind data analyses, whereby site codes and species codes were anonymized in all analyses (Figure 10.7). Still, the preliminary (statistical) analyses yielded preliminary diagnostics that made sense. In particular, the first diagnosis of apparent metal mixture impacts in region C was not considered valid by specialists from the Ohio EPA, since the area is not known for its smelters or industrial activities. Later, however, the diagnosis was confirmed by determining that the area is rich in metal mines abandoned in the eighteenth century.

Apart from this type of validation (is the outcome in line with some kind of expectation on water quality or habitat problems, though now quantified?), the ecoepidemiological analyses also yielded a "validation" of the mixture model itself. That is, the data allow for a direct comparison between the toxic pressure of mixtures (msPAF) and the fractional impact on the species assemblages in the field, given all other stressors and natural variability.

For the fish data on Ohio, this associative analysis resulted in the association depicted in Figure 10.8 (Posthuma and De Zwart, 2006). The increase of toxic pressure (x-axis, model result based on concentrations of chemicals) appeared to be associated with the fraction of fish species affected in the field. This association, on a log-log scale, was $y = 0.92\ x - 1$; that is, the slope of the acute-toxic pressure prediction to the fraction affected was nearly 1 (higher predicted impact results in higher impacts in the field), though the fraction of species affected in the field was lower than expected (represented by the "–1" in the formula). In another study with benthic invertebrates the association between toxic pressure of mixtures and species abundances was further elaborated and expanded, resulting in Figure 10.9

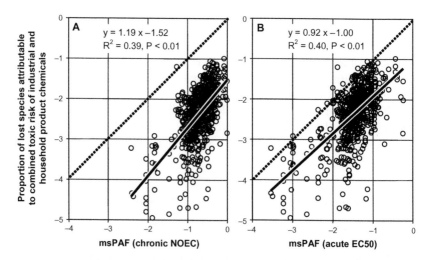

Figure 10.8. When mixture toxic pressure on fish in Ohio surface waters increases (x-axis), the impact on local species assemblages (Y) does relate, so that increasing toxic pressures of mixtures implies increased local impacts on the fish assemblages (Posthuma and De Zwart, 2006). Dashed lines represent an ideal fit (predicted impact equals observed impacts, while the solid lines represent the line fitted through the data).

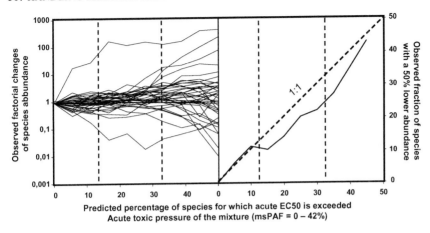

Figure 10.9. (Left) Example graph showing how the abundance of a subset of benthic invertebrate species changes when mixture toxic pressure (*X*) increases. For ease of interpretation, each species has an abundance depicted as "1" at the lowest toxic pressure value. (Right) fraction affected in the field (abundance change –50% or more, or +50% or more [broken line]) relates to predicted msPAF (Posthuma and de Zwart, 2012).

(Posthuma and De Zwart, 2012). The latter study shows that species indeed show a large variety of response patterns, varying between opportunistic, neutral, and sensitive (as also shown in Figure 10.6).

Figure 10.9 shows that the expected impact of toxic mixture exposure can actually be correlated one-to-one with field observations (Posthuma and De Zwart, 2012). That is: the predicted and observed fractions of species affected are similar in the latter study. Note, however, that the abundance responses of the individual species from which this 1:1 relationship was derived are highly variable among individual species. That is, toxic pressure quantitatively predicts a level of impact in terms of fraction of species affected, not what those species' specific response patterns are.

6. CONCLUSIONS, OVERVIEW, AND SUMMARY

While massive fish kills and depauperate fish assemblages were common in the earlier days of industrialization, the situation is generally considered as being improved today. All the same, natural fish assemblages are still exposed to mixtures of contaminants whereby the exposure types and levels vary over sites. Evidence from tissue residue studies, like the U.S. EPA (2009) inventory report, clearly shows the presence of such exposures in the field. Furthermore, mixture tests with individual species suggested that multiple exposure asks for cumulative risk assessments. This warrants attention for mixture risk and impact assessment for fish and other species groups, superimposed on single-compound assessment approaches.

Sections 3 and 4 summarize the principles of risk assessment, showing the availability of different types of problem definitions and different options (tiers) to address single compounds as well as mixtures. Methods exist to address mixtures of known composition as well as for mixtures of unknown composition. Next to criterion-based (ratio-type) approaches (RCR, HI, and so forth), quantitative approaches have been proposed, especially to quantify toxic pressure on fish assemblages by means of SSD modeling in conjunction with mixture and bioavailability models and whole-sample assays that resulted in the pT-msPAF. These methods are able to summarize, at least in a relative way, mixture-related hazard differences among field samples, so that spatiotemporal trends in mixture pressures can be analyzed. These would be helpful in setting the focus in river basin management planning.

Specific risk assessments thereby ask for different degrees of specificity, as conceptually summarized in Figure 10.1. The principle of tiering is the option for use in simple risk assessment (exposure and effects assessment) when possible and in more complex ones when needed. Various proposals have been described in order to provide methods tailored to the assessment problem, such as Posthuma et al. (2008) and Ragas et al. (2010). These authors developed a tiered ranking of mixture risk assessment methods for ecosystems and further substantiated the scheme for ecological and human mixture risk assessment, respectively. These proposed tiering schemes can be used as a guideline to select (mixture) risk assessment approaches tailored to the assessment problem.

Section 5 demonstrated that using these techniques presents evidence for current impacts of ambient mixtures on fish species assemblages under field conditions. That section made clear that impacts of toxic mixtures in fish and other species groups are commonly masked by the influences of other stressors. Mixture impacts may not be overt on first sight, but they are demonstrated when taking account of multiple stressors. Similar analyses for various species groups (e.g., invertebrates) clearly showed that substantial fractions of species currently exhibit significant abundance associations to mixture toxic pressure in the field.

As an overall conclusion, mixture exposures occur, mixture risk and impact methods exist, these methods unveil mixture issues in the natural environment, and the methods yield useful results for planning risk reduction strategies.

REFERENCES

Aldenberg, T. and Slob, W. (1991). Confidence limits for hazardous concentrations based on logistically distributed NOEC toxicity data. *Ecotoxicology and Environmental Safety* 18, 221S–251S.

Altenburger, R., Backhaus, T., et al. (2000). Predictability of the toxicity of multiple chemical mixtures to *Vibrio fischeri*: Mixtures composed of similarly acting chemicals. *Environmental Toxicology and Chemistry* 19, 2341–2347.

Backhaus, T., Altenburger, R., et al. (2000). Predictability of the toxicity of a multiple mixture dissimilarly acting chemicals to *Vibrio Fischeri*. *Environmental Toxicology and Chemistry* 19 (9), 2348–2356.

Belden, J. B., Gilliom, R. J., et al. (2007). How well can we predict the toxicity of pesticide mixtures to aquatic life? *Integrated Environmental Assessment and Management* 3 (3), 364–372.

Berenbaum, M. C. (1989). What is synergy? *Pharmacological Reviews* 41 (2), 93–141.

Bliss, C. I. (1939). The toxicity of poisons applied jointly. *Annals of Applied Biology* 26, 585–615.

Bro-Rasmussen, F. and Lokke, H. (1984). Ecoepidemiology—a casuistic discipline describing ecological disturbances and damages in relation to their specific causes; exemplified by chlorinated phenols and chlorophenoxy acids. *Regulatory Toxicology and Pharmacology* 4, 391–399.

Coors, A. and Frische, T. (2011). Predicting the aquatic toxicity of commercial pesticide mixtures. *Environmental Sciences Europe* 23.

De Lange, H. J., van der Pol, J. J. C., et al. (2006). *Ecological vulnerability in wildlife: A conceptual approach to assess impact of environmental stressors*. Wageningen, Netherlands: Alterra, 112.

De Zwart, D. (2002). Observed regularities in species sensitivity distributions for aquatic species. In *Species Sensitivity Distributions in Ecotoxicology* (eds. L. Posthuma, G. W. Suter and T. P. Traas), pp. 133–154. Boca Raton, FL: Lewis Publishers.

De Zwart, D. (2005). Ecological effects of pesticide use in the Netherlands: Modeled and observed effects in the field ditch. *Integrated Environmental Assessment and Management* 1 (2), 123–134.

De Zwart, D., Dyer, S. D., et al. (2006). Predictive models attribute effects on fish assemblages to toxicity and habitat alteration. *Ecological Applications* 16 (4), 1295–1310.

De Zwart, D. and Posthuma, L. (2005). Complex mixture toxicity for single and multiple species: proposed methodologies. *Environmental Toxicology and Chemistry* 24 (10), 2665–2676.

De Zwart, D., Posthuma, L., et al. (2009). Diagnosis of ecosystem impairment in a multiple stress context—how to formulate effective river basin management plans. *Integrated Environmental Assessment and Management* 5 (1), 38–49. DOI: 10.1897/IEAM_2008-1030.1891.

De Zwart, D., Warne, A., et al. (2008). Matrix and media extrapolation. In *Extrapolation Practice for Ecotoxicological Effect Characterization of Chemicals* (K. R. Solomon, T. Brock, D. De Zwart, et al.), pp. 33–74. Boca Raton, FL: CRC-Press.

EC (2003). *Technical Guidance Documents on Risk Assessment, Part II. EUR 20418 EN/2 (http://ecb.jrc.it/tgdoc)*. Ispra, Italy: European Commission, Joint Research Centre.

EC (2012). *Opinion on the Toxicity and Assessment of Chemical Mixtures*. Brussels, Belgium: European Commission—SCHER, SCCS, SCENIHR, 50.

EIFAC (1987). Water quality criteria for European freshwater fish - Revised report on combined effects on freshwater fish and other aquatic life of mixtures of toxicants in water. EIFAC technical paper 37, Rome, Italy. FAO, (75).

Escher, B. I. and Hermans, J. L. M. (2002). Modes of action in ecotoxicology: Their role in body burdens, species sensitivity, QSARs and mixture effects. A critical review. *Environmental Science and Technology* **36**, 4201–4216.

Forbes, V. E. and Calow, P. (2002). Species sensitivity distributions revisited: A critical appraisal. *Human and Ecological Risk Assessment* 8, 473–492.

Giddings, J. M. (2011). *The Relative Sensitivity of Macrophyte and Algal Species to Herbicides and Fungicides: An Analysis Using Species Sensitivity Distributions.* Rochester, MA: Compliance Services International, 77.

Giddings, J. M., Solomon, K. R., et al. (2001). Probabilistic risk assessment of cotton pyrethroids: II. Aquatic mesocosm and field studies. *Environmental Toxicology and Chemistry* 20, 660–668.

Haddad, S. and Krishnan., K. (1998). Physiological modeling of toxicokinetic interactions: implications for mixture risk assessment. *Environmental Health Perspectives* 106 (supplement 6), 1377–1384.

Hamers, T., Aldenberg, T., et al. (1996). *Definition Report–Indicator Effects Toxic Substances (Itox).* Bilthoven, Netherlands. RIVM (95).

Harbers, J. V., Huijbregts, M. A. J., et al. (2006). Estimating the impact of high-production-volume chemicals on remote ecosystems by toxic pressure calculation. *Environmental Science and Technology* 40 (5), 1573–1580.

Hewlett, P. S. and Plackett, R. L. (1959). A unified theory for quantal responses to mixtures of drugs: non-interactive action. *Biometrics* 15, 591–610.

Hewlett, P. S. and Plackett, R. L. (1979). *An Introduction to the Interpretation of Quantal Responses in Biology.* London, UK: Edward Arnold Ltd.

Junghans, M., Backhaus, T., et al. (2006). Application and validation of approaches for the predictive hazard assessment of realistic pesticide mixtures. *Aquatic Toxicology* 76, 93–110.

Kapo, K. E., Burton, G. A., Jr., et al. (2008). Quantitative lines of evidence for screening-level diagnostic assessment of regional fish community impacts: A comparison of spatial database evaluation methods. *Environmental Science and Technology* 42 (24), 9412–9418.

Könemann, H. (1980). Structure-activity relationships and additivity in fish toxicities of environmental pollutants. *Ecotoxicology and Environmental Safety* 4, 415–421.

Könemann, H. (1981). Fish toxicity tests with mixtures of more than two chemicals: A proposal for a quantitative approach and experimental results. *Toxicology* 19, 229–238.

Kooijman, S. A. L. M. (1987). A safety factor for LC50 values allowing for differences in sensitivity among species. *Water Research* 21, 269–276.

Kortenkamp, A., Backhaus, T., et al. (2009). *State of the Art Report on Mixture Toxicity: Final Report.* Brussels, Belgium: European Commission, Directorate General for the Environment, 391.

Mebane, C. A. (2010). Relevance of risk predictions derived from a chronic species-sensitivity distribution with cadmium to aquatic populations and ecosystems. *Risk Analysis* 30 (2), 203–223.

Niyogia, S. and Wood, C. M. (2004). Biotic ligand model, a flexible tool for developing site-specific water quality guidelines for metals. *Environmental Science and Technology* 38 (23), 6177–6192.

NRC (2008). *Phthalates and Cumulative Risk Assessment—The Task Ahead.* Washington, DC, USA: National Research Council, Committee on the Health Risks of Phthalates, The National Academies Press, 188.

OECD (1981). *Guidelines for Testing of Chemicals.* Paris, France: Organization for Economic Cooperation and Development.

OECD (1992). *Report of the OECD Workshop on the extrapolation of laboratory aquatic toxicity data to the real environment.* Paris, France: Organisation for Economic Co-Operation and Development.

Posthuma, L. and de Zwart, D. (2006). Predicted effects of toxicant mixtures are confirmed by changes in fish species assemblages in Ohio, USA rivers. *Environmental Toxicology and Chemistry* 25 (4), 1094–1105.

Posthuma, L. and de Zwart, D. (2012). Predicted mixture toxic pressure relates to observed fraction of benthic macrofauna species impacted by contaminant mixtures. *Environmental Toxicology and Chemistry* 31 (9), 2175–2188.

Posthuma, L., Richards, S. M., et al. (2008). Mixture extrapolation approaches. In *Extrapolation Practice for Ecological Effect Characterization of Chemicals (EXPECT)* (K. R. Solomon, T. C. M. Brock, S. D. Dyer, et al.), Pensacola, FL: SETAC Press.

Posthuma, L. and Suter, G. W. (eds). *Ecological Risk Assessment of Diffuse and Local Soil Contamination Using Species Sensitivity Distributions*. Dealing with contaminated sites. From theory towards practical applications. Dordrecht, Netherlands: Springer.

Posthuma, L., Traas, T. P., et al. (2002). Conceptual and technical outlook on species sensitivity distributions. In *Species Sensitivity Distributions in Ecotoxicology* (eds. L. Posthuma, G. W. Suter, II and T. P. Traas), pp. 475–510. Boca Raton, FL: Lewis Publishers.

Ragas, A. M. J., L. K. P. Teuschler, L., et al. (2010). Human and Ecological Risk Assessment of Chemical Mixtures. In *Mixture Toxicity—Linking Approaches from Ecological and Human Toxicology* (eds. C. A. M. Van Gestel, M. J. Jonker, J. E. Kammenga, R. Laskowski and C. Svendsen), pp. 157–215. Boca Raton, FL: CRC Press.

Scott-Fordsmand, J. J. and Jensen, J. (2002). Ecotoxicological soil quality criteria in Denmark. In *Species Sensitivity Distributions in Ecotoxicology* (eds. L. Posthuma, G. W. Suter, II and T. P. Traas), 275–284. Boca Raton, FL: Lewis Publishers.

Sijm, D. T. H. M., Van Wezel, A. P., et al. (2002). Environmental risk limits in the Netherlands. In *Species Sensitivity Distributions in Ecotoxicology* (eds. L. Posthuma, G. W. Suter, II and T. P. Traas), 221–253. Boca Raton, FL: Lewis Publishers.

Solomon, K. R., Brock, T. C. M., et al. (2008). Extrapolation in the context of criteria setting and risk assessment. EXPECT: *Extrapolation Practice for Ecological Effects and Exposure Characterization of Chemicals.*

Solomon, K. R. and Takacs, P. (2002). Probabilistic risk assessment using species sensitivity distributions. In *Species Sensitivity Distributions in Ecotoxicology* (eds. L. Posthuma, G. W. Suter, II and T. P. Traas), 285–313. Boca Raton, FL: Lewis Publishers.

Sprague, J. B. (1970). Measurement of pollutant toxicity to fish. II. Utilizing and applying bioassay results. *Water Research* 4, 3–32.

Stephan, C. E. (2002). Use of species sensitivity distributions in the derivation of water quality criteria for aquatic life by the U.S. Environmental Protection Agency. In *Species Sensitivity Distributions in Ecotoxicology* (eds. L. Posthuma, G. W. Suter, II and T. P. Traas), pp. 211–254. Boca Raton, FL: Lewis Publishers.

Stephan, C. E., Mount, D. I., et al. (1985). *Guidelines for deriving numerical national water quality criteria for the protection of aquatic organisms and their uses*. Duluth MN: U.S. EPA ORD ERL, 1–97.

Struijs, J., Van der Grinten, E., et al. (2010). *Toxic Pressure in the Dutch Delta Measured with Bioassays—Trends over the Years 2000–2009*. Bilthoven, Netherlands. RIVM (78).

Suter, G. W., Barnthouse, L. W., et al. (1993). *Ecological Risk Assessment*. Boca Raton, FL: Lewis Publishers.

Teuschler, L. K. and Hertzberg, R. C. (1995). Current and future risk assessment guidelines, policy, and methods development for chemical mixtures. *Toxicology* 105 (2–3), 137–144.

Traas, T. P., Van de Meent, D., et al. (2002). The potentially affected fraction as a measure of ecological risk. In *Species Sensitivity Distributions in Ecotoxicology* (eds. L. Posthuma, G. W. Suter, II and T. P. Traas), pp. 315–344. Boca Raton, FL: Lewis Publishers.

U.S. EPA (2000). *Supplementary guidance for conducting health risk assessment of chemical mixtures.* Washington, DC: U.S. Environmental Protection Agency.

U.S. EPA (2007). ECOTOX User Guide: ECOTOXicology Database System. Version 4.0. Available: http:/www.epa.gov/ecotox/ Accessed September 25, 2012.

U.S., EPA (2009). *The National Study of Chemical Residues in Lake Fish Tissue.* Washington, DC: U.S. Environmental Protection Agency, Office of Water.

Van den Brink, P. J., Blake, N., et al. (2006). Predictive value of species sensitivity distributions for effects of herbicides in freshwater ecosystems. *Human and Ecological Risk Assessment* 12 (4), 645–674.

Van den Brink, P. J., Brock, T. C. M., et al. (2002). The value of the species sensitivity distribution concept for predicting field effects: (Non-)confirmation of the concept using semi-field experiments. In *Species Sensitivity Distributions in Ecotoxicology* (eds. L. Posthuma, G. W. Suter, II and T. P. Traas), pp. 155–198. Boca Raton, FL: Lewis Publishers.

Van Straalen, N. M. and Denneman, C. A. J. (1989). Ecotoxicological evaluation of soil quality criteria. *Ecotoxicology and Environmental Safety* 18, 241–251.

Verhaar, H. J. M., Van Leeuwen, C. J., et al. (1992). Classifying environmental pollutants 1: Structure-activity relationships for prediction of aquatic inherent toxicity. *Chemosphere* 25, 471–491.

Wagner, C. and Løkke, H. (1991). Estimation of ecotoxicological protection levels from NOEC toxicity data. *Water Research* 25, 1237–1242.

Warne, M. S. and Hawker, D. W. (1995). The number of components in a mixture determines whether synergistic and antagonistic or additive toxicity predominate: the funnel hypothesis.". *Ecotoxicoloy and Environmental Safety* 31 (1), 23–28.

Warne, M. S. J. (1991). *Mechanism and Prediction of the Non-specific Toxicity of Individual Compounds and Mixtures.* Griffith University.

Warne, M. S. J. and Hawker, D. W. (1995). The number of components in a mixture determines whether synergistic, antagonistic or additive toxicity predominate: The Funnell Hypothesis. *Ecotoxicology and Environmental Safety* 31, 23–28.

Warren-Hicks, W. J., Parkhurst, B. J., et al. (2002). In *Methodology for aquatic ecological risk assessment.* Species Sensitivity Distributions in Ecotoxicology (eds. L. Posthuma, G. W. Suter, II and T. P. Traas), pp. 345–382. Boca Raton, FL: Lewis Publishers.

Wilkinson, C. F., Christoph, G. R., et al. (2000). Assessing the risks of exposures to multiple chemicals with a common mechanism of toxicity: how to cumulate? *Regulatory Toxicology and Pharmacology* 31 (1), 30–43.

Wright, J. F. and Sutcliffe, D. W. (eds). *Assessing the Biological Quality of Fresh Waters: RIVPACS and Other Techniques.* Ambleside, UK: The Freshwater Biological Association.

INDEX

525

538

OTHER VOLUMES IN THE FISH PHYSIOLOGY SERIES

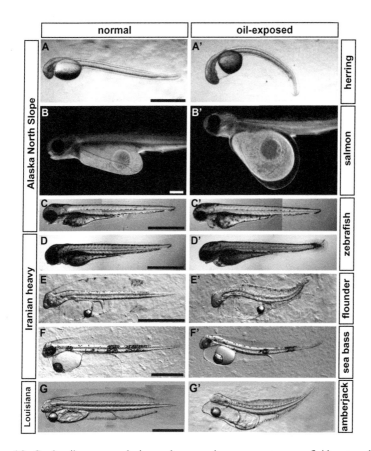

Figure 4.3. Crude oil exposure during embryogenesis causes a common fluid accumulation syndrome across multiple species of fish, geological sources of oil, and exposure methods. Left panels show control embryos with the egg shell dissected or newly hatched larvae; right panels are corresponding oil-exposed clutch mates. All exposures occurred from shortly after fertilization to the hatching stage. (A) Pacific herring and (B) pink salmon exposed to ANSCO using oiled gravel columns; (C) zebrafish exposed to a mechanically dispersed water-accommodated fraction of ANSCO; (D) zebrafish exposed to Iranian heavy crude oil using an oiled gravel column; (E) olive flounder (*Paralichthys olivaceus*) and (F) Japanese sea bass (*Lateolabrax japonicas*) exposed to Iranian heavy crude oil using an oiled gravel column; (G) Yellowtail amberjack (*Seriola lalandi*) exposed to a mechanically dispersed water-accommodated fraction of a Louisiana crude oil. (A) Incardona et al., 2009; (B) Carls and Incardona, unpublished; (C) and (D) Jung et al., 2013; (E) and (F) Jung and Incardona, unpublished; (G) Incardona et al., unpublished.

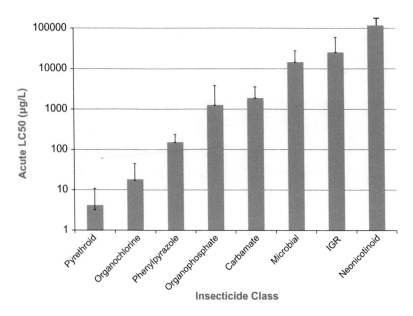

Figure 6.2. Acute toxicity values for fish by insecticide class. Values are average of rainbow trout, bluegill sunfish, and sheepshead minnow LC50 values on a logarithmic scale. Error bars denote standard deviation.

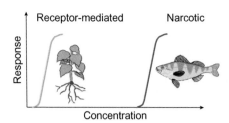

Figure 7.1. Illustration of the difference in toxicity of an herbicide to a plant and a fish and the difference in mode of action in these two groups of organisms.

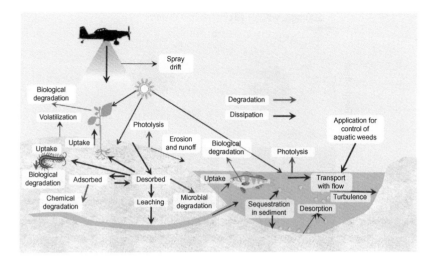

Figure 7.2. Graphical illustration of sources, routes of dissipation, and degradation of herbicides in the surface waters.

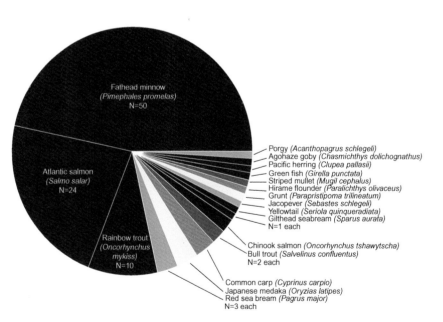

Figure 7.4. Number of sublethal herbicide toxicity data points by fish species.

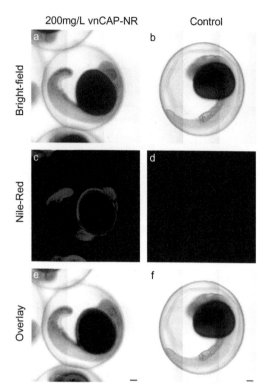

Figure 9.2. Nile red nanocapsules are able to transverse the chorion of zebrafish. Bright field (a, b), Nile red nanocapsule fluorescence (c, d) and overlay (e, f) confocal laser scanning micrographs of 24-hour postfertilization zebrafish exposed to 200 mg/L dialyzed nanocapsules containing Nile red (vnCAP-NR; left) or dechlorinated tap water control (right). Scale bars are 100 μm. Photos courtesy of Lindsey Felix and James Ede.

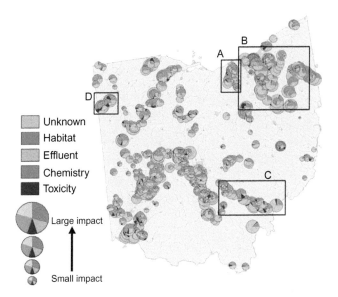

Figure 10.7. A GIS map of Ohio, USA, of magnitudes of impacts (fractions of fish species lost depicted by pie size differences) and probable causes (colors); mixture ecotoxicity summarized in red (via msPAF) (De Zwart, Dyer et al., 2006).

Printed and bound by CPI Group (UK) Ltd, Croydon, CR0 4YY

08/05/2025

01864957-0004